Stochastic Modelling and Applied Probability

(Formerly:
Applications of Mathematics)

38

T0215923

Amir Dembo · Ofer Zeitouni

Large Deviations Techniques and Applications

Corrected printing of the 1998 Edition

 Springer

Amir Dembo
Department of Statistics and
Department of Mathematics
Stanford University
Stanford, CA 94305
USA
amir@math.stanford.edu

Ofer Zeitouni
School of Mathematics
University of Minnesota
206 Church St. SE
Minneapolis, MN 55455
USA

and

Faculty of Mathematics
Weizmann Institute of Science
PoB 26, Rehovot 76100
Israel
zeitouni@math.umn.edu

Managing Editors

Boris Rozovskiĭ
Division of Applied Mathematics
Brown University
182 George St
Providence, RI 02912
USA
rozovsky@dam.brown.edu

Geoffrey Grimmett
Centre for Mathematical Sciences
University of Cambridge
Wilberforce Road
Cambridge CB3 0WB
UK
g.r.grimmett@statslab.cam.ac.uk

ISSN 0172-4568
ISBN 978-3-642-03310-0   e-ISBN 978-3-642-03311-7
DOI 10.1007/978-3-642-03311-7
Springer Heidelberg Dordrecht London New York

Library of Congress Control Number: 97045236

Mathematics Subject Classification (1991): 60F10, 60E15, 60G57, 60H10, 60J15, 93E10

1st edition: © Jones & Bartlett 1993
2nd edition: © Springer-Verlag New York, Inc., 1998 (hard cover)

Cover design: WMXDesign, Heidelberg

Printed on acid-free paper

Springer is part of Springer Science+Business Media (www.springer.com)

To Daphne and Naomi

Preface to the Second Edition

This edition does not involve any major reorganization of the basic plan of the book; however, there are still a substantial number of changes. The inaccuracies and typos that were pointed out, or detected by us, and that were previously posted on our web site, have been corrected. Here and there, clarifying remarks have been added. Some new exercises have been added, often to reflect a result we consider interesting that did not find its way into the main body of the text. Some exercises have been dropped, either because the new presentation covers them, or because they were too difficult or unclear. The general principles of Chapter 4 have been updated by the addition of Theorem 4.4.13 and Lemmas 4.1.23, 4.1.24, and 4.6.5.

More substantial changes have also been incorporated in the text.

1. A new section on concentration inequalities (Section 2.4) has been added. It overviews techniques, ranging from martingale methods to Talagrand's inequalities, to obtain upper bound on exponentially negligible events.

2. A new section dealing with a metric framework for large deviations (Section 4.7) has been added.

3. A new section explaining the basic ingredients of a weak convergence approach to large deviations (Section 6.6) has been added. This section largely follows the recent text of Dupuis and Ellis, and provides yet another approach to the proof of Sanov's theorem.

4. A new subsection with refinements of the Gibbs conditioning principle (Section 7.3.3) has been added.

5. Section 7.2 dealing with sampling without replacement has been completely rewritten. This is a much stronger version of the results, which

vii

also provides an alternative proof of Mogulskii's theorem. This advance was possible by introducing an appropriate coupling.

The added material preserves the numbering of the first edition. In particular, theorems, lemmas and definitions in the first edition have retained the same numbers, although some exercises may now be labeled differently.

Another change concerns the bibliography: The historical notes have been rewritten with more than 100 entries added to the bibliography, both to rectify some omissions in the first edition and to reflect some advances that have been made since then. As in the first edition, no claim is being made for completeness.

The web site http://www-ee.technion.ac.il/~zeitouni/cor.ps will contain corrections, additions, etc. related to this edition. Readers are strongly encouraged to send us their corrections or suggestions.

We thank Tiefeng Jiang for a preprint of [Jia95], on which Section 4.7 is based. The help of Alex de Acosta, Peter Eichelsbacher, Ioannis Kontoyiannis, Stephen Turner, and Tim Zajic in suggesting improvements to this edition is gratefully acknowledged. We conclude this preface by thanking our editor, John Kimmel, and his staff at Springer for their help in producing this edition.

STANFORD, CALIFORNIA
HAIFA, ISRAEL

AMIR DEMBO
OFER ZEITOUNI
DECEMBER 1997

Preface to the First Edition

In recent years, there has been renewed interest in the (old) topic of large deviations, namely, the asymptotic computation of small probabilities on an exponential scale. (Although the term *large deviations* historically was also used for asymptotic expositions off the CLT regime, we always take large deviations to mean the evaluation of small probabilities on an exponential scale). The reasons for this interest are twofold. On the one hand, starting with Donsker and Varadhan, a general foundation was laid that allowed one to point out several "general" tricks that seem to work in diverse situations. On the other hand, large deviations estimates have proved to be the crucial tool required to handle many questions in statistics, engineering, statistical mechanics, and applied probability.

The field of large deviations is now developed enough to enable one to expose the basic ideas and representative applications in a systematic way. Indeed, such treatises exist; see, e.g., the books of Ellis and Deuschel–Stroock [Ell85, DeuS89b]. However, in view of the diversity of the applications, there is a wide range in the backgrounds of those who need to apply the theory. This book is an attempt to provide a rigorous exposition of the theory, which is geared towards such different audiences. We believe that a field as technical as ours calls for a rigorous presentation. Running the risk of satisfying nobody, we tried to expose large deviations in such a way that the principles are first discussed in a relatively simple, finite dimensional setting, and the abstraction that follows is motivated and based on it and on real applications that make use of the "simple" estimates. This is also the reason for our putting our emphasis on the projective limit approach, which is the natural tool to pass from simple finite dimensional statements to abstract ones.

With the recent explosion in the variety of problems in which large deviations estimates have been used, it is only natural that the collection of applications discussed in this book reflects our taste and interest, as well as applications in which we have been involved. Obviously, it does not represent the most important or the deepest possible ones.

The material in this book can serve as a basis for two types of courses: The first, geared mainly towards the finite dimensional application, could be centered around the material of Chapters 2 and 3 (excluding Section 2.1.3 and the proof of Lemma 2.3.12). A more extensive, semester-long course would cover the first four chapters (possibly excluding Section 4.5.3) and either Chapter 5 or Chapter 6, which are independent of each other. The mathematical sophistication required from the reader runs from a senior undergraduate level in mathematics/statistics/engineering (for Chapters 2 and 3) to advanced graduate level for the latter parts of the book.

Each section ends with exercises. While some of those are routine applications of the material described in the section, most of them provide new insight (in the form of related computations, counterexamples, or refinements of the core material) or new applications, and thus form an integral part of our exposition. Many "hinted" exercises are actually theorems with a sketch of the proof.

Each chapter ends with historical notes and references. While a complete bibliography of the large deviations literature would require a separate volume, we have tried to give due credit to authors whose results are related to our exposition. Although we were in no doubt that our efforts could not be completely successful, we believe that an incomplete historical overview of the field is better than no overview at all. We have not hesitated to ignore references that deal with large deviations problems other than those we deal with, and even for the latter, we provide an indication to the literature rather than an exhaustive list. We apologize in advance to those authors who are not given due credit.

Any reader of this book will recognize immediately the immense impact of the Deuschel–Stroock book [DeuS89b] on our exposition. We are grateful to Dan Stroock for teaching one of us (O.Z.) large deviations, for providing us with an early copy of [DeuS89b], and for his advice. O.Z. is also indebted to Sanjoy Mitter for his hospitality at the Laboratory for Information and Decision Systems at MIT, where this project was initiated. A course based on preliminary drafts of this book was taught at Stanford and at the Technion. The comments of people who attended these courses—in particular, the comments and suggestions of Andrew Nobel, Yuval Peres, and Tim Zajic—contributed much to correct mistakes and omissions. We wish to thank Sam Karlin for motivating us to derive the results of Sections 3.2 and 5.5 by suggesting their application in molecular biology. We thank Tom Cover and Joy Thomas for a preprint of [CT91], which influenced our treatment of Sections 2.1.1 and 3.4. The help of Wlodek Bryc, Marty

Day, Gerald Edgar, Alex Ioffe, Dima Ioffe, Sam Karlin, Eddy Mayer-Wolf, and Adam Shwartz in suggesting improvements, clarifying omissions, and correcting outright mistakes is gratefully acknowledged. We thank Alex de Acosta, Richard Ellis, Richard Olshen, Zeev Schuss and Sandy Zabell for helping us to put things in their correct historical perspective. Finally, we were fortunate to benefit from the superb typing and editing job of Lesley Price, who helped us with the intricacies of LaTeX and the English language.

STANFORD, CALIFORNIA
HAIFA, ISRAEL

AMIR DEMBO
OFER ZEITOUNI
AUGUST 1992

Contents

Preface to the Second Edition vii

Preface to the First Edition ix

1 Introduction 1

 1.1 Rare Events and Large Deviations 1

 1.2 The Large Deviation Principle 4

 1.3 Historical Notes and References 9

2 LDP for Finite Dimensional Spaces 11

 2.1 Combinatorial Techniques for Finite Alphabets 11

  2.1.1 The Method of Types and Sanov's Theorem 12

  2.1.2 Cramér's Theorem for Finite Alphabets in $\mathbb{R}$ 18

  2.1.3 Large Deviations for Sampling Without Replacement . 20

 2.2 Cramér's Theorem 26

  2.2.1 Cramér's Theorem in $\mathbb{R}$ 26

  2.2.2 Cramér's Theorem in $\mathbb{R}^d$ 36

 2.3 The Gärtner–Ellis Theorem 43

 2.4 Concentration Inequalities 55

  2.4.1 Inequalities for Bounded Martingale Differences 55

  2.4.2 Talagrand's Concentration Inequalities 60

 2.5 Historical Notes and References 68

3 Applications—The Finite Dimensional Case   **71**

  3.1 Large Deviations for Finite State Markov Chains 72

    3.1.1 LDP for Additive Functionals of Markov Chains . . . 73

    3.1.2 Sanov's Theorem for the Empirical Measure of Markov
           Chains 76

    3.1.3 Sanov's Theorem for the Pair Empirical Measure of
           Markov Chains 78

  3.2 Long Rare Segments in Random Walks 82

  3.3 The Gibbs Conditioning Principle for Finite Alphabets 87

  3.4 The Hypothesis Testing Problem 90

  3.5 Generalized Likelihood Ratio Test for Finite Alphabets . . . 96

  3.6 Rate Distortion Theory 101

  3.7 Moderate Deviations and Exact Asymptotics in $\mathbb{R}^d$ 108

  3.8 Historical Notes and References 113

4 General Principles   **115**

  4.1 Existence of an LDP and Related Properties 116

    4.1.1 Properties of the LDP 117

    4.1.2 The Existence of an LDP 120

  4.2 Transformations of LDPs 126

    4.2.1 Contraction Principles 126

    4.2.2 Exponential Approximations 130

  4.3 Varadhan's Integral Lemma 137

  4.4 Bryc's Inverse Varadhan Lemma 141

  4.5 LDP in Topological Vector Spaces 148

    4.5.1 A General Upper Bound 149

    4.5.2 Convexity Considerations 151

    4.5.3 Abstract Gärtner–Ellis Theorem 157

  4.6 Large Deviations for Projective Limits 161

  4.7 The LDP and Weak Convergence in Metric Spaces 168

  4.8 Historical Notes and References 173

5 Sample Path Large Deviations   **175**

5.1   Sample Path Large Deviations for Random Walks 176

5.2   Brownian Motion Sample Path Large Deviations 185

5.3   Multivariate Random Walk and Brownian Sheet 188

5.4   Performance Analysis of DMPSK Modulation 193

5.5   Large Exceedances in $\mathbb{R}^d$ 200

5.6   The Freidlin–Wentzell Theory 212

5.7   The Problem of Diffusion Exit from a Domain 220

5.8   The Performance of Tracking Loops 238

    5.8.1   An Angular Tracking Loop Analysis 238

    5.8.2   The Analysis of Range Tracking Loops 242

5.9   Historical Notes and References 248

6 The LDP for Abstract Empirical Measures   **251**

6.1   Cramér's Theorem in Polish Spaces 251

6.2   Sanov's Theorem 260

6.3   LDP for the Empirical Measure—The Uniform Markov
      Case 272

6.4   Mixing Conditions and LDP 278

    6.4.1   LDP for the Empirical Mean in $\mathbb{R}^d$ 279

    6.4.2   Empirical Measure LDP for Mixing Processes 285

6.5   LDP for Empirical Measures of Markov Chains 289

    6.5.1   LDP for Occupation Times 289

    6.5.2   LDP for the k-Empirical Measures 295

    6.5.3   Process Level LDP for Markov Chains 298

6.6   A Weak Convergence Approach to Large Deviations 302

6.7   Historical Notes and References 306

7 Applications of Empirical Measures LDP   **311**

7.1   Universal Hypothesis Testing 311

    7.1.1   A General Statement of Test Optimality 311

    7.1.2   Independent and Identically Distributed
            Observations 317

7.2   Sampling Without Replacement 318

7.3 The Gibbs Conditioning Principle 323

    7.3.1 The Non-Interacting Case 327

    7.3.2 The Interacting Case 330

    7.3.3 Refinements of the Gibbs Conditioning Principle . . . 335

7.4 Historical Notes and References 338

Appendix   **341**

A Convex Analysis Considerations in $\mathbb{R}^d$ 341

B Topological Preliminaries 343

    B.1 Generalities 343

    B.2 Topological Vector Spaces and Weak Topologies . . . 346

    B.3 Banach and Polish Spaces 347

    B.4 Mazur's Theorem 349

C Integration and Function Spaces 350

    C.1 Additive Set Functions 350

    C.2 Integration and Spaces of Functions 352

D Probability Measures on Polish Spaces 354

    D.1 Generalities 354

    D.2 Weak Topology 355

    D.3 Product Space and Relative Entropy
         Decompositions 357

E Stochastic Analysis 359

Bibliography   **363**

General Conventions   **385**

Glossary   **387**

Index   **391**

Chapter 1

Introduction

1.1 Rare Events and Large Deviations

This book is concerned with the study of the probabilities of very rare
events. To understand why rare events are important at all, one only has
to think of a lottery to be convinced that rare events (such as hitting the
jackpot) can have an enormous impact.

If any mathematics is to be involved, it must be quantified what is meant
by *rare*. Having done so, a *theory* of rare events should provide an analysis
of the rarity of these events. It is the scope of the theory of large deviations
to answer both these questions. Unfortunately, as Deuschel and Stroock
pointed out in the introduction of [DeuS89b], there is no real "theory" of
large deviations. Rather, besides the basic definitions that by now are standard, a variety of tools are available that allow analysis of small probability
events. Often, the same answer may be reached by using different paths
that seem completely unrelated. It is the goal of this book to explore some
of these tools and show their strength in a variety of applications. The
approach taken here emphasizes making *probabilistic* estimates in a finite
dimensional setting and using *analytical* considerations whenever necessary
to lift up these estimates to the particular situation of interest. In so doing, a particular device, namely, the projective limit approach of Dawson
and Gärtner, will play an important role in our presentation. Although the
reader is exposed to the beautiful convex analysis ideas that have been the
driving power behind the development of the large deviations theory, it is
the projective limit approach that often allows sharp results to be obtained
in general situations. To emphasize this point, derivations for many of the
large deviations theorems using this approach have been provided.

A. Dembo, O. Zeitouni, *Large Deviations Techniques and Applications*,
Stochastic Modelling and Applied Probability 38,
DOI 10.1007/978-3-642-03311-7_1,
© Springer-Verlag Berlin Heidelberg 1998, corrected printing 2010

The uninitiated reader must wonder, at this point, what exactly is meant by *large deviations*. Although precise definitions and statements are postponed to the next section, a particular example is discussed here to provide both motivation and some insights as to what this book is about. Let us begin with the most classical topic of probability theory, namely, the behavior of the empirical mean of independent, identically distributed random variables. Let $X_1, X_2, \ldots, X_n$ be a sequence of independent, standard Normal, real-valued random variables, and consider the empirical mean $\hat{S}_n = \frac{1}{n} \sum_{i=1}^{n} X_i$. Since $\hat{S}_n$ is again a Normal random variable with zero mean and variance $1/n$, it follows that for any $\delta > 0$,

$$P(|\hat{S}_n| \geq \delta) \xrightarrow[n \to \infty]{} 0, \tag{1.1.1}$$

and, for any interval A,

$$P(\sqrt{n}\hat{S}_n \in A) \xrightarrow[n \to \infty]{} \frac{1}{\sqrt{2\pi}} \int_A e^{-x^2/2} dx. \tag{1.1.2}$$

Note now that

$$P(|\hat{S}_n| \geq \delta) = 1 - \frac{1}{\sqrt{2\pi}} \int_{-\delta\sqrt{n}}^{\delta\sqrt{n}} e^{-x^2/2} dx;$$

therefore,

$$\frac{1}{n} \log P(|\hat{S}_n| \geq \delta) \xrightarrow[n \to \infty]{} -\frac{\delta^2}{2}. \tag{1.1.3}$$

Equation (1.1.3) is an example of a large deviations statement: The "typical" value of $\hat{S}_n$ is, by (1.1.2), of the order $1/\sqrt{n}$, but with small probability (of the order of $e^{-n\delta^2/2}$), $|\hat{S}_n|$ takes relatively large values.

Since both (1.1.1) and (1.1.2) remain valid as long as $\{X_i\}$ are independent, identically distributed (i.i.d.) random variables of zero mean and unit variance, it could be asked whether (1.1.3) also holds for non-Normal $\{X_i\}$. The answer is that while the limit of $n^{-1} \log P(|\hat{S}_n| \geq \delta)$ always exists, its value depends on the distribution of X_i. This is precisely the content of Cramér's theorem derived in Chapter 2.

The preceding analysis is not limited to the case of real-valued random variables. With a somewhat more elaborate proof, a similar result holds for d-dimensional, i.i.d. random vectors. Moreover, the independence assumption can be replaced by appropriate notions of *weak dependence*. For example, $\{X_i\}$ may be a realization of a Markov chain. This is discussed in Chapter 2 and more generally in Chapter 6. However, some restriction on the dependence must be made, for examples abound in which the rate of convergence in the law of large numbers is not exponential.

Once the asymptotic rate of convergence of the probabilities $P\left(\left|\frac{1}{n}\sum_{i=1}^{n} X_i\right| \geq \delta\right)$ is available for every distribution of X_i satisfying certain moment conditions, it may be computed in particular for $P\left(\left|\frac{1}{n}\sum_{i=1}^{n} f(X_i)\right| \geq \delta\right)$, where f is an arbitrary bounded measurable function. Similarly, from the corresponding results in $\mathbb{R}^d$, tight bounds may be obtained on the asymptotic decay rate of

$$P\left(\left|\frac{1}{n}\sum_{i=1}^{n} f_1(X_i)\right| \geq \delta, \ldots, \left|\frac{1}{n}\sum_{i=1}^{n} f_d(X_i)\right| \geq \delta\right),$$

where $f_1, \ldots, f_d$ are arbitrary bounded and measurable functions. From here, it is only a relatively small logical step to ask about the rate of convergence of the empirical measure $\frac{1}{n}\sum_{i=1}^{n} \delta_{X_i}$, where δ_{X_i} denotes the (random) measure concentrated at X_i, to the distribution of X_1. This is the content of Sanov's impressive theorem and its several extensions discussed in Chapter 6. It should be noted here that Sanov's theorem provides a quite unexpected link between Large Deviations, Statistical Mechanics, and Information Theory.

Another class of large deviations questions involves the sample path of stochastic processes. Specifically, if $X^\epsilon(t)$ denotes a family of processes that converge, as $\epsilon \to 0$, to some deterministic limit, it may be asked what the rate of this convergence is. This question, treated first by Mogulskii and Schilder in the context, respectively, of a random walk and of the Brownian motion, is explored in Chapter 5, which culminates in the Freidlin–Wentzell theory for the analysis of dynamical systems. This theory has implications to the study of partial differential equations with small parameters.

It is appropriate at this point to return to the *applications* part of the title of this book, in the context of the simple example described before. As a first application, suppose that the mean of the Normal random variables X_i is unknown and, based on the observation $(X_1, X_2, \ldots, X_n)$, one tries to decide whether the mean is -1 or 1. A reasonable, and commonly used decision rule is as follows: Decide that the mean is 1 whenever $\hat{S}_n \geq 0$. The probability of error when using this rule is the probability that, when the mean is -1, the empirical mean is nonnegative. This is exactly the computation encountered in the context of Cramér's theorem. This application is addressed in Chapters 3 and 7 along with its generalization to more than two alternatives and to weakly dependent random variables.

Another important application concerns conditioning on rare events. The best known example of such a conditioning is related to Gibbs conditioning in statistical mechanics, which has found many applications in the seemingly unrelated areas of image processing, computer vision, VLSI design, and nonlinear programming. To illustrate this application, we return to the example where $\{X_i\}$ are i.i.d. standard Normal random variables, and assume that $\hat{S}_n \geq 1$. To find the conditional distribution of X_1 given this rare event, observe that it may be expressed as $P(X_1|X_1 \geq Y)$, where $Y = n - \sum_{i=2}^{n} X_i$ is independent of X_1 and has a Normal distribution with mean n and variance $(n - 1)$. By an asymptotic evaluation of the relevant integrals, it can be deduced that as $n \to \infty$, the conditional distribution converges to a Normal distribution of mean 1 and unit variance. When the marginal distribution of the X_i is not Normal, such a direct computation becomes difficult, and it is reassuring to learn that the limiting behavior of the conditional distribution may be found using large deviations bounds. These results are first obtained in Chapter 3 for X_i taking values in a finite set, whereas the general case is presented in Chapter 7.

A good deal of the preliminary material required to be able to follow the proofs in the book is provided in the Appendix section. These appendices are not intended to replace textbooks on analysis, topology, measure theory, or differential equations. Their inclusion is to allow readers needing a reminder of basic results to find them in this book instead of having to look elsewhere.

1.2   The Large Deviation Principle

The *large deviation principle* (LDP) characterizes the limiting behavior, as $\epsilon \to 0$, of a family of probability measures $\{\mu_\epsilon\}$ on $(\mathcal{X}, \mathcal{B})$ in terms of a *rate function*. This characterization is via asymptotic upper and lower exponential bounds on the values that μ_ϵ assigns to measurable subsets of $\mathcal{X}$. Throughout, $\mathcal{X}$ is a topological space so that open and closed subsets of $\mathcal{X}$ are well-defined, and the simplest situation is when elements of $\mathcal{B}_\mathcal{X}$, the Borel σ-field on $\mathcal{X}$, are of interest. To reduce possible measurability questions, all probability spaces in this book are assumed to have been completed, and, with some abuse of notations, $\mathcal{B}_\mathcal{X}$ always denotes the thus completed Borel σ-field.

Definitions *A rate function I is a lower semicontinuous mapping $I : \mathcal{X} \to [0, \infty]$ (such that for all $\alpha \in [0, \infty)$, the level set $\Psi_I(\alpha) \triangleq \{x : I(x) \leq \alpha\}$ is a closed subset of $\mathcal{X}$). A good rate function is a rate function for which all the level sets $\Psi_I(\alpha)$ are compact subsets of $\mathcal{X}$. The effective domain of I, denoted $\mathcal{D}_I$, is the set of points in $\mathcal{X}$ of finite rate, namely, $\mathcal{D}_I \triangleq \{x : I(x) < \infty\}$. When no confusion occurs, we refer to $\mathcal{D}_I$ as the domain of I.*

Note that if $\mathcal{X}$ is a metric space, the lower semicontinuity property may be checked on sequences, i.e., I is lower semicontinuous if and only if $\liminf_{x_n \to x} I(x_n) \geq I(x)$ for all $x \in \mathcal{X}$. A consequence of a rate function being good is that its infimum is achieved over closed sets.

The following standard notation is used throughout this book. For any set Γ, $\overline{\Gamma}$ denotes the closure of Γ, Γ^o the interior of Γ, and Γ^c the complement of Γ. The infimum of a function over an empty set is interpreted as ∞.

Definition $\{\mu_\epsilon\}$ *satisfies the large deviation principle with a rate function* I *if, for all* $\Gamma \in \mathcal{B}$,

$$- \inf_{x \in \Gamma^o} I(x) \le \liminf_{\epsilon \to 0} \epsilon \log \mu_\epsilon(\Gamma) \le \limsup_{\epsilon \to 0} \epsilon \log \mu_\epsilon(\Gamma) \le - \inf_{x \in \overline{\Gamma}} I(x) . \quad (1.2.4)$$

The right- and left-hand sides of (1.2.4) are referred to as the upper and lower bounds, respectively.

Remark: Note that in (1.2.4), $\mathcal{B}$ need not necessarily be the Borel σ-field. Thus, there can be a separation between the sets on which probability may be assigned and the values of the bounds. In particular, (1.2.4) makes sense even if some open sets are not measurable. Except for this section, we always assume that $\mathcal{B}_\mathcal{X} \subseteq \mathcal{B}$ unless explicitly stated otherwise.

The sentence "μ_ϵ satisfies the LDP" is used as shorthand for "$\{\mu_\epsilon\}$ satisfies the large deviation principle with rate function I." It is obvious that if μ_ϵ satisfies the LDP and $\Gamma \in \mathcal{B}$ is such that

$$\inf_{x \in \Gamma^o} I(x) = \inf_{x \in \overline{\Gamma}} I(x) \stackrel{\triangle}{=} I_\Gamma, \quad (1.2.5)$$

then

$$\lim_{\epsilon \to 0} \epsilon \log \mu_\epsilon(\Gamma) = -I_\Gamma . \quad (1.2.6)$$

A set Γ that satisfies (1.2.5) is called an I *continuity set*. In general, the LDP implies a precise limit in (1.2.6) only for I continuity sets. Finer results may well be derived on a case-by-case basis for specific families of measures $\{\mu_\epsilon\}$ and particular sets. While such results do not fall within our definition of the LDP, a few illustrative examples are included in this book. (See Sections 2.1 and 3.7.)

Some remarks on the definition now seem in order. Note first that in any situation involving non-atomic measures, $\mu_\epsilon(\{x\}) = 0$ for every x in $\mathcal{X}$. Thus, if the lower bound of (1.2.4) was to hold with the infimum over Γ instead of Γ^o, it would have to be concluded that $I(x) \equiv \infty$, contradicting the upper bound of (1.2.4) because $\mu_\epsilon(\mathcal{X}) = 1$ for all ϵ. Thus, some topological restrictions are necessary, and the definition of the LDP codifies a particularly convenient way of stating asymptotic results that, on the one hand, are accurate enough to be useful and, on the other hand, are loose enough to be correct.

Since $\mu_\epsilon(\mathcal{X}) = 1$ for all ϵ, it is necessary that $\inf_{x \in \mathcal{X}} I(x) = 0$ for the upper bound to hold. When I is a good rate function, this means that there exists at least one point x for which $I(x) = 0$. Next, the upper bound trivially holds whenever $\inf_{x \in \overline{\Gamma}} I(x) = 0$, while the lower bound trivially holds whenever $\inf_{x \in \Gamma^o} I(x) = \infty$. This leads to an alternative formulation of the LDP which is actually more useful when proving it. Suppose I is a rate function and $\Psi_I(\alpha)$ its level set. Then (1.2.4) is equivalent to the following bounds.

(a) (Upper bound) For every $\alpha < \infty$ and every measurable set Γ with $\overline{\Gamma} \subset \Psi_I(\alpha)^c$,

$$\limsup_{\epsilon \to 0} \epsilon \log \mu_\epsilon(\Gamma) \leq -\alpha. \tag{1.2.7}$$

(b) (Lower bound) For any $x \in \mathcal{D}_I$ and any measurable Γ with $x \in \Gamma^o$,

$$\liminf_{\epsilon \to 0} \epsilon \log \mu_\epsilon(\Gamma) \geq -I(x). \tag{1.2.8}$$

Inequality (1.2.8) emphasizes the local nature of the lower bound.

In proving the upper bound, it is often convenient to avoid rate functions whose range is unbounded.

Definition *For any rate function I and any $\delta > 0$, the δ-rate function is defined as*

$$I^\delta(x) \triangleq \min\{I(x) - \delta, \frac{1}{\delta}\}. \tag{1.2.9}$$

While in general I^δ is not a rate function, its usefulness stems from the fact that for any set Γ,

$$\lim_{\delta \to 0} \inf_{x \in \Gamma} I^\delta(x) = \inf_{x \in \Gamma} I(x). \tag{1.2.10}$$

Consequently, the upper bound in (1.2.4) is equivalent to the statement that for any $\delta > 0$ and for any measurable set Γ,

$$\limsup_{\epsilon \to 0} \epsilon \log \mu_\epsilon(\Gamma) \leq - \inf_{x \in \overline{\Gamma}} I^\delta(x). \tag{1.2.11}$$

When $\mathcal{B}_\mathcal{X} \subseteq \mathcal{B}$, the LDP is equivalent to the following bounds:

(a) (Upper bound) For any closed set $F \subseteq \mathcal{X}$,

$$\limsup_{\epsilon \to 0} \epsilon \log \mu_\epsilon(F) \leq - \inf_{x \in F} I(x). \tag{1.2.12}$$

(b) (Lower bound) For any open set $G \subseteq \mathcal{X}$,

$$\liminf_{\epsilon \to 0} \epsilon \log \mu_\epsilon(G) \geq - \inf_{x \in G} I(x). \tag{1.2.13}$$

In many cases, a countable family of measures μ_n is considered (for example, when μ_n is the law governing the empirical mean of n random variables). Then the LDP corresponds to the statement

$$- \inf_{x \in \Gamma^o} I(x) \ \leq \ \liminf_{n \to \infty} a_n \log \mu_n(\Gamma) \leq \limsup_{n \to \infty} a_n \log \mu_n(\Gamma)$$

$$\leq - \inf_{x \in \overline{\Gamma}} I(x) \qquad\qquad\qquad (1.2.14)$$

for some sequence $a_n \to 0$. Note that here a_n replaces ϵ of (1.2.4) and similarly, the statements (1.2.7)–(1.2.13) are appropriately modified. For consistency, the convention $a_n = 1/n$ is used throughout and μ_n is renamed accordingly to mean $\mu_{a^{-1}(1/n)}$, where a^{-1} denotes the inverse of $n \mapsto a_n$.

Having defined what is meant by an LDP, the rest of this section is devoted to proving two elementary properties related to the upper bound that will be put to good use throughout this book.

Lemma 1.2.15 *Let N be a fixed integer. Then, for every $a_\epsilon^i \geq 0$,*

$$\limsup_{\epsilon \to 0} \epsilon \log \left(\sum_{i=1}^{N} a_\epsilon^i \right) = \max_{i=1}^{N} \limsup_{\epsilon \to 0} \epsilon \log a_\epsilon^i . \qquad (1.2.16)$$

Proof: First note that for all ϵ,

$$0 \leq \epsilon \log \left(\sum_{i=1}^{N} a_\epsilon^i \right) - \max_{i=1}^{N} \epsilon \log a_\epsilon^i \leq \epsilon \log N .$$

Since N is fixed, $\epsilon \log N \to 0$ as $\epsilon \to 0$ and

$$\limsup_{\epsilon \to 0} \max_{i=1}^{N} \epsilon \log a_\epsilon^i = \max_{i=1}^{N} \limsup_{\epsilon \to 0} \epsilon \log a_\epsilon^i . \qquad \square$$

Often, a natural approach to proving the large deviations upper bound is to prove it first for compact sets. This motivates the following:

Definition *Suppose that all the compact subsets of $\mathcal{X}$ belong to $\mathcal{B}$. A family of probability measures $\{\mu_\epsilon\}$ is said to satisfy the weak LDP with the rate function I if the upper bound (1.2.7) holds for every $\alpha < \infty$ and all compact subsets of $\Psi_I(\alpha)^c$, and the lower bound (1.2.8) holds for all measurable sets.*

It is important to realize that there are families of probability measures that satisfy the weak LDP with a good rate function but do not satisfy the full LDP. For example, let μ_ϵ be the probability measures degenerate at $1/\epsilon$. This family satisfies the weak LDP in $\mathbb{R}$ with the good rate function $I(x) = \infty$. On the other hand, it is not hard to prove that μ_ϵ can not satisfy the LDP with this or any other rate function.

In view of the preceding example, strengthening the weak LDP to a full LDP requires a way of showing that most of the probability mass (at least on an exponential scale) is concentrated on compact sets. The tool for doing that is described next. Here, and in the rest of this book, all topological spaces are assumed to be Hausdorff.

Definition *Suppose that all the compact subsets of $\mathcal{X}$ belong to $\mathcal{B}$. A family of probability measures $\{\mu_\epsilon\}$ on $\mathcal{X}$ is exponentially tight if for every $\alpha < \infty$, there exists a compact set $K_\alpha \subset \mathcal{X}$ such that*

$$\limsup_{\epsilon \to 0} \; \epsilon \log \mu_\epsilon(K_\alpha^c) < -\alpha. \tag{1.2.17}$$

Remarks:
(a) Beware of the logical mistake that consists of identifying exponential tightness and the goodness of the rate function: The measures $\{\mu_\epsilon\}$ need not be exponentially tight in order to satisfy a LDP with a good rate function (you will see such an example in Exercise 6.2.24). In some situations, however, and in particular whenever $\mathcal{X}$ is locally compact or, alternatively, Polish, exponential tightness is implied by the goodness of the rate function. For details, *c.f.* Exercises 1.2.19 and 4.1.10.
(b) Whenever it is stated that μ_ϵ satisfies the weak LDP or μ_ϵ is exponentially tight, it will be implicitly assumed that all the compact subsets of $\mathcal{X}$ belong to $\mathcal{B}$.
(c) Obviously, for $\{\mu_\epsilon\}$ to be exponentially tight, it suffices to have pre-compact K_α for which (1.2.17) holds.

In the following lemma, exponential tightness is applied to strengthen a weak LDP.

Lemma 1.2.18 *Let $\{\mu_\epsilon\}$ be an exponentially tight family.*
(a) If the upper bound (1.2.7) holds for some $\alpha < \infty$ and all compact subsets of $\Psi_I(\alpha)^c$, then it also holds for all measurable sets Γ with $\overline{\Gamma} \subset \Psi_I(\alpha)^c$. In particular, if $\mathcal{B}_{\mathcal{X}} \subseteq \mathcal{B}$ and the upper bound (1.2.12) holds for all compact sets, then it also holds for all closed sets.
(b) If the lower bound (1.2.8) (the lower bound (1.2.13) in case $\mathcal{B}_{\mathcal{X}} \subseteq \mathcal{B}$) holds for all measurable sets (all open sets), then $I(\cdot)$ is a good rate function.

Thus, when an exponentially tight family of probability measures satisfies the weak LDP with a rate function $I(\cdot)$, then I is a good rate function and the LDP holds.

Proof: We consider the general situation, the case where $\mathcal{B}_{\mathcal{X}} \subseteq \mathcal{B}$ being included in it.
(a) To establish (1.2.7), fix a set $\Gamma \in \mathcal{B}$ and $\alpha < \infty$ such that $\overline{\Gamma} \subset \Psi_I(\alpha)^c$. Let K_α be the compact set in (1.2.17), noting that both $\overline{\Gamma} \cap K_\alpha \in \mathcal{B}$ and $K_\alpha^c \in \mathcal{B}$. Clearly,

$$\mu_\epsilon(\Gamma) \leq \mu_\epsilon(\overline{\Gamma} \cap K_\alpha) + \mu_\epsilon(K_\alpha^c) .$$

Note that $\overline{\Gamma} \cap K_\alpha \subset \Psi_I(\alpha)^c$, so $\inf_{x \in \overline{\Gamma} \cap K_\alpha} I(x) \geq \alpha$. Combining the inequality (1.2.17), the upper bound (1.2.7) for the compact set $\overline{\Gamma} \cap K_\alpha$, and Lemma 1.2.15, results in $\limsup_{\epsilon \to 0} \epsilon \log \mu_\epsilon(\Gamma) \leq -\alpha$ as claimed.

(b) Applying the lower bound (1.2.8) to the open set $K_\alpha^c \in \mathcal{B}$, it is concluded from (1.2.17) that $\inf_{x \in K_\alpha^c} I(x) > \alpha$. Therefore, $\Psi_I(\alpha) \subseteq K_\alpha$ yields the compactness of the closed level set $\Psi_I(\alpha)$. As this argument holds for any $\alpha < \infty$, it follows that $I(\cdot)$ is a good rate function. $\qquad\square$

Exercise 1.2.19 Assume that $\mathcal{X}$ is a locally compact topological space, i.e., every point possesses a neighborhood whose closure is compact (a particularly useful example is $\mathcal{X} = \mathbb{R}^d$). Prove that if $\{\mu_\epsilon\}$ satisfies the LDP with a good rate function I, then $\{\mu_\epsilon\}$ is exponentially tight.
Hint: Cover a level set of I by the neighborhoods from the definition of local compactness.

Exercise 1.2.20 Assume $\mathcal{X}$ is a metric space. Further assume that there exists a sequence of constants $\delta_n \to 0$ and a point $x_0 \in \mathcal{X}$ such that $\mu_n(B_{x_0,\delta_n}) = 1$, where B_{x_0,δ_n} denotes the ball of radius δ_n and center x_0. Prove that the sequence $\{\mu_n\}$ satisfies the LDP with the good rate function $I(x_0) = 0$ and $I(x) = \infty$ for $x \neq x_0$.

Exercise 1.2.21 In the example following the definition of the weak LDP, the probability measures μ_ϵ do not even converge weakly. Modify this example to yield a sequence of probability measures that converges weakly to the probability measure degenerate at 0, satisfies the weak LDP, but does not satisfy the full LDP.

1.3 Historical Notes and References

While much of the credit for the modern theory of large deviations and its various applications goes to Donsker and Varadhan (in the West) and Freidlin and Wentzell (in the East), the topic is much older and may be traced back to the early 1900s. General treatments, in book form, of various aspects of the theory of large deviations have already appeared. Varadhan provides a clear and concise description of the main results up to 1984 in his lecture notes [Var84]. Freidlin and Wentzell describe their theory of small random perturbations of dynamical systems in [FW84] (originally published, in Russian, in 1979). A systematic application of large deviations to statistical mechanics and an introduction to the theory may be found in Ellis's [Ell85]. An introduction to the theory of large deviations together with a thorough treatment of the relation between empirical measure LDP and analytical properties of Markov semigroups may be found in [St84]. The latter was expanded in [DeuS89b] and forms the basis for the theory as presented in this book. Finally, engineering applications and a description of the basic results of large deviations theory may be found in [Buc90].

It seems that the general abstract framework for the LDP was proposed by Varadhan [Var66]. Earlier work was mainly concerned with the exponential decay of specific events (for example, the probability of an appropriately normalized random walk to lie in a tube, as in [Mog76]) and did not emphasize the topological structure of the LDP and the different bounds for open and closed sets. (See however [Lan73] and [Var66] where limits of the form of the general theory appear.) It should be noted that many authors use the term LDP only when the rate function is good. (See [FW84, Var84, Ell85].) Our definition follows [DeuS89b]. The term *exponential tightness* was coined in [DeuS89b], although it was implicitly used, in conjunction with the goodness of the rate function, in almost all early work on the LDP. See also [LyS87] for a version of Lemma 1.2.18. Many, but not all, of the results stated in this book for Borel measures carry over to situations where $\mathcal{B}_\mathcal{X} \not\subseteq \mathcal{B}$. For a treatment of such a situation, see for example [deA94b, EicS96].

One may find in the literature results more precise than the LDP. Although the case may be made that the LDP provides only some rough information on the asymptotic probabilities, its scope and ease of application have made it a popular tool. For more refined results, see Wentzell [Wen90] and the historical notes of Chapter 3.

Chapter 2

LDP for Finite Dimensional Spaces

This chapter is devoted to the study of the LDP in a framework that is not yet encumbered with technical details. The main example studied is the empirical mean of a sequence of random variables taking values in $\mathbb{R}^d$. The concreteness of this situation enables the LDP to be obtained under conditions that are much weaker than those that will be imposed in the "general" theory. Many of the results presented here have counterparts in the infinite dimensional context dealt with later, starting in Chapter 4.

2.1 Combinatorial Techniques for Finite Alphabets

Throughout this section, all random variables assume values in a finite set $\Sigma = \{a_1, a_2, \ldots, a_N\}$; Σ, which is also called the *underlying alphabet*, satisfies $|\Sigma| = N$, where for any set A, $|A|$ denotes its cardinality, or size. Combinatorial methods are then applicable for deriving LDPs for the empirical measures of Σ-valued processes (Sections 2.1.1 and 2.1.3), and for the corresponding empirical means (Section 2.1.2). While the scope of these methods is limited to finite alphabets, they illustrate the results one can hope to obtain for more abstract alphabets. It will be seen that some of the latter are actually direct consequences of the LDP derived next via the combinatorial method. Unlike other approaches, this method for deriving the LDP is based on point estimates and thus yields more information than the LDP statement.

A. Dembo, O. Zeitouni, *Large Deviations Techniques and Applications*, 11
Stochastic Modelling and Applied Probability 38,
DOI 10.1007/978-3-642-03311-7_2,
© Springer-Verlag Berlin Heidelberg 1998, corrected printing 2010

2.1.1   The Method of Types and Sanov's Theorem

Throughout, $M_1(\Sigma)$ denotes the space of all probability measures (laws) on the alphabet Σ. Here $M_1(\Sigma)$ is identified with the standard probability simplex in $\mathbb{R}^{|\Sigma|}$, i.e., the set of all $|\Sigma|$-dimensional real vectors with nonnegative components that sum to 1. Open sets in $M_1(\Sigma)$ are obviously induced by the open sets in $\mathbb{R}^{|\Sigma|}$.

Let $Y_1, Y_2, \ldots, Y_n$ be a sequence of random variables that are independent and identically distributed according to the law $\mu \in M_1(\Sigma)$. Let Σ_μ denote the support of the law μ, i.e., $\Sigma_\mu = \{a_i : \mu(a_i) > 0\}$. In general, Σ_μ could be a strict subset of Σ; When considering a single measure μ, it may be assumed, without loss of generality, that $\Sigma_\mu = \Sigma$ by ignoring those symbols that appear with zero probability. For example, this is assumed throughout Section 2.1.2, while in Section 2.1.3 as well as Section 3.5, one has to keep track of various support sets of the form of Σ_μ.

Definition 2.1.1 *The type $L_n^{\mathbf{y}}$ of a finite sequence $\mathbf{y} = (y_1, \ldots, y_n) \in \Sigma^n$ is the empirical measure (law) induced by this sequence. Explicitly, $L_n^{\mathbf{y}} = (L_n^{\mathbf{y}}(a_1), \ldots, L_n^{\mathbf{y}}(a_{|\Sigma|}))$ is the element of $M_1(\Sigma)$ where*

$$L_n^{\mathbf{y}}(a_i) = \frac{1}{n} \sum_{j=1}^{n} 1_{a_i}(y_j), \quad i = 1, \ldots, |\Sigma|,$$

i.e., $L_n^{\mathbf{y}}(a_i)$ is the fraction of occurrences of a_i in the sequence $y_1, \ldots, y_n$.

Let $\mathcal{L}_n$ denote the set of all possible types of sequences of length n. Thus, $\mathcal{L}_n \triangleq \{\nu : \nu = L_n^{\mathbf{y}} \text{ for some } \mathbf{y}\} \subset \mathbb{R}^{|\Sigma|}$, and the empirical measure $L_n^{\mathbf{Y}}$ associated with the sequence $\mathbf{Y} \triangleq (Y_1, \ldots, Y_n)$ is a *random* element of the set $\mathcal{L}_n$. These concepts are useful for finite alphabets because of the following volume and approximation distance estimates.

Lemma 2.1.2 *(a) $|\mathcal{L}_n| \leq (n+1)^{|\Sigma|}$.*
(b) For any probability vector $\nu \in M_1(\Sigma)$,

$$d_V(\nu, \mathcal{L}_n) \triangleq \inf_{\nu' \in \mathcal{L}_n} d_V(\nu, \nu') \leq \frac{|\Sigma|}{2n}, \qquad (2.1.3)$$

where $d_V(\nu, \nu') \triangleq \sup_{A \subset \Sigma}[\nu(A) - \nu'(A)]$ is the variational distance between the measures ν and ν'.

Proof: Note that every component of the vector $L_n^{\mathbf{y}}$ belongs to the set $\{\frac{0}{n}, \frac{1}{n}, \ldots, \frac{n}{n}\}$, whose cardinality is $(n+1)$. Part (a) of the lemma follows, since the vector $L_n^{\mathbf{y}}$ is specified by at most $|\Sigma|$ such quantities.

To prove part (b), observe that $\mathcal{L}_n$ contains all probability vectors composed of $|\Sigma|$ coordinates from the set $\{\frac{0}{n}, \frac{1}{n}, \ldots, \frac{n}{n}\}$. Thus, for any $\nu \in M_1(\Sigma)$, there exists a $\nu' \in \mathcal{L}_n$ with $|\nu(a_i) - \nu'(a_i)| \leq 1/n$ for $i = 1, \ldots, |\Sigma|$. The bound of (2.1.3) now follows, since for finite Σ,

$$d_V(\nu, \nu') = \frac{1}{2} \sum_{i=1}^{|\Sigma|} |\nu(a_i) - \nu'(a_i)| . \qquad \square$$

Remarks:
(a) Since $L_n^{\mathbf{y}}$ is a probability vector, at most $|\Sigma| - 1$ of its components need to be specified and so $|\mathcal{L}_n| \leq (n+1)^{|\Sigma|-1}$.
(b) Lemma 2.1.2 states that the cardinality of $\mathcal{L}_n$, the support of the random empirical measures $L_n^{\mathbf{Y}}$, grows polynomially in n and further that for large enough n, the sets $\mathcal{L}_n$ approximate uniformly and arbitrarily well (in the sense of variational distance) any measure in $M_1(\Sigma)$. Both properties fail to hold when $|\Sigma| = \infty$.

Definition 2.1.4 *The type class $T_n(\nu)$ of a probability law $\nu \in \mathcal{L}_n$ is the set $T_n(\nu) = \{\mathbf{y} \in \Sigma^n : L_n^{\mathbf{y}} = \nu\}$.*

Note that a type class consists of all permutations of a given vector in this set. In the definitions to follow, $0 \log 0 \stackrel{\triangle}{=} 0$ and $0 \log(0/0) \stackrel{\triangle}{=} 0$.

Definition 2.1.5 *(a) The entropy of a probability vector ν is*

$$H(\nu) \stackrel{\triangle}{=} - \sum_{i=1}^{|\Sigma|} \nu(a_i) \log \nu(a_i) .$$

(b) The relative entropy of a probability vector ν with respect to another probability vector μ is

$$H(\nu|\mu) \stackrel{\triangle}{=} \sum_{i=1}^{|\Sigma|} \nu(a_i) \log \frac{\nu(a_i)}{\mu(a_i)} .$$

Remark: By applying Jensen's inequality to the convex function $x \log x$, it follows that the function $H(\cdot|\mu)$ is nonnegative. Note that $H(\cdot|\mu)$ is finite and continuous on the compact set $\{\nu \in M_1(\Sigma) : \Sigma_\nu \subseteq \Sigma_\mu\}$, because $x \log x$ is continuous for $0 \leq x \leq 1$. Moreover, $H(\cdot|\mu) = \infty$ outside this set, and hence $H(\cdot|\mu)$ is a good rate function.

The probabilities of the events $\{L_n^{\mathbf{Y}} = \nu\}$, $\nu \in \mathcal{L}_n$, are estimated in the following three lemmas. First, it is shown that outcomes belonging to the

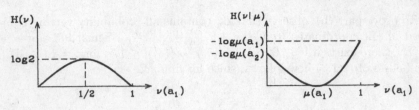

Figure 2.1.1: $H(\nu)$ and $H(\nu|\mu)$ for $|\Sigma| = 2$.

same type class are equally likely, and then the exponential growth rate of each type class is estimated.

Let P_μ denote the probability law $\mu^{\mathbb{Z}_+}$ associated with an infinite sequence of i.i.d. random variables $\{Y_j\}$ distributed following $\mu \in M_1(\Sigma)$.

Lemma 2.1.6 *If* $\mathbf{y} \in T_n(\nu)$ *for* $\nu \in \mathcal{L}_n$, *then*

$$P_\mu((Y_1, \ldots, Y_n) = \mathbf{y}) = e^{-n[H(\nu) + H(\nu|\mu)]}.$$

Proof: The random empirical measure $L_n^{\mathbf{Y}}$ concentrates on types $\nu \in \mathcal{L}_n$ for which $\Sigma_\nu \subseteq \Sigma_\mu$ i.e., $H(\nu|\mu) < \infty$. Therefore, assume without loss of generality that $L_n^{\mathbf{y}} = \nu$ and $\Sigma_\nu \subseteq \Sigma_\mu$. Then

$$P_\mu((Y_1, \ldots, Y_n) = \mathbf{y}) = \prod_{i=1}^{|\Sigma|} \mu(a_i)^{n\nu(a_i)} = e^{-n[H(\nu) + H(\nu|\mu)]},$$

where the last equality follows by the identity

$$H(\nu) + H(\nu|\mu) = -\sum_{i=1}^{|\Sigma|} \nu(a_i) \log \mu(a_i).\qquad\qquad\square$$

In particular, since $H(\mu|\mu) = 0$, it follows that for all $\mu \in \mathcal{L}_n$ and $\mathbf{y} \in T_n(\mu)$,

$$P_\mu((Y_1, \ldots, Y_n) = \mathbf{y}) = e^{-nH(\mu)}. \qquad\qquad (2.1.7)$$

Lemma 2.1.8 *For every* $\nu \in \mathcal{L}_n$,

$$(n+1)^{-|\Sigma|} e^{nH(\nu)} \leq |T_n(\nu)| \leq e^{nH(\nu)}.$$

Remark: Since $|T_n(\nu)| = n!/(n\nu(a_1))! \cdots (n\nu(a_{|\Sigma|}))!$, a good estimate of $|T_n(\nu)|$ can be obtained from Stirling's approximation. (See [Fel71, page 48].) Here, a different route, with an information theory flavor, is taken.

Proof: Under P_ν, any type class has probability one at most and all its realizations are of equal probability. Therefore, for every $\nu \in \mathcal{L}_n$, by (2.1.7),

$$1 \geq P_\nu(L_n^{\mathbf{Y}} = \nu) = P_\nu((Y_1, \ldots, Y_n) \in T_n(\nu)) = e^{-nH(\nu)}|T_n(\nu)|$$

and the upper bound on $|T_n(\nu)|$ follows.

Turning now to prove the lower bound, let $\nu' \in \mathcal{L}_n$ be such that $\Sigma_{\nu'} \subseteq \Sigma_\nu$, and for convenience of notations, reduce Σ so that $\Sigma_\nu = \Sigma$. Then

$$\frac{P_\nu(L_n^{\mathbf{Y}} = \nu)}{P_\nu(L_n^{\mathbf{Y}} = \nu')} = \frac{|T_n(\nu)| \displaystyle\prod_{i=1}^{|\Sigma|} \nu(a_i)^{n\nu(a_i)}}{|T_n(\nu')| \displaystyle\prod_{i=1}^{|\Sigma|} \nu(a_i)^{n\nu'(a_i)}}$$

$$= \prod_{i=1}^{|\Sigma|} \frac{(n\nu'(a_i))!}{(n\nu(a_i))!} \nu(a_i)^{n\nu(a_i)-n\nu'(a_i)}.$$

This last expression is a product of terms of the form $\frac{m!}{\ell!} \left(\frac{\ell}{n}\right)^{\ell-m}$. Considering separately the cases $m \geq \ell$ and $m < \ell$, it is easy to verify that $m!/\ell! \geq \ell^{(m-\ell)}$ for every $m, \ell \in \mathbb{Z}_+$. Hence, the preceding equality yields

$$\frac{P_\nu(L_n^{\mathbf{Y}} = \nu)}{P_\nu(L_n^{\mathbf{Y}} = \nu')} \geq \prod_{i=1}^{|\Sigma|} n^{n\nu'(a_i)-n\nu(a_i)} = n^{n\left[\sum_{i=1}^{|\Sigma|} \nu'(a_i)-\sum_{i=1}^{|\Sigma|} \nu(a_i)\right]} = 1.$$

Note that $P_\nu(L_n^{\mathbf{Y}} = \nu') > 0$ only when $\Sigma_{\nu'} \subseteq \Sigma_\nu$ and $\nu' \in \mathcal{L}_n$. Therefore, the preceding implies that, for all $\nu, \nu' \in \mathcal{L}_n$,

$$P_\nu(L_n^{\mathbf{Y}} = \nu) \geq P_\nu(L_n^{\mathbf{Y}} = \nu').$$

Thus,

$$1 = \sum_{\nu' \in \mathcal{L}_n} P_\nu(L_n^{\mathbf{Y}} = \nu') \leq |\mathcal{L}_n| P_\nu(L_n^{\mathbf{Y}} = \nu)$$

$$= |\mathcal{L}_n| e^{-nH(\nu)} |T_n(\nu)|,$$

and the lower bound on $|T_n(\nu)|$ follows by part (a) of Lemma 2.1.2. $\square$

Lemma 2.1.9 (Large deviations probabilities) *For any $\nu \in \mathcal{L}_n$,*

$$(n+1)^{-|\Sigma|} e^{-nH(\nu|\mu)} \leq P_\mu(L_n^{\mathbf{Y}} = \nu) \leq e^{-nH(\nu|\mu)}.$$

Proof: By Lemma 2.1.6,

$$
\begin{aligned}
P_\mu(L_n^{\mathbf{Y}} = \nu) &= |T_n(\nu)|\, P_\mu((Y_1, \ldots, Y_n) = \mathbf{y}\,,\, L_n^{\mathbf{y}} = \nu) \\
&= |T_n(\nu)|\, e^{-n[H(\nu) + H(\nu|\mu)]}\,.
\end{aligned}
$$

The proof is completed by applying Lemma 2.1.8.   □

Combining Lemmas 2.1.2 and 2.1.9, Sanov's theorem is proved for the finite alphabet context.

Theorem 2.1.10 (Sanov) *For every set Γ of probability vectors in $M_1(\Sigma)$,*

$$
- \inf_{\nu \in \Gamma^o} H(\nu|\mu) \ \le\ \liminf_{n \to \infty} \frac{1}{n} \log P_\mu(L_n^{\mathbf{Y}} \in \Gamma) \tag{2.1.11}
$$

$$
\le\ \limsup_{n \to \infty} \frac{1}{n} \log P_\mu(L_n^{\mathbf{Y}} \in \Gamma) \le - \inf_{\nu \in \Gamma} H(\nu|\mu)\,,
$$

where Γ^o is the interior of Γ considered as a subset of $M_1(\Sigma)$.

Remark: Comparing (2.1.11) and (1.2.14), it follows that Sanov's theorem states that the family of laws $P_\mu(L_n^{\mathbf{Y}} \in \cdot)$ satisfies the LDP with the rate function $H(\cdot|\mu)$. Moreover, in the case of a finite alphabet Σ, there is no need for a closure operation in the upper bound. For a few other improvements that are specific to this case, see Exercises 2.1.16, 2.1.18, and 2.1.19. Note that there are closed sets for which the upper and lower bounds of (2.1.11) are distinct. Moreover, there are closed sets Γ for which the limit of $\frac{1}{n} \log P_\mu(L_n^{\mathbf{Y}} \in \Gamma)$ does not exist. (See Exercises 2.1.20 and 2.1.21.)

Proof: First, from Lemma 2.1.9, upper and lower bounds for all finite n are deduced. By the upper bound of Lemma 2.1.9,

$$
\begin{aligned}
P_\mu(L_n^{\mathbf{Y}} \in \Gamma) &= \sum_{\nu \in \Gamma \cap \mathcal{L}_n} P_\mu(L_n^{\mathbf{Y}} = \nu) \le \sum_{\nu \in \Gamma \cap \mathcal{L}_n} e^{-nH(\nu|\mu)} \\
&\le |\Gamma \cap \mathcal{L}_n| e^{-n \inf_{\nu \in \Gamma \cap \mathcal{L}_n} H(\nu|\mu)} \\
&\le (n+1)^{|\Sigma|} e^{-n \inf_{\nu \in \Gamma \cap \mathcal{L}_n} H(\nu|\mu)}\,. \tag{2.1.12}
\end{aligned}
$$

The accompanying lower bound is

$$
\begin{aligned}
P_\mu(L_n^{\mathbf{Y}} \in \Gamma) &= \sum_{\nu \in \Gamma \cap \mathcal{L}_n} P_\mu(L_n^{\mathbf{Y}} = \nu) \ge \sum_{\nu \in \Gamma \cap \mathcal{L}_n} (n+1)^{-|\Sigma|} e^{-nH(\nu|\mu)} \\
&\ge (n+1)^{-|\Sigma|} e^{-n \inf_{\nu \in \Gamma \cap \mathcal{L}_n} H(\nu|\mu)}\,. \tag{2.1.13}
\end{aligned}
$$

Since $\lim_{n \to \infty} \frac{1}{n} \log(n+1)^{|\Sigma|} = 0$, the normalized logarithmic limits of (2.1.12) and (2.1.13) yield

$$
\limsup_{n \to \infty} \frac{1}{n} \log P_\mu(L_n^{\mathbf{Y}} \in \Gamma) = - \liminf_{n \to \infty} \{ \inf_{\nu \in \Gamma \cap \mathcal{L}_n} H(\nu|\mu) \} \tag{2.1.14}
$$

and

$$\liminf_{n\to\infty} \frac{1}{n}\, \log\, P_\mu(L_n^{\mathbf{Y}} \in \Gamma) = -\limsup_{n\to\infty} \{ \inf_{\nu\in\Gamma\cap\mathcal{L}_n} H(\nu|\mu)\}. \qquad (2.1.15)$$

The upper bound of (2.1.11) follows, since $\Gamma\cap\mathcal{L}_n \subset \Gamma$ for all n.

Turning now to complete the proof of the lower bound of (2.1.11), fix an arbitrary point ν in the interior of Γ such that $\Sigma_\nu \subseteq \Sigma_\mu$. Then, for some $\delta > 0$ small enough, $\{\nu' : d_V(\nu, \nu') < \delta\}$ is contained in Γ. Thus, by part (b) of Lemma 2.1.2, there exists a sequence $\nu_n \in \Gamma\cap\mathcal{L}_n$ such that $\nu_n \to \nu$ as $n \to \infty$. Moreover, without loss of generality, it may be assumed that $\Sigma_{\nu_n} \subseteq \Sigma_\mu$, and hence

$$-\limsup_{n\to\infty}\{\inf_{\nu'\in\Gamma\cap\mathcal{L}_n} H(\nu'|\mu)\} \geq -\lim_{n\to\infty} H(\nu_n|\mu) = -H(\nu|\mu)\,.$$

Recall that $H(\nu|\mu) = \infty$ whenever, for some $i \in \{1, 2, \ldots, |\Sigma|\}$, $\nu(a_i) > 0$ while $\mu(a_i) = 0$. Therefore, by the preceding inequality,

$$-\limsup_{n\to\infty}\{\inf_{\nu\in\Gamma\cap\mathcal{L}_n} H(\nu|\mu)\} \geq -\inf_{\nu\in\Gamma^o} H(\nu|\mu)\,,$$

and the lower bound of (2.1.11) follows by (2.1.15).   □

Exercise 2.1.16 Prove that for every open set Γ,

$$-\lim_{n\to\infty} \{\inf_{\nu\in\Gamma\cap\mathcal{L}_n} H(\nu|\mu)\} = \lim_{n\to\infty} \frac{1}{n} \log P_\mu (L_n^{\mathbf{Y}} \in \Gamma)$$

$$= -\inf_{\nu\in\Gamma} H(\nu|\mu) \stackrel{\triangle}{=} -I_\Gamma\,. \qquad (2.1.17)$$

Exercise 2.1.18 (a) Extend the conclusions of Exercise 2.1.16 to any subset Γ of $\{\nu \in M_1(\Sigma) : \Sigma_\nu \subseteq \Sigma_\mu\}$ that is contained in the closure of its interior.
(b) Prove that for any such set, $I_\Gamma < \infty$ and $I_\Gamma = H(\nu^*|\mu)$ for some $\nu^* \in \overline{\Gamma^o}$.
Hint: Use the continuity of $H(\cdot|\mu)$ on the compact set $\overline{\Gamma^o}$.

Exercise 2.1.19 Assume $\Sigma_\mu = \Sigma$ and that Γ is a *convex* subset of $M_1(\Sigma)$ of non-empty interior. Prove that all the conclusions of Exercise 2.1.18 apply. Moreover, prove that $I_\Gamma = H(\nu^*|\mu)$ for a *unique* $\nu^* \in \overline{\Gamma^o}$.
Hint: If $\nu \in \Gamma$ and $\nu' \in \Gamma^o$, then the entire line segment between ν and ν' is in Γ^o, except perhaps the end point ν. Deduce that $\Gamma \subset \overline{\Gamma^o}$, and prove that $H(\cdot|\mu)$ is a strictly convex function on $M_1(\Sigma)$.

Exercise 2.1.20 Find a closed set Γ for which the two limits in (2.1.17) do not exist.
Hint: Any set $\Gamma = \{\nu\}$, where $\nu \in \mathcal{L}_n$ for some n and $\Sigma_\nu \subseteq \Sigma_\mu$, will do.

Exercise 2.1.21 Find a closed set Γ such that $\inf_{\nu \in \Gamma} H(\nu|\mu) < \infty$ and $\Gamma = \overline{\Gamma^o}$ while $\inf_{\nu \in \Gamma^o} H(\nu|\mu) = \infty$.
Hint: For this construction, you need $\Sigma_\mu \neq \Sigma$. Try $\Gamma = M_1(\Sigma)$, where $|\Sigma| = 2$ and $\mu(a_1) = 0$.

Exercise 2.1.22 Let G be an open subset of $M_1(\Sigma)$ and suppose that μ is chosen at random, uniformly on G. Let $Y_1, \ldots, Y_n$ be i.i.d. random variables taking values in the finite set Σ, distributed according to the law μ. Prove that the LDP holds for the sequence of empirical measures $L_n^{\mathbf{Y}}$ with the good rate function $I(\nu) = \inf_{\mu \in G} H(\nu|\mu)$.
Hint: Show that $\nu_n \to \nu$, $\mu_n \to \mu$, and $H(\nu_n|\mu_n) \leq \alpha$ imply that $H(\nu|\mu) \leq \alpha$. Prove that $H(\nu|\cdot)$ is a continuous function for every fixed $\nu \in M_1(\Sigma)$, and use it for proving the large deviations lower bound and the lower semicontinuity of the rate function $I(\cdot)$.

2.1.2 Cramér's Theorem for Finite Alphabets in $\mathbb{R}$

As an application of Sanov's theorem, a version of Cramér's theorem about the large deviations of the empirical mean of i.i.d. random variables is proved. Specifically, throughout this section the sequence of empirical means $\hat{S}_n \triangleq \frac{1}{n} \sum_{j=1}^n X_j$ is considered, where $X_j = f(Y_j)$, $f : \Sigma \to \mathbb{R}$, and $Y_j \in \Sigma$ are i.i.d. with law μ as in Section 2.1.1. Without loss of generality, it is further assumed that $\Sigma = \Sigma_\mu$ and that $f(a_1) < f(a_2) < \cdots < f(a_{|\Sigma|})$. Cramér's theorem deals with the LDP associated with the real-valued random variables $\hat{S}_n$. Sections 2.2 and 2.3 are devoted to successive generalizations of this result, first to $\mathbb{R}^d$ (Section 2.2), and then to weakly dependent random vectors in $\mathbb{R}^d$ (Section 2.3).

Note that in the case considered here, the random variables $\hat{S}_n$ assume values in the compact interval $K \triangleq [f(a_1), f(a_{|\Sigma|})]$. Moreover, $\hat{S}_n = \sum_{i=1}^{|\Sigma|} f(a_i) L_n^{\mathbf{Y}}(a_i) \triangleq \langle \mathbf{f}, L_n^{\mathbf{Y}} \rangle$, where $\mathbf{f} \triangleq (f(a_1), \ldots, f(a_{|\Sigma|}))$. Therefore, for every set A and every integer n,

$$\hat{S}_n \in A \iff L_n^{\mathbf{Y}} \in \{\nu : \langle \mathbf{f}, \nu \rangle \in A\} \triangleq \Gamma. \qquad (2.1.23)$$

Thus, the following version of Cramér's theorem is a direct consequence of Sanov's theorem (Theorem 2.1.10).

Theorem 2.1.24 (Cramér's theorem for finite subsets of $\mathbb{R}$) *For any set $A \subset \mathbb{R}$,*

$$-\inf_{x \in A^o} I(x) \leq \liminf_{n \to \infty} \frac{1}{n} \log \mathrm{P}_\mu(\hat{S}_n \in A) \qquad (2.1.25)$$

$$\leq \limsup_{n \to \infty} \frac{1}{n} \log \mathrm{P}_\mu(\hat{S}_n \in A) \leq -\inf_{x \in A} I(x),$$

where A^o is the interior of A and $I(x) \triangleq \inf_{\{\nu:\langle \mathbf{f},\nu\rangle = x\}} H(\nu|\mu)$. The rate function $I(x)$ is continuous at $x \in K$ and satisfies there

$$I(x) = \sup_{\lambda \in \mathbb{R}} \{\lambda x - \Lambda(\lambda)\}, \qquad (2.1.26)$$

where

$$\Lambda(\lambda) = \log \sum_{i=1}^{|\Sigma|} \mu(a_i) e^{\lambda f(a_i)} .$$

Remark: Since the rate function $I(\cdot)$ is continuous on K, it follows from (2.1.25) that whenever $A \subset \overline{A^o} \subseteq K$,

$$\lim_{n \to \infty} \frac{1}{n} \log P_\mu(\hat{S}_n \in A) = -\inf_{x \in A} I(x) .$$

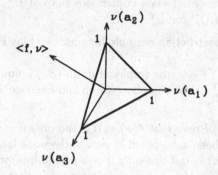

Figure 2.1.2: $M_1(\Sigma)$ and $\langle \mathbf{f}, \nu \rangle$ for $|\Sigma| = 3$.

Proof: When the set A is open, so is the set Γ of (2.1.23), and the bounds of (2.1.25) are simply the bounds of (2.1.11) for Γ. By Jensen's inequality, for every $\nu \in M_1(\Sigma)$ and every $\lambda \in \mathbb{R}$,

$$
\begin{aligned}
\Lambda(\lambda) &= \log \sum_{i=1}^{|\Sigma|} \mu(a_i) e^{\lambda f(a_i)} \geq \sum_{i=1}^{|\Sigma|} \nu(a_i) \log \frac{\mu(a_i) e^{\lambda f(a_i)}}{\nu(a_i)} \\
&= \lambda \langle \mathbf{f}, \nu \rangle - H(\nu|\mu) ,
\end{aligned}
$$

with equality for $\nu_\lambda(a_i) \triangleq \mu(a_i) e^{\lambda f(a_i) - \Lambda(\lambda)}$. Thus, for all λ and all x,

$$\lambda x - \Lambda(\lambda) \leq \inf_{\{\nu:\langle \mathbf{f},\nu\rangle = x\}} H(\nu|\mu) = I(x) , \qquad (2.1.27)$$

with equality when $x = \langle \mathbf{f}, \nu_\lambda \rangle$. The function $\Lambda(\lambda)$ is differentiable with $\Lambda'(\lambda) = \langle \mathbf{f}, \nu_\lambda \rangle$. Therefore, (2.1.26) holds for all $x \in \{\Lambda'(\lambda) : \lambda \in \mathbb{R}\}$.

Observe that $\Lambda'(\cdot)$ is strictly increasing, since $\Lambda(\cdot)$ is strictly convex, and moreover, $f(a_1) = \inf_\lambda \Lambda'(\lambda)$ and $f(a_{|\Sigma|}) = \sup_\lambda \Lambda'(\lambda)$. Hence, (2.1.26) holds for all $x \in K^o$. Consider now the end point $x = f(a_1)$ of K, and let $\nu^*(a_1) = 1$ so that $\langle \mathbf{f}, \nu^* \rangle = x$. Then

$$-\log \mu(a_1) = H(\nu^*|\mu) \quad \geq \quad I(x) \geq \sup_\lambda \{\lambda x - \Lambda(\lambda)\}$$
$$\geq \quad \lim_{\lambda \to -\infty} [\lambda x - \Lambda(\lambda)] = -\log \mu(a_1) \, .$$

The proof for the other end point of K, i.e., $x = f(a_{|\Sigma|})$, is similar. The continuity of $I(x)$ for $x \in K$ is a direct consequence of the continuity of the relative entropy $H(\cdot|\mu)$.   □

It is interesting to note that Cramér's theorem was derived from Sanov's theorem following a pattern that will be useful in the sequel and that is referred to as the *contraction principle*. In this perspective, the random variables $\hat{S}_n$ are represented via a continuous map of $L_n^{\mathbf{Y}}$, and the LDP for $\hat{S}_n$ follows from the LDP for $L_n^{\mathbf{Y}}$.

Exercise 2.1.28 Construct an example for which $n^{-1} \log P_\mu(\hat{S}_n = x)$ has no limit as $n \to \infty$.
Hint: Note that for $|\Sigma| = 2$, the empirical mean $(\hat{S}_n)$ uniquely determines the empirical measure $(L_n^{\mathbf{Y}})$. Use this observation and Exercise 2.1.20 to construct the example.

Exercise 2.1.29 (a) Prove that $I(x) = 0$ if and only if $x = E(X_1)$. Explain why this should have been anticipated in view of the weak law of large numbers.
(b) Check that $H(\nu|\mu) = 0$ if and only if $\nu = \mu$, and interpret this result.
(c) Prove the strong law of large numbers by showing that, for all $\epsilon > 0$,

$$\sum_{n=1}^\infty P(|\hat{S}_n - E(X_1)| > \epsilon) < \infty \, .$$

Exercise 2.1.30 Guess the value of $\lim_{n \to \infty} P_\mu(X_1 = f(a_i)|\hat{S}_n \geq q)$ for $q \in (E(X_1), f(a_{|\Sigma|}))$. Try to justify your guess, at least heuristically.

Exercise 2.1.31 Extend Theorem 2.1.24 to the empirical means of $X_j = f(Y_j)$, where $f : \Sigma \to \mathbb{R}^d$, $d > 1$. In particular, determine the shape of the set K and find the appropriate extension of the formula (2.1.26).

2.1.3 Large Deviations for Sampling Without Replacement

The scope of the method of types is not limited to the large deviations of the empirical measure of i.i.d. random variables. For example, consider the setup of sampling without replacement, a common procedure in many statistical problems. From an initial deterministic pool of m distinct items, $\mathbf{y} \triangleq (y_1, \ldots, y_m)$, an n-tuple $\mathbf{Y} \triangleq (y_{i_1}, y_{i_2}, \ldots, y_{i_n})$ is sampled without replacement, namely, the indices $\{i_1, i_2, i_3, \ldots, i_n\}$ are chosen at random, such that each subset of n distinct elements of $\{1, 2, \ldots, m\}$ is equally likely.

Suppose that for all m, $(y_1^{(m)}, \ldots, y_m^{(m)})$ are elements of the finite set $\Sigma = \{a_1, \ldots, a_{|\Sigma|}\}$. Moreover, suppose that $m = m(n)$ and as $n \to \infty$, the deterministic relative frequency vectors $L_m^{\mathbf{y}} = (L_m^{\mathbf{y}}(a_1), \ldots, L_m^{\mathbf{y}}(a_{|\Sigma|}))$ converge to a probability measure $\mu \in M_1(\Sigma)$. Recall that

$$L_m^{\mathbf{y}}(a_i) = \frac{1}{m} \sum_{j=1}^{m} 1_{a_i}(y_j^{(m)}), \quad i = 1, \ldots, |\Sigma| \ .$$

Suppose further that $\mathbf{Y}$ is a random vector obtained by the sampling without replacement of n out of m elements as described before. An investigation is made next of the LDP for the random empirical measures $L_n^{\mathbf{Y}}$ associated with the vectors $\mathbf{Y}$. In particular, the analog of Theorem 2.1.10, is established for $m = m(n)$ and $\lim_{n \to \infty} (n/m(n)) = \beta$, $0 < \beta < 1$. To this end, consider the following candidate rate function

$$I(\nu|\beta, \mu) \triangleq \begin{cases} H(\nu|\mu) + \frac{1-\beta}{\beta} H\left(\frac{\mu - \beta\nu}{1-\beta}\Big|\mu\right) & \text{if } \mu(a_i) \geq \beta\nu(a_i) \text{ for all } i \\ \infty & \text{otherwise.} \end{cases}$$

$$(2.1.32)$$

Observe that as $\beta \to 0$, the function $I(\cdot|\beta, \mu)$ approaches $H(\cdot|\mu)$, while as $\beta \to 1$, the domain of ν for which $I(\nu|\beta, \mu) < \infty$ coalesces to a single measure $\nu = \mu$. This reflects the reduction in the amount of "randomness" as β increases. Note that $L_n^{\mathbf{Y}}$ belongs to the set $\mathcal{L}_n$ whose size grows polynomially in n by Lemma 2.1.2. Further, the following estimates of large deviations probabilities for $L_n^{\mathbf{Y}}$ are obtained by elementary combinatorics.

Lemma 2.1.33 *For every probability vector $\nu \in \mathcal{L}_n$:*
(a) If $I(\nu| \frac{n}{m}, L_m^{\mathbf{y}}) < \infty$, then

$$\left| \frac{1}{n} \log P(L_n^{\mathbf{Y}} = \nu) + I(\nu| \tfrac{n}{m}, L_m^{\mathbf{y}}) \right| \leq 2(|\Sigma| + 1) \left(\frac{\log(m+1)}{n} \right) . \quad (2.1.34)$$

(b) If $I(\nu| \frac{n}{m}, L_m^{\mathbf{y}}) = \infty$, then $P(L_n^{\mathbf{Y}} = \nu) = 0$.

Proof: (a) Under sampling without replacement, the probability of the event $\{L_n^{\mathbf{Y}} = \nu\}$ for $\nu \in \mathcal{L}_n$ is exactly the number of n-tuples $i_1 \neq i_2 \neq \cdots \neq i_n$ resulting in type ν, compared to the overall number of n-tuples,

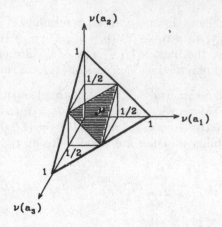

Figure 2.1.3: Domain of $I(\cdot|\beta, \mu)$ for $\beta = \frac{2}{3}$ and $\mu = (\frac{1}{3}, \frac{1}{3}, \frac{1}{3})$.

that is,

$$
P(L_n^{\mathbf{Y}} = \nu) = \frac{\prod_{i=1}^{|\Sigma|} \binom{m\, L_m^{\mathbf{y}}(a_i)}{n\, \nu(a_i)}}{\binom{m}{n}}. \tag{2.1.35}
$$

An application of Lemma 2.1.8 for $|\Sigma| = 2$, where $|T_n(\nu)| = \binom{n}{k}$ when $\nu(a_1) = k/n$, $\nu(a_2) = 1 - k/n$, results in the estimate

$$
\max_{0 \le k \le n} \left| \log \binom{n}{k} - nH\left(\frac{k}{n}\right) \right| \le 2\log(n+1), \tag{2.1.36}
$$

where

$$
H(p) \overset{\triangle}{=} -p\, \log p - (1-p)\, \log(1-p).
$$

Alternatively, (2.1.36) follows by Stirling's formula. (See [Fel71, page 48].)
Combining the exact expression (2.1.35) and the bound (2.1.36) results in

$$
\left| \frac{1}{n} \log P(L_n^{\mathbf{Y}} = \nu) - \sum_{i=1}^{|\Sigma|} \frac{m\, L_m^{\mathbf{y}}(a_i)}{n} H\left(\frac{n\nu(a_i)}{mL_m^{\mathbf{y}}(a_i)}\right) + \frac{m}{n} H\left(\frac{n}{m}\right) \right|
$$

$$
\le 2(|\Sigma| + 1)\left(\frac{\log(m+1)}{n}\right). \tag{2.1.37}
$$

The inequality (2.1.34) follows by rearranging the left side of (2.1.37).

(b) Note that $I(\nu| \frac{n}{m}, L_m^{\mathbf{y}}) = \infty$ if and only if $n\nu(a_i) > mL_m^{\mathbf{y}}(a_i)$ for some $a_i \in \Sigma$. It is then impossible, in sampling without replacement, to have $L_n^{\mathbf{Y}} = \nu$, since $nL_n^{\mathbf{Y}}(a_i) \le mL_m^{\mathbf{y}}(a_i)$ for every $a_i \in \Sigma$. $\qquad\square$

As in the proof of Theorem 2.1.10, the preceding point estimates give the analogs of (2.1.14) and (2.1.15).

Corollary 2.1.38 *With $m = m(n)$,*

$$\limsup_{n \to \infty} \frac{1}{n} \log P(L_n^{\mathbf{Y}} \in \Gamma) = -\liminf_{n \to \infty} \{ \inf_{\nu \in \Gamma \cap \mathcal{L}_n} I(\nu | \tfrac{n}{m}, L_m^{\mathbf{Y}}) \}, \quad (2.1.39)$$

and

$$\liminf_{n \to \infty} \frac{1}{n} \log P(L_n^{\mathbf{Y}} \in \Gamma) = -\limsup_{n \to \infty} \{ \inf_{\nu \in \Gamma \cap \mathcal{L}_n} I(\nu | \tfrac{n}{m}, L_m^{\mathbf{Y}}) \}. \quad (2.1.40)$$

The proof of Corollary 2.1.38 is left as Exercise 2.1.46.

The following is the desired analog of Sanov's theorem (Theorem 2.1.10).

Theorem 2.1.41 *Suppose $L_m^{\mathbf{y}}$ converges to μ and $n/m \to \beta \in (0, 1)$ as $n \to \infty$. Then the random empirical measures $L_n^{\mathbf{Y}}$ satisfy the LDP with the good rate function $I(\nu | \beta, \mu)$. Explicitly, for every set Γ of probability vectors in $M_1(\Sigma) \subset \mathbb{R}^{|\Sigma|}$,*

$$-\inf_{\nu \in \Gamma^\circ} I(\nu | \beta, \mu) \leq \liminf_{n \to \infty} \frac{1}{n} \log P(L_n^{\mathbf{Y}} \in \Gamma) \quad (2.1.42)$$

$$\leq \limsup_{n \to \infty} \frac{1}{n} \log P(L_n^{\mathbf{Y}} \in \Gamma) \leq -\inf_{\nu \in \overline{\Gamma}} I(\nu | \beta, \mu).$$

Remark: Note that the upper bound of (2.1.42) is weaker than the upper bound of (2.1.11) in the sense that the infimum of the rate function is taken over the closure of Γ. See Exercise 2.1.47 for examples of sets Γ where the above lower and upper bounds coincide.

The following lemma is needed for the proof of Theorem 2.1.41.

Lemma 2.1.43 *Let $\beta_n \in (0, 1)$, $\mu_n, \nu_n \in M_1(\Sigma)$ be such that $\beta_n \to \beta \in (0, 1)$ and $\mu_n \to \mu$ as $n \to \infty$.*
(a) If $\nu_n \to \nu$ and $I(\nu_n | \beta_n, \mu_n) < \infty$ for all n large enough, then

$$\lim_{n \to \infty} I(\nu_n | \beta_n, \mu_n) = I(\nu | \beta, \mu).$$

(b) If $I(\nu | \beta, \mu) < \infty$, then there exists a sequence $\{\nu_n\}$ such that $\nu_n \in \mathcal{L}_n$, $\nu_n \to \nu$, and

$$\lim_{n \to \infty} I(\nu_n | \beta_n, \mu_n) = I(\nu | \beta, \mu).$$

Proof: (a) Since $I(\nu_n|\beta_n,\mu_n) < \infty$ for all n large enough, it follows that $\mu_n(a_i) \geq \beta_n\nu_n(a_i)$ for every $a_i \in \Sigma$. Hence, by rearranging (2.1.32),

$$I(\nu_n|\beta_n,\mu_n) = \frac{1}{\beta_n}H(\mu_n) - H(\nu_n) - \frac{1-\beta_n}{\beta_n}H\left(\frac{\mu_n - \beta_n\nu_n}{1-\beta_n}\right),$$

and $I(\nu_n|\beta_n,\mu_n) \to I(\nu|\beta,\mu)$, since $H(\cdot)$ is continuous on $M_1(\Sigma)$ and $\{\beta_n\}$ is bounded away from 0 and 1.

(b) Consider first $\nu \in M_1(\Sigma)$ for which

$$\min_{a_i \in \Sigma}\{\mu(a_i) - \beta\nu(a_i)\} > 0. \tag{2.1.44}$$

By part (b) of Lemma 2.1.2, there exist $\nu_n \in \mathcal{L}_n$ such that $\nu_n \to \nu$ as $n \to \infty$. Thus, the *strict* inequality (2.1.44) implies that for all n large,

$$\min_{a_i \in \Sigma}\{\mu_n(a_i) - \beta_n\nu_n(a_i)\} \geq 0.$$

Consequently, $I(\nu_n|\beta_n,\mu_n) < \infty$ for all n large enough and the desired conclusion follows by part (a).

Suppose now that $\Sigma_\mu = \Sigma$, but possibly (2.1.44) does not hold. Then, since $\beta < 1$ and $I(\nu|\beta,\mu) < \infty$, there exist $\nu_k \to \nu$ such that (2.1.44) holds for all k. By the preceding argument, there exist $\{\nu_{n,k}\}_{n=1}^\infty$ such that for all k, $\nu_{n,k} \in \mathcal{L}_n$, $\nu_{n,k} \to \nu_k$ and

$$\lim_{n\to\infty} I(\nu_{n,k}|\beta_n,\mu_n) = I(\nu_k|\beta,\mu).$$

By the standard Cantor diagonalization argument, there exist $\nu_n \in \mathcal{L}_n$ such that $\nu_n \to \nu$ and

$$\lim_{n\to\infty} I(\nu_n|\beta_n,\mu_n) = \lim_{k\to\infty} I(\nu_k|\beta,\mu) = I(\nu|\beta,\mu),$$

where the last equality is due to part (a) of the lemma.

Finally, even if $\Sigma_\mu \neq \Sigma$, it still holds that $\Sigma_\nu \subseteq \Sigma_\mu$, since $I(\nu|\beta,\mu) < \infty$. Hence, by repeating the preceding argument with Σ_μ instead of Σ, there exist $\nu_n \in \mathcal{L}_n$ such that $\Sigma_{\nu_n} \subseteq \Sigma_\mu$, $\nu_n \to \nu$, and $I(\nu_n|\beta_n,\mu_n) < \infty$ for all n large enough. The proof is completed by applying part (a) of the lemma. $\square$

Proof of Theorem 2.1.41: It follows from part (a) of Lemma 2.1.43 that $I(\nu|\beta,\mu)$ is lower semicontinuous jointly in $\beta \in (0,1)$, μ, and ν. Since it is a nonnegative function, and the probability simplex is a compact set, $I(\cdot|\beta,\mu)$ is a good rate function.

Turning now to prove the upper bound of (2.1.42), first deduce from (2.1.39) that for some infinite subsequence n_k, there exists a sequence $\{\nu_k\} \subset \Gamma$ such that

$$\limsup_{n \to \infty} \frac{1}{n} \log \mathrm{P}(L_n^{\mathbf{Y}} \in \Gamma) = -\lim_{k \to \infty} I\left(\nu_k \Big| \frac{n_k}{m_k}, L_{m_k}^{\mathbf{y}}\right) \stackrel{\triangle}{=} -I^*, \qquad (2.1.45)$$

where possibly $I^* = \infty$. The sequence $\{\nu_k\}$ has a limit point ν^* in the *compact* set $\overline{\Gamma}$. Passing to a convergent subsequence, the lower semicontinuity of I jointly in β, μ, and ν implies that

$$I^* \geq I(\nu^* | \beta, \mu) \geq \inf_{\nu \in \overline{\Gamma}} I(\nu | \beta, \mu) .$$

The upper bound follows by combining this inequality and (2.1.45).

Finally, in order to prove the lower bound, consider an arbitrary $\nu \in \Gamma^o$ such that $I(\nu | \beta, \mu) < \infty$. Then, by part (b) of Lemma 2.1.43, there exist $\nu_n \in \mathcal{L}_n$ such that $\nu_n \to \nu$ and

$$\lim_{n \to \infty} I\left(\nu_n \big| \tfrac{n}{m}, L_m^{\mathbf{y}}\right) = I(\nu | \beta, \mu) .$$

Since for all n large enough, $\nu_n \in \Gamma \cap \mathcal{L}_n$, it follows that

$$-\limsup_{n \to \infty} \left\{ \inf_{\nu' \in \Gamma \cap \mathcal{L}_n} I\left(\nu' \big| \tfrac{n}{m}, L_m^{\mathbf{y}}\right) \right\} \geq -I(\nu | \beta, \mu) .$$

Combining the preceding inequality with (2.1.40), one concludes that for each such ν,

$$\liminf_{n \to \infty} \frac{1}{n} \log \mathrm{P}(L_n^{\mathbf{Y}} \in \Gamma) \geq -I(\nu | \beta, \mu) .$$

The proof of the theorem is now complete in view of the formulation (1.2.8) of the large deviations lower bound. $\qquad \square$

Exercise 2.1.46 Prove Corollary 2.1.38.

Exercise 2.1.47 Let $I_\Gamma \stackrel{\triangle}{=} \inf_{\nu \in \Gamma^o} I(\nu | \beta, \mu)$. Prove that when $\Gamma \subset \mathcal{D}_I$ and Γ is contained in the closure of its interior, then

$$I_\Gamma = \lim_{n \to \infty} \frac{1}{n} \log \mathrm{P}(L_n^{\mathbf{Y}} \in \Gamma) .$$

Hint: Use Exercise 2.1.18 and the continuity of $I(\cdot | \beta, \mu)$ on its level sets.

Exercise 2.1.48 Prove that the rate function $I(\cdot | \beta, \mu)$ is convex.

Exercise 2.1.49 Prove that the LDP of Theorem 2.1.41 holds for $\beta = 0$ with the good rate function $I(\cdot | 0, \mu) \stackrel{\triangle}{=} H(\cdot | \mu)$.

Hint: First prove that the left side of (2.1.37) goes to zero as $n \to \infty$ (i.e., even when $n^{-1} \log(m+1) \to \infty$). Then, show that Lemma 2.1.43 holds for $\beta_n > 0$ and $\beta = 0$.

2.2   Cramér's Theorem

Cramér's theorem about the large deviations associated with the empirical mean of i.i.d. random variables taking values in a finite set is presented in Section 2.1.2 as an application of the method of types. However, a direct application of the method of types is limited to finite alphabets. In this section, Theorem 2.1.24 is extended to the case of i.i.d. random variables taking values in $\mathbb{R}^d$.

Specifically, consider the empirical means $\hat{S}_n \stackrel{\triangle}{=} \frac{1}{n} \sum_{j=1}^n X_j$, for i.i.d., d-dimensional random vectors $X_1, \ldots, X_n, \ldots$, with X_1 distributed according to the probability law $\mu \in M_1(\mathbb{R}^d)$. The *logarithmic moment generating function* associated with the law μ is defined as

$$\Lambda(\lambda) \stackrel{\triangle}{=} \log M(\lambda) \stackrel{\triangle}{=} \log E[e^{\langle \lambda, X_1 \rangle}] , \qquad (2.2.1)$$

where $\langle \lambda, x \rangle \stackrel{\triangle}{=} \sum_{j=1}^d \lambda^j x^j$ is the usual scalar product in $\mathbb{R}^d$, and x^j the jth coordinate of x. Another common name for $\Lambda(\cdot)$ is the *cumulant generating function*. In what follows, $|x| \stackrel{\triangle}{=} \sqrt{\langle x, x \rangle}$, is the usual Euclidean norm. Note that $\Lambda(0) = 0$, and while $\Lambda(\lambda) > -\infty$ for all λ, it is possible to have $\Lambda(\lambda) = \infty$. Let μ_n denote the law of $\hat{S}_n$ and $\overline{x} \stackrel{\triangle}{=} E[X_1]$. When $\overline{x}$ exists and is finite, and $E[|X_1 - \overline{x}|^2] < \infty$, then $\hat{S}_n \xrightarrow[n \to \infty]{\text{Prob}} \overline{x}$, since

$$E\left[|\hat{S}_n - \overline{x}|^2\right] = \frac{1}{n^2} \sum_{j=1}^n E\left[|X_j - \overline{x}|^2\right] = \frac{1}{n} E\left[|X_1 - \overline{x}|^2\right] \xrightarrow[n \to \infty]{} 0 .$$

Hence, in this situation, $\mu_n(F) \xrightarrow[n \to \infty]{} 0$ for any closed set F such that $\overline{x} \notin F$. Cramér's theorem characterizes the logarithmic rate of this convergence by the following (rate) function.

Definition 2.2.2 *The Fenchel–Legendre transform of $\Lambda(\lambda)$ is*

$$\Lambda^*(x) \stackrel{\triangle}{=} \sup_{\lambda \in \mathbb{R}^d} \{\langle \lambda, x \rangle - \Lambda(\lambda)\} .$$

It is instructive to consider first the case $d = 1$, followed by the additional work necessary for handling the general case.

2.2.1   Cramér's Theorem in $\mathbb{R}$

Let $\mathcal{D}_\Lambda \stackrel{\triangle}{=} \{\lambda : \Lambda(\lambda) < \infty\}$ and $\mathcal{D}_{\Lambda^*} \stackrel{\triangle}{=} \{x : \Lambda^*(x) < \infty\}$. Cramér's theorem in $\mathbb{R}$, which is stated next, is applicable even when $\overline{x}$ does not exist.

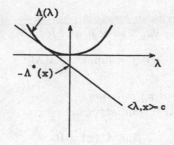

Figure 2.2.1: Geometrical interpretation of Λ^*.

Theorem 2.2.3 (Cramér) *When $X_i \in \mathbb{R}$, the sequence of measures $\{\mu_n\}$ satisfies the LDP with the convex rate function $\Lambda^*(\cdot)$, namely:*
(a) For any closed set $F \subset \mathbb{R}$,

$$\limsup_{n \to \infty} \frac{1}{n} \log \mu_n(F) \leq - \inf_{x \in F} \Lambda^*(x) . \qquad (2.2.4)$$

(b) For any open set $G \subset \mathbb{R}$,

$$\liminf_{n \to \infty} \frac{1}{n} \log \mu_n(G) \geq - \inf_{x \in G} \Lambda^*(x) .$$

Remarks:
(a) The definition of the Fenchel–Legendre transform for (topological) vector spaces and some of its properties are presented in Section 4.5. It is also shown there that the Fenchel–Legendre transform is a natural candidate for the rate function, since the upper bound (2.2.4) holds for compact sets in a general setup.
(b) As follows from part (b) of Lemma 2.2.5 below, Λ^* need not in general be a good rate function.
(c) A close inspection of the proof reveals that, actually, (2.2.4) may be strengthened to the statement that, for all n,

$$\mu_n(F) \leq 2e^{-n \inf_{x \in F} \Lambda^*(x)} .$$

The following lemma states the properties of $\Lambda^*(\cdot)$ and $\Lambda(\cdot)$ that are needed for proving Theorem 2.2.3.

Lemma 2.2.5 *(a) Λ is a convex function and Λ^* is a convex rate function.*
(b) If $\mathcal{D}_\Lambda = \{0\}$, then Λ^* is identically zero. If $\Lambda(\lambda) < \infty$ for some $\lambda > 0$, then $\bar{x} < \infty$ (possibly $\bar{x} = -\infty$), and for all $x \geq \bar{x}$,

$$\Lambda^*(x) = \sup_{\lambda \geq 0}[\lambda x - \Lambda(\lambda)] \qquad (2.2.6)$$

is, for $x > \overline{x}$, a nondecreasing function. Similarly, if $\Lambda(\lambda) < \infty$ for some $\lambda < 0$, then $\overline{x} > -\infty$ (possibly $\overline{x} = \infty$), and for all $x \leq \overline{x}$,

$$\Lambda^*(x) = \sup_{\lambda \leq 0}[\lambda x - \Lambda(\lambda)] \tag{2.2.7}$$

is, for $x < \overline{x}$, a nonincreasing function.

When $\overline{x}$ is finite, $\Lambda^*(\overline{x}) = 0$, and always,

$$\inf_{x \in \mathbb{R}} \Lambda^*(x) = 0 \,. \tag{2.2.8}$$

(c) $\Lambda(\cdot)$ is differentiable in $\mathcal{D}_\Lambda^o$ with

$$\Lambda'(\eta) = \frac{1}{M(\eta)} E[X_1 e^{\eta X_1}] \tag{2.2.9}$$

and

$$\Lambda'(\eta) = y \implies \Lambda^*(y) = \eta y - \Lambda(\eta) \,. \tag{2.2.10}$$

Proof: (a) The convexity of Λ follows by Hölder's inequality, since

$$\Lambda(\theta\lambda_1 + (1 - \theta)\lambda_2) = \log E[(e^{\lambda_1 X_1})^\theta (e^{\lambda_2 X_1})^{(1-\theta)}]$$
$$\leq \log\{E[e^{\lambda_1 X_1}]^\theta E[e^{\lambda_2 X_1}]^{(1-\theta)}\} = \theta\Lambda(\lambda_1) + (1 - \theta)\Lambda(\lambda_2)$$

for any $\theta \in [0, 1]$. The convexity of Λ^* follows from its definition, since

$$\theta\Lambda^*(x_1) + (1 - \theta)\Lambda^*(x_2)$$
$$= \sup_{\lambda \in \mathbb{R}}\{\theta\lambda x_1 - \theta\Lambda(\lambda)\} + \sup_{\lambda \in \mathbb{R}}\{(1 - \theta)\lambda x_2 - (1 - \theta)\Lambda(\lambda)\}$$
$$\geq \sup_{\lambda \in \mathbb{R}}\{(\theta x_1 + (1 - \theta)x_2)\lambda - \Lambda(\lambda)\} = \Lambda^*(\theta x_1 + (1 - \theta)x_2) \,.$$

Recall that $\Lambda(0) = \log E[1] = 0$, so $\Lambda^*(x) \geq 0x - \Lambda(0) = 0$ is nonnegative. In order to establish that Λ^* is lower semicontinuous and hence a rate function, fix a sequence $x_n \to x$. Then, for every $\lambda \in \mathbb{R}$,

$$\liminf_{x_n \to x} \Lambda^*(x_n) \geq \liminf_{x_n \to x}[\lambda x_n - \Lambda(\lambda)] = \lambda x - \Lambda(\lambda) \,.$$

Thus,

$$\liminf_{x_n \to x} \Lambda^*(x_n) \geq \sup_{\lambda \in \mathbb{R}}[\lambda x - \Lambda(\lambda)] = \Lambda^*(x) \,.$$

(b) If $\mathcal{D}_\Lambda = \{0\}$, then $\Lambda^*(x) = \Lambda(0) = 0$ for all $x \in \mathbb{R}$. If $\Lambda(\lambda) = \log M(\lambda) < \infty$ for some $\lambda > 0$, then $\int_0^\infty x d\mu < M(\lambda)/\lambda < \infty$, implying that $\overline{x} < \infty$ (possibly $\overline{x} = -\infty$). Now, for all $\lambda \in \mathbb{R}$, by Jensen's inequality,

$$\Lambda(\lambda) = \log E[e^{\lambda X_1}] \geq E[\log e^{\lambda X_1}] = \lambda\overline{x} \,.$$

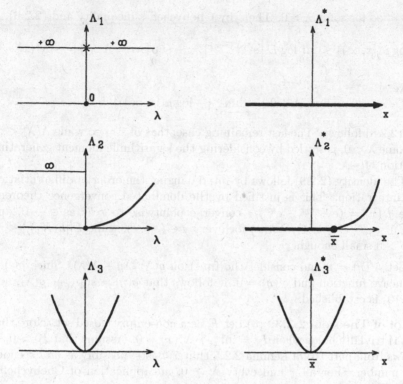

Figure 2.2.2: Pairs of Λ and Λ^*.

If $\overline{x} = -\infty$, then $\Lambda(\lambda) = \infty$ for λ negative, and (2.2.6) trivially holds. When $\overline{x}$ is finite, it follows from the preceding inequality that $\Lambda^*(\overline{x}) = 0$. In this case, for every $x \geq \overline{x}$ and every $\lambda < 0$,

$$\lambda x - \Lambda(\lambda) \leq \lambda\overline{x} - \Lambda(\lambda) \leq \Lambda^*(\overline{x}) = 0 ,$$

and (2.2.6) follows. Observe that (2.2.6) implies the monotonicity of $\Lambda^*(x)$ on $(\overline{x}, \infty)$, since for every $\lambda \geq 0$, the function $\lambda x - \Lambda(\lambda)$ is nondecreasing as a function of x.

When $\Lambda(\lambda) < \infty$ for some $\lambda < 0$, then both (2.2.7) and the monotonicity of Λ^* on $(-\infty, \overline{x})$ follow by considering the logarithmic moment generating function of $-X$, for which the preceding proof applies.

It remains to prove that $\inf_{x \in \mathbb{R}} \Lambda^*(x) = 0$. This is already established for $\mathcal{D}_\Lambda = \{0\}$, in which case $\Lambda^* \equiv 0$, and when $\overline{x}$ is finite, in which case, as shown before, $\Lambda^*(\overline{x}) = 0$. Now, consider the case when $\overline{x} = -\infty$ while

$\Lambda(\lambda) < \infty$ for some $\lambda > 0$. Then, by Chebycheff's inequality and (2.2.6),

$$\log \mu([x, \infty)) \le \inf_{\lambda \ge 0} \log E\left[e^{\lambda(X_1 - x)}\right] = -\sup_{\lambda \ge 0}\{\lambda x - \Lambda(\lambda)\} = -\Lambda^*(x) .$$

Hence,

$$\lim_{x \to -\infty} \Lambda^*(x) \le \lim_{x \to -\infty} \{-\log \mu([x, \infty))\} = 0 ,$$

and (2.2.8) follows. The last remaining case, that of $\overline{x} = \infty$ while $\Lambda(\lambda) < \infty$ for some $\lambda < 0$, is settled by considering the logarithmic moment generating function of $-X$.

(c) The identity (2.2.9) follows by interchanging the order of differentiation and integration. This is justified by the dominated convergence theorem, since $f_\epsilon(x) = (e^{(\eta + \epsilon)x} - e^{\eta x})/\epsilon$ converge pointwise to $xe^{\eta x}$ as $\epsilon \to 0$, and $|f_\epsilon(x)| \le e^{\eta x}(e^{\delta|x|} - 1)/\delta \triangleq h(x)$ for every $\epsilon \in (-\delta, \delta)$, while $E[|h(X_1)|] < \infty$ for $\delta > 0$ small enough.

Let $\Lambda'(\eta) = y$ and consider the function $g(\lambda) \triangleq \lambda y - \Lambda(\lambda)$. Since $g(\cdot)$ is a concave function and $g'(\eta) = 0$, it follows that $g(\eta) = \sup_{\lambda \in \mathbb{R}} g(\lambda)$, and (2.2.10) is established.   □

Proof of Theorem 2.2.3: (a) Let F be a non-empty closed set. Note that (2.2.4) trivially holds when $I_F = \inf_{x \in F} \Lambda^*(x) = 0$. Assume that $I_F > 0$. It follows from part (b) of Lemma 2.2.5 that $\overline{x}$ exists, possibly as an extended real number. For all x and every $\lambda \ge 0$, an application of Chebycheff's inequality yields

$$\mu_n([x, \infty)) = E\left[1_{\hat{S}_n - x \ge 0}\right] \le E\left[e^{n\lambda(\hat{S}_n - x)}\right] \qquad (2.2.11)$$

$$= e^{-n\lambda x} \prod_{i=1}^{n} E\left[e^{\lambda X_i}\right] = e^{-n[\lambda x - \Lambda(\lambda)]} .$$

Therefore, if $\overline{x} < \infty$, then by (2.2.6), for every $x > \overline{x}$,

$$\mu_n([x, \infty)) \le e^{-n\Lambda^*(x)} . \qquad (2.2.12)$$

By a similar argument, if $\overline{x} > -\infty$ and $x < \overline{x}$, then

$$\mu_n((-\infty, x]) \le e^{-n\Lambda^*(x)} . \qquad (2.2.13)$$

First, consider the case of $\overline{x}$ finite. Then $\Lambda^*(\overline{x}) = 0$, and because by assumption $I_F > 0$, $\overline{x}$ must be contained in the open set F^c. Let (x_-, x_+) be the union of all the open intervals $(a, b) \in F^c$ that contain $\overline{x}$. Note that $x_- < x_+$ and that either x_- or x_+ must be finite since F is non-empty. If x_- is finite, then $x_- \in F$, and consequently $\Lambda^*(x_-) \ge I_F$. Likewise,

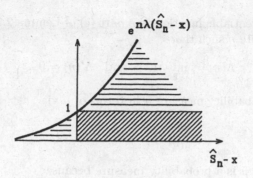

Figure 2.2.3: Chebycheff's bound.

$\Lambda^*(x_+) \geq I_F$ whenever x_+ is finite. Applying (2.2.12) for $x = x_+$ and (2.2.13) for $x = x_-$, the union of events bound ensures that

$$\mu_n(F) \leq \mu_n((-\infty, x_-]) + \mu_n([x_+, \infty)) \leq 2e^{-nI_F} ,$$

and the upper bound follows when the normalized logarithmic limit as $n \to \infty$ is considered.

Suppose now that $\overline{x} = -\infty$. Then, since Λ^* is nondecreasing, it follows from (2.2.8) that $\lim_{x \to -\infty} \Lambda^*(x) = 0$, and hence $x_+ = \inf\{x : x \in F\}$ is finite for otherwise $I_F = 0$. Since F is a closed set, $x_+ \in F$ and consequently $\Lambda^*(x_+) \geq I_F$. Moreover, $F \subset [x_+, \infty)$ and, therefore, the large deviations upper bound follows by applying (2.2.12) for $x = x_+$.

The case of $\overline{x} = \infty$ is handled analogously.

(b) We prove next that for every $\delta > 0$ and every marginal law $\mu \in M_1(\mathbb{R})$,

$$\liminf_{n \to \infty} \frac{1}{n} \log \mu_n((-\delta, \delta)) \geq \inf_{\lambda \in \mathbb{R}} \Lambda(\lambda) = -\Lambda^*(0) . \qquad (2.2.14)$$

Since the transformation $Y = X - x$ results with $\Lambda_Y(\lambda) = \Lambda(\lambda) - \lambda x$, and hence with $\Lambda_Y^*(\cdot) = \Lambda^*(\cdot + x)$, it follows from the preceding inequality that for every x and every $\delta > 0$,

$$\liminf_{n \to \infty} \frac{1}{n} \log \mu_n((x - \delta, x + \delta)) \geq -\Lambda^*(x) . \qquad (2.2.15)$$

For any open set G, any $x \in G$, and any $\delta > 0$ small enough, $(x - \delta, x + \delta) \subset G$. Thus, the large deviations lower bound follows from (2.2.15).

Turning to the proof of the key inequality (2.2.14), first suppose that $\mu((-\infty, 0)) > 0$, $\mu((0, \infty)) > 0$, and that μ is supported on a bounded subset of $\mathbb{R}$. By the former assumption, $\Lambda(\lambda) \to \infty$ as $|\lambda| \to \infty$, and by the latter assumption, $\Lambda(\cdot)$ is finite everywhere. Accordingly, $\Lambda(\cdot)$ is a continuous, differentiable function (see part (c) of Lemma 2.2.5), and hence there exists a finite η such that

$$\Lambda(\eta) = \inf_{\lambda \in \mathbb{R}} \Lambda(\lambda) \quad \text{and} \quad \Lambda'(\eta) = 0 .$$

Define a new probability measure $\tilde{\mu}$ in terms of μ via

$$\frac{d\tilde{\mu}}{d\mu}(x) = e^{\eta x - \Lambda(\eta)} ,$$

and observe that $\tilde{\mu}$ is a probability measure because

$$\int_{\mathbb{R}} d\tilde{\mu} = \frac{1}{M(\eta)} \int_{\mathbb{R}} e^{\eta x} d\mu = 1.$$

Let $\tilde{\mu}_n$ be the law governing $\hat{S}_n$ when X_i are i.i.d. random variables of law $\tilde{\mu}$. Note that for every $\epsilon > 0$,

$$
\begin{aligned}
\mu_n((-\epsilon, \epsilon)) &= \int_{|\sum_{i=1}^{n} x_i| < n\epsilon} \mu(dx_1) \cdots \mu(dx_n) \\
&\geq e^{-n\epsilon |\eta|} \int_{|\sum_{i=1}^{n} x_i| < n\epsilon} \exp(\eta \sum_{i=1}^{n} x_i) \, \mu(dx_1) \cdots \mu(dx_n) \\
&= e^{-n\epsilon |\eta|} e^{n\Lambda(\eta)} \tilde{\mu}_n((-\epsilon, \epsilon)) .
\end{aligned}
\tag{2.2.16}
$$

By (2.2.9) and the choice of η,

$$E_{\tilde{\mu}}[X_1] = \frac{1}{M(\eta)} \int_{\mathbb{R}} x e^{\eta x} d\mu = \Lambda'(\eta) = 0 .$$

Hence, by the law of large numbers,

$$\lim_{n \to \infty} \tilde{\mu}_n((-\epsilon, \epsilon)) = 1 .
\tag{2.2.17}$$

It now follows from (2.2.16) that for every $0 < \epsilon < \delta$,

$$\liminf_{n \to \infty} \frac{1}{n} \log \mu_n((-\delta, \delta)) \geq \liminf_{n \to \infty} \frac{1}{n} \log \mu_n((-\epsilon, \epsilon)) \geq \Lambda(\eta) - \epsilon |\eta| ,$$

and (2.2.14) follows by considering the limit $\epsilon \to 0$.

Suppose now that μ is of unbounded support, while both $\mu((-\infty, 0)) > 0$ and $\mu((0, \infty)) > 0$. Fix M large enough such that $\mu([-M, 0)) > 0$ as well as $\mu((0, M]) > 0$, and let

$$\Lambda_M(\lambda) = \log \int_{-M}^{M} e^{\lambda x} d\mu .$$

Let ν denote the law of X_1 conditioned on $\{|X_1| \leq M\}$, and let ν_n be the law of $\hat{S}_n$ conditioned on $\{|X_i| \leq M, i = 1, \ldots, n\}$. Then, for all n and every $\delta > 0$,

$$\mu_n((-\delta, \delta)) \geq \nu_n((-\delta, \delta))\mu([-M, M])^n .$$

Observe that by the preceding proof, (2.2.14) holds for ν_n. Therefore, with the logarithmic moment generating function associated with ν being merely $\Lambda_M(\lambda) - \log\mu([-M, M])$,

$$\liminf_{n \to \infty} \frac{1}{n} \log\mu_n((-\delta, \delta)) \geq$$

$$\log\mu([-M, M]) + \liminf_{n \to \infty} \frac{1}{n} \log\nu_n((-\delta, \delta)) \geq \inf_{\lambda \in \mathbb{R}} \Lambda_M(\lambda) .$$

With $I_M = -\inf_{\lambda \in \mathbb{R}} \Lambda_M(\lambda)$ and $I^* = \limsup_{M \to \infty} I_M$, it follows that

$$\liminf_{n \to \infty} \frac{1}{n} \log\mu_n((-\delta, \delta)) \geq -I^* . \qquad (2.2.18)$$

Note that $\Lambda_M(\cdot)$ is nondecreasing in M, and thus so is $-I_M$. Moreover, $-I_M \leq \Lambda_M(0) \leq \Lambda(0) = 0$, and hence $-I^* \leq 0$. Now, since $-I_M$ is finite for all M large enough, $-I^* > -\infty$. Therefore, the level sets $\{\lambda : \Lambda_M(\lambda) \leq -I^*\}$ are non-empty, compact sets that are nested with respect to M, and hence there exists at least one point, denoted λ_0, in their intersection. By Lesbegue's monotone convergence theorem, $\Lambda(\lambda_0) = \lim_{M \to \infty} \Lambda_M(\lambda_0) \leq -I^*$, and consequently the bound (2.2.18) yields (2.2.14), now for μ of unbounded support.

The proof of (2.2.14) for an arbitrary probability law μ is completed by observing that if either $\mu((-\infty, 0)) = 0$ or $\mu((0, \infty)) = 0$, then $\Lambda(\cdot)$ is a monotone function with $\inf_{\lambda \in \mathbb{R}} \Lambda(\lambda) = \log\mu(\{0\})$. Hence, in this case, (2.2.14) follows from

$$\mu_n((-\delta, \delta)) \geq \mu_n(\{0\}) = \mu(\{0\})^n . \qquad \square$$

Remarks:
(a) The crucial step in the proof of the upper bound is based on Chebycheff's inequality, combined with the independence assumption. For weakly dependent random variables, a similar approach is to use the logarithmic limit of the right side of (2.2.11), instead of the logarithmic moment generating function for a single random variable. This is further explored in Sections 2.3 and 4.5.
(b) The essential step in the proof of the lower bound is the exponential change of measure that is used to define $\tilde{\mu}$. This is particularly well-suited to problems where, even if the random variables involved are not directly

independent, some form of underlying independence exists, e.g., when as in Exercise 5.2.11, a Girsanov-type formula is used.

In the preceding proof, the change of measure is coupled with the law of large numbers to obtain the lower bound. One may obtain tighter bounds for semi-infinite intervals by using the Central Limit Theorem (CLT), as the next corollary demonstrates. (See also Exercise 2.2.25 for extensions.)

Corollary 2.2.19 *For any $y \in \mathbb{R}$,*

$$\lim_{n \to \infty} \frac{1}{n} \log \mu_n([y, \infty)) = - \inf_{x \geq y} \Lambda^*(x) .$$

Proof: Since $[x, x + \delta) \subset [y, \infty)$ for all $x \geq y$ and all $\delta > 0$, it follows that

$$\liminf_{n \to \infty} \frac{1}{n} \log \mu_n([y, \infty)) \geq \sup_{x \geq y} \liminf_{n \to \infty} \frac{1}{n} \log \mu_n([x, x + \delta)).$$

The corollary is thus a consequence of the following strengthened version of (2.2.15):

$$\liminf_{n \to \infty} \frac{1}{n} \log \mu_n([x, x + \delta)) \geq -\Lambda^*(x) .$$

The proof paraphrases the proof of (2.2.15), where it is sufficient to consider $x = 0$, and everywhere $[0, \delta)$ and $[0, \epsilon)$ replace $(-\delta, \delta)$ and $(-\epsilon, \epsilon)$, respectively. This is possible, since (2.2.17) is replaced by the CLT statement

$$\lim_{n \to \infty} \tilde{\mu}_n([0, \epsilon)) = \frac{1}{2} . \qquad \square$$

Another strengthening of Cramér's theorem (Theorem 2.2.3) is related to the goodness of the rate function.

Lemma 2.2.20 *If $0 \in \mathcal{D}_\Lambda^o$ then Λ^* is a good rate function. Moreover, if $\mathcal{D}_\Lambda = \mathbb{R}$, then*

$$\lim_{|x| \to \infty} \Lambda^*(x)/|x| = \infty. \qquad (2.2.21)$$

Proof: As $0 \in \mathcal{D}_\Lambda^o$, there exist $\lambda_- < 0$ and $\lambda_+ > 0$ that are both in $\mathcal{D}_\Lambda$. Since for any $\lambda \in \mathbb{R}$,

$$\frac{\Lambda^*(x)}{|x|} \geq \lambda \operatorname{sign}(x) - \frac{\Lambda(\lambda)}{|x|} ,$$

it follows that

$$\liminf_{|x| \to \infty} \frac{\Lambda^*(x)}{|x|} \geq \min\{\lambda_+, -\lambda_-\} > 0 .$$

In particular, $\Lambda^*(x) \longrightarrow \infty$ as $|x| \to \infty$, and its level sets are closed and bounded, hence compact. Thus, Λ^* is a good rate function. Note that (2.2.21) follows for $\mathcal{D}_\Lambda = \mathbb{R}$ by considering $-\lambda_- = \lambda_+ \to \infty$.   □

Exercise 2.2.22 Prove by an application of Fatou's lemma that $\Lambda(\cdot)$ is lower semicontinuous.

Exercise 2.2.23 Show that:
(a) For $X_1 \sim$ Poisson(θ), $\Lambda^*(x) = \theta - x + x\log(x/\theta)$ for nonnegative x, and $\Lambda^*(x) = \infty$ otherwise.
(b) For $X_1 \sim$ Bernoulli(p), $\Lambda^*(x) = x\log(\frac{x}{p}) + (1-x)\log(\frac{1-x}{1-p})$ for $x \in [0,1]$ and $\Lambda^*(x) = \infty$ otherwise. Note that $\mathcal{D}_\Lambda = \mathbb{R}$, but $\Lambda^*(\cdot)$ is discontinuous.
(c) For $X_1 \sim$ Exponential(θ), $\Lambda^*(x) = \theta x - 1 - \log(\theta x)$ for $x > 0$ and $\Lambda^*(x) = \infty$ otherwise.
(d) For $X_1 \sim$ Normal$(0, \sigma^2)$, $\Lambda^*(x) = x^2/2\sigma^2$.

Exercise 2.2.24 Prove that $\Lambda(\lambda)$ is C^∞ in $\mathcal{D}_\Lambda^o$ and that $\Lambda^*(x)$ is strictly convex, and C^∞ in the interior of the set $\mathcal{F} \triangleq \{\Lambda'(\lambda) : \lambda \in \mathcal{D}_\Lambda^o\}$.
Hint: Use (2.2.9) to show that $x = \Lambda'(\eta) \in \mathcal{F}^o$ implies that $\Lambda''(\eta) > 0$ and then apply (2.2.10).

Exercise 2.2.25 (a) Suppose A is a Borel measurable set such that $[y, z) \subset A \subset [y, \infty)$ for some $y < z$ and either $\mathcal{D}_\Lambda = \{0\}$ or $\bar{x} < z$. Prove that

$$\lim_{n\to\infty} \frac{1}{n} \log \mu_n(A) = -\inf_{x \in A} \Lambda^*(x).$$

(b) Use Exercise 2.2.24 to prove that the conclusion of Corollary 2.2.19 holds for $A = (y, \infty)$ when $y \in \mathcal{F}^o$ and $y > \bar{x}$.
Hint: $[y + \delta, \infty) \subset A \subset [y, \infty)$ while Λ^* is continuous at y.

Exercise 2.2.26 Suppose for some $a < b$, $\bar{x} \in [a, b]$ and any $\lambda \in \mathbb{R}$,

$$M(\lambda) \leq \frac{b - \bar{x}}{b - a} e^{\lambda a} + \frac{\bar{x} - a}{b - a} e^{\lambda b}. \qquad (2.2.27)$$

(a) Show that then for any $a \leq x \leq b$,

$$\Lambda^*(x) \geq H\left(\frac{x - a}{b - a} \Big| \frac{\bar{x} - a}{b - a}\right), \qquad (2.2.28)$$

where $H(p|p_0) \triangleq p \log(p/p_0) + (1 - p)\log((1 - p)/(1 - p_0))$.
(b) Find a distribution of X_1, with support on $[a, b]$, for which equality is achieved in (2.2.27) and (2.2.28).

Remark: See also Corollary 2.4.5 to appreciate the significance of this result.

Exercise 2.2.29 Suppose that for some $b \geq \overline{x}$, $\sigma > 0$ and any $\lambda \geq 0$,

$$M(\lambda) \leq e^{\lambda \overline{x}} \left\{ \frac{(b - \overline{x})^2}{(b - \overline{x})^2 + \sigma^2} e^{-\lambda \sigma^2 / (b - \overline{x})} + \frac{\sigma^2}{(b - \overline{x})^2 + \sigma^2} e^{\lambda (b - \overline{x})} \right\} .$$

Show that $\Lambda^*(x) \geq H(p_x | p_{\overline{x}})$, for $p_x \triangleq \left((b - \overline{x})(x - \overline{x}) + \sigma^2 \right) / \left((b - \overline{x})^2 + \sigma^2 \right)$ and $\overline{x} \leq x \leq b$.

Remark: See also Lemma 2.4.1 to appreciate the significance of this result.

2.2.2 Cramér's Theorem in $\mathbb{R}^d$

Cramér's theorem (Theorem 2.2.3) possesses a multivariate counterpart dealing with the large deviations of the empirical means of i.i.d. random vectors in $\mathbb{R}^d$. Our analysis emphasizes the new points in which the proof for $\mathbb{R}^d$ differs from the proof for $\mathbb{R}$, with an eye towards infinite dimensional extensions. An interesting consequence is Sanov's theorem for finite alphabets. (See Corollary 2.2.35 and Exercise 2.2.36.) To ease the proofs, it is assumed throughout that $\mathcal{D}_\Lambda = \mathbb{R}^d$, i.e., $\Lambda(\lambda) < \infty$ for all λ. In particular, $E(|X_1|^2) < \infty$ and $\hat{S}_n \xrightarrow[n \to \infty]{\text{Prob}} \overline{x}$.

The following theorem is the counterpart of Theorem 2.2.3 for $\mathbb{R}^d$, $d > 1$.

Theorem 2.2.30 (Cramér) *Assume $\mathcal{D}_\Lambda = \mathbb{R}^d$. Then $\{\mu_n\}$ satisfies the LDP on $\mathbb{R}^d$ with the good convex rate function $\Lambda^*(\cdot)$.*

Remarks:
(a) A stronger version of this theorem is proved in Section 6.1 via a more sophisticated sub-additivity argument. In particular, it is shown in Corollary 6.1.6 that the assumption $0 \in \mathcal{D}_\Lambda^o$ suffices for the conclusion of Theorem 2.2.30. Even without this assumption, for every open convex $A \subset \mathbb{R}^d$,

$$\lim_{n \to \infty} \frac{1}{n} \log \mu_n(A) = - \inf_{x \in A} \Lambda^*(x) .$$

(b) For the extension of Theorem 2.2.30 to dependent random vectors that do not necessarily satisfy the condition $\mathcal{D}_\Lambda = \mathbb{R}^d$, see Section 2.3.

The following lemma summarizes the properties of $\Lambda(\cdot)$ and $\Lambda^*(\cdot)$ that are needed to prove Theorem 2.2.30.

Lemma 2.2.31 *(a) $\Lambda(\cdot)$ is convex and differentiable everywhere, and $\Lambda^*(\cdot)$ is a good convex rate function.*
(b) $\qquad y = \nabla \Lambda(\eta) \implies \Lambda^*(y) = \langle \eta, y \rangle - \Lambda(\eta).$

Proof: (a) The convexity of Λ follows by Hölder's inequality. Its differentiability follows by dominated convergence. (See the proof of Lemma 2.2.5.) Convexity and lower semicontinuity of Λ^* follow from Definition 2.2.2 by an argument similar to the proof of part (a) of Lemma 2.2.5. Since $\Lambda(0) = 0$, Λ^* is nonnegative, and hence it is a rate function. Observe that for all $x \in \mathbb{R}^d$ and all $\rho > 0$,

$$\Lambda^*(x) \geq \rho|x| - \sup_{|\lambda|=\rho} \{\Lambda(\lambda)\}.$$

In particular, all level sets of Λ^* are bounded and Λ^* is a good rate function.
(b) Let $y = \nabla\Lambda(\eta)$. Fix an arbitrary point $\lambda \in \mathbb{R}^d$ and let

$$g(\alpha) \overset{\triangle}{=} \alpha\langle \lambda - \eta, y \rangle - \Lambda(\eta + \alpha(\lambda - \eta)) + \langle \eta, y \rangle, \quad \alpha \in [0,1].$$

Since Λ is convex and $\Lambda(\eta)$ is finite, $g(\cdot)$ is concave, and $|g(0)| < \infty$. Thus,

$$g(1) - g(0) \leq \liminf_{\alpha \searrow 0} \frac{g(\alpha) - g(0)}{\alpha} = \langle \lambda - \eta, y - \nabla\Lambda(\eta) \rangle = 0,$$

where the last equality follows by the assumed identity $y = \nabla\Lambda(\eta)$. Therefore, for all λ,

$$g(1) = [\langle \lambda, y \rangle - \Lambda(\lambda)] \leq g(0) = [\langle \eta, y \rangle - \Lambda(\eta)] \leq \Lambda^*(y).$$

The conclusion follows by taking the supremum over λ in the preceding inequality. □

Proof of Theorem 2.2.30:　The first step of the proof is to establish the large deviations upper bound. In $\mathbb{R}^d$, the monotonicity of Λ^* stated in part (b) of Lemma 2.2.5 is somewhat lost. Thus, the method of containing each closed set F by two half-spaces is not as useful as it is in $\mathbb{R}$. Instead, upper bounds on the probabilities that $\{\mu_n\}$ assign to balls are deduced by Chebycheff's inequality. Compact sets are then covered by an appropriate finite collection of small enough balls and the upper bound for compact sets follows by the union of events bound.

As mentioned in Section 1.2, establishing the upper bound is equivalent to proving that for every $\delta > 0$ and every closed set $F \subset \mathbb{R}^d$,

$$\limsup_{n \to \infty} \frac{1}{n} \log \mu_n(F) \leq \delta - \inf_{x \in F} I^\delta(x), \qquad (2.2.32)$$

where I^δ is the δ-rate function (as defined in (1.2.9)) associated with Λ^*. Fix a *compact* set $\Gamma \subset \mathbb{R}^d$. For every $q \in \Gamma$, choose $\lambda_q \in \mathbb{R}^d$ for which

$$\langle \lambda_q, q \rangle - \Lambda(\lambda_q) \geq I^\delta(q).$$

This is feasible on account of the definitions of Λ^* and I^δ. For each q, choose $\rho_q > 0$ such that $\rho_q |\lambda_q| \leq \delta$, and let $B_{q,\rho_q} = \{x : |x - q| < \rho_q\}$ be the ball with center at q and radius ρ_q. Observe for every n, $\lambda \in \mathbb{R}^d$, and measurable $G \subset \mathbb{R}^d$, that

$$\mu_n(G) = E\left[1_{\hat{S}_n \in G}\right] \leq E\left[\exp\left(\langle \lambda, \hat{S}_n \rangle - \inf_{x \in G}\{\langle \lambda, x \rangle\}\right)\right] .$$

In particular, for each n and $q \in \Gamma$,

$$\mu_n(B_{q,\rho_q}) \leq E\left[\exp(n\langle \lambda_q, \hat{S}_n \rangle)\right] \exp\left(-\inf_{x \in B_{q,\rho_q}}\{n\langle \lambda_q, x \rangle\}\right) .$$

Also, for any $q \in \Gamma$,

$$-\inf_{x \in B_{q,\rho_q}} \langle \lambda_q, x \rangle \leq \rho_q |\lambda_q| - \langle \lambda_q, q \rangle \leq \delta - \langle \lambda_q, q \rangle ,$$

and therefore,

$$\frac{1}{n} \log \mu_n(B_{q,\rho_q}) \leq -\inf_{x \in B_{q,\rho_q}} \langle \lambda_q, x \rangle + \Lambda(\lambda_q) \leq \delta - \langle \lambda_q, q \rangle + \Lambda(\lambda_q) .$$

Since Γ is compact, one may extract from the open covering $\cup_{q \in \Gamma} B_{q,\rho_q}$ of Γ a finite covering that consists of $N = N(\Gamma, \delta) < \infty$ such balls with centers $q_1, \ldots, q_N$ in Γ. By the union of events bound and the preceding inequality,

$$\frac{1}{n} \log \mu_n(\Gamma) \leq \frac{1}{n} \log N + \delta - \min_{i=1,\ldots,N}\{\langle \lambda_{q_i}, q_i \rangle - \Lambda(\lambda_{q_i})\} .$$

Hence, by our choice of λ_q,

$$\limsup_{n \to \infty} \frac{1}{n} \log \mu_n(\Gamma) \leq \delta - \min_{i=1,\ldots,N} I^\delta(q_i) .$$

Since $q_i \in \Gamma$, the upper bound (2.2.32) is established for all compact sets.

The large deviations upper bound is extended to all closed subsets of $\mathbb{R}^d$ by showing that μ_n is an exponentially tight family of probability measures and applying Lemma 1.2.18. Let $H_\rho \triangleq [-\rho, \rho]^d$. Since $H_\rho^c = \cup_{j=1}^d \{x : |x^j| > \rho\}$, the union of events bound yields

$$\mu_n(H_\rho^c) \leq \sum_{j=1}^d \mu_n^j([\rho, \infty)) + \sum_{j=1}^d \mu_n^j((-\infty, -\rho]) , \qquad (2.2.33)$$

where μ_n^j, $j = 1, \ldots, d$ are the laws of the coordinates of the random vector $\hat{S}_n$, namely, the laws governing $\frac{1}{n} \sum_{i=1}^n X_i^j$. By applying (2.2.12) and (2.2.13), one has for any $\rho \geq |\bar{x}|$,

$$\mu_n^j((-\infty, -\rho]) \leq e^{-n\Lambda_j^*(-\rho)} , \quad \mu_n^j([\rho, \infty)) \leq e^{-n\Lambda_j^*(\rho)} ,$$

where Λ_j^* denote the Fenchel–Legendre transform of $\log E[e^{\lambda X_1^j}]$, $j = 1, \ldots,$ d. As shown in Lemma 2.2.20, $\Lambda_j^*(x) \to \infty$ when $|x| \to \infty$. Therefore, by combining the preceding bounds with (2.2.33) and considering the limits, first as $n \to \infty$ and then as $\rho \to \infty$, one obtains the identity

$$\lim_{\rho \to \infty} \limsup_{n \to \infty} \frac{1}{n} \log \mu_n(H_\rho^c) = -\infty \ .$$

Consequently, $\{\mu_n\}$ is an exponentially tight sequence of probability measures, since the hypercubes H_ρ are compact.

The large deviations lower bound is next established. To this end, it suffices to prove that for every $y \in \mathcal{D}_{\Lambda^*}$ and every $\delta > 0$,

$$\liminf_{n \to \infty} \frac{1}{n} \log \mu_n(B_{y,\delta}) \geq -\Lambda^*(y) \ . \tag{2.2.34}$$

Suppose first that $y = \nabla \Lambda(\eta)$ for some $\eta \in \mathbb{R}^d$. Define the probability measure $\tilde{\mu}$ via

$$\frac{d\tilde{\mu}}{d\mu}(z) = e^{\langle \eta, z \rangle - \Lambda(\eta)} \ ,$$

and let $\tilde{\mu}_n$ denote the law of $\hat{S}_n$ when X_i are i.i.d. with law $\tilde{\mu}$. Then

$$\begin{aligned}
\frac{1}{n} \cdot \log \mu_n(B_{y,\delta}) &= \Lambda(\eta) - \langle \eta, y \rangle + \frac{1}{n} \log \int_{z \in B_{y,\delta}} e^{n\langle \eta, y-z \rangle} \tilde{\mu}_n(dz) \\
&\geq \Lambda(\eta) - \langle \eta, y \rangle - |\eta|\delta + \frac{1}{n} \log \tilde{\mu}_n(B_{y,\delta}) \ .
\end{aligned}$$

Note that by the dominated convergence theorem,

$$E_{\tilde{\mu}}(X_1) = \frac{1}{M(\eta)} \int_{\mathbb{R}^d} x e^{\langle \eta, x \rangle} d\mu = \nabla \Lambda(\eta) = y \ ,$$

and by the weak law of large numbers, $\lim_{n \to \infty} \tilde{\mu}_n(B_{y,\delta}) = 1$ for all $\delta > 0$. Moreover, since $\Lambda(\eta) - \langle \eta, y \rangle \geq -\Lambda^*(y)$, the preceding inequality implies the lower bound

$$\liminf_{n \to \infty} \frac{1}{n} \log \mu_n(B_{y,\delta}) \geq -\Lambda^*(y) - |\eta|\delta \ .$$

Hence,

$$\liminf_{n \to \infty} \frac{1}{n} \log \mu_n(B_{y,\delta}) \geq \liminf_{\delta \to 0} \liminf_{n \to \infty} \frac{1}{n} \log \mu_n(B_{y,\delta}) \geq -\Lambda^*(y) \ .$$

To extend the lower bound (2.2.34) to also cover $y \in \mathcal{D}_{\Lambda^*}$ such that $y \notin \{\nabla \Lambda(\lambda) : \lambda \in \mathbb{R}^d\}$, we now regularize μ by adding to each X_j a small

Normal random variable. Specifically, fix $M < \infty$ and let ν denote the marginal law of the i.i.d. random vectors $Y_j \triangleq X_j + V_j/\sqrt{M}$, where $V_1, \ldots, V_n$ are i.i.d. standard multivariate Normal random variables independent of $X_1, \ldots, X_n$. Let $\Lambda_M(\cdot)$ denote the logarithmic moment generating function of Y_1, while ν_n denotes the law governing $\hat{S}_n^{(M)} \triangleq \frac{1}{n} \sum_{j=1}^n Y_j$.

Since the logarithmic moment generating function of a standard multivariate Normal is $|\lambda|^2/2$, it follows that

$$\Lambda_M(\lambda) = \Lambda(\lambda) + \frac{1}{2M}|\lambda|^2 \geq \Lambda(\lambda),$$

and hence

$$\Lambda^*(y) = \sup_{\lambda \in \mathbb{R}^d} \{\langle \lambda, y \rangle - \Lambda(\lambda)\} \geq \sup_{\lambda \in \mathbb{R}^d} \{\langle \lambda, y \rangle - \Lambda_M(\lambda)\}.$$

By assumption, $\bar{x} = E(X_1)$ is finite; thus, Jensen's inequality implies that $\Lambda(\lambda) \geq \langle \lambda, \bar{x} \rangle$ for all $\lambda \in \mathbb{R}^d$. Hence, the finite and differentiable function

$$g(\lambda) \triangleq \langle \lambda, y \rangle - \Lambda_M(\lambda)$$

satisfies

$$\lim_{\rho \to \infty} \sup_{|\lambda| > \rho} g(\lambda) = -\infty.$$

Consequently, the supremum of $g(\cdot)$ over $\mathbb{R}^d$ is obtained at some finite η, for which

$$0 = \nabla g(\eta) = y - \nabla \Lambda_M(\eta),$$

namely, $y = \nabla \Lambda_M(\eta)$. Thus, by the preceding proof, the large deviations lower bound (2.2.34) applies for $\{\nu_n\}$ yielding for all $\delta > 0$,

$$\liminf_{n \to \infty} \frac{1}{n} \log \nu_n(B_{y,\delta}) \geq -\Lambda^*(y) > -\infty.$$

Observe that $\hat{S}_n^{(M)}$ possesses the same distribution as $\hat{S}_n + V/\sqrt{Mn}$, where $V \sim$ Normal$(0, I)$ and is independent of $\hat{S}_n$. Therefore,

$$\mu_n(B_{y,2\delta}) \geq \nu_n(B_{y,\delta}) - P(|V| \geq \sqrt{Mn}\delta).$$

Finally, since V is a standard multivariate Normal random vector,

$$\limsup_{n \to \infty} \frac{1}{n} \log P(|V| \geq \sqrt{Mn}\delta) \leq -\frac{M\delta^2}{2}.$$

The proof of (2.2.34) is now completed by combining the preceding three inequalities and considering first $n \to \infty$ and then $M \to \infty$. □

Remark: An inspection of the last proof reveals that the upper bound for compact sets holds with no assumptions on $\mathcal{D}_\Lambda$. The condition $0 \in \mathcal{D}_\Lambda^o$ suffices to ensure that Λ^* is a good rate function and that $\{\mu_n\}$ is exponentially tight. However, the proof of the lower bound is based on the strong assumption that $\mathcal{D}_\Lambda = \mathbb{R}^d$.

Sanov's theorem for finite alphabets, Theorem 2.1.10, may be deduced as a consequence of Cramér's theorem in $\mathbb{R}^d$. Indeed, note that the empirical mean of the random vectors $X_i \triangleq [1_{a_1}(Y_i), 1_{a_2}(Y_i), \ldots, 1_{a_{|\Sigma|}}(Y_i)]$ equals $L_n^\mathbf{Y}$, the empirical measure of the i.i.d. random variables $Y_1, \ldots, Y_n$ that take values in the finite alphabet Σ. Moreover, as X_i are bounded, $\mathcal{D}_\Lambda = \mathbb{R}^{|\Sigma|}$, and the following corollary of Cramér's theorem is obtained.

Corollary 2.2.35 *For any set Γ of probability vectors in $\mathbb{R}^{|\Sigma|}$,*

$$
-\inf_{\nu \in \Gamma^o} \Lambda^*(\nu) \leq \liminf_{n \to \infty} \frac{1}{n} \log \mathrm{P}_\mu \left(L_n^\mathbf{Y} \in \Gamma \right)
$$

$$
\leq \limsup_{n \to \infty} \frac{1}{n} \log \mathrm{P}_\mu \left(L_n^\mathbf{Y} \in \Gamma \right) \leq -\inf_{\nu \in \overline{\Gamma}} \Lambda^*(\nu) ,
$$

where Λ^* is the Fenchel–Legendre transform of the logarithmic moment generating function

$$
\Lambda(\lambda) = \log E \left(e^{\langle \lambda, X_1 \rangle} \right) = \log \sum_{i=1}^{|\Sigma|} e^{\lambda_i} \mu(a_i) ,
$$

with $\lambda = (\lambda_1, \lambda_2, \ldots, \lambda_{|\Sigma|}) \in \mathbb{R}^{|\Sigma|}$.

Remark: Comparing this corollary to Theorem 2.1.10, it is tempting to conjecture that, on the probability simplex $M_1(\Sigma)$, $\Lambda^*(\cdot) = H(\cdot|\mu)$. Indeed, this is proved in Exercise 2.2.36. Actually, as shown in Section 4.1, the rate function controlling an LDP in $\mathbb{R}^d$ is always unique.

Exercise 2.2.36 For the function $\Lambda^*(\cdot)$ defined in Corollary 2.2.35, show that $\Lambda^*(x) = H(x|\mu)$.
Hint: Prove that $\mathcal{D}_{\Lambda^*} = M_1(\Sigma)$. Show that for $\nu \in M_1(\Sigma)$, the value of $\Lambda^*(\nu)$ is obtained by taking $\lambda_i = \log[\nu(a_i)/\mu(a_i)]$ when $\nu(a_i) > 0$, and $\lambda_i \to -\infty$ when $\nu(a_i) = 0$.

Exercise 2.2.37 [From [GOR79], Example 5.1]. Corollary 2.2.19 implies that for $\{X_i\}$ i.i.d. d-dimensional random vectors with finite logarithmic moment generating function Λ, the laws μ_n of $\hat{S}_n$ satisfy

$$
\lim_{n \to \infty} \frac{1}{n} \log \mu_n(F) = -\inf_{x \in F} \Lambda^*(x)
$$

for every closed half-space F. Find a quadrant $F = [a, \infty) \times [b, \infty)$ for which the preceding identity is false.

Hint: Consider $a = b = 0.5$ and μ which is supported on the points $(0, 1)$ and $(1, 0)$.

Exercise 2.2.38 Let μ_n denote the law of $\hat{S}_n$, the empirical mean of the i.i.d. random vectors $X_i \in \mathbb{R}^d$, and $\Lambda(\cdot)$ denote the logarithmic moment generating function associated with the law of X_1. Do not assume that $\mathcal{D}_{\Lambda^*} = \mathbb{R}^d$.
(a) Use Chebycheff's inequality to prove that for any measurable $C \subset \mathbb{R}^d$, any n, and any $\lambda \in \mathbb{R}^d$,

$$\frac{1}{n} \log \mu_n(C) \leq - \inf_{y \in C} \langle \lambda, y \rangle + \Lambda(\lambda) .$$

(b) Recall the following version of the min–max theorem: Let $g(\theta, y)$ be convex and lower semicontinuous in y, concave and upper semicontinuous in θ. Let $C \subset \mathbb{R}^d$ be convex and compact. Then

$$\inf_{y \in C} \sup_{\theta} g(\theta, y) = \sup_{\theta} \inf_{y \in C} g(\theta, y)$$

(*c.f.* [ET76, page 174].) Apply this theorem to justify the upper bound

$$\frac{1}{n} \log \mu_n(C) \leq - \sup_{\lambda \in \mathbb{R}^d} \inf_{y \in C} [\langle \lambda, y \rangle - \Lambda(\lambda)] = - \inf_{y \in C} \Lambda^*(y)$$

for every n and every *convex, compact* set C.
(c) Show that the preceding upper bound holds for every n and every *convex, closed* set C by considering the convex, compact sets $C \cap [-\rho, \rho]^d$ with $\rho \to \infty$.
(d) Use this bound to show that the large deviations upper bound holds for all compact sets (with the rate function Λ^*).

Exercise 2.2.39 In this exercise, you examine the conditions needed for $\Lambda^*(x) < \infty$ when x belongs to the support of the law μ.
(a) Suppose that $\mu \in M_1(\mathbb{R}^d)$ possesses a density $f(\cdot)$ with respect to Lebesgue measure, that $f(x) > 0$, and that $f(\cdot)$ is continuous at x. Prove that under these conditions, $\Lambda^*(x) < \infty$.
Hint: Without loss of generality, you may consider only $x = 0$. Then the preceding conditions imply that $f(\cdot) \geq \epsilon > 0$ in some small ball around the origin. As a result, the moment generating function at λ is bounded below by $c \sinh(|\lambda|a)/(|\lambda|a)$ for some $a, c > 0$.
(b) Suppose that $\mu \in M_1(\mathbb{R})$ possesses the density $f(x) = x$ for $x \in [0, 1]$, $f(x) = 2 - x$ for $x \in [1, 2]$, and zero otherwise. Prove that for this measure, $\Lambda^*(0) = \infty$ while $\mu((-\delta, \delta)) > 0$ for all $\delta > 0$.

Exercise 2.2.40 Let $(w_{t_1}, \ldots, w_{t_d})$ be samples of a Brownian motion at the fixed times $t_1, \ldots, t_d$; so $\{w_{t_{j+1}} - w_{t_j}\}$ are zero-mean, independent Normal random variables of variances $\{t_{j+1} - t_j\}$, respectively. Find the rate function for the empirical mean $\hat{S}_n$ of $X_i \stackrel{\triangle}{=} (w_{t_1}^i, \ldots, w_{t_d}^i)$, where $w_{t_j}^i$, $i = 1, \ldots, n$ are samples of independent Brownian motions at time instances t_j.

Remark: Note that the law of $\hat{S}_n$ is the same as that of $(1/\sqrt{n})\,(w_{t_1}, \ldots, w_{t_d})$, and compare your result with Theorem 5.2.3.

2.3 The Gärtner–Ellis Theorem

Cramér's theorem (Theorem 2.2.30) is limited to the i.i.d. case. However, a glance at the proof should be enough to convince the reader that some extension to the non-i.i.d. case is possible. It is the goal of this section to present such an extension. As a by-product, a somewhat stronger version of Theorem 2.2.30 follows. (See Exercise 2.3.16.) Simple applications are left as Exercises 2.3.19 and 2.3.23, whereas Section 3.1 is devoted to a class of important applications, the large deviations of the empirical measure for finite state Markov chains.

Consider a sequence of random vectors $Z_n \in \mathbb{R}^d$, where Z_n possesses the law μ_n and logarithmic moment generating function

$$\Lambda_n(\lambda) \stackrel{\triangle}{=} \log E\left[e^{\langle \lambda, Z_n \rangle}\right]. \tag{2.3.1}$$

The existence of a limit of properly scaled logarithmic moment generating functions indicates that μ_n may satisfy the LDP. Specifically, the following assumption is imposed throughout this section.

Assumption 2.3.2 *For each $\lambda \in \mathbb{R}^d$, the logarithmic moment generating function, defined as the limit*

$$\Lambda(\lambda) \stackrel{\triangle}{=} \lim_{n \to \infty} \frac{1}{n} \Lambda_n(n\lambda)$$

exists as an extended real number. Further, the origin belongs to the interior of $\mathcal{D}_\Lambda \stackrel{\triangle}{=} \{\lambda \in \mathbb{R}^d : \Lambda(\lambda) < \infty\}$.

In particular, if μ_n is the law governing the empirical mean $\hat{S}_n$ of i.i.d. random vectors $X_i \in \mathbb{R}^d$, then for every $n \in \mathbb{Z}_+$,

$$\frac{1}{n}\Lambda_n(n\lambda) = \Lambda(\lambda) \stackrel{\triangle}{=} \log E[e^{\langle \lambda, X_1 \rangle}],$$

and Assumption 2.3.2 holds whenever $0 \in \mathcal{D}_\Lambda^\circ$.

Let $\Lambda^*(\cdot)$ be the Fenchel–Legendre transform of $\Lambda(\cdot)$, with $\mathcal{D}_{\Lambda^*} = \{x \in \mathbb{R}^d : \Lambda^*(x) < \infty\}$. Motivated by Theorem 2.2.30, it is our goal to state conditions under which the sequence μ_n satisfies the LDP with the rate function Λ^*.

Definition 2.3.3 $y \in \mathbb{R}^d$ *is an exposed point of* Λ^* *if for some* $\lambda \in \mathbb{R}^d$ *and all* $x \neq y$,

$$\langle \lambda, y \rangle - \Lambda^*(y) > \langle \lambda, x \rangle - \Lambda^*(x). \tag{2.3.4}$$

λ *in (2.3.4) is called an exposing hyperplane.*

Definition 2.3.5 *A convex function* $\Lambda : \mathbb{R}^d \to (-\infty, \infty]$ *is essentially smooth if:*
(a) $\mathcal{D}_\Lambda^o$ *is non-empty.*
(b) $\Lambda(\cdot)$ *is differentiable throughout* $\mathcal{D}_\Lambda^o$.
(c) $\Lambda(\cdot)$ *is steep, namely,* $\lim_{n \to \infty} |\nabla \Lambda(\lambda_n)| = \infty$ *whenever* $\{\lambda_n\}$ *is a sequence in* $\mathcal{D}_\Lambda^o$ *converging to a boundary point of* $\mathcal{D}_\Lambda^o$.

The following theorem is the main result of this section.

Theorem 2.3.6 (Gärtner–Ellis) *Let Assumption 2.3.2 hold.*
(a) For any closed set F,

$$\limsup_{n \to \infty} \frac{1}{n} \log \mu_n(F) \leq - \inf_{x \in F} \Lambda^*(x). \tag{2.3.7}$$

(b) For any open set G,

$$\liminf_{n \to \infty} \frac{1}{n} \log \mu_n(G) \geq - \inf_{x \in G \cap \mathcal{F}} \Lambda^*(x); \tag{2.3.8}$$

where $\mathcal{F}$ *is the set of exposed points of* Λ^* *whose exposing hyperplane belongs to* $\mathcal{D}_\Lambda^o$.
(c) If Λ *is an essentially smooth, lower semicontinuous function, then the LDP holds with the good rate function* $\Lambda^*(\cdot)$.

Remarks:
(a) All results developed in this section are valid, as in the statement (1.2.14) of the LDP, when $1/n$ is replaced by a sequence of constants $a_n \to 0$, or even when a continuous parameter family $\{\mu_\epsilon\}$ is considered, with Assumption 2.3.2 properly modified.
(b) The essential ingredients for the proof of parts (a) and (b) of the Gärtner–Ellis theorem are those presented in the course of proving Cramér's theorem in $\mathbb{R}^d$; namely, Chebycheff's inequality is applied for deriving the upper bound and an exponential change of measure is used for deriving the

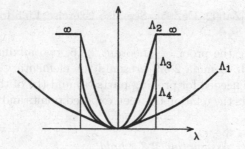

Figure 2.3.1: $\Lambda_1 - \Lambda_3$ are steep; Λ_4 is not.

lower bound. However, since the law of large numbers is no longer available *a priori*, the large deviations *upper* bound is used in order to prove the lower bound.

(c) The proof of part (c) of the Gärtner–Ellis theorem depends on rather intricate convex analysis considerations that are summarized in Lemma 2.3.12. A proof of this part for the case of $\mathcal{D}_\Lambda = \mathbb{R}^d$, which avoids these convex analysis considerations, is outlined in Exercise 2.3.20. This proof, which is similar to the proof of Cramér's theorem in $\mathbb{R}^d$, is based on a regularization of the random variables Z_n by adding asymptotically negligible Normal random variables.

(d) Although the Gärtner–Ellis theorem is quite general in its scope, it does not cover all $\mathbb{R}^d$ cases in which an LDP exists. As an illustrative example, consider $Z_n \sim$ Exponential(n). Assumption 2.3.2 then holds with $\Lambda(\lambda) = 0$ for $\lambda < 1$ and $\Lambda(\lambda) = \infty$ otherwise. Moreover, the law of Z_n possesses the density $ne^{-nz}1_{[0,\infty)}(z)$, and consequently the LDP holds with the good rate function $I(x) = x$ for $x \geq 0$ and $I(x) = \infty$ otherwise. A direct computation reveals that $I(\cdot) = \Lambda^*(\cdot)$. Hence, $\mathcal{F} = \{0\}$ while $\mathcal{D}_{\Lambda^*} = [0, \infty)$, and therefore the Gärtner–Ellis theorem yields a trivial lower bound for sets that do not contain the origin. A non-trivial example of the same phenomenon is presented in Exercise 2.3.24, motivated by the problem of non-coherent detection in digital communication. In that exercise, the Gärtner–Ellis method is refined, and by using a change of measure that depends on n, the LDP is proved. See also [DeZ95] for a more general exposition of this approach and [BryD97] for its application to quadratic forms of stationary Gaussian processes.

(e) Assumption 2.3.2 implies that $\Lambda^*(x) \leq \liminf_{n\to\infty} \Lambda_n^*(x)$ for $\Lambda_n^*(x) = \sup_\lambda\{\langle\lambda, x\rangle - n^{-1}\Lambda_n(n\lambda)\}$. However, pointwise convergence of $\Lambda_n^*(x)$ to $\Lambda^*(x)$ is not guaranteed. For example, when $P(Z_n = n^{-1}) = 1$, we have $\Lambda_n(\lambda) = \lambda/n \to 0 = \Lambda(\lambda)$, while $\Lambda_n^*(0) = \infty$ and $\Lambda^*(0) = 0$. This phenomenon is relevant when trying to go beyond the Gärtner–Ellis theorem, as for example in [Zab92, DeZ95]. See also Exercise 4.5.5 for another motivation.

Before bringing the proof of Theorem 2.3.6, two auxiliary lemmas are stated and proved. Lemma 2.3.9 presents the elementary properties of Λ and Λ^*, which are needed for proving parts (a) and (b) of the theorem, and moreover highlights the relation between exposed points and differentiability properties.

Lemma 2.3.9 *Let Assumption 2.3.2 hold.*
(a) $\Lambda(\lambda)$ is a convex function, $\Lambda(\lambda) > -\infty$ everywhere, and $\Lambda^*(x)$ is a good convex rate function.
(b) Suppose that $y = \nabla\Lambda(\eta)$ for some $\eta \in \mathcal{D}_\Lambda^o$. Then

$$\Lambda^*(y) = \langle \eta, y \rangle - \Lambda(\eta) \,. \tag{2.3.10}$$

Moreover $y \in \mathcal{F}$, with η being the exposing hyperplane for y.

Proof: (a) Since Λ_n are convex functions (see the proof of part (a) of Lemma 2.2.5), so are $\Lambda_n(n\cdot)/n$, and their limit $\Lambda(\cdot)$ is convex as well. Moreover, $\Lambda_n(0) = 0$ and, therefore, $\Lambda(0) = 0$, implying that Λ^* is nonnegative.

If $\Lambda(\lambda) = -\infty$ for some $\lambda \in \mathbb{R}^d$, then by convexity $\Lambda(\alpha\lambda) = -\infty$ for all $\alpha \in (0,1]$. Since $\Lambda(0) = 0$, it follows by convexity that $\Lambda(-\alpha\lambda) = \infty$ for all $\alpha \in (0,1]$, contradicting the assumption that $0 \in \mathcal{D}_\Lambda^o$. Thus, $\Lambda > -\infty$ everywhere.

Since $0 \in \mathcal{D}_\Lambda^o$, it follows that $\overline{B}_{0,\delta} \subset \mathcal{D}_\Lambda^o$ for some $\delta > 0$, and $c = \sup_{\lambda \in \overline{B}_{0,\delta}} \Lambda(\lambda) < \infty$ because the convex function Λ is continuous in $\mathcal{D}_\Lambda^o$. (See Appendix A.) Therefore,

$$\Lambda^*(x) \geq \sup_{\lambda \in \overline{B}_{0,\delta}} \{\langle \lambda, x \rangle - \Lambda(\lambda)\}$$

$$\geq \sup_{\lambda \in \overline{B}_{0,\delta}} \langle \lambda, x \rangle - \sup_{\lambda \in \overline{B}_{0,\delta}} \Lambda(\lambda) = \delta|x| - c \,. \tag{2.3.11}$$

Thus, for every $\alpha < \infty$, the level set $\{x : \Lambda^*(x) \leq \alpha\}$ is bounded. The function Λ^* is both convex and lower semicontinuous, by an argument similar to the proof of part (a) of Lemma 2.2.5. Hence, Λ^* is indeed a good convex rate function.
(b) The proof of (2.3.10) is a repeat of the proof of part (b) of Lemma 2.2.31. Suppose now that for some $x \in \mathbb{R}^d$,

$$\Lambda(\eta) = \langle \eta, y \rangle - \Lambda^*(y) \leq \langle \eta, x \rangle - \Lambda^*(x) \,.$$

Then, for every $\theta \in \mathbb{R}^d$,

$$\langle \theta, x \rangle \leq \Lambda(\eta + \theta) - \Lambda(\eta) \,.$$

In particular,

$$\langle \theta, x \rangle \leq \lim_{\epsilon \to 0} \frac{1}{\epsilon} [\Lambda(\eta + \epsilon\theta) - \Lambda(\eta)] = \langle \theta, \nabla\Lambda(\eta) \rangle .$$

Since this inequality holds for all $\theta \in \mathbb{R}^d$, necessarily $x = \nabla\Lambda(\eta) = y$. Hence, y is an exposed point of Λ^*, with exposing hyperplane $\eta \in \mathcal{D}_\Lambda^o$. $\square$

For every non-empty convex set C, the *relative interior* of C, denoted ri C, is defined as the set

$$\text{ri } C \triangleq \{y \in C : x \in C \Rightarrow y - \epsilon(x - y) \in C \text{ for some } \epsilon > 0\} .$$

Various properties of the relative interiors of convex sets are collected in Appendix A.

The following lemma, which is Corollary 26.4.1 of [Roc70], is needed for proving part (c) of the Gärtner–Ellis theorem.

Lemma 2.3.12 (Rockafellar) *If* $\Lambda : \mathbb{R}^d \to (-\infty, \infty]$ *is an essentially smooth, lower semicontinuous, convex function, then* ri $\mathcal{D}_{\Lambda^*} \subseteq \mathcal{F}$.

Proof: The proof is based on the results of Appendix A. Note first that there is nothing to prove if $\mathcal{D}_{\Lambda^*}$ is empty. Hence, it is assumed hereafter that $\mathcal{D}_{\Lambda^*}$ is non-empty. Fix a point $x \in \text{ri } \mathcal{D}_{\Lambda^*}$ and define the function

$$f(\lambda) \triangleq \Lambda(\lambda) - \langle \lambda, x \rangle + \Lambda^*(x) .$$

If $f(\lambda) = 0$, then clearly λ belongs to the *subdifferential* of $\Lambda^*(\cdot)$ at x. (Recall that for any convex function $g(x)$, the subdifferential at x is just the set $\{\lambda : g(y) \geq g(x) + \langle \lambda, y - x \rangle \; \forall y \in \mathbb{R}^d\}$.) The proof that $x \in \mathcal{F}$ is based on showing that such a λ exists and that it belongs to $\mathcal{D}_\Lambda^o$.

Observe that $f : \mathbb{R}^d \to [0, \infty]$ is a convex, lower semicontinuous function, and $\inf_{\lambda \in \mathbb{R}^d} f(\lambda) = 0$. It is easy to check that the Fenchel–Legendre transform of $f(\cdot)$ is $f^*(\cdot) = \Lambda^*(\cdot + x) - \Lambda^*(x)$. Therefore, with $x \in \text{ri } \mathcal{D}_{\Lambda^*}$, it follows that $0 \in \text{ri } \mathcal{D}_{f^*}$. By Lemma A.2, there exists an $\eta \in \mathcal{D}_\Lambda$ such that $f(\eta) = 0$. Let $\tilde{\Lambda}(\cdot) = \Lambda(\cdot + \eta) - \Lambda(\eta)$. By our assumptions, $\tilde{\Lambda}(\cdot)$ is an essentially smooth, convex function and $\tilde{\Lambda}(0) = 0$. Moreover, it is easy to check that $\tilde{\Lambda}^*(x) = f(\eta) = 0$. Consequently, by Lemma A.5, $\tilde{\Lambda}(\cdot)$ is finite in a neighborhood of the origin. Hence, $\eta \in \mathcal{D}_\Lambda^o$ and by our assumptions $f(\cdot)$ is differentiable at η. Moreover, $f(\eta) = \inf_{\lambda \in \mathbb{R}^d} f(\lambda)$, implying that $\nabla f(\eta) = 0$, i.e., $x = \nabla\Lambda(\eta)$. It now follows by part (b) of Lemma 2.3.9 that $x \in \mathcal{F}$. Since this holds true for every $x \in \text{ri } \mathcal{D}_{\Lambda^*}$, the proof of the lemma is complete. $\square$

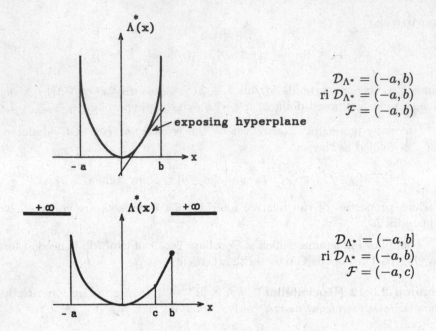

Figure 2.3.2: $\mathcal{D}_{\Lambda^*}$, ri $\mathcal{D}_{\Lambda^*}$, and $\mathcal{F}$.

Proof of the Gärtner–Ellis theorem: (a) The upper bound (2.3.7) for compact sets is established by the same argument as in the proof of Cramér's theorem in $\mathbb{R}^d$. The details are omitted, as they are presented for a more general setup in the course of the proof of Theorem 4.5.3. The extension to all closed sets follows by proving that the sequence of measures $\{\mu_n\}$ is exponentially tight. To this end, let $\mathbf{u}_j$ denote the jth unit vector in $\mathbb{R}^d$ for $j = 1,\ldots,d$. Since $0 \in \mathcal{D}_\Lambda^o$, there exist $\theta_j > 0$, $\eta_j > 0$ such that $\Lambda(\theta_j \mathbf{u}_j) < \infty$ and $\Lambda(-\eta_j \mathbf{u}_j) < \infty$ for $j = 1,\ldots,d$. Therefore, by Chebycheff's inequality,

$$\mu_n^j((-\infty,-\rho]) \le \exp(-n\eta_j\rho + \Lambda_n(-n\eta_j\mathbf{u}_j)),$$
$$\mu_n^j([\rho,\infty)) \le \exp(-n\theta_j\rho + \Lambda_n(n\theta_j\mathbf{u}_j)), \quad j = 1,\ldots,d,$$

where μ_n^j, $j = 1,\ldots,d$ are the laws of the coordinates of the random vector Z_n. Hence, for $j = 1,\ldots,d$,

$$\lim_{\rho\to\infty}\limsup_{n\to\infty}\frac{1}{n}\log\mu_n^j((-\infty,-\rho]) = -\infty,$$

$$\lim_{\rho\to\infty}\limsup_{n\to\infty}\frac{1}{n}\log\mu_n^j([\rho,\infty)) = -\infty.$$

Consequently, by the union of events bound, and Lemma 1.2.15, $\lim_{\rho\to\infty}$ $\limsup_{n\to\infty} \frac{1}{n} \log \mu_n(([-\rho,\rho]^d)^c) = -\infty$, i.e., $\{\mu_n\}$ is an exponentially tight sequence of probability measures.

(b) In order to establish the lower bound (2.3.8) for every open set, it suffices to prove that for all $y \in \mathcal{F}$,

$$\lim_{\delta\to 0}\liminf_{n\to\infty} \frac{1}{n} \log \mu_n(B_{y,\delta}) \geq -\Lambda^*(y). \qquad (2.3.13)$$

Fix $y \in \mathcal{F}$ and let $\eta \in \mathcal{D}_\Lambda^o$ denote an exposing hyperplane for y. Then, for all n large enough, $\Lambda_n(n\eta) < \infty$ and the associated probability measures $\tilde{\mu}_n$ are well-defined via

$$\frac{d\tilde{\mu}_n}{d\mu_n}(z) = \exp\left[n\langle\eta, z\rangle - \Lambda_n(n\eta)\right].$$

After some calculations,

$$\frac{1}{n} \log \mu_n(B_{y,\delta}) = \frac{1}{n}\Lambda_n(n\eta) - \langle\eta, y\rangle + \frac{1}{n} \log \int_{z\in B_{y,\delta}} e^{n\langle\eta, y-z\rangle}\, d\tilde{\mu}_n(z)$$

$$\geq \frac{1}{n}\Lambda_n(n\eta) - \langle\eta, y\rangle - |\eta|\delta + \frac{1}{n} \log \tilde{\mu}_n(B_{y,\delta}).$$

Therefore,

$$\lim_{\delta\to 0}\liminf_{n\to\infty} \frac{1}{n} \log \mu_n(B_{y,\delta})$$

$$\geq \Lambda(\eta) - \langle\eta, y\rangle + \lim_{\delta\to 0}\liminf_{n\to\infty} \frac{1}{n} \log \tilde{\mu}_n(B_{y,\delta})$$

$$\geq -\Lambda^*(y) + \lim_{\delta\to 0}\liminf_{n\to\infty} \frac{1}{n} \log \tilde{\mu}_n(B_{y,\delta}), \qquad (2.3.14)$$

where the second inequality follows by the definition of Λ^* (and actually holds with equality due to (2.3.10)).

Here, a new obstacle stems from the removal of the independence assumption, since the weak law of large numbers no longer applies. Our strategy is to use the upper bound proved in part (a). For that purpose, it is first verified that $\tilde{\mu}_n$ satisfies Assumption 2.3.2 with the limiting logarithmic moment generating function $\tilde{\Lambda}(\cdot) \triangleq \Lambda(\cdot + \eta) - \Lambda(\eta)$. Indeed, $\tilde{\Lambda}(0) = 0$, and since $\eta \in \mathcal{D}_\Lambda^o$, it follows that $\tilde{\Lambda}(\lambda) < \infty$ for all $|\lambda|$ small enough. Let $\tilde{\Lambda}_n(\cdot)$ denote the logarithmic moment generating function corresponding to the law $\tilde{\mu}_n$. Then for every $\lambda \in \mathbb{R}^d$,

$$\frac{1}{n}\tilde{\Lambda}_n(n\lambda) \triangleq \frac{1}{n} \log\left[\int_{\mathbb{R}^d} e^{n\langle\lambda, z\rangle}\, d\tilde{\mu}_n(z)\right]$$

$$= \frac{1}{n}\Lambda_n(n(\lambda + \eta)) - \frac{1}{n}\Lambda_n(n\eta) \to \tilde{\Lambda}(\lambda).$$

because $\Lambda_n(n\eta) < \infty$ for all n large enough. Define

$$\tilde{\Lambda}^*(x) \triangleq \sup_{\lambda \in \mathbb{R}^d} \{\langle \lambda, x \rangle - \tilde{\Lambda}(\lambda)\} = \Lambda^*(x) - \langle \eta, x \rangle + \Lambda(\eta) . \tag{2.3.15}$$

Since Assumption 2.3.2 also holds for $\tilde{\mu}_n$, it follows by applying Lemma 2.3.9 to $\tilde{\Lambda}$ that $\tilde{\Lambda}^*$ is a good rate function. Moreover, by part (a) earlier, a large deviations *upper* bound of the form of (2.3.7) holds for the sequence of measures $\tilde{\mu}_n$, and with the good rate function $\tilde{\Lambda}^*$. In particular, for the closed set $B_{y,\delta}^c$, it yields

$$\limsup_{n \to \infty} \frac{1}{n} \log \tilde{\mu}_n (B_{y,\delta}^c) \leq - \inf_{x \in B_{y,\delta}^c} \tilde{\Lambda}^*(x) = -\tilde{\Lambda}^*(x_0)$$

for some $x_0 \neq y$, where the equality follows from the goodness of $\tilde{\Lambda}^*(\cdot)$. Recall that y is an exposed point of Λ^*, with η being the exposing hyperplane. Hence, since $\Lambda^*(y) \geq [\langle \eta, y \rangle - \Lambda(\eta)]$, and $x_0 \neq y$,

$$\tilde{\Lambda}^*(x_0) \geq [\Lambda^*(x_0) - \langle \eta, x_0 \rangle] - [\Lambda^*(y) - \langle \eta, y \rangle] > 0 .$$

Thus, for every $\delta > 0$,

$$\limsup_{n \to \infty} \frac{1}{n} \log \tilde{\mu}_n(B_{y,\delta}^c) < 0 .$$

This inequality implies that $\tilde{\mu}_n(B_{y,\delta}^c) \to 0$ and hence $\tilde{\mu}_n(B_{y,\delta}) \to 1$ for all $\delta > 0$. In particular,

$$\lim_{\delta \to 0} \liminf_{n \to \infty} \frac{1}{n} \log \tilde{\mu}_n(B_{y,\delta}) = 0 ,$$

and the lower bound (2.3.13) follows by (2.3.14).

(c) In view of parts (a) and (b) and Lemma 2.3.12, it suffices to show that for any open set G,

$$\inf_{x \in G \cap \mathrm{ri}\, \mathcal{D}_{\Lambda^*}} \Lambda^*(x) \leq \inf_{x \in G} \Lambda^*(x) .$$

There is nothing to prove if $G \cap \mathcal{D}_{\Lambda^*} = \emptyset$. Thus, we may, and will, assume that $\mathcal{D}_{\Lambda^*}$ is non-empty, which implies that there exists some $z \in \mathrm{ri}\, \mathcal{D}_{\Lambda^*}$. Fix $y \in G \cap \mathcal{D}_{\Lambda^*}$. Then, for all sufficiently small $\alpha > 0$,

$$\alpha z + (1 - \alpha)y \in G \cap \mathrm{ri}\, \mathcal{D}_{\Lambda^*} .$$

Hence,

$$\inf_{x \in G \cap \mathrm{ri}\, \mathcal{D}_{\Lambda^*}} \Lambda^*(x) \leq \lim_{\alpha \searrow 0} \Lambda^*(\alpha z + (1 - \alpha)y) \leq \Lambda^*(y) .$$

(Consult Appendix A for the preceding convex analysis results.) The arbitrariness of y completes the proof.   □

Remark: As shown in Chapter 4, the preceding proof actually extends under certain restrictions to general topological vector spaces. However, two points of caution are that the exponential tightness has to be proved on a case-by-case basis, and that in infinite dimensional spaces, the convex analysis issues are more subtle.

Exercise 2.3.16 (a) Prove by an application of Fatou's lemma that any logarithmic moment generating function is lower semicontinuous.
(b) Prove that the conclusions of Theorem 2.2.30 hold whenever the logarithmic moment generating function $\Lambda(\cdot)$ of (2.2.1) is a steep function that is finite in some ball centered at the origin.
Hint: Assumption 2.3.2 holds with $\Lambda(\cdot)$ being the logarithmic moment generating function of (2.2.1). This function is lower semicontinuous by part (a). Check that it is differentiable in $\mathcal{D}_\Lambda^o$ and apply the Gärtner–Ellis theorem.

Exercise 2.3.17 (a) Find a non-steep logarithmic moment generating function for which $0 \in \mathcal{D}_\Lambda^o$.
Hint: Try a distribution with density $g(x) = Ce^{-|x|}/(1 + |x|^{d+2})$.
(b) Find a logarithmic moment generating function for which $\overline{\mathcal{F}} = \overline{B}_{0,a}$ for some $a < \infty$ while $\mathcal{D}_{\Lambda^*} = \mathbb{R}^d$.
Hint: Show that for the density $g(x)$, $\Lambda(\lambda)$ depends only on $|\lambda|$ and $\mathcal{D}_\Lambda = \overline{B}_{0,1}$. Hence, the limit of $|\nabla\Lambda(\lambda)|$ as $|\lambda| \nearrow 1$, denoted by a, is finite, while $\Lambda^*(x) = |x| - \Lambda(1)$ for all $x \notin B_{0,a}$.

Exercise 2.3.18 (a) Prove that if $\Lambda(\cdot)$ is a steep logarithmic moment generating function, then $\exp(\Lambda(\cdot)) - 1$ is also a steep function.
Hint: Recall that Λ is lower semicontinuous.
(b) Let X_j be $\mathbb{R}^d$-valued i.i.d. random variables with a steep logarithmic moment generating function Λ such that $0 \in \mathcal{D}_\Lambda^o$. Let $N(t)$ be a Poisson process of unit rate that is independent of the X_j variables, and consider the random variables

$$\hat{S}_n \triangleq \frac{1}{n} \sum_{j=1}^{N(n)} X_j .$$

Let μ_n denote the law of $\hat{S}_n$ and prove that μ_n satisfies the LDP, with the rate function being the Fenchel–Legendre transform of $e^{\Lambda(\lambda)} - 1$.
Hint: You can apply part (b) of Exercise 2.3.16 as $N(n) = \sum_{j=1}^n N_j$, where N_j are i.i.d. Poisson(1) random variables.

Exercise 2.3.19 Let $N(n)$ be a sequence of nonnegative integer-valued random variables such that the limit $\Lambda(\lambda) = \lim_{n\to\infty} \frac{1}{n} \log E[e^{\lambda N(n)}]$ exists, and $0 \in \mathcal{D}_\Lambda^o$. Let X_j be $\mathbb{R}^d$-valued i.i.d. random variables, independent of $\{N(n)\}$, with *finite* logarithmic moment generating function Λ_X. Let μ_n denote the law of

$$Z_n \stackrel{\triangle}{=} \frac{1}{n} \sum_{j=1}^{N(n)} X_j.$$

(a) Prove that if the convex function $\Lambda(\cdot)$ is essentially smooth and lower semicontinuous, then so is $\Lambda(\Lambda_X(\cdot))$, and moreover, $\Lambda(\Lambda_X(\cdot))$ is finite in some ball around the origin.

Hint: Show that either $\mathcal{D}_\Lambda = (-\infty, a]$ or $\mathcal{D}_\Lambda = (-\infty, a)$ for some $a > 0$. Moreover, if $a < \infty$ and $\Lambda_X(\cdot) \leq a$ in some ball around λ, then $\Lambda_X(\lambda) < a$.

(b) Deduce that $\{\mu_n\}$ then satisfies the LDP with the rate function being the Fenchel–Legendre transform of $\Lambda(\Lambda_X(\cdot))$.

Exercise 2.3.20 Suppose that Assumption 2.3.2 holds for Z_n and that $\Lambda(\cdot)$ is finite and differentiable everywhere. For all $\delta > 0$, let $Z_{n,\delta} = Z_n + \sqrt{\delta/n}V$, where V is a standard multivariate Normal random variable independent of Z_n.

(a) Check that Assumption 2.3.2 holds for $Z_{n,\delta}$ with the finite and differentiable limiting logarithmic moment generating function $\Lambda_\delta(\lambda) = \Lambda(\lambda) + \delta|\lambda|^2/2$.

(b) Show that for any $x \in \mathbb{R}^d$, the value of the Fenchel–Legendre transform of Λ_δ does not exceed $\Lambda^*(x)$.

(c) Observe that the upper bound (2.3.7) for $F = \mathbb{R}^d$ implies that $\mathcal{D}_{\Lambda^*}$ is non-empty, and deduce that every $x \in \mathbb{R}^d$ is an exposed point of the Fenchel–Legendre transform of $\Lambda_\delta(\cdot)$.

(d) By applying part (b) of the Gärtner–Ellis theorem for $Z_{n,\delta}$ and (a)–(c) of this exercise, deduce that for any $x \in \mathbb{R}^d$ and any $\epsilon > 0$,

$$\liminf_{n\to\infty} \frac{1}{n} \log \mathrm{P}(Z_{n,\delta} \in B_{x,\epsilon/2}) \geq -\Lambda^*(x). \qquad (2.3.21)$$

(e) Prove that

$$\limsup_{n\to\infty} \frac{1}{n} \log \mathrm{P}\left(\sqrt{\delta/n}|V| \geq \epsilon/2\right) \leq -\frac{\epsilon^2}{8\delta}. \qquad (2.3.22)$$

(f) By combining (2.3.21) and (2.3.22), prove that the large deviations lower bound holds for the laws μ_n corresponding to Z_n.

(g) Deduce now by part (a) of the Gärtner–Ellis theorem that $\{\mu_n\}$ satisfies the LDP with rate function Λ^*.

Remark: This derivation of the LDP when $\mathcal{D}_\Lambda = \mathbb{R}^d$ avoids Rockafellar's lemma (Lemma 2.3.12).

Exercise 2.3.23 Let $X_1, \ldots, X_n, \ldots$ be a real-valued, zero mean, stationary Gaussian process with covariance sequence $R_i \stackrel{\triangle}{=} E(X_n X_{n+i})$. Suppose the process has a finite power P defined via $P \stackrel{\triangle}{=} \lim_{n\to\infty} \sum_{i=-n}^{n} R_i \left(1 - \frac{|i|}{n}\right)$. Let μ_n be the law of the empirical mean $\hat{S}_n$ of the first n samples of this process. Prove that $\{\mu_n\}$ satisfy the LDP with the good rate function $\Lambda^*(x) = x^2/2P$.

Exercise 2.3.24 [Suggested by Y. Kofman]. This exercise presents yet another example of an LDP in $\mathbb{R}$ that is not covered by the Gärtner–Ellis theorem, but that may be proved by perturbing the change of measure arguments, allowing the change of measure parameter to depend on n.

The motivation for the exercise comes from the problem of non-coherent detection in digital communication. The optimal receiver for the detection of orthogonal signals in Gaussian white noise forms from the signal received the random variable

$$Z_n = \frac{N_2^2/2 - (N_1/\sqrt{2} + \sqrt{cn})^2}{n}$$

where N_1, N_2 are independent standard Normal random variables (quadrature noise), c (signal) is some deterministic positive constant, and n is related to the signal-to-noise ratio. (See [Pro83, section 4.3.1, pages 205–209].) The optimal receiver then makes the decision "signal is present" if $Z_n \leq 0$.

The error probability, in this situation, is the probability $P(Z_n > 0)$. More generally, the probabilities $P(Z_n > z)$ are of interest.
(a) Show that

$$E(e^{n\lambda Z_n}) = \begin{cases} \frac{1}{\sqrt{1-\lambda^2}} e^{-\lambda cn/(1+\lambda)} & \text{if } \lambda \in (-1,1) \\ \infty & \text{otherwise,} \end{cases}$$

$$\Lambda(\lambda) = \begin{cases} -\frac{\lambda c}{1+\lambda} & \text{if } \lambda \in (-1,1) \\ \infty & \text{otherwise,} \end{cases}$$

and

$$\Lambda^*(x) = \begin{cases} (\sqrt{c} - \sqrt{-x})^2 & , \quad x \leq -c/4 \\ (x + c/2) & , \quad x > -c/4 \,. \end{cases}$$

Thus, $\mathcal{D}_{\Lambda^*} = \mathbb{R}$ while $\mathcal{F} = (-\infty, -c/4)$.
(b) Check that the Gärtner–Ellis theorem yields both the large deviations upper bound for arbitrary sets and the correct lower bound for open sets $G \subset \mathcal{F}$.
(c) Generalize the lower bound to arbitrary open sets. To this end, fix $z > -c/4$ and define the measures $\tilde{\mu}_n$ as in the proof of the Gärtner–Ellis theorem, except that instead of a fixed η consider $\eta_n = 1 - 1/[2n(z + c/4 + r_n)]$, where $r_n \to 0$ is a sequence of positive constants for which $nr_n \to \infty$. Show that $n^{-1} \log E(\exp(n\eta_n Z_n)) \to -c/2$ and that under $\tilde{\mu}_n$, $N_1/\sqrt{n}$ (and $N_2/\sqrt{n}$) are independent Normal random variables with mean $-\sqrt{c/2} + o(r_n)$ (respectively, 0) and variance $o(r_n)$ (respectively, $2(z + c/4 + r_n)$). Use this fact to show that for any δ,

$$\tilde{\mu}_n(z - \delta, z + \delta)$$
$$\geq \tilde{\mu}_n \left(N_1/\sqrt{n} \in \left(\sqrt{c/2 - \delta/2} - \sqrt{2c}, \sqrt{c/2 + \delta/2} - \sqrt{2c} \right) \right)$$
$$\cdot \tilde{\mu}_n \left(N_2/\sqrt{n} \in \left(\sqrt{2z + c/2 - \delta}, \sqrt{2z + c/2 + \delta} \right) \right) \geq c_1 \,,$$

where c_1 is a constant that depends on c and δ but does not depend on n. Deduce that for every $\delta > 0$, $\liminf_{n\to\infty} \tilde{\mu}_n((z-\delta, z+\delta)) > 0$, and complete the construction of the lower bound.

Remarks:
(a) By a direct computation (*c.f.* [Pro83]),

$$P(Z_n > 0) = 2\Phi(\sqrt{cn})(1 - \Phi(\sqrt{cn})),$$

where $\Phi(x) = (1/\sqrt{2\pi}) \int_{-\infty}^{x} e^{-\theta^2/2} d\theta$.
(b) The lower bound derived here completes the upper bounds of [SOSL85, page 208] and may be extended to M-ary channels.

Exercise 2.3.25 Suppose that $0 \in \mathcal{D}_\Lambda^o$ for $\Lambda(\lambda) = \limsup_{n\to\infty} n^{-1}\Lambda_n(n\lambda)$.
(a) Show that Lemma 2.3.9 and part (a) of the Gärtner–Ellis theorem hold under this weaker form of Assumption 2.3.2.
(b) Show that if $z = \nabla\Lambda(0)$ then $P(|Z_n - z| \geq \delta) \to 0$ exponentially in n for any fixed $\delta > 0$ (sometimes called *the exponential convergence of Z_n to z*).
Hint: Check that $\Lambda^*(x) > \Lambda^*(z) = 0$ for all $x \neq z$.

Exercise 2.3.26 Let $Z_n = \sum_{i=1}^{n} \eta_i^{(n)} Y_i^2$, where Y_i are real-valued i.i.d. $N(0,1)$ random variables and $\{\eta_1^{(n)} \geq \eta_2^{(n)} \geq \cdots \geq \eta_n^{(n)} \geq 0\}$ are non-random such that the real-valued sequence $\{\eta_1^{(n)}\}$ converges to $M < \infty$ and the empirical measure $n^{-1}\sum_{i=1}^{n} \delta_{\eta_i^{(n)}}$ converges in law to a probability measure μ. Define $\Lambda(\theta) = -0.5 \int \log(1 - 2\theta\eta)\mu(d\eta)$ for $\theta \leq 1/(2M)$ and $\Lambda(\theta) = \infty$ otherwise.
(a) Check that the logarithmic moment generating functions Λ_n of Z_n satisfy $n^{-1}\Lambda_n(n\theta) \to \Lambda(\theta)$ for any $\theta \neq 1/(2M)$.
(b) Let $x_0 = \lim_{\theta \nearrow 1/(2M)} \Lambda'(\theta)$. Check that the Gärtner–Ellis theorem yields for Z_n and the good rate function $\Lambda^*(\cdot)$ both the large deviations upper bound for arbitrary sets and the lower bound for any open set intersecting $\mathcal{F} = (-\infty, x_0)$. Show that, moreover, the same lower bounds apply to $\tilde{Z}_n = Z_n - n^{-1}\eta_1^{(n)} Y_1^2$.
(c) Verify that $\Lambda^*(x) = \Lambda^*(x_0) + (x - x_0)/(2M)$ for any $x > x_0$.
(d) Complete the large deviations lower bound for $G = (x, \infty)$, any $x > x_0$, and conclude that Z_n satisfies the LDP with the good rate function $\Lambda^*(\cdot)$.
Hint: $P(n^{-1}Z_n > x) \geq P(n^{-1}\tilde{Z}_n > x_0 - \delta)P(n^{-1}\eta_1^{(n)} Y_1^2 > x - x_0 + \delta)$ for all $\delta > 0$.

Remark: Let $X_1, \ldots, X_n, \ldots$ be a real-valued, zero mean, stationary Gaussian process with covariance $E(X_n X_{n+k}) = (2\pi)^{-1} \int_0^{2\pi} e^{iks} f(s) ds$ for some $f : [0, 2\pi] \to [0, M]$. Then, $Z_n = n^{-1}\sum_{i=1}^{n} X_i^2$ is of the form considered here with $\eta_i^{(n)}$ being the eigenvalues of the n-dimensional covariance Toeplitz matrix associated with the process $\{X_i\}$. By the limiting distribution of Toeplitz matrices (see [GS58]) we have $\mu(\Gamma) = (2\pi)^{-1} \int_{\{s: f(s) \in \Gamma\}} ds$ with the corresponding LDP for Z_n (c.f. [BGR97]).

Exercise 2.3.27 Let X_j be $\mathbb{R}^d$-valued i.i.d. random variables with an everywhere finite logarithmic moment generating function Λ and a_j an absolutely summable sequence of real numbers such that $\sum_{i=-\infty}^{\infty} a_i = 1$. Consider the normalized partial sums $Z_n = n^{-1} \sum_{j=1}^{n} Y_j$ of the moving average process $Y_j = \sum_{i=-\infty}^{\infty} a_{j+i} X_i$. Show that Assumption 2.3.2 holds, hence by the Gärtner–Ellis theorem Z_n satisfy the LDP with good rate function Λ^*.
Hint: Show that $n^{-1} \sum_{i=-\infty}^{\infty} \phi(\sum_{j=i+1}^{i+n} a_j) \to \phi(1)$ for any $\phi : \mathbb{R} \to \mathbb{R}$ continuously differentiable with $\phi(0) = 0$.

2.4 Concentration Inequalities

The precise large deviations estimates presented in this chapter are all related to rather simple functionals of an independent sequence of random variables, namely to empirical means of such a sequence. We digress in this section from this theme by, while still keeping the independence structure, allowing for more complicated functionals. In such a situation, it is often hopeless to have a full LDP, and one is content with the rough concentration properties of the random variables under investigation. While such concentration properties have diverse applications in computer science, combinatorics, operations research and geometry, only a few simple examples are presented here. See the historical notes for partial references to the extensive literature dealing with such results and their applications.

In Section 2.4.1, we present concentration inequalities for discrete time martingales of bounded differences and show how these may apply for certain functionals of independent variables. Section 2.4.2 is devoted to Talagrand's far reaching extensions of this idea, extensions whose root can be traced back to isoperimetric inequalities for product measures.

In order not to be distracted by measureability concerns, we assume throughout this section that Σ is a Polish space, that is, a complete separable metric space. To understand the main ideas, suffices to take $\Sigma = \{0, 1\}$. In applications, Σ is often either a finite set or a subset of $\mathbb{R}$.

2.4.1 Inequalities for Bounded Martingale Differences

Our starting point is a bound on the moment generating function of a random variable in terms of its maximal possible value and first two moments.

Lemma 2.4.1 (Bennett) *Suppose $X \leq b$ is a real-valued random variable with $\overline{x} = E(X)$ and $E[(X - \overline{x})^2] \leq \sigma^2$ for some $\sigma > 0$. Then, for any $\lambda \geq 0$,*

$$E(e^{\lambda X}) \leq e^{\lambda \overline{x}} \left\{ \frac{(b - \overline{x})^2}{(b - \overline{x})^2 + \sigma^2} e^{-\frac{\lambda \sigma^2}{b - \overline{x}}} + \frac{\sigma^2}{(b - \overline{x})^2 + \sigma^2} e^{\lambda(b - \overline{x})} \right\} . \qquad (2.4.2)$$

Proof: In case $b = \overline{x}$, it follows that $X = \overline{x}$ almost surely and (2.4.2) trivially holds with equality for any $\sigma > 0$. Similarly, (2.4.2) trivially holds with equality when $\lambda = 0$. Turning to deal with the general case of $b > \overline{x}$ and $\lambda > 0$, let $Y \triangleq \lambda(X - \overline{x})$, noting that $Y \leq \lambda(b - \overline{x}) \triangleq m$ is of zero mean and $EY^2 \leq \lambda^2 \sigma^2 \triangleq v$. Thus, (2.4.2) amounts to showing that

$$E(e^Y) \leq \frac{m^2}{m^2 + v} e^{-v/m} + \frac{v}{m^2 + v} e^m = E(e^{Y_o}) , \qquad (2.4.3)$$

for any random variable $Y \leq m$ of zero mean and $E(Y^2) \leq v$, where the random variable Y_o takes values in $\{-v/m, m\}$, with $P(Y_o = m) = v/(m^2 + v)$. To this end, fix $m, v > 0$ and let $\phi(\cdot)$ be the (unique) quadratic function such that $f(y) \triangleq \phi(y) - e^y$ is zero at $y = m$, and $f(y) = f'(y) = 0$ at $y = -v/m$. Note that $f''(y) = 0$ at exactly one value of y, say y_0. Since $f(-v/m) = f(m)$ and $f(\cdot)$ is not constant on $[-v/m, m]$, it follows that $f'(y) = 0$ at some $y_1 \in (-v/m, m)$. By the same argument, now applied to $f'(\cdot)$ on $[-v/m, y_1]$, also $y_0 \in (-v/m, y_1)$. With $f(\cdot)$ convex on $(-\infty, y_0]$, minimal at $-v/m$, and $f(\cdot)$ concave on $[y_0, \infty)$, maximal at y_1, it follows that $f(y) \geq 0$ for any $y \in (-\infty, m]$. Thus, $E(f(Y)) \geq 0$ for any random variable $Y \leq m$, that is,

$$E(e^Y) \leq E(\phi(Y)) , \qquad (2.4.4)$$

with equality whenever $P(Y \in \{-v/m, m\}) = 1$. Since $f(y) \geq 0$ for $y \to -\infty$, it follows that $\phi''(0) = \phi''(y) \geq 0$ (recall that $\phi(\cdot)$ is a quadratic function). Hence, for Y of zero mean, $E(\phi(Y))$, which depends only upon $E(Y^2)$, is a non-decreasing function of $E(Y^2)$. It is not hard to check that Y_o, taking values in $\{-v/m, m\}$, is of zero mean and such that $E(Y_o^2) = v > 0$. So, (2.4.4) implies that

$$E(e^Y) \leq E(\phi(Y)) \leq E(\phi(Y_o)) = E(e^{Y_o}) ,$$

establishing (2.4.3), as needed. $\square$

Specializing Lemma 2.4.1 we next bound the moment generating function of a random variable in terms of its mean and support.

Corollary 2.4.5 *Fix $a < b$. Suppose that $a \leq X \leq b$ is a real-valued random variable with $\overline{x} = E(X)$. Then, for any $\lambda \in \mathbb{R}$,*

$$E(e^{\lambda X}) \leq \frac{\overline{x} - a}{b - a} e^{\lambda b} + \frac{b - \overline{x}}{b - a} e^{\lambda a} . \qquad (2.4.6)$$

Proof: Set $\sigma^2 = (b-\overline{x})(\overline{x}-a)$. If $\sigma = 0$ then either $X = \overline{x} = a$ almost surely or $X = \overline{x} = b$ almost surely, with (2.4.6) trivially holding with equality in both cases. Assuming hereafter that $\sigma > 0$, since $x^2 - (a+b)x + ab \leq 0$ for any $x \in [a,b]$, integrating with respect to the law of X we conclude that $E[(X - \overline{x})^2] \leq (b - \overline{x})(\overline{x} - a) = \sigma^2$. Setting this value of σ^2 in (2.4.2) we recover (2.4.6) for any $\lambda \geq 0$. In case $\lambda \leq 0$ we apply (2.4.6) for $-\lambda \geq 0$ and the random variable $-X \in [-b, -a]$ of mean $-\overline{x}$ for which also $E[(X - \overline{x})^2] \leq \sigma^2$. $\qquad\square$

As we see next, Lemma 2.4.1 readily provides concentration inequalities for discrete time martingales of bounded differences, null at 0.

Corollary 2.4.7 *Suppose $v > 0$ and the real valued random variables $\{Y_n : n = 1, 2, \ldots\}$ are such that both $Y_n \leq 1$ almost surely, and $E[Y_n | S_{n-1}] = 0$, $E[Y_n^2 | S_{n-1}] \leq v$ for $S_n \triangleq \sum_{j=1}^n Y_j$, $S_0 = 0$. Then, for any $\lambda \geq 0$,*

$$E\left[e^{\lambda S_n}\right] \leq \left(\frac{e^{-v\lambda} + v e^{\lambda}}{1 + v}\right)^n. \tag{2.4.8}$$

Moreover, for all $x \geq 0$,

$$P(n^{-1} S_n \geq x) \leq \exp\left(-n H\left(\frac{x+v}{1+v}\Big|\frac{v}{1+v}\right)\right), \tag{2.4.9}$$

where $H(p|p_0) \triangleq p \log(p/p_0) + (1-p) \log((1-p)/(1-p_0))$ for $p \in [0,1]$ and $H(p|p_0) = \infty$ otherwise. Finally, for all $y \geq 0$,

$$P(n^{-1/2} S_n \geq y) \leq e^{-2y^2/(1+v)^2}. \tag{2.4.10}$$

Proof: Applying Lemma 2.4.1 for the conditional law of Y_k given S_{k-1}, $k = 1, 2, \ldots$, for which $\overline{x} = 0$, $b = 1$ and $\sigma^2 = v$, it follows that almost surely

$$E(e^{\lambda Y_k} | S_{k-1}) \leq \frac{e^{-v\lambda} + v e^{\lambda}}{1 + v}. \tag{2.4.11}$$

In particular, $S_0 = 0$ is non-random and hence (2.4.11) yields (2.4.8) in case of $k = n = 1$. Multiplying (2.4.11) by $e^{\lambda S_{k-1}}$ and taking its expectation, results with

$$E[e^{\lambda S_k}] = E[e^{\lambda S_{k-1}} E[e^{\lambda Y_k} | S_{k-1}]] \leq E[e^{\lambda S_{k-1}}]\frac{e^{-v\lambda} + v e^{\lambda}}{1 + v}. \tag{2.4.12}$$

Iterating (2.4.12) from $k = n$ to $k = 1$, establishes (2.4.8).

Applying Chebycheff's inequality we get by (2.4.8) that for any $x, \lambda \geq 0$,

$$P(n^{-1} S_n \geq x) \leq e^{-\lambda n x} E\left[e^{\lambda S_n}\right] \leq e^{-\lambda n x}\left(\frac{e^{-v\lambda} + v e^{\lambda}}{1 + v}\right)^n. \tag{2.4.13}$$

For $x \in [0,1)$ the inequality (2.4.9) follows when considering $\lambda = (1 + v)^{-1} \log((x+v)/v(1-x))$ in (2.4.13). The case of $x = 1$ in (2.4.9) is similar, now considering $\lambda \to \infty$ in (2.4.13). Since $n^{-1}S_n \leq 1$ almost surely, (2.4.9) trivially holds for any $x > 1$.

Note that $H(p_0|p_0) = 0$ and $dH(p|p_0)/dp = \log(p(1-p_0)) - \log(p_0(1-p))$ is zero for $p = p_0$, whereas $d^2 H(p|p_0)/dp^2 = 1/(p(1-p)) \geq 4$ for any $p, p_0 \in (0,1)$. Consequently, $H(p|p_0) \geq 2(p - p_0)^2$ for any $p, p_0 \in [0,1]$, and setting $x = n^{-1/2}y$, (2.4.10) directly follows from (2.4.9). $\qquad\square$

A typical application of Corollary 2.4.7 is as follows.

Corollary 2.4.14 *Let $Z_n = g_n(X_1, \ldots, X_n)$ for independent Σ-valued random variables $\{X_i\}$ and real-valued, measurable $g_n(\cdot)$. Let $\{\hat{X}_i\}$ be an independent copy of $\{X_i\}$. Suppose that for $k = 1, \ldots, n$,*

$$|g_n(X_1, \ldots, X_n) - g_n(X_1, \ldots, X_{k-1}, \hat{X}_k, X_{k+1}, \ldots, X_n)| \leq 1, \qquad (2.4.15)$$

almost surely. Then, for all $x \geq 0$,

$$\mathrm{P}(n^{-1}(Z_n - EZ_n) \geq x) \leq \exp\left(-nH\left(\frac{x+1}{2}\Big|\frac{1}{2}\right)\right), \qquad (2.4.16)$$

and for all $y \geq 0$,

$$\mathrm{P}(n^{-1/2}(Z_n - EZ_n) \geq y) \leq e^{-y^2/2}. \qquad (2.4.17)$$

Proof: For $k = 1, \ldots, n$ let $S_k \triangleq E[Z_n|X_1, \ldots, X_k] - EZ_n$ with $S_0 \triangleq 0$. In particular, $S_n = Z_n - EZ_n$ and $Y_k \triangleq S_k - S_{k-1}$ is such that

$$E[Y_k|S_{k-1}] = E[E[Y_k|X_1, \ldots, X_{k-1}]|S_{k-1}] = 0.$$

Moreover, integrating (2.4.15) with respect to $\hat{X}_k, X_{k+1}, \ldots, X_n$ results with

$$|E[g_n(X_1, \ldots, X_n) - g_n(X_1, \ldots, \hat{X}_k, \ldots, X_n)|X_1, \ldots, X_k]| \leq 1, \qquad (2.4.18)$$

almost surely. The mutual independence of $\{X_i\}$ and $\hat{X}_k$ implies in particular that

$$E[g_n(X_1, \ldots, X_n)|X_1, \ldots, X_{k-1}] = E[g_n(X_1, \ldots, \hat{X}_k, \ldots, X_n)|X_1, \ldots, X_{k-1}]$$
$$= E[g_n(X_1, \ldots, \hat{X}_k, \ldots, X_n)|X_1, \ldots, X_k],$$

and hence $|Y_k| \leq 1$ almost surely by (2.4.18). Clearly, then $E[Y_k^2|S_{k-1}] \leq 1$, with (2.4.16) and (2.4.17) established by applying Corollary 2.4.7 (for $v = 1$). $\qquad\square$

Remark: It is instructive to note that a slightly stronger version of (2.4.17) can be obtained while bypassing Bennett's lemma (Lemma 2.4.1). See Exercise 2.4.22 for more details about this derivation, starting with a direct proof of (2.4.6).

We conclude this section with a representative example for the application of Corollary 2.4.14. It arises in the context of bin-packing problems in computer science and operations research (See also Exercise 2.4.44 for an improvement using Talagrand's inequalities). To this end, let

$$B_n(\mathbf{x}) \;=\; \min\{N : \exists \{i_\ell\}_{\ell=1}^N, \{j_\ell^m\}_{m=1}^{i_\ell} \text{ such that}$$

$$\sum_{\ell=1}^N i_\ell = n, \; \bigcup_{\ell=1}^N \bigcup_{m=1}^{i_\ell} \{j_\ell^m\} = \{1,\ldots,n\}, \; \sum_{m=1}^{i_\ell} x_{j_\ell^m} \le 1\},$$

denote the minimal number of unit size bins (intervals) needed to store $\mathbf{x} = \{x_k : 1 \le k \le n\}$, and $E_n \triangleq E(B_n(\mathbf{X}))$ for $\{X_k\}$ independent $[0,1]$-valued random variables.

Corollary 2.4.19 *For any* $n, t > 0$,

$$P(|B_n(\mathbf{X}) - E_n| \ge t) \le 2\exp(-t^2/2n).$$

Proof: Clearly, $|B_n(\mathbf{x}) - B_n(\mathbf{x}')| \le 1$ when $\mathbf{x}$ and $\mathbf{x}'$ differ only in one coordinate. It follows that $B_n(\mathbf{X})$ satisfies (2.4.15). The claim follows by an application of (2.4.17), first for $Z_n = B_n(\mathbf{X})$, $y = n^{-1/2}t$, then for $Z_n = -B_n(\mathbf{X})$. □

Exercise 2.4.20 Check that the conclusions of Corollary 2.4.14 apply for $Z_n = |\sum_{j=1}^n \epsilon_j x_j|$ where $\{\epsilon_i\}$ are i.i.d. such that $P(\epsilon_i = 1) = P(\epsilon_i = -1) = 1/2$ and $\{x_i\}$ are non-random vectors in the unit ball of a normed space $(\mathcal{X}, |\cdot|)$.
Hint: Verify that $|E[Z_n|\epsilon_1,\ldots,\epsilon_k] - E[Z_n|\epsilon_1,\ldots,\epsilon_{k-1}]| \le 1$.

Exercise 2.4.21 Let $B(u) \triangleq 2u^{-2}[(1+u)\log(1+u) - u]$ for $u > 0$.
(a) Show that for any $x, v > 0$,

$$H\Big(\frac{x+v}{1+v}\Big|\frac{v}{1+v}\Big) \ge \frac{x^2}{2v} B\Big(\frac{x}{v}\Big),$$

hence (2.4.9) implies that for any $z > 0$,

$$P(S_n \ge z) \le \exp\Big(-\frac{z^2}{2nv} B\Big(\frac{z}{v}\Big)\Big).$$

(b) Suppose that $(S_n, \mathcal{F}_n)$ is a discrete time martingale such that $S_0 = 0$ and $Y_k \triangleq S_k - S_{k-1} \le 1$ almost surely. Let $Q_n \triangleq \sum_{j=1}^n E[Y_j^2|\mathcal{F}_{j-1}]$ and show that for any $z, r > 0$,

$$P(S_n \ge z, Q_n \le r) \le \exp\Big(-\frac{z^2}{2r} B\Big(\frac{z}{r}\Big)\Big).$$

Hint: $\exp(\lambda S_n - \theta Q_n)$ is an $\mathcal{F}_n$ super-martingale for $\theta = e^\lambda - \lambda - 1 \geq \lambda^2/2$ and any $\lambda \geq 0$.

Exercise 2.4.22 Suppose in the setting of Corollary 2.4.14 that assumption (2.4.15) is replaced by

$$|g_n(X_1,\ldots,X_n) - g_n(X_1,\ldots,X_{k-1},\hat{X}_k,X_{k+1},\ldots,X_n)| \leq c_k \,,$$

for some $c_k \geq 0$, $k = 1,\ldots,n$.
(a) Applying Corollary 2.4.5, show that for any $\lambda \in \mathbb{R}$,

$$E[e^{\lambda Y_k}|S_{k-1}] \leq e^{\lambda^2 c_k^2/8} \,.$$

Hint: Check that for any $a < b$ and $\overline{x} \in [a,b]$,

$$\frac{\overline{x}-a}{b-a}e^{\lambda b} + \frac{b-\overline{x}}{b-a}e^{\lambda a} \leq e^{\lambda \overline{x}}e^{\lambda^2(b-a)^2/8} \,.$$

Note that given $\{X_1,\ldots,X_{k-1}\}$, the martingale difference Y_k as a function of X_k, is of zero mean and supported on $[a_k, a_k+c_k]$ for some $a_k(X_1,\ldots,X_{k-1})$.
(b) Deduce from part (a), along the line of proof of Corollary 2.4.7, that for any $z \geq 0$,

$$P(Z_n - EZ_n \geq z) \leq e^{-2z^2/\sum_{k=1}^n c_k^2} \,.$$

Remark: Even when $c_k = 1$ for all k, this is an improvement upon (2.4.17) by a factor of 4 in the exponent. Moreover, both are sharper than what follows by setting $r = \sum_{k=1}^n c_k^2$ in Exercise 2.4.21 (as $B(\cdot) \geq 1$).

2.4.2 Talagrand's Concentration Inequalities

The Azuma-Hoeffding-Bennett inequalities of Corollary 2.4.14 apply for coordinate wise Lipschitz functions of independent random variables. Talagrand's concentration inequalities to which this section is devoted, also rely on the product structure of the underlying probability space, but allow for a much wider class of functions to be considered. As such, they are extremely useful in combinatorial applications such as Corollary 2.4.36, in statistical physics applications such as the study of spin glass models, and in areas touching upon functional analysis such as probability in Banach spaces (for details and references see the historical notes at the end of this chapter).

For any $n \in \mathbb{Z}_+$, let $(\mathbf{y},\mathbf{x}) = (y_1,\ldots,y_n,x_1,\ldots,x_n)$ denote a generic point in $(\Sigma^n)^2$. Let $\mathcal{M}_n(\mathbf{Q},\mathbf{P})$ denote the set of all probability measures on $(\Sigma^n)^2$ whose marginals are the prescribed probability measures $\mathbf{Q}$ and $\mathbf{P}$ on Σ^n. That is, if $\mathbf{Y}, \mathbf{X} \in \Sigma^n$ are such that $(\mathbf{Y}, \mathbf{X})$ has the joint law $\pi \in \mathcal{M}_n(\mathbf{Q}, \mathbf{P})$, then $\mathbf{Q}$ is the law of $\mathbf{Y}$ and $\mathbf{P}$ is the law of $\mathbf{X}$.

For any $\alpha > 0$ let $\phi_\alpha : [0, \infty) \to [0, \infty)$ be such that

$$\phi_\alpha(x) = \alpha x \log x - (1 + \alpha x) \log\left(\frac{1 + \alpha x}{1 + \alpha}\right).$$

Consider the "coupling distance"

$$d_\alpha(\mathbf{Q}, \mathbf{P}) = \inf_{\pi \in \mathcal{M}_n(\mathbf{Q}, \mathbf{P})} \int_{\Sigma^n} \sum_{k=1}^{n} \phi_\alpha(\pi^{\mathbf{x}}(\{y_k = x_k\})) d\mathbf{P}(\mathbf{x}),$$

between probability measures $\mathbf{Q}$ and $\mathbf{P}$ on Σ^n. Here and throughout, for $\mathbf{z} = (y_{i_1}, \ldots, y_{i_m}, x_{j_1}, \ldots, x_{j_\ell})$, $\pi^{\mathbf{z}}(\cdot)$ denotes the regular conditional probability distribution of π given the σ-field generated by the restriction of $(\Sigma^n)^2$ to the coordinates specified by $\mathbf{z}$. (Such a conditional probability exists because $(\Sigma^n)^2$ is Polish; *c.f.* Appendix D.) We also use the notations $\mathbf{y}^k \triangleq (y_1, \ldots, y_k)$ and $\mathbf{x}^k \triangleq (x_1, \ldots, x_k)$, $k = 1, \ldots, n$.

Define the *relative entropy* of the probability measure ν with respect to $\mu \in M_1(\Sigma^n)$ as

$$H(\nu|\mu) \triangleq \begin{cases} \int_{\Sigma^n} f \log f \, d\mu & \text{if } f \triangleq \frac{d\nu}{d\mu} \text{ exists} \\ \infty & \text{otherwise}, \end{cases}$$

where $d\nu/d\mu$ stands for the Radon-Nikodym derivative of ν with respect to μ when it exists. As already hinted in Section 2.1.2, the relative entropy plays a crucial role in large deviations theory. See Section 6.5.3 and Appendix D.3 for some of its properties.

The next theorem, the proof of which is deferred to the end of this section, provides an upper bound on $d_\alpha(\cdot, \cdot)$ in terms of $H(\cdot|\cdot)$.

Theorem 2.4.23 *Suppose* $\mathbf{R} = \prod_k R_k$ *is a product probability measure on* Σ^n. *Then, for any* $\alpha > 0$ *and any probability measures* $\mathbf{P}, \mathbf{Q}$ *on* Σ^n,

$$d_\alpha(\mathbf{Q}, \mathbf{P}) \leq H(\mathbf{P}|\mathbf{R}) + \alpha H(\mathbf{Q}|\mathbf{R}). \tag{2.4.24}$$

For $\alpha > 0$ and $A \in \mathcal{B}_{\Sigma^n}$ consider the function

$$f_\alpha(A, \mathbf{x}) \triangleq \inf_{\{\nu \in M_1(\Sigma^n) : \nu(A) = 1\}} \sum_{k=1}^{n} \phi_\alpha(\nu(\{\mathbf{y} : y_k = x_k\})). \tag{2.4.25}$$

For any $J \subset \{1, \ldots, n\}$ let $p_J(A) = \{\mathbf{z} \in \Sigma^n : \exists \mathbf{y} \in A, \text{ such that } y_k = z_k$ for all $k \in J\}$, noting that $p_J(A) \in \mathcal{B}_{\Sigma^n}$ for any such J (since Σ is Polish, see Theorem D.4). As the value of $f_\alpha(A, \cdot)$ is uniquely determined by the finite length binary sequence $\{1_{p_J(A)}(\cdot) : J \subset \{1, \ldots, n\}\}$, it follows that $f_\alpha(A, \cdot)$ is Borel measureable. See also Exercise 2.4.43 for an alternative representation of $f_\alpha(A, \mathbf{x})$.

We next show that concentration inequalities for $f_\alpha(\cdot, \cdot)$ are a direct consequence of Theorem 2.4.23.

Corollary 2.4.26 *For any product probability measure* $\mathbf{R}$ *on* Σ^n, *any* $\alpha > 0$, $t > 0$ *and* $A \in \mathcal{B}_{\Sigma^n}$,

$$\mathbf{R}(\{\mathbf{x} : f_\alpha(A, \mathbf{x}) \geq t\})e^t \leq \int_{\Sigma^n} e^{f_\alpha(A, \mathbf{x})} d\mathbf{R}(\mathbf{x}) \leq \mathbf{R}(A)^{-\alpha} . \qquad (2.4.27)$$

Proof: Fix a product probability measure $\mathbf{R}$, $\alpha > 0$ and $A \in \mathcal{B}_{\Sigma^n}$. The right inequality in (2.4.27) trivially holds when $\mathbf{R}(A) = 0$. When $\mathbf{R}(A) > 0$, set $\mathbf{Q}(\cdot) = \mathbf{R}(\cdot \cap A)/\mathbf{R}(A)$ for which $H(\mathbf{Q}|\mathbf{R}) = -\log \mathbf{R}(A)$, and $\mathbf{P}$ such that

$$\frac{d\mathbf{P}}{d\mathbf{R}} = \frac{e^{f_\alpha(A, \mathbf{x})}}{\int_{\Sigma^n} e^{f_\alpha(A, \mathbf{x})} d\mathbf{R}(\mathbf{x})} ,$$

for which

$$H(\mathbf{P}|\mathbf{R}) = \int_{\Sigma^n} f_\alpha(A, \mathbf{x}) d\mathbf{P}(\mathbf{x}) - \log \int_{\Sigma^n} e^{f_\alpha(A, \mathbf{x})} d\mathbf{R}(\mathbf{x}) . \qquad (2.4.28)$$

Since $\pi^{\mathbf{x}}(A \times \{\mathbf{x}\}) = 1$ for any $\pi \in \mathcal{M}_n(\mathbf{Q}, \mathbf{P})$ and $\mathbf{P}$ almost every $\mathbf{x} \in \Sigma^n$, it follows that

$$\int_{\Sigma^n} f_\alpha(A, \mathbf{x}) d\mathbf{P}(\mathbf{x}) \leq d_\alpha(\mathbf{Q}, \mathbf{P}) . \qquad (2.4.29)$$

Combining (2.4.24) with (2.4.28) and (2.4.29) gives the right inequality in (2.4.27). For any $t > 0$, the left inequality in (2.4.27) is a direct application of Chebycheff's inequality. $\qquad \square$

We next deduce concentration inequalities for

$$g(A, \mathbf{x}) \stackrel{\triangle}{=} \sup_{\{\beta : |\beta| \leq 1\}} \inf_{\mathbf{y} \in A} \sum_{k=1}^n \beta_k 1_{y_k \neq x_k} , \qquad (2.4.30)$$

where $\beta = (\beta_1, \ldots, \beta_n) \in \mathbb{R}^n$, out of those for $f_\alpha(\cdot, \cdot)$. The function $g(A, \cdot)$ is Borel measurable by an argument similar to the one leading to the measurability of $f_\alpha(A, \cdot)$.

Corollary 2.4.31 *For any product probability measure* $\mathbf{R}$ *on* Σ^n, $A \in \mathcal{B}_{\Sigma^n}$, *and any* $u > 0$,

$$\mathbf{R}(\{\mathbf{x} : g(A, \mathbf{x}) \geq u\})\mathbf{R}(A) \leq e^{-u^2/4} . \qquad (2.4.32)$$

If $\mathbf{R}(A) > 0$ and $u > \sqrt{2\log(1/\mathbf{R}(A))}$, then also

$$\mathbf{R}(\{\mathbf{x} : g(A, \mathbf{x}) \geq u\}) \leq \exp\left(-\frac{1}{2}(u - \sqrt{2\log(1/\mathbf{R}(A))})^2\right) . \qquad (2.4.33)$$

Remarks:
(a) For any $A \in \mathcal{B}_{\Sigma^n}$ and $\mathbf{x} \in \Sigma^n$, the choice of $\beta_k = n^{-1/2}$, $k = 1, \ldots, n$, in (2.4.30), leads to $g(A, \mathbf{x}) \geq n^{-1/2}h(A, \mathbf{x})$, where

$$h(A, \mathbf{x}) \stackrel{\triangle}{=} \inf_{\mathbf{y} \in A} \sum_{k=1}^{n} 1_{y_k \neq x_k} .$$

Inequalities similar to (2.4.33) can be derived for $n^{-1/2}h(\cdot, \cdot)$ directly out of Corollary 2.4.14. However, Corollary 2.4.36 and Exercise 2.4.44 demonstrate the advantage of using $g(\cdot, \cdot)$ over $n^{-1/2}h(\cdot, \cdot)$ in certain applications.
(b) By the Cauchy-Schwartz inequality, $g(A, \mathbf{x}) \leq \sqrt{h(A, \mathbf{x})}$ for any $A \in \mathcal{B}_{\Sigma^n}$ and any $\mathbf{x} \in \Sigma^n$. It is however shown in Exercise 2.4.45 that inequalities such as (2.4.33) do not hold in general for $\sqrt{h(\cdot, \cdot)}$.

Proof: Note first that

$$\inf_{\{\nu \in M_1(\Sigma^n) : \nu(A) = 1\}} \int_{\Sigma^n} \left(\sum_{k=1}^{n} \beta_k 1_{y_k \neq x_k}\right) \nu(d\mathbf{y}) = \inf_{\mathbf{y} \in A} \sum_{k=1}^{n} \beta_k 1_{y_k \neq x_k} ,$$

hence,

$$g(A, \mathbf{x}) = \sup_{\{\beta : |\beta| \leq 1\}} \inf_{\{\nu \in M_1(\Sigma^n) : \nu(A) = 1\}} \sum_{k=1}^{n} \beta_k \nu(\{\mathbf{y} : y_k \neq x_k\}) .$$

By the Cauchy-Schwartz inequality,

$$g(A, \mathbf{x})^2 \leq \inf_{\{\nu \in M_1(\Sigma^n) : \nu(A) = 1\}} \sum_{k=1}^{n} \nu(\{\mathbf{y} : y_k \neq x_k\})^2 . \qquad (2.4.34)$$

Since $\phi_\alpha(1) = 0$, $\phi'_\alpha(1) = 0$ and $\phi''_\alpha(x) = \frac{\alpha}{x(1+\alpha x)} \geq \frac{\alpha}{1+\alpha}$ for $x \in [0, 1]$, it follows that $\phi_\alpha(x) \geq \frac{\alpha}{2(1+\alpha)}(1-x)^2$ for any $\alpha > 0$ and any $x \in [0, 1]$. Hence, by (2.4.25) and (2.4.34), for any $\alpha > 0$, $A \in \mathcal{B}_{\Sigma^n}$ and $\mathbf{x} \in \Sigma^n$,

$$\frac{\alpha}{2(1+\alpha)}g(A, \mathbf{x})^2 \leq f_\alpha(A, \mathbf{x}) . \qquad (2.4.35)$$

Set $t = \frac{\alpha}{2(1+\alpha)}u^2$ in (2.4.27) and apply (2.4.35) to get (2.4.32) by choosing $\alpha = 1$. For $\mathbf{R}(A) < 1$, the inequality (2.4.33) is similarly obtained by choosing $\alpha = u/\sqrt{2\log(1/\mathbf{R}(A))} - 1 \in (0, \infty)$, whereas it trivially holds for $\mathbf{R}(A) = 1$. $\qquad \square$

The next example demonstrates how concentration inequalities for $g(\cdot, \cdot)$ are typically applied. To this end, let

$$Z_n(\mathbf{x}) \stackrel{\triangle}{=} \max\{m : x_{k_1} < \cdots < x_{k_m} \text{ for some } 1 \leq k_1 < \cdots < k_m \leq n\},$$

denote the length of the longest increasing subsequence of $\mathbf{x} = \{x_k : 1 \leq k \leq n\}$, and $M_n \stackrel{\triangle}{=} \text{Median } (Z_n(\mathbf{X}))$ for $\{X_k\}$ independent random variables, each distributed uniformly on $[0, 1]$.

Corollary 2.4.36 *For any $v \in \mathbb{Z}_+$,*

$$P(Z_n(\mathbf{X}) \geq M_n + v) \leq 2\exp\left(-\frac{v^2}{4(M_n + v)}\right), \qquad (2.4.37)$$

$$P(Z_n(\mathbf{X}) \leq M_n - v) \leq 2\exp\left(-\frac{v^2}{4M_n}\right). \qquad (2.4.38)$$

Proof: Fix $v \in \mathbb{Z}_+$, and let $\Sigma = [0, 1]$, with $A(j) \stackrel{\triangle}{=} \{\mathbf{y} : Z_n(\mathbf{y}) \leq j\} \subset \Sigma^n$ for $j \in \{1, 2, \ldots, n\}$. Suppose $\mathbf{x} \in \Sigma^n$ is such that $Z_n(\mathbf{x}) \geq j + v$, so there exist $1 \leq k_1 < k_2 \cdots < k_{j+v} \leq n$ with $x_{k_1} < x_{k_2} \cdots < x_{k_{j+v}}$. Hence, $\sum_{i=1}^{j+v} 1_{y_{k_i} \neq x_{k_i}} \geq v$ for any $\mathbf{y} \in A(j)$. Thus, setting $\beta_{k_i} = 1/\sqrt{j+v}$ for $i = 1, \ldots, j+v$ and $\beta_k = 0$ for $k \notin \{k_1, \ldots, k_{j+v}\}$, it follows that $g(A(j), \mathbf{x}) \geq v/\sqrt{j+v}$. To establish (2.4.37) apply (2.4.32) for $A = A(M_n)$ and $\mathbf{R}$ being Lebesgue's measure, so that $\mathbf{R}(A) = P(Z_n(\mathbf{X}) \leq M_n) \geq 1/2$, while

$$P(Z_n(\mathbf{X}) \geq M_n + v) \leq \mathbf{R}(\{\mathbf{x} : g(A, \mathbf{x}) \geq v/\sqrt{M_n + v}\}).$$

Noting that $\mathbf{R}(A) = P(Z_n(\mathbf{X}) \leq M_n - v)$ for $A = A(M_n - v)$, and moreover,

$$\frac{1}{2} \leq P(Z_n(\mathbf{X}) \geq M_n) \leq \mathbf{R}(\{\mathbf{x} : g(A, \mathbf{x}) \geq v/\sqrt{M_n}\}),$$

we establish (2.4.38) by applying (2.4.32), this time for $A = A(M_n - v)$. $\square$

The following lemma, which is of independent interest, is key to the proof of Theorem 2.4.23.

Lemma 2.4.39 *For any $\alpha > 0$ and probability measures P, Q on Σ, let*

$$\Delta_\alpha(Q, P) \stackrel{\triangle}{=} \int_{\tilde{\Sigma}} \phi_\alpha(\min\{\frac{dQ}{dP}, 1\})dP,$$

where $\tilde{\Sigma}$ is such that $P(\tilde{\Sigma}^c) = 0$ and $\frac{dQ}{dP}$ exists on $\tilde{\Sigma}$.
(a) There exists $\pi \in \mathcal{M}_1(Q, P)$ such that

$$\int_\Sigma \phi_\alpha(\pi^x(\{y = x\}))dP(x) = \Delta_\alpha(Q, P). \qquad (2.4.40)$$

(b) For any probability measure R on Σ,

$$\Delta_\alpha(Q, P) \le H(P|R) + \alpha H(Q|R) . \tag{2.4.41}$$

Proof: (a) Let $(P - Q)_+$ denote the positive part of the finite (signed) measure $P - Q$ while $Q \wedge P$ denotes the positive measure $P - (P - Q)_+ = Q - (Q - P)_+$. Let $q = (P - Q)_+ (\Sigma) = (Q - P)_+ (\Sigma)$. Suppose $q \in (0, 1)$ and enlarging the probability space if needed, define the independent random variables W_1, W_2 and Z with $W_1 \sim (1 - q)^{-1} (Q \wedge P)$, $W_2 \sim q^{-1} (P - Q)_+$, and $Z \sim q^{-1} (Q - P)_+$. Let $I \in \{1, 2\}$ be chosen independently of these variables such that $I = 2$ with probability q. Set $X = Y = W_1$ when $I = 1$, whereas $X = W_2 \ne Y = Z$ otherwise. If $q = 1$ then we do not need the variable W_1 for the construction of Y, X, whereas for $q = 0$ we never use Z and W_2. This coupling $(Y, X) \sim \pi \in \mathcal{M}_1(Q, P)$ is such that $\pi(\{(y, x) : y = x, x \in \cdot\}) = (Q \wedge P)(\cdot)$. Hence, by the definition of regular conditional probability distribution (see Appendix D.3), for every $\Gamma \in \mathcal{B}_\Sigma$, $\Gamma \subset \tilde{\Sigma}$,

$$\begin{aligned}
\int_\Gamma \pi^x(\{y = x\}) P(dx) &= \int_\Gamma \pi^x(\{(y, x) : y = x\}) P(dx) \\
&= \pi(\bigcup_{x \in \Gamma} \{(y, x) : y = x\}) \\
&= Q \wedge P(\Gamma) = \int_\Gamma \frac{dQ \wedge P}{dP}(x) P(dx),
\end{aligned}$$

implying that $\pi^x(\{y = x\}) = d(Q \wedge P)/dP = \min\{\frac{dQ}{dP}, 1\}$ for P-almost every $x \in \tilde{\Sigma}$. With $P(\tilde{\Sigma}^c) = 0$, (2.4.40) follows.

(b) Suffices to consider R such that $f = dP/dR$ and $g = dQ/dR$ exist. Let $R_\alpha = (P + \alpha Q)/(1 + \alpha)$ so that $dR_\alpha/dR \triangleq h = (f + \alpha g)/(1 + \alpha)$. Thus,

$$\begin{aligned}
H(P|R) + \alpha H(Q|R) &= \int_\Sigma [f \log f + \alpha g \log g] dR \\
&\ge \int_\Sigma \left[f \log \frac{f}{h} + \alpha g \log \frac{g}{h} \right] dR, \quad (2.4.42)
\end{aligned}$$

since $\int_\Sigma h \log h \, dR = H(R_\alpha | R) \ge 0$. Let $\rho \triangleq dQ/dP$ on $\tilde{\Sigma}$. Then, on $\tilde{\Sigma}$, $f/h = (1 + \alpha)/(1 + \alpha\rho)$ and $g/f = \rho$, whereas $P(\tilde{\Sigma}^c) = 0$ implying that $g/h = (1 + \alpha)/\alpha$ on $\tilde{\Sigma}^c$. Hence, the inequality (2.4.42) implies that

$$H(P|R) + \alpha H(Q|R) \ge \int_{\tilde{\Sigma}} \phi_\alpha(\rho) dP + Q(\tilde{\Sigma}^c) v_\alpha$$

for $v_\alpha \triangleq \alpha \log((1 + \alpha)/\alpha) > 0$. With $\phi_\alpha(x) \ge \phi_\alpha(1) = 0$ for all $x \ge 0$, we thus obtain (2.4.41). $\qquad \square$

Proof of Theorem 2.4.23: Fix $\alpha > 0$, $\mathbf{P}$, $\mathbf{Q}$ and $\mathbf{R} = \prod_{k=1}^{n} R_k$. For $k = 1, \ldots, n$, let $P_k^{\mathbf{x}^{k-1}}(\cdot)$ denote the regular conditional probability distribution induced by $\mathbf{P}$ on the k-th coordinate given the σ-field generated by the restriction of Σ^n to the first $(k-1)$ coordinates. Similarly, let $Q_k^{\mathbf{y}^{k-1}}(\cdot)$ be the corresponding regular conditional probability distribution induced by $\mathbf{Q}$. By part (a) of Lemma 2.4.39, for any $k = 1, \ldots, n$ there exists a $\pi_k = \pi_k^{\mathbf{y}^{k-1}, \mathbf{x}^{k-1}} \in \mathcal{M}_1(Q_k^{\mathbf{y}^{k-1}}, P_k^{\mathbf{x}^{k-1}})$ such that

$$\int_{\Sigma} \phi_\alpha(\pi_k^x(\{y = x\})) dP_k^{\mathbf{x}^{k-1}}(x) = \Delta_\alpha(Q_k^{\mathbf{y}^{k-1}}, P_k^{\mathbf{x}^{k-1}})$$

(recall that π_k^x denotes the regular conditional probability distribution of π_k given the σ-field generated by the x coordinate). Let $\pi \in M_1(\Sigma^{2n})$ denote the surgery of the π_k, $k = 1, \ldots, n$, that is, for any $\Gamma \in \mathcal{B}_{\Sigma^{2n}}$,

$$\pi(\Gamma) =$$
$$\int_{\Sigma^2} \cdots \int_{\Sigma^2} 1_{(\mathbf{y},\mathbf{x}) \in \Gamma} \, \pi_1(dy_1, dx_1) \, \pi_2^{\mathbf{y}^1, \mathbf{x}^1}(dy_2, dx_2) \cdots \pi_n^{\mathbf{y}^{n-1}, \mathbf{x}^{n-1}}(dy_n, dx_n).$$

Note that $\pi \in \mathcal{M}_n(\mathbf{Q}, \mathbf{P})$. Moreover, for $k = 1, \ldots, n$, since $\pi_k = \pi_k^{\mathbf{y}^{k-1}, \mathbf{x}^{k-1}} \in \mathcal{M}_1(Q_k^{\mathbf{y}^{k-1}}, P_k^{\mathbf{x}^{k-1}})$, the restriction of the regular conditional probability distribution $\pi^{\mathbf{y}^{k-1}, \mathbf{x}^n}(\cdot)$ to the coordinates (y_k, x_k) coincides with that of $\pi^{\mathbf{y}^{k-1}, \mathbf{x}^k}(\cdot)$, i.e., with $\pi_k^{x_k}$. Consequently, with E denoting expectations with respect to π, the convexity of $\phi_\alpha(\cdot)$ implies that

$$\begin{aligned}
E\phi_\alpha(\pi^{\mathbf{X}}(\{y_k = x_k\})) &\leq E\phi_\alpha(\pi^{\mathbf{Y}^{k-1}, \mathbf{X}}(\{y_k = x_k\})) \\
&= E\phi_\alpha(\pi^{\mathbf{Y}^{k-1}, \mathbf{X}^k}(\{y_k = x_k\})) \\
&= E\phi_\alpha(\pi_k^{X_k}(\{y_k = x_k\})) \\
&= E\Delta_\alpha(Q_k^{\mathbf{Y}^{k-1}}, P_k^{\mathbf{X}^{k-1}}).
\end{aligned}$$

Hence, for any n, by (2.4.41),

$$d_\alpha(\mathbf{Q}, \mathbf{P}) \leq \sum_{k=1}^{n} E\Delta_\alpha(Q_k^{\mathbf{Y}^{k-1}}, P_k^{\mathbf{X}^{k-1}})$$

$$\leq E\left[\sum_{k=1}^{n} H(P_k^{\mathbf{X}^{k-1}} | R_k)\right] + \alpha E\left[\sum_{k=1}^{n} H(Q_k^{\mathbf{Y}^{k-1}} | R_k)\right].$$

Apply Theorem D.13 successively $(n-1)$ times, to obtain the so-called *chain-rule for relative entropies*,

$$H(\mathbf{P} | \mathbf{R}) = \sum_{k=1}^{n} EH(P_k^{\mathbf{X}^{k-1}} | R_k).$$

The corresponding identity holds for $H(\mathbf{Q}|\mathbf{R})$, hence the inequality (2.4.24) follows. $\square$

Exercise 2.4.43 Fix $\alpha > 0$ and $A \in \mathcal{B}_{\Sigma^n}$. For any $\mathbf{x} \in \Sigma^n$ let $V_A(\mathbf{x})$ denote the closed convex hull of $\{(1_{y_1 = x_1}, \ldots, 1_{y_n = x_n}) : \mathbf{y} \in A\} \subset [0,1]^n$. Show that the measurable function $f_\alpha(A, \cdot)$ of (2.4.25) can be represented also as

$$f_\alpha(A, \mathbf{x}) = \inf_{\mathbf{s} \in V_A(\mathbf{x})} \sum_{k=1}^n \phi_\alpha(s_k) .$$

Exercise 2.4.44 [From [Tal95], Section 6]
In this exercise you improve upon Corollary 2.4.19 in case $v \triangleq EX_1^2$ is small by applying Talagrand's concentration inequalities.
(a) Let $A(j) \triangleq \{\mathbf{y} : B_n(\mathbf{y}) \le j\}$. Check that $B_n(\mathbf{x}) \le 2\sum_{k=1}^n x_k + 1$, and hence that

$$B_n(\mathbf{x}) \le j + 2\|\mathbf{x}\|_2 g(A(j), \mathbf{x}) + 1 ,$$

where $\|\mathbf{x}\|_2 \triangleq (\sum_{k=1}^n x_k^2)^{1/2}$.
(b) Check that $E(\exp\|\mathbf{X}\|_2^2) \le \exp(2nv)$ and hence

$$P(\|\mathbf{X}\|_2 \ge 2\sqrt{nv}) \le \exp(-2nv) .$$

(c) Conclude from Corollary 2.4.31 that for any $u > 0$

$$P(B_n(\mathbf{X}) \le j) P(B_n(\mathbf{X}) \ge j + 4u\sqrt{nv} + 1) \le \exp(-u^2/4) + \exp(-2nv) .$$

(d) Deduce from part (c) that, for $M_n \triangleq \text{Median}(B_n(\mathbf{X}))$ and $t \in (0, 8nv)$,

$$P(|B_n(\mathbf{X}) - M_n| \ge t + 1) \le 8\exp(-t^2/(64nv)) .$$

Exercise 2.4.45 [From [Tal95], Section 4]
In this exercise you check the sharpness of Corollary 2.4.31.
(a) Let $\Sigma = \{0, 1\}$, and for $j \in \mathbb{Z}_+$, let $A(j) \triangleq \{\mathbf{y} : \|\mathbf{y}\| \le j\}$, where $\|\mathbf{x}\| \triangleq \sum_{k=1}^n x_k$. Check that then $h(A(j), \mathbf{x}) = \max\{\|\mathbf{x}\| - j, 0\}$ and $g(A(j), \mathbf{x}) = h(A(j), \mathbf{x})/\sqrt{\|\mathbf{x}\|}$.
(b) Suppose $\{X_k\}$ are i.i.d. Bernoulli(j/n) random variables and $j, n \to \infty$ such that $j/n \to p \in (0, 1)$. Check that $P(\mathbf{X} \in A(j)) \to 1/2$ and $P(h(A(j), \mathbf{X}) \ge u) \to 1/2$ for any fixed $u > 0$. Conclude that inequalities such as (2.4.33) do not hold in this case for $\sqrt{h(\cdot, \cdot)}$.
(c) Check that

$$P(g(A(j), \mathbf{X}) \ge u) \to \frac{1}{2\pi} \int_{u/\sqrt{1-p}}^\infty e^{-x^2/2} dx .$$

Conclude that for $p > 0$ arbitrarily small, the coefficient $1/2$ in the exponent in the right side of (2.4.33) is optimal.
Hint: $n^{-1}\|\mathbf{X}\| \to p$ in probability and $(np)^{-1/2}(\|\mathbf{X}\| - j)$ converges in distribution to a Normal$(0, 1 - p)$ variable.

2.5   Historical Notes and References

The early development of large deviation bounds did not follow the order of our presentation. Statisticians, starting with Khinchin [Khi29], have analyzed various forms of Cramér's theorem for special random variables. See [Smi33], [Lin61], and [Pet75] for additional references on this early work. It should be mentioned in this context that in the early literature, particularly in the Russian literature, the term *large deviations* often refers to refinements of the CLT when an expansion is made not at the mean but at some other point. For representative examples, see [BoR65], the survey article by Nagaev [Nag79] and the references there, and the historical notes of Chapter 3.

Although Stirling's formula, which is at the heart of the combinatorial estimates of Section 2.1, dates back at least to the 19th century, the notion of types and bounds of the form of Lemmas 2.1.2–2.1.9 had to wait until information theorists discovered that they are useful tools for analyzing the efficiency of codes. For early references, see the excellent book by Gallager [Ga68], while an extensive use of combinatorial estimates in the context of information theory may be found in [CsK81]. An excellent source of applications of exponential inequalities to discrete mathematics is [AS91]. Non-asymptotic computations of moments and some exponential upper bounds for the problem of sampling without replacement can be found in [Kem73].

The first statement of Cramér's theorem for distributions on $\mathbb{R}$ possessing densities is due, of course, to Cramér [Cra38], who introduced the change of measure argument to this context. An extension to general distributions was done by Chernoff [Che52], who introduced the upper bound that was to carry his name. Bahadur [Bah71] was apparently the first to use the truncation argument to prove Cramér's theorem in $\mathbb{R}$ with no exponential moments conditions. That some finite exponential moments condition is necessary for Cramér's theorem in $\mathbb{R}^d$, $d \geq 3$, is amply demonstrated in the counterexamples of [Din91], building upon the I continuity sets examples of [Sla88]. Even in the absence of any finite moment, the large deviations bounds of Cramér's theorem may apply to *certain* subsets of $\mathbb{R}^d$, as demonstrated in [DeS98]. Finally, the $\mathbb{R}^d$ case admits a sharp result, due to Bahadur and Zabell [BaZ79], which is described in Corollary 6.1.6.

The credit for the extension of Cramér's theorem to the dependent case goes to Plachky and Steinebach [PS75], who considered the one-dimensional case, and to Gärtner [Gär77], who considered $\mathcal{D}_\Lambda = \mathbb{R}^d$. Ellis [Ell84] extended this result to the steep setup. Section 2.3 is an embellishment of his results. Sharper results in the i.i.d. case are presented in [Ney83]. In the one-dimensional, zero mean case, [Bry93] shows that the existence of limiting logarithmic moment generating function in a centered disk of the complex plane implies the CLT.

Exercise 2.3.24 uses the same methods as [MWZ93, DeZ95], and is based on a detection scheme described in [Pro83]. Related computations and extensions may be found in [Ko92]. The same approach is applied by [BryD97] to study quadratic forms of stationary Gaussian processes. Exercise 2.3.26 provides an alternative method to deal with the latter objects, borrowed from [BGR97].

Exercise 2.3.27 is taken from [BD90]. See also [JiWR92, JiRW95] for the LDP in different scales and the extension to Banach space valued moving average processes.

For more references to the literature dealing with the dependent case, see the historical notes of Chapter 6.

Hoeffding, in [Hoe63], derives Corollary 2.4.5 and the tail estimates of Corollary 2.4.7 in the case of sums of independent variables. For the same purpose, Bennett derives Lemma 2.4.1 in [Benn62]. Both Lemma 2.4.1 and Corollary 2.4.5 are special cases of the theory of Chebycheff systems; see, for example, Section 2 of Chapter XII of [KaS66]. Azuma, in [Azu67], extends Hoeffding's tail estimates to the more general context of bounded martingale differences as in Corollary 2.4.7. For other variants of Corollary 2.4.7 and more applications along the lines of Corollary 2.4.14, see [McD89]. See also [AS91], Chapter 7, for applications in the study of random graphs, and [MiS86] for applications in the local theory of Banach spaces. The bound of Exercise 2.4.21 goes back at least to [Frd75]. For more on the relation between moderate deviations of a martingale of bounded jumps (possibly in continuous time), and its quadratic variation, see [Puk94b, Dem96] and the references therein.

The concentration inequalities of Corollaries 2.4.26 and 2.4.31 are taken from Talagrand's monograph [Tal95]. The latter contains references to earlier works as well as other concentration inequalities and a variety of applications, including that of Corollary 2.4.36. Talagrand proves these concentration inequalities and those of [Tal96a] by a clever induction on the dimension n of the product space. Our proof, via the entropy bounds of Theorem 2.4.23, is taken from [Dem97], where other concentration inequalities are proved by the same approach. See also [DeZ96c, Mar96a, Mar96b, Tal96b] for similar results and extensions to a certain class of Markov chains. For a proof of such concentration inequalities starting from log-Sobolev or Poincaré's inequalities, see [Led96, BoL97].

The problem of the longest increasing subsequence presented in Corollary 2.4.36 is the same as Ulam's problem of finding the longest increasing subsequence of a random permutation, with deep connections to combinatorics and group theory. See [LoS77, VK77, AD95] for more information and references. It is interesting to note that a naive application of bounded martingale differences in this problem yields only poor results, and to get meaningful results by this approach requires some ingenuity. See [BB92] for details. Finally, for some full LDPs for this problem, see [Sep97, DeuZ98].

Chapter 3

Applications—The Finite Dimensional Case

This chapter consists of applications of the theory presented in Chapter 2. The LDPs associated with finite state irreducible Markov chains are derived in Section 3.1 as a corollary of the Gärtner–Ellis theorem. Varadhan's characterization of the spectral radius of nonnegative irreducible matrices is derived along the way. (See Exercise 3.1.19.) The asymptotic size of long rare segments in random walks is found by combining, in Section 3.2, the basic large deviations estimates of Cramér's theorem with the Borel–Cantelli lemma. The Gibbs conditioning principle is of fundamental importance in statistical mechanics. It is derived in Section 3.3, for finite alphabet, as a direct result of Sanov's theorem. The asymptotics of the probability of error in hypothesis testing problems are analyzed in Sections 3.4 and 3.5 for testing between two *a priori* known product measures and for universal testing, respectively. Shannon's source coding theorem is proved in Section 3.6 by combining the classical random coding argument with the large deviations lower bound of the Gärtner–Ellis theorem. Finally, Section 3.7 is devoted to refinements of Cramér's theorem in $\mathbb{R}$. Specifically, it is shown that for $\beta \in (0, 1/2)$, $\{n^\beta \hat{S}_n\}$ satisfies the LDP with a Normal-like rate function, and the pre-exponent associated with $P(\hat{S}_n \geq q)$ is computed for appropriate values of q.

A. Dembo, O. Zeitouni, *Large Deviations Techniques and Applications*, 71
Stochastic Modelling and Applied Probability 38,
DOI 10.1007/978-3-642-03311-7_3,
© Springer-Verlag Berlin Heidelberg 1998, corrected printing 2010

3.1 Large Deviations for Finite State Markov Chains

The results of Section 2.1 are extended in this section to Markov chains $Y_1, \ldots, Y_n, \ldots$ taking values in a finite alphabet Σ, where without loss of generality Σ is identified with the set $\{1, \ldots, N\}$ and $|\Sigma| = N$. Although these results may be derived by the method of types presented in Section 2.1, the combinatorial arguments involved are quite elaborate. (See Exercise 3.1.21.) An alternative derivation of these results via an application of the Gärtner–Ellis theorem is given here.

Let $\mathbf{\Pi} = \{\pi(i,j)\}_{i,j=1}^{|\Sigma|}$ be a stochastic matrix (i.e., a matrix whose elements are nonnegative and such that each row–sum is one). Let P_σ^π denote the Markov probability measure associated with the transition probability $\mathbf{\Pi}$ and with the initial state $\sigma \in \Sigma$, i.e.,

$$P_\sigma^\pi(Y_1 = y_1, \ldots, Y_n = y_n) = \pi(\sigma, y_1) \prod_{i=1}^{n-1} \pi(y_i, y_{i+1}).$$

Expectations with respect to P_σ^π are denoted by $E_\sigma^\pi(\cdot)$.

Let $\mathbf{B}^m$ denote the mth power of the matrix $\mathbf{B}$. A matrix $\mathbf{B}$ with nonnegative entries is called *irreducible*, if for any pair of indices i, j there exists an $m = m(i, j)$ such that $B^m(i, j) > 0$. Irreducibility is equivalent to the condition that one may find for each i, j a sequence of indices $i_1, \ldots, i_m$ such that $i_1 = i$, $i_m = j$ and $B(i_k, i_{k+1}) > 0$ for all $k = 1, \ldots, m-1$. The following theorem describes some properties of irreducible matrices.

Theorem 3.1.1 (Perron–Frobenius) *Let* $\mathbf{B} = \{B(i,j)\}_{i,j=1}^{|\Sigma|}$ *be an irreducible matrix. Then* $\mathbf{B}$ *possesses an eigenvalue* ρ *(called the Perron–Frobenius eigenvalue) such that:*
(a) $\rho > 0$ *is real.*
(b) For any eigenvalue λ *of* $\mathbf{B}$, $|\lambda| \leq \rho$.
(c) There exist left and right eigenvectors corresponding to the eigenvalue ρ *that have strictly positive coordinates.*
(d) The left and right eigenvectors μ, ϑ *corresponding to the eigenvalue* ρ *are unique up to a constant multiple.*
(e) For every $i \in \Sigma$ *and every* $\phi = (\phi_1, \ldots, \phi_{|\Sigma|})$ *such that* $\phi_j > 0$ *for all* j,

$$\lim_{n \to \infty} \frac{1}{n} \log \left[\sum_{j=1}^{|\Sigma|} B^n(i,j) \, \phi_j \right]$$

$$= \lim_{n \to \infty} \frac{1}{n} \log \left[\sum_{j=1}^{|\Sigma|} \phi_j \, B^n(j,i) \right] = \log \rho.$$

Proof: The proofs of parts (a)–(d) can be found in [Sen81], Theorem 1.5. To prove part (e), let $\alpha \triangleq \sup_i \vartheta_i$, $\beta \triangleq \inf_i \vartheta_i > 0$, $\gamma \triangleq \sup_j \phi_j$, and $\delta \triangleq \inf_j \phi_j > 0$, where ϑ is the right eigenvector corresponding to ρ as before. Then, for all $i, j \in \Sigma$,

$$\frac{\gamma}{\beta} B^n(i,j)\, \vartheta_j \geq B^n(i,j)\, \phi_j \geq \frac{\delta}{\alpha} B^n(i,j)\, \vartheta_j \, .$$

Therefore,

$$\lim_{n \to \infty} \frac{1}{n} \log \left[\sum_{j=1}^{|\Sigma|} B^n(i,j)\, \phi_j \right] \;=\; \lim_{n \to \infty} \frac{1}{n} \log \left[\sum_{j=1}^{|\Sigma|} B^n(i,j)\, \vartheta_j \right]$$

$$=\; \lim_{n \to \infty} \frac{1}{n} \log \left(\rho^n\, \vartheta_i \right) \;=\; \log \rho \; .$$

A similar argument leads to

$$\lim_{n \to \infty} \frac{1}{n} \log \left[\sum_{j=1}^{|\Sigma|} \phi_j B^n(j,i) \right] = \log \rho .$$

$\square$

Throughout this section, $\phi \gg 0$ denotes strictly positive vectors, i.e., $\phi_j > 0$ for all $j \in \Sigma$.

3.1.1 LDP for Additive Functionals of Markov Chains

The subject of this section is the large deviations of the empirical means

$$Z_n = \frac{1}{n} \sum_{k=1}^{n} X_k \, ,$$

where $X_k = f(Y_k)$ and $f : \Sigma \to \mathbb{R}^d$ is a given deterministic function; for an extension to random functions, see Exercise 3.1.4. If the random variables Y_k, and hence X_k, are independent, it follows from Cramér's theorem that the Fenchel–Legendre transform of the logarithmic moment generating function of X_1 is the rate function for the LDP associated with $\{Z_n\}$. The Gärtner–Ellis theorem hints that the rate function may still be expressed in terms of a Fenchel–Legendre transform, even when the random variables Y_k obey a Markov dependence.

To find an alternative representation for the logarithmic moment generating function $\Lambda(\lambda)$, associate with every $\lambda \in \mathbb{R}^d$ a nonnegative matrix $\mathbf{\Pi}_\lambda$, whose elements are

$$\pi_\lambda(i,j) = \pi(i,j)\, e^{\langle \lambda, f(j) \rangle} \quad i,j \in \Sigma \ .$$

Because the quantities $e^{\langle \lambda, f(j) \rangle}$ are always positive, π_λ is irreducible as soon as π is. For each $\lambda \in \mathbb{R}^d$, let $\rho(\mathbf{\Pi}_\lambda)$ denote the Perron–Frobenius eigenvalue of the matrix $\mathbf{\Pi}_\lambda$. It will be shown next that $\log \rho(\mathbf{\Pi}_\lambda)$ plays the role of the logarithmic moment generating function $\Lambda(\lambda)$.

Theorem 3.1.2 *Let $\{Y_k\}$ be a finite state Markov chain possessing an irreducible transition matrix $\mathbf{\Pi}$. For every $z \in \mathbb{R}^d$, define*

$$I(z) \stackrel{\triangle}{=} \sup_{\lambda \in \mathbb{R}^d} \left\{ \langle \lambda, z \rangle - \log \rho(\mathbf{\Pi}_\lambda) \right\} \ .$$

Then the empirical mean Z_n satisfies the LDP with the convex, good rate function $I(\cdot)$. Explicitly, for any set $\Gamma \subseteq \mathbb{R}^d$, and any initial state $\sigma \in \Sigma$,

$$- \inf_{z \in \Gamma^o} I(z) \ \leq \ \liminf_{n \to \infty} \frac{1}{n} \log P_\sigma^\pi (Z_n \in \Gamma) \tag{3.1.3}$$

$$\leq \ \limsup_{n \to \infty} \frac{1}{n} \log P_\sigma^\pi (Z_n \in \Gamma) \leq - \inf_{z \in \overline{\Gamma}} I(z) \ .$$

Proof: Define

$$\Lambda_n(\lambda) \stackrel{\triangle}{=} \log E_\sigma^\pi \left[e^{\langle \lambda, Z_n \rangle} \right] \ .$$

In view of the Gärtner–Ellis theorem (Theorem 2.3.6), it is enough to check that the limit

$$\Lambda(\lambda) \stackrel{\triangle}{=} \lim_{n \to \infty} \frac{1}{n} \Lambda_n(n\lambda) = \lim_{n \to \infty} \frac{1}{n} \log E_\sigma^\pi \left[e^{n\langle \lambda, Z_n \rangle} \right]$$

exists for every $\lambda \in \mathbb{R}^d$, that $\Lambda(\cdot)$ is finite and differentiable everywhere in $\mathbb{R}^d$, and that $\Lambda(\lambda) = \log \rho(\mathbf{\Pi}_\lambda)$. To begin, note that

$$
\begin{aligned}
\Lambda_n(n\lambda) &= \log E_\sigma^\pi \left[e^{\langle \lambda, \sum_{k=1}^n X_k \rangle} \right] \\
&= \log \sum_{y_1, \ldots, y_n} P_\sigma^\pi (Y_1 = y_1, \ldots, Y_n = y_n) \prod_{k=1}^n e^{\langle \lambda, f(y_k) \rangle} \\
&= \log \sum_{y_1, \ldots, y_n} \pi(\sigma, y_1) e^{\langle \lambda, f(y_1) \rangle} \cdots \pi(y_{n-1}, y_n) e^{\langle \lambda, f(y_n) \rangle} \\
&= \log \sum_{y_n = 1}^{|\Sigma|} (\mathbf{\Pi}_\lambda)^n (\sigma, y_n) \ .
\end{aligned}
$$

Since $\mathbf{\Pi}_\lambda$ is irreducible, part (e) of the Perron–Frobenius theorem yields (with $\phi = (1, \dots, 1)$)

$$\Lambda(\lambda) = \lim_{n \to \infty} \frac{1}{n} \Lambda_n(n\lambda) = \log \rho\,(\mathbf{\Pi}_\lambda)\,.$$

Moreover, since $|\Sigma|$ is finite, $\rho(\mathbf{\Pi}_\lambda)$, being an isolated root of the characteristic equation for the matrix $\mathbf{\Pi}_\lambda$, is positive, finite and differentiable with respect to λ. (See [Lanc69], Theorem 7.7.1.)   □

Remarks:

(a) The preceding proof relies on two properties of the Markov chain—namely, part (e) of the Perron–Frobenius theorem and the differentiability of $\rho(\mathbf{\Pi}_\lambda)$ with respect to λ. Thus, the theorem holds as long as the Markov chain possesses these two properties. In particular, the finiteness of Σ is not crucial; the LDP for the general Markov chain setup is presented in Sections 6.3 and 6.5.

(b) The good rate function $I(\cdot)$ in the LDP of Theorem 3.1.2 is convex and does not depend on the initial state $\sigma \in \Sigma$. Both properties might be lost when the transition matrix $\mathbf{\Pi}$ is reducible, even when Z_n satisfy the law of large numbers (and the central limit theorem). For an example with dependence on the initial state, consider $\Sigma = \{1, 2\}$, with $\mathbf{\Pi}$ such that $\pi(2,1) = \pi(2,2) = 1/2$, $\pi(1,1) = 1$ and $f(j) = 1_2(j)$. Then, $\rho(\mathbf{\Pi}_\lambda) = \exp(\max\{\lambda - \log 2, 0\})$ and $Z_n = L_n^{\mathbf{Y}}(2) \to 0$ almost surely. For $\sigma = 1$, obviously $Y_k = 1$ for all k, hence $Z_n = 0$ for all n satisfies the LDP with the convex good rate function $I(0) = 0$ and $I(z) = \infty$ for $z \neq 0$. In contrast, the good rate function in Theorem 3.1.2 is $I(z) = z \log 2$ for $z \in [0,1]$ and $I(z) = \infty$ otherwise. Indeed, this is the rate function for the LDP in case of $\sigma = 2$, for which $P_\sigma^\pi(Z_n = k/n) = 2^{-(k+1)}(1 + 1_n(k))$. For an example with a non-convex rate function, add to this chain the states $\{3, 4\}$ with $\pi(4,3) = \pi(4,2) = \pi(3,3) = \pi(3,1) = 1/2$. Consider the initial state $\sigma = 4$ and $f(j) = (1_1(j), 1_2(j), 1_3(j)) \in \mathbb{R}^3$ so that $Z_n \to (1, 0, 0)$ almost surely. Computing directly $P_\sigma^\pi(Z_n \in \cdot)$, it is not hard to verify that Z_n satisfies the LDP with the non-convex, good rate function $I(1 - z, z, 0) = I(1 - z, 0, z) = z \log 2$ for $z \in [0,1]$, and $I(\cdot) = \infty$ otherwise.

Exercise 3.1.4 Assume that $Y_1, \dots, Y_n$ are distributed according to the joint law P_σ^π determined by the irreducible stochastic matrix $\mathbf{\Pi}$. Let the conditional law of $\{X_k\}$ for each realization $\{Y_k = j_k\}_{k=1}^n$ be the product of the measures $\mu_{j_k} \in M_1(\mathbb{R}^d)$; i.e., the variables X_k are conditionally independent. Suppose that the logarithmic moment generating functions Λ_j associated with μ_j are finite everywhere (for all $j \in \Sigma$). Consider the empirical mean $Z_n = \frac{1}{n} \sum_{k=1}^n X_k$, and prove that Theorem 3.1.2 holds for Borel measurable sets Γ with

$$\pi_\lambda(i,j) \overset{\triangle}{=} \pi(i,j)\, e^{\Lambda_j(\lambda)}, \quad i, j \in \Sigma \ .$$

3.1.2 Sanov's Theorem for the Empirical Measure of Markov Chains

A particularly important application of Theorem 3.1.2 yields the LDP satisfied by the empirical measures $L_n^{\mathbf{Y}} = (L_n^{\mathbf{Y}}(1), \ldots, L_n^{\mathbf{Y}}(|\Sigma|))$ of Markov chains. Here, $L_n^{\mathbf{Y}}$ denotes the vector of frequencies in which the Markov chain visits the different states, namely,

$$L_n^{\mathbf{Y}}(i) = \frac{1}{n} \sum_{k=1}^{n} 1_i(Y_k), \quad i = 1, \ldots, |\Sigma|.$$

Suppose that $\mathbf{\Pi}$ is an irreducible matrix, and let μ be the stationary distribution of the Markov chain, i.e., the unique non negative left eigenvector of $\mathbf{\Pi}$ whose components sum to 1. The ergodic theorem then implies that $L_n^{\mathbf{Y}} \to \mu$ in probability as $n \to \infty$, at least when $\mathbf{\Pi}$ is aperiodic and the initial state Y_0 is distributed according to μ. It is not hard to check that the same holds for any irreducible $\mathbf{\Pi}$ and any distribution of the initial state. Hence, the sequence $\{L_n^{\mathbf{Y}}\}$ is a good candidate for an LDP in $M_1(\Sigma)$.

It is clear that $L_n^{\mathbf{Y}}$ fits into the framework of the previous section if we take $f(y) = (1_1(y), \ldots, 1_{|\Sigma|}(y))$. Therefore, by Theorem 3.1.2, the LDP holds for $\{L_n^{\mathbf{Y}}\}$ with the rate function

$$I(q) = \sup_{\lambda \in \mathbb{R}^d} \{\langle \lambda, q \rangle - \log \rho(\mathbf{\Pi}_\lambda)\}, \tag{3.1.5}$$

where $\pi_\lambda(i,j) \triangleq \pi(i,j) e^{\lambda_j}$. The following alternative characterization of $I(q)$ is sometimes more useful.

Theorem 3.1.6

$$I(q) = J(q) \triangleq \begin{cases} \sup_{\mathbf{u} \gg 0} \sum_{j=1}^{|\Sigma|} q_j \log\left[\frac{u_j}{(\mathbf{u}\mathbf{\Pi})_j}\right], & q \in M_1(\Sigma) \\ \infty & q \notin M_1(\Sigma). \end{cases} \tag{3.1.7}$$

Remarks:
(a) If the random variables $\{Y_k\}$ are i.i.d., then the rows of $\mathbf{\Pi}$ are identical, in which case $J(q)$ is the relative entropy $H(q|\pi(1,\cdot))$. (See Exercise 3.1.8.)
(b) The preceding identity actually also holds for non-stochastic matrices $\mathbf{\Pi}$. (See Exercise 3.1.9.)

Proof: Note that for every n and every realization of the Markov chain, $L_n^{\mathbf{Y}} \in M_1(\Sigma)$, which is a closed subset of $\mathbb{R}^{|\Sigma|}$. Hence, the lower bound of (3.1.3) yields for the open set $M_1(\Sigma)^c$

$$-\infty = \liminf_{n \to \infty} \frac{1}{n} \log P_\sigma^\pi \left(\{L_n^{\mathbf{Y}} \in M_1(\Sigma)^c\}\right) \geq - \inf_{q \notin M_1(\Sigma)} I(q).$$

Consequently, $I(q) = \infty$ for every $q \notin M_1(\Sigma)$.

Fix a probability vector $q \in M_1(\Sigma)$, a strictly positive vector $\mathbf{u} \gg 0$, and for $j = 1, \ldots, |\Sigma|$ set $\lambda_j = \log[u_j/(\mathbf{u}\mathbf{\Pi})_j]$. Note that since $\mathbf{u} \gg 0$ and $\mathbf{\Pi}$ is irreducible, it follows that $\mathbf{u}\mathbf{\Pi} \gg 0$. Observe that $\mathbf{u}\mathbf{\Pi}_\lambda = \mathbf{u}$, hence $\mathbf{u}\mathbf{\Pi}_\lambda^n = \mathbf{u}$ and thus $\rho(\mathbf{\Pi}_\lambda) = 1$ by part (e) of the Perron–Frobenius theorem (with $\phi_i = u_i > 0$). Therefore, by definition,

$$I(q) \geq \sum_{j=1}^{|\Sigma|} q_j \log\left[\frac{u_j}{(\mathbf{u}\mathbf{\Pi})_j}\right].$$

Since $\mathbf{u} \gg 0$ is arbitrary, this inequality implies that $I(q) \geq J(q)$.

To establish the reverse inequality, fix an arbitrary vector $\lambda \in \mathbb{R}^{|\Sigma|}$ and let $\mathbf{u}^* \gg 0$ be a left eigenvector corresponding to the eigenvalue $\rho(\mathbf{\Pi}_\lambda)$ of the irreducible matrix $\mathbf{\Pi}_\lambda$. Then, $\mathbf{u}^*\mathbf{\Pi}_\lambda = \rho(\mathbf{\Pi}_\lambda)\mathbf{u}^*$, and by the definition of $\mathbf{\Pi}_\lambda$,

$$\langle \lambda, q \rangle + \sum_{j=1}^{|\Sigma|} q_j \log\left[\frac{(\mathbf{u}^*\mathbf{\Pi})_j}{u_j^*}\right] = \sum_{j=1}^{|\Sigma|} q_j \log\left[\frac{(\mathbf{u}^*\mathbf{\Pi}_\lambda)_j}{u_j^*}\right]$$

$$= \sum_{j=1}^{|\Sigma|} q_j \log \rho(\mathbf{\Pi}_\lambda) = \log \rho(\mathbf{\Pi}_\lambda).$$

Therefore,

$$\langle \lambda, q \rangle - \log \rho(\mathbf{\Pi}_\lambda) \leq \sup_{\mathbf{u} \gg 0} \sum_{j=1}^{|\Sigma|} q_j \log\left[\frac{u_j}{(\mathbf{u}\mathbf{\Pi})_j}\right] = J(q).$$

Since λ is arbitrary, $I(q) \leq J(q)$ and the proof is complete. □

Exercise 3.1.8 Suppose that for every $i, j \in \Sigma$, $\pi(i, j) = \mu(j)$, where $\mu \in M_1(\Sigma)$. Show that $J(\cdot) = H(\cdot|\mu)$ (the relative entropy with respect to μ), and that $I(\cdot)$ is the Fenchel–Legendre transform of $\log[\sum_j e^{\lambda_j} \mu(j)]$. Thus, Theorem 3.1.6 is a natural extension of Exercise 2.2.36 to the Markov setup.

Exercise 3.1.9 (a) Show that the relation in Theorem 3.1.6 holds for any nonnegative irreducible matrix $\mathbf{B} = \{b(i, j)\}$ (not necessarily stochastic).
Hint: Let $\phi(i) = \sum_j b(i, j)$. Clearly, $\phi \gg 0$, and the matrix $\mathbf{\Pi}$ with $\pi(i, j) = b(i, j)/\phi(i)$ is stochastic. Let I_B and J_B denote the rate functions I and J associated with the matrix $\mathbf{B}$ via (3.1.5) and (3.1.7), respectively. Now prove that $J_\Pi(q) = J_B(q) + \sum_j q_j \log \phi(j)$ for all $q \in \mathbb{R}^{|\Sigma|}$, and likewise,

$I_\Pi(q) = I_B(q) + \sum_j q_j \log \phi(j)$.

(b) Show that for any irreducible, nonnegative matrix $\mathbf{B}$,

$$\log \rho(\mathbf{B}) = \sup_{\nu \in M_1(\Sigma)} \{-J_B(\nu)\}. \qquad (3.1.10)$$

This characterization is useful when looking for bounds on the spectral radius of nonnegative matrices. (For an alternative characterization of the spectral radius, see Exercise 3.1.19.)

Exercise 3.1.11 Show that for any nonnegative irreducible matrix $\mathbf{B}$,

$$J(q) = \begin{cases} \displaystyle\sup_{\mathbf{u} \gg 0} \sum_{j=1}^{|\Sigma|} q_j \log \left[\frac{u_j}{(\mathbf{Bu})_j} \right], & q \in M_1(\Sigma) \\ \infty & q \notin M_1(\Sigma). \end{cases}$$

Hint: Prove that the matrices $\{b(j,i)e^{\lambda_j}\}$ and $\mathbf{B}_\lambda$ have the same eigenvalues, and use part (a) of Exercise 3.1.9.

3.1.3 Sanov's Theorem for the Pair Empirical Measure of Markov Chains

The rate function governing the LDP for the empirical measure of a Markov chain is still in the form of an optimization problem. Moreover, the elegant interpretation in terms of relative entropy (recall Section 2.1.1 where the i.i.d. case is presented) has disappeared. It is interesting to note that by considering a somewhat different random variable, from which the large deviations for L_n^Y may be recovered (see Exercise 3.1.17), an LDP may be obtained with a rate function that is an appropriate relative entropy.

Consider the space $\Sigma^2 \triangleq \Sigma \times \Sigma$, which corresponds to consecutive pairs of elements from the sequence $\mathbf{Y}$. Note that by considering the pairs formed by $Y_1, \ldots, Y_n$, i.e., the sequence $Y_0 Y_1, Y_1 Y_2, Y_2 Y_3, \ldots, Y_i Y_{i+1}, \ldots, Y_{n-1} Y_n$, where $Y_0 = \sigma$, a Markov chain is recovered with state space Σ^2 and transition matrix $\mathbf{\Pi}^{(2)}$ specified via

$$\pi^{(2)}(k \times \ell, \ i \times j) = 1_\ell(i)\, \pi(i,j). \qquad (3.1.12)$$

For simplicity, it is assumed throughout this section that $\mathbf{\Pi}$ is strictly positive (i.e., $\pi(i,j) > 0$ for all i, j). Then $\mathbf{\Pi}^{(2)}$ is an irreducible transition matrix, and therefore, the results of Section 3.1.2 may be applied to yield the rate function $I_2(q)$ associated with the large deviations of the pair

$$\pi = \begin{bmatrix} 2/3 & 1/3 \\ 1/3 & 2/3 \end{bmatrix}, \quad \pi^{(2)} = \begin{bmatrix} 2/3 & 1/3 & 0 & 0 \\ 0 & 0 & 1/3 & 2/3 \\ 2/3 & 1/3 & 0 & 0 \\ 0 & 0 & 1/3 & 2/3 \end{bmatrix}$$

Figure 3.1.1: Example of π and π^2, $|\Sigma| = 2$.

empirical measures

$$L_{n,2}^{\mathbf{Y}}(y) \overset{\triangle}{=} \frac{1}{n} \sum_{i=1}^{n} 1_y(Y_{i-1}Y_i), \quad y \in \Sigma^2 .$$

Note that $L_{n,2}^{\mathbf{Y}} \in M_1(\Sigma^2)$ and, therefore, $I_2(\cdot)$ is a good, convex rate function on this space. To characterize $I_2(\cdot)$ as an appropriate relative entropy, the following definitions are needed. For any $q \in M_1(\Sigma^2)$, let

$$q_1(i) \overset{\triangle}{=} \sum_{j=1}^{|\Sigma|} q(i,j) \quad \text{and} \quad q_2(i) \overset{\triangle}{=} \sum_{j=1}^{|\Sigma|} q(j,i)$$

be its marginals. Whenever $q_1(i) > 0$, let $q_f(j|i) \overset{\triangle}{=} q(i,j)/q_1(i)$. A probability measure $q \in M_1(\Sigma^2)$ is *shift invariant* if $q_1 = q_2$, i.e., both marginals of q are identical.

Theorem 3.1.13 *Assume that* $\mathbf{\Pi}$ *is strictly positive. Then for every probability measure* $q \in M_1(\Sigma^2)$,

$$I_2(q) = \begin{cases} \sum_i q_1(i) H(q_f(\cdot|i) \mid \pi(i,\cdot)) & , \quad \text{if } q \text{ is shift invariant} \\ \infty & , \quad \text{otherwise}, \end{cases}$$

(3.1.14)

where $H(\cdot|\cdot)$ *is the relative entropy function defined in Section 2.1.1, i.e.,*

$$H(q_f(\cdot|i) \mid \pi(i,\cdot)) = \sum_{j=1}^{|\Sigma|} q_f(j|i) \, \log \frac{q_f(j|i)}{\pi(i,j)} .$$

Remarks:
(a) When $\mathbf{\Pi}$ is not strictly positive (but is irreducible), the theorem still applies, with Σ^2 replaced by $\{(i,j) : \pi(i,j) > 0\}$, and a similar proof.
(b) The preceding representation of $I_2(q)$ is useful in characterizing the spectral radius of nonnegative matrices. (See Exercise 3.1.19.) It is also useful because bounds on the relative entropy are readily available and may be used to obtain bounds on the rate function.

Proof: By Theorem 3.1.6,

$$
\begin{aligned}
I_2(q) &= \sup_{\mathbf{u} \gg 0} \sum_{j=1}^{|\Sigma|} \sum_{i=1}^{|\Sigma|} q(i,j) \, \log \frac{u(i,j)}{(\mathbf{u}\Pi^{(2)})(i,j)} \\
&= \sup_{\mathbf{u} \gg 0} \sum_{j=1}^{|\Sigma|} \sum_{i=1}^{|\Sigma|} q(i,j) \, \log \frac{u(i,j)}{[\sum_k u(k,i)]\pi(i,j)}, \quad (3.1.15)
\end{aligned}
$$

where the last equality follows by the definition of $\Pi^{(2)}$.

Assume first that q is not shift invariant. Then $q_1(j_0) < q_2(j_0)$ for some j_0. For $\mathbf{u}$ such that $u(\cdot, j) = 1$ when $j \neq j_0$ and $u(\cdot, j_0) = e^\alpha$,

$$
\begin{aligned}
& \sum_{j=1}^{|\Sigma|} \sum_{i=1}^{|\Sigma|} q(i,j) \, \log \frac{u(i,j)}{[\sum_k u(k,i)] \, \pi(i,j)} \\
= \; & \sum_{j=1}^{|\Sigma|} \sum_{i=1}^{|\Sigma|} q(i,j) \, \log \frac{u(1,j)}{|\Sigma| u(1,i) \pi(i,j)} \\
= \; & -\sum_{j=1}^{|\Sigma|} \sum_{i=1}^{|\Sigma|} q(i,j) \, \log \{|\Sigma| \pi(i,j)\} + \alpha[q_2(j_0) - q_1(j_0)].
\end{aligned}
$$

Letting $\alpha \to \infty$, we find that $I_2(q) = \infty$.

Finally, if q is shift invariant, then for every $\mathbf{u} \gg 0$,

$$
\sum_{i=1}^{|\Sigma|} \sum_{j=1}^{|\Sigma|} q(i,j) \, \log \frac{\sum_k u(k,i) q_2(j)}{\sum_k u(k,j) q_1(i)} = 0. \quad (3.1.16)
$$

Let $u(i|j) = u(i,j)/\sum_k u(k,j)$ and $q_b(i|j) \overset{\triangle}{=} q(i,j)/q_2(j) = q(i,j)/q_1(j)$ (as q is shift invariant). By (3.1.15) and (3.1.16),

$$
\begin{aligned}
& I_2(q) - \sum_{i=1}^{|\Sigma|} q_1(i) \, H(q_f(\cdot|i) \mid \pi(i, \cdot)) \\
= \; & \sup_{\mathbf{u} \gg 0} \sum_{i=1}^{|\Sigma|} \sum_{j=1}^{|\Sigma|} q(i,j) \, \log \frac{u(i,j) q_1(i)}{[\sum_k u(k,i)] q(i,j)} \\
= \; & \sup_{\mathbf{u} \gg 0} \sum_{i=1}^{|\Sigma|} \sum_{j=1}^{|\Sigma|} q(i,j) \, \log \frac{u(i|j)}{q_b(i|j)} \\
= \; & \sup_{\mathbf{u} \gg 0} \left\{ -\sum_{j=1}^{|\Sigma|} q_2(j) H(q_b(\cdot|j) \mid u(\cdot|j)) \right\}.
\end{aligned}
$$

Note that always $I_2(q) \leq \sum_i q_1(i) H(q_f(\cdot|i) \,|\, \pi(i,\cdot))$ because $H(\cdot|\cdot)$ is a nonnegative function, whereas if $q \gg 0$, then the choice $\mathbf{u} = q$ yields equality in the preceding. The proof is completed for q, which is not strictly positive, by considering a sequence $\mathbf{u}_n \gg 0$ such that $\mathbf{u}_n \to q$ (so that $q_2(j) H(q_b(\cdot|j) \,|\, u_{\bar{n}}(\cdot|j)) \to 0$ for each j). $\qquad \square$

Exercise 3.1.17 Prove that for any strictly positive stochastic matrix $\mathbf{\Pi}$

$$J(\nu) = \inf_{\{q : q_2 = \nu\}} I_2(q) , \qquad (3.1.18)$$

where $J(\cdot)$ is the rate function defined in (3.1.7), while $I_2(\cdot)$ is as specified in Theorem 3.1.13.

Hint: There is no need to prove the preceding identity directly. Instead, for $Y_0 = \sigma$, observe that $L_n^{\mathbf{Y}} \in A$ iff $L_{n,2}^{\mathbf{Y}} \in \{q : q_2 \in A\}$. Since the projection of any measure $q \in M_1(\Sigma^2)$ to its marginal q_2 is continuous and $I_2(\cdot)$ controls the LDP of $L_{n,2}^{\mathbf{Y}}$, deduce that the right side of (3.1.18) is a rate function governing the LDP of $L_n^{\mathbf{Y}}$. Conclude by proving the uniqueness of such a function.

Exercise 3.1.19 (a) Extend the validity of the identity (3.1.18) to any irreducible nonnegative matrix $\mathbf{B}$.

Hint: Consider the remark following Theorem 3.1.13 and the transformation $\mathbf{B} \mapsto \mathbf{\Pi}$ used in Exercise 3.1.9.
(b) Deduce by applying the identities (3.1.10) and (3.1.18) that for any nonnegative irreducible matrix $\mathbf{B}$,

$$
\begin{aligned}
-\log \rho(\mathbf{B}) &= \inf_{q \in M_1(\Sigma_B)} I_2(q) \\
&= \inf_{q \in M_1(\Sigma_B),\, q_1 = q_2} \sum_{(i,j) \in \Sigma_B} q(i,j) \log \frac{q_f(j|i)}{b(i,j)} ,
\end{aligned}
$$

where $\Sigma_B = \{(i,j) : b(i,j) > 0\}$.
Remark: This is Varadhan's characterization of the spectral radius of nonnegative irreducible matrices.

Exercise 3.1.20 Suppose that $Y_1, \ldots, Y_n, \ldots$ are Σ-valued i.i.d. random variables, with $p(j)$ denoting the probability that $Y = j$ for $j = 1, 2, \ldots, |\Sigma|$. A typical k-scan process is defined via $X_i = f_1(Y_i) + \cdots + f_k(Y_{i+k-1})$, where $i \in \mathbb{Z}_+$ and f_ℓ, $\ell = 1, \ldots, k$, are deterministic real valued functions. The LDP of the empirical measures $L_{n,k}^{\mathbf{Y}}$ of the k-tuples $(Y_1 Y_2 \cdots Y_k)$, $(Y_2 Y_3 \cdots Y_{k+1})$, $\ldots, (Y_n Y_{n+1} \cdots Y_{n+k})$ is instrumental for the large deviations of these processes as $n \to \infty$ (while k is fixed).
(a) Show that the rate function governing the LDP of $L_{n,2}^{\mathbf{Y}}$ is

$$
I_2(q) = \begin{cases} \sum_{i=1}^{|\Sigma|} \sum_{j=1}^{|\Sigma|} q(i,j) \log \frac{q_f(j|i)}{p(j)} , & q \text{ shift invariant} \\ \infty , & \text{otherwise.} \end{cases}
$$

(b) Show that the rate function governing the LDP of $L_{n,k}^Y$ is

$$
I_k(q) = \begin{cases} \sum_{j_1=1}^{|\Sigma|} \cdots \sum_{j_k=1}^{|\Sigma|} q(j_1,\ldots,j_k) \log \frac{q_f(j_k|j_1,\ldots,j_{k-1})}{p(j_k)}, \\ \qquad\qquad\qquad\qquad\qquad\qquad\qquad q \text{ shift invariant} \\ \infty \qquad\qquad\qquad\qquad\qquad\quad, \quad \text{otherwise}, \end{cases}
$$

where $q \in M_1(\Sigma^k)$ is shift invariant if

$$
\sum_i q(i,j_1,\ldots,j_{k-1}) = \sum_i q(j_1,\ldots,j_{k-1},i).
$$

Exercise 3.1.21 (a) Prove that for any sequence $\mathbf{y} = (y_1,\ldots,y_n) \in \Sigma^n$ of nonzero P_σ^π probability,

$$
\frac{1}{n} \log P_\sigma^\pi(Y_1 = y_1,\ldots,Y_n = y_n) = \sum_{i=1}^{|\Sigma|} \sum_{j=1}^{|\Sigma|} L_{n,2}^\mathbf{y}(i,j) \log \pi(i,j) .
$$

(b) Let

$$
\mathcal{L}_n \triangleq \{q : q = L_{n,2}^\mathbf{y}, P_\sigma^\pi(Y_1 = y_1,\ldots,Y_n = y_n) > 0 \text{ for some } \mathbf{y} \in \Sigma^n\}
$$

be the set of possible types of pairs of states of the Markov chain. Prove that $\mathcal{L}_n$ can be identified with a subset of $M_1(\Sigma_\Pi)$, where $\Sigma_\Pi = \{(i,j) : \pi(i,j) > 0\}$, and that $|\mathcal{L}_n| \leq (n+1)^{|\Sigma|^2}$.
(c) Let $T_n(q)$ be the type class of $q \in \mathcal{L}_n$, namely, the set of sequences $\mathbf{y}$ of positive P_σ^π probability for which $L_{n,2}^\mathbf{y} = q$, and let $H(q) \triangleq -\sum_{i,j} q(i,j) \log q_f(j|i)$. Suppose that for any $q \in \mathcal{L}_n$,

$$
(n+1)^{-(|\Sigma|^2+|\Sigma|)} e^{nH(q)} \leq |T_n(q)| \leq e^{nH(q)} , \qquad (3.1.22)
$$

and moreover that for all $q \in M_1(\Sigma_\Pi)$,

$$
\lim_{n\to\infty} d_V(q, \mathcal{L}_n) = 0 \quad \text{iff} \quad q \text{ is shift invariant} . \qquad (3.1.23)
$$

Prove by adapting the method of types of Section 2.1.1 that $L_{n,2}^Y$ satisfies the LDP with the rate function $I_2(\cdot)$ specified in (3.1.14).

Remark: Using Stirling's formula, the estimates (3.1.22) and (3.1.23) are consequences of a somewhat involved combinatorial estimate of $|T_n(q)|$. (See, for example, [CsCC87], Eqs. (35)–(37), Lemma 3, and the references therein.)

3.2   Long Rare Segments in Random Walks

Consider the random walk $S_0 = 0$, $S_k = \sum_{i=1}^k X_i$, $k = 1, 2, \ldots$, where X_i are i.i.d. random variables taking values in $\mathbb{R}^d$. Let R_m be segments of maximal length of the random walk up to time m whose empirical mean belongs to a measurable set $A \subset \mathbb{R}^d$, i.e.,

$$R_m \stackrel{\triangle}{=} \max \left\{ \ell - k : 0 \le k < \ell \le m, \ \frac{S_\ell - S_k}{\ell - k} \in A \right\}.$$

Associated with $\{R_m\}$ are the stopping times

$$T_r \stackrel{\triangle}{=} \inf \left\{ \ell : \frac{S_\ell - S_k}{\ell - k} \in A \quad \text{for some} \quad 0 \le k \le \ell - r \right\},$$

so that $\{R_m \ge r\}$ if and only if $\{T_r \le m\}$.

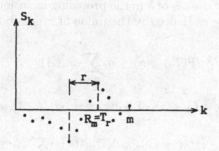

Figure 3.2.1: Maximal length segments.

The random variables R_m and T_r appear in comparative analysis of DNA sequences and in the analysis of computer search algorithms. The following theorem gives estimates on the asymptotics of R_m (and T_r) as $m \to \infty$ ($r \to \infty$, respectively). For some applications and refinements of these estimates, see [ArGW90] and the exercises at the end of this section.

Theorem 3.2.1 *Suppose A is such that the limit*

$$I_A \stackrel{\triangle}{=} - \lim_{n \to \infty} \frac{1}{n} \ \log \mu_n(A) \tag{3.2.2}$$

exists, where μ_n is the law of $\hat{S}_n = \frac{1}{n} S_n$. Then, almost surely,

$$\lim_{m \to \infty} (R_m / \log m) = \lim_{r \to \infty} (r / \log T_r) = 1/I_A.$$

Remark: The condition (3.2.2) is typically established as a result of a LDP. For example, if $\Lambda(\cdot)$, the logarithmic moment generating function of X_1, is finite everywhere, then by Cramér's theorem, condition (3.2.2) holds whenever

$$I_A = \inf_{x \in \overline{A}} \Lambda^*(x) = \inf_{x \in A^\circ} \Lambda^*(x). \tag{3.2.3}$$

Examples for which (3.2.2) holds are presented in Exercises 3.2.5 and 3.2.6.

Proof of Theorem 3.2.1: First, the left tail of the distribution of T_r is bounded by the inclusion of events

$$\{T_r \leq m\} \subset \bigcup_{k=0}^{m-r} \bigcup_{\ell=k+r}^{m} C_{k,\ell} \subset \bigcup_{k=0}^{m-1} \bigcup_{\ell=k+r}^{\infty} C_{k,\ell} ,$$

where

$$C_{k,\ell} \triangleq \left\{ \frac{S_\ell - S_k}{\ell - k} \in A \right\} .$$

There are m possible choices of k in the preceding inclusion, while $\mathrm{P}\,(C_{k,\ell}) = \mu_{\ell-k}(A)$ and $\ell - k \geq r$. Hence, by the union of events bound,

$$\mathrm{P}(T_r \leq m) \leq m \sum_{n=r}^{\infty} \mu_n(A) .$$

Suppose first that $0 < I_A < \infty$. Taking $m = \lfloor e^{r(I_A-\epsilon)} \rfloor$, and using (3.2.2),

$$\sum_{r=1}^{\infty} \mathrm{P}\,(T_r \leq e^{r(I_A-\epsilon)}) \;\leq\; \sum_{r=1}^{\infty} e^{r(I_A-\epsilon)} \sum_{n=r}^{\infty} c e^{-n(I_A-\epsilon/2)}$$

$$\leq\; c' \sum_{r=1}^{\infty} e^{-r\epsilon/2} < \infty$$

for every $\epsilon > 0$ and some positive constants c, c' (possibly ϵ dependent). When $I_A = \infty$, the same argument with $m = \lfloor e^{r/\epsilon} \rfloor$ implies that

$$\sum_{r=1}^{\infty} \mathrm{P}\,(T_r \leq e^{r/\epsilon}) < \infty \qquad \text{for all } \epsilon > 0.$$

By the Borel–Cantelli lemma, these estimates result in

$$\liminf_{r \to \infty} \frac{1}{r} \log T_r \geq I_A \quad \text{almost surely}.$$

Using the duality of events $\{R_m \geq r\} \equiv \{T_r \leq m\}$, it then follows that

$$\limsup_{m \to \infty} \frac{R_m}{\log m} \leq \frac{1}{I_A} \quad \text{almost surely}.$$

Note that for $I_A = \infty$, the proof of the theorem is complete. To establish the opposite inequality (when $I_A < \infty$), the right tail of the distribution of T_r needs to be bounded. Let $B_\ell \triangleq \left\{ \frac{1}{r}(S_{\ell r} - S_{(\ell-1)r}) \in A \right\}$. Note that $\{B_\ell\}_{\ell=1}^{\infty}$

are independent events (related to disjoint segments of the random walk) of equal probabilities $P(B_\ell) = \mu_r(A)$. Therefore, the inclusion

$$\bigcup_{\ell=1}^{\lfloor m/r \rfloor} B_\ell \subset \{T_r \leq m\}$$

yields

$$P(T_r > m) \leq 1 - P\left(\bigcup_{\ell=1}^{\lfloor m/r \rfloor} B_\ell\right) = (1 - P(B_1))^{\lfloor m/r \rfloor}$$

$$\leq e^{-\lfloor m/r \rfloor P(B_1)} = e^{-\lfloor m/r \rfloor \mu_r(A)} .$$

Combining this inequality (for $m = \lfloor e^{r(I_A + \epsilon)} \rfloor$) with (3.2.2), it follows that for every $\epsilon > 0$,

$$\sum_{r=1}^{\infty} P\left(T_r > e^{r(I_A + \epsilon)}\right) \leq \sum_{r=1}^{\infty} \exp\left(-\frac{c_1}{r} e^{r(I_A + \epsilon)} e^{-r(I_A + \epsilon/2)}\right)$$

$$\leq \sum_{r=1}^{\infty} \exp(-c_2 e^{c_3 r}) < \infty ,$$

where c_1, c_2, c_3 are some positive constants (which may depend on ϵ). By the Borel–Cantelli lemma and the duality of events $\{R_m < r\} \equiv \{T_r > m\}$,

$$\liminf_{m \to \infty} \frac{R_m}{\log m} = \liminf_{r \to \infty} \frac{r}{\log T_r} \geq \frac{1}{I_A} \quad \text{almost surely} ,$$

and the proof of the theorem is complete. $\square$

Exercise 3.2.4 Suppose that $I_A < \infty$ and that the identity (3.2.2) may be refined to

$$\lim_{r \to \infty} [\mu_r(A) r^{d/2} e^{r I_A}] = a$$

for some $a \in (0, \infty)$. (Such an example is presented in Section 3.7 for $d = 1$.) Let

$$\hat{R}_m \triangleq \frac{I_A R_m - \log m}{\log \log m} + \frac{d}{2} .$$

(a) Prove that $\limsup\limits_{m \to \infty} |\hat{R}_m| \leq 1$ almost surely.

(b) Deduce that $\lim\limits_{m \to \infty} P(\hat{R}_m \geq \epsilon) = 0$ for all $\epsilon > 0$.

Exercise 3.2.5 (a) Consider a sequence $X_1, \ldots, X_m$ of i.i.d. random variables taking values in a finite alphabet Σ, distributed according to the strictly positive marginal law μ. Let R_m be the maximal length segment of this sequence up to time m, whose empirical measure is in the non-empty, open set $\Gamma \subset M_1(\Sigma)$. Apply Theorem 3.2.1 to this situation.

(b) Assume further that Γ is convex. Let ν^* be the unique minimizer of $H(\cdot|\mu)$ in $\bar{\Gamma}$. Prove that as $m \to \infty$, the empirical measures associated with the segments contributing to R_m converge almost surely to ν^*.

Hint: Let $R_m^{(\delta)}$ denote the maximal length segment up to time m whose empirical measure is in the set $\{\mu : \mu \in \Gamma, d_V(\mu, \nu^*) > \delta\}$. Prove that

$$\limsup_{m \to \infty} \frac{R_m^{(\delta)}}{R_m} < 1, \quad \text{almost surely.}$$

Exercise 3.2.6 Prove that Theorem 3.2.1 holds when $X_1, \ldots, X_n$ and $Y_1, \ldots, Y_n$ are as in Exercise 3.1.4. Specifically, Y_k are the states of a Markov chain on the finite set $\{1, 2, \ldots, |\Sigma|\}$ with an irreducible transition matrix $\mathbf{\Pi}$, and the conditional law of X_k when $Y_k = j$ is $\mu_j \in M_1(\mathbb{R}^d)$; the random variables $\{X_k\}$ are independent given any realization of the Markov chain states, and the logarithmic moment generating functions Λ_j associated with μ_j are finite everywhere.

Remark: Here, $\Lambda^*(\cdot)$ of (3.2.3) is replaced by the Fenchel–Legendre transform of $\log \rho(\mathbf{\Pi}_\lambda)$, where $\pi_\lambda(i,j) \triangleq \pi(i,j) \, e^{\Lambda_j(\lambda)}$.

Exercise 3.2.7 Consider the i.i.d. random variables $\{\tilde{Y}_j\}$, $\{Y_j\}$ all distributed following $\tilde{\mu} \in M_1(\tilde{\Sigma})$ for $\tilde{\Sigma}$ a finite set. Let $\Sigma = \tilde{\Sigma}^2$ and $\mu = \tilde{\mu}^2 \in M_1(\Sigma)$. For any integers $s, r \geq 0$ let $L_k^{T^s \tilde{\mathbf{y}}, T^r \mathbf{y}}$ denote the empirical measure of the sequence $((\tilde{y}_{s+1}, y_{r+1}), \ldots, (\tilde{y}_{s+k}, y_{r+k}))$.

(a) Using Lemma 2.1.9 show that for any $\nu \in \mathcal{L}_k$, $k \in \{1, \ldots, n\}$,

$$P\left(\bigcup_{s, r \leq n-k} \{L_k^{T^s \tilde{\mathbf{Y}}, T^r \mathbf{Y}} = \nu\} \right) \leq n^2 e^{-kH(\nu|\mu)} .$$

(b) Fix $f : \Sigma \to \mathbb{R}$ such that $E(f(\tilde{Y}_1, Y_1)) < 0$ and $P(f(\tilde{Y}_1, Y_1) > 0) > 0$. Consider

$$M_n = \max_{k, 0 \leq s, r \leq n-k} \sum_{j=1}^{k} f(\tilde{Y}_{s+j}, Y_{r+j}) .$$

Prove that almost surely,

$$\limsup_{n \to \infty} \frac{M_n}{\log n} \leq \sup_{\nu \in M_1(\Sigma)} \frac{2 \int f d\nu}{H(\nu|\mu)} .$$

Hint: Apply (2.2.12) for $X_j = f(\tilde{Y}_j, Y_j)$ and $x = 0 > \bar{x}$ to show that

$k \geq (5/\Lambda^*(0)) \log n$ is negligible when considering M_n. Use (a) to bound the contribution of each $k \leq (5/\Lambda^*(0)) \log n$. Then, apply the Borel-Cantelli lemma along the skeleton $n_k = \lfloor e^k \rfloor$ and conclude the proof using the monotonicity of $n \mapsto M_n$.

3.3 The Gibbs Conditioning Principle for Finite Alphabets

Let $Y_1, Y_2, \ldots, Y_n$ be a sequence of i.i.d. random variables with strictly positive law μ on the finite alphabet Σ. Let $X_k = f(Y_k)$ for some deterministic $f : \Sigma \to \mathbb{R}$. The following question is of fundamental importance in statistical mechanics. Given a set $A \in \mathbb{R}$ and a constraint of the type $\hat{S}_n \in A$, what is the conditional law of Y_1 when n is large? In other words, what are the limit points, as $n \to \infty$, of the conditional probability vector

$$\mu_n^*(a_i) \triangleq P_\mu(Y_1 = a_i \mid \hat{S}_n \in A), \quad i = 1, \ldots, |\Sigma| \ .$$

Recall that $\hat{S}_n \triangleq \frac{1}{n} \sum_{j=1}^n X_j = \langle \mathbf{f}, L_n^{\mathbf{Y}} \rangle$, where $\mathbf{f} = (f(a_1), \ldots, f(a_{|\Sigma|}))$, and note that under the conditioning $\hat{S}_n \in A$, Y_j are identically distributed, although not independent. Therefore, for every function $\phi : \Sigma \to \mathbb{R}$,

$$\begin{aligned}
\langle \phi, \mu_n^* \rangle &= E[\phi(Y_1) \mid \hat{S}_n \in A] = E[\phi(Y_2) \mid \hat{S}_n \in A] \\
&= E[\frac{1}{n} \sum_{j=1}^n \phi(Y_j) \mid \hat{S}_n \in A] = E[\langle \phi, L_n^{\mathbf{Y}} \rangle \mid \langle \mathbf{f}, L_n^{\mathbf{Y}} \rangle \in A] ,
\end{aligned}$$

where $\phi = (\phi(a_1), \ldots, \phi(a_{|\Sigma|}))$. Hence, with $\Gamma \triangleq \{\nu : \langle \mathbf{f}, \nu \rangle \in A\}$,

$$\mu_n^* = E[L_n^{\mathbf{Y}} \mid L_n^{\mathbf{Y}} \in \Gamma] . \tag{3.3.1}$$

Using this identity, the following characterization of the limit points of $\{\mu_n^*\}$ applies to any non-empty set Γ for which

$$I_\Gamma \triangleq \inf_{\nu \in \Gamma^o} H(\nu|\mu) = \inf_{\nu \in \overline{\Gamma}} H(\nu|\mu) . \tag{3.3.2}$$

Theorem 3.3.3 (Gibbs's principle) *Let*

$$\mathcal{M} \triangleq \{\nu \in \overline{\Gamma} : H(\nu|\mu) = I_\Gamma\} . \tag{3.3.4}$$

(a) All the limit points of $\{\mu_n^*\}$ belong to $\overline{co}(\mathcal{M})$—the closure of the convex hull of $\mathcal{M}$.
(b) When Γ is a convex set of non-empty interior, the set $\mathcal{M}$ consists of a single point to which μ_n^* converge as $n \to \infty$.

Remarks:
(a) For conditions on Γ (alternatively, on A) under which (3.3.2) holds, see Exercises 2.1.16–2.1.19.
(b) Since $\mathcal{M}$ is compact, it always holds true that $\mathrm{co}(\mathcal{M}) = \overline{\mathrm{co}}(\mathcal{M})$.
(c) To see why the limit distribution may be in $\mathrm{co}(\mathcal{M})$ and not just in $\mathcal{M}$, let $\{Y_i\}$ be i.i.d. Bernoulli($\frac{1}{2}$), let $X_i = Y_i$, and take $\Gamma = [0, \alpha] \cup [1 - \alpha, 1]$ for some small $\alpha > 0$. It is easy to see that while $\mathcal{M}$ consists of the probability distributions $\{$Bernoulli(α), Bernoulli($1 - \alpha$)$\}$, the symmetry of the problem implies that the only possible limit point of $\{\mu_n^*\}$ is the probability distribution Bernoulli($\frac{1}{2}$).

Proof: Since $|\Sigma| < \infty$, $\overline{\Gamma}$ is a compact set and thus $\mathcal{M}$ is non-empty. Moreover, part (b) of the theorem follows from part (a) by Exercise 2.1.19 and the compactness of $M_1(\Sigma)$. Next, for every $U \subset M_1(\Sigma)$,

$$E[L_n^{\mathbf{Y}} \mid L_n^{\mathbf{Y}} \in \Gamma] - E[L_n^{\mathbf{Y}} \mid L_n^{\mathbf{Y}} \in U \cap \Gamma]$$
$$= P_\mu(L_n^{\mathbf{Y}} \in U^c \mid L_n^{\mathbf{Y}} \in \Gamma)$$
$$\left\{ E[L_n^{\mathbf{Y}} \mid L_n^{\mathbf{Y}} \in U^c \cap \Gamma] - E[L_n^{\mathbf{Y}} \mid L_n^{\mathbf{Y}} \in U \cap \Gamma] \right\} .$$

Since $E[L_n^{\mathbf{Y}} \mid L_n^{\mathbf{Y}} \in U \cap \Gamma]$ belongs to $\mathrm{co}(U)$, while $\mu_n^* = E[L_n^{\mathbf{Y}} \mid L_n^{\mathbf{Y}} \in \Gamma]$, it follows that

$$d_V(\mu_n^*, \mathrm{co}(U))$$
$$\leq P_\mu(L_n^{\mathbf{Y}} \in U^c \mid L_n^{\mathbf{Y}} \in \Gamma) d_V\left(E[L_n^{\mathbf{Y}} \mid L_n^{\mathbf{Y}} \in U^c \cap \Gamma], E[L_n^{\mathbf{Y}} \mid L_n^{\mathbf{Y}} \in U \cap \Gamma] \right)$$
$$\leq P_\mu(L_n^{\mathbf{Y}} \in U^c \mid L_n^{\mathbf{Y}} \in \Gamma) \tag{3.3.5}$$

where the last inequality is due to the bound $d_V(\cdot, \cdot) \leq 1$. With $\mathcal{M}^\delta \triangleq \{\nu : d_V(\nu, \mathcal{M}) < \delta\}$, it is proved shortly that for every $\delta > 0$,

$$\lim_{n \to \infty} P_\mu(L_n^{\mathbf{Y}} \in \mathcal{M}^\delta \mid L_n^{\mathbf{Y}} \in \Gamma) = 1 , \tag{3.3.6}$$

with an exponential (in n) rate of convergence. Consequently, (3.3.5) applied to $U = \mathcal{M}^\delta$ results in $d_V(\mu_n^*, \mathrm{co}(\mathcal{M}^\delta)) \to 0$. Since d_V is a convex function on $M_1(\Sigma) \times M_1(\Sigma)$, each point in $\mathrm{co}(\mathcal{M}^\delta)$ is within variational distance δ of some point in $\mathrm{co}(\mathcal{M})$. With $\delta > 0$ being arbitrarily small, limit points of μ_n^* are necessarily in the closure of $\mathrm{co}(\mathcal{M})$.

To prove (3.3.6), apply Sanov's theorem (Theorem 2.1.10) and (3.3.2) to obtain

$$I_\Gamma = -\lim_{n \to \infty} \frac{1}{n} \log P_\mu(L_n^{\mathbf{Y}} \in \Gamma) \tag{3.3.7}$$

and

$$\limsup_{n \to \infty} \frac{1}{n} \log P_\mu(L_n^{\mathbf{Y}} \in (\mathcal{M}^\delta)^c \cap \Gamma) \leq - \inf_{\nu \in (\mathcal{M}^\delta)^c \cap \Gamma} H(\nu|\mu)$$
$$\leq - \inf_{\nu \in (\mathcal{M}^\delta)^c \cap \overline{\Gamma}} H(\nu|\mu) . \tag{3.3.8}$$

Observe that $\mathcal{M}^\delta$ are open sets and, therefore, $(\mathcal{M}^\delta)^c \cap \overline{\Gamma}$ are compact sets. Thus, for some $\tilde{\nu} \in (\mathcal{M}^\delta)^c \cap \overline{\Gamma}$,

$$\inf_{\nu \in (\mathcal{M}^\delta)^c \cap \overline{\Gamma}} H(\nu|\mu) = H(\tilde{\nu}|\mu) > I_\Gamma . \qquad (3.3.9)$$

Now, (3.3.6) follows from (3.3.7)–(3.3.9) because

$$\limsup_{n \to \infty} \frac{1}{n} \log \mathbf{P}_\mu(L_n^{\mathbf{Y}} \in (\mathcal{M}^\delta)^c \mid L_n^{\mathbf{Y}} \in \Gamma)$$

$$= \limsup_{n \to \infty} \left\{ \frac{1}{n} \log \mathbf{P}_\mu(L_n^{\mathbf{Y}} \in (\mathcal{M}^\delta)^c \cap \Gamma) - \frac{1}{n} \log \mathbf{P}_\mu(L_n^{\mathbf{Y}} \in \Gamma) \right\} < 0.$$

$\square$

Remarks:
(a) Intuitively, one expects $Y_1, \ldots, Y_k$ to be asymptotically independent (as $n \to \infty$) for any fixed k, when the conditioning event is $\{L_n^{\mathbf{Y}} \in \Gamma\}$. This is indeed shown in Exercise 3.3.12 by considering "super-symbols" from the enlarged alphabet Σ^k.
(b) The preceding theorem holds whenever the set Γ satisfies (3.3.2). The particular conditioning set $\{\nu : \langle \mathbf{f}, \nu \rangle \in A\}$ has an important significance in statistical mechanics because it represents an energy-like constraint.
(c) Recall that by the discussion preceding (2.1.27), if A is a non-empty, convex, open subset of K, the interval supporting $\{f(a_i)\}$, then the unique limit of μ_n^* is of the form

$$\nu_\lambda(a_i) = \mu(a_i) e^{\lambda f(a_i) - \Lambda(\lambda)}$$

for some appropriately chosen $\lambda \in \mathbb{R}$, which is called the *Gibbs parameter* associated with A. In particular, for any $x \in K^o$, the Gibbs parameter associated with the set $(x - \delta, x + \delta)$ converges as $\delta \to 0$ to the unique solution of the equation $\Lambda'(\lambda) = x$.
(d) A Gibbs conditioning principle holds beyond the i.i.d. case. All that is needed is that Y_i are exchangeable conditionally upon any given value of $L_n^{\mathbf{Y}}$ (so that (3.3.1) holds). For such an example, consider Exercise 3.3.11.

Exercise 3.3.10 Using Lemma 2.1.9, prove Gibbs's principle (Theorem 3.3.3) by the method of types.

Exercise 3.3.11 Prove the Gibbs conditioning principle for sampling without replacement, i.e., under the assumptions of Section 2.1.3.
(a) Observe that Y_j again are identically distributed even under conditioning on their empirical measures. Conclude that (3.3.1) holds.
(b) Assume that Γ is such that

$$I_\Gamma = \inf_{\nu \in \Gamma^o} I(\nu|\beta, \mu) = \inf_{\nu \in \overline{\Gamma}} I(\nu|\beta, \mu) < \infty .$$

Define $\mathcal{M} \triangleq \{\nu \in \overline{\Gamma} : I(\nu|\beta, \mu) = I_\Gamma\}$, and prove that both parts of Theorem 3.3.3 hold. (For part (b), you may rely on Exercise 2.1.48.)

Exercise 3.3.12 (a) Suppose that $\Sigma = (\Sigma')^k$ and $\mu = (\mu')^k$ are, respectively, a kth product alphabet and a kth product probability measure on it, and assume that μ' is strictly positive on Σ'. For any law $\nu \in M_1(\Sigma)$ and $j \in \{1, \ldots, k\}$, let $\nu^{(j)} \in M_1(\Sigma')$, denote the jth marginal of ν on Σ'. Prove that

$$\frac{1}{k} H(\nu|\mu) \geq \frac{1}{k} \sum_{j=1}^{k} H(\nu^{(j)} | \mu') \geq H\left(\frac{1}{k} \sum_{j=1}^{k} \nu^{(j)} | \mu'\right),$$

with equality if and only if $\nu = (\nu')^k$ for some $\nu' \in M_1(\Sigma')$.
(b) Assume that

$$\Gamma \triangleq \{\nu : \frac{1}{k} \sum_{j=1}^{k} \nu^{(j)} \in \Gamma'\} \qquad (3.3.13)$$

for some $\Gamma' \subset M_1(\Sigma')$, which satisfies (3.3.2) with respect to μ'. Let $\mathcal{M}' = \{\nu' \in \overline{\Gamma}' : H(\nu'|\mu') = I_{\Gamma'}\}$ and prove that $\mathcal{M} = \{\nu : \nu = (\nu')^k, \nu' \in \mathcal{M}'\}$.
(c) Consider the kth joint conditional law

$$\mu_n^*(a'_{i_1}, \ldots, a'_{i_k}) \triangleq P_{\mu'}(Y_1 = a'_{i_1}, \ldots, Y_k = a'_{i_k} \mid L_n^{\mathbf{Y}} \in \Gamma'),$$

where Y_i are i.i.d. with marginal law $\mu' \in M_1(\Sigma')$ and $\Gamma' \subset M_1(\Sigma')$ satisfies (3.3.2), with $\mathcal{M}'$ being a single point. Let $\mu = (\mu')^k$ be the law of $X_i = (Y_{k(i-1)+1}, \ldots, Y_{ki})$ on a new alphabet Σ. Prove that for any $n \in \mathbb{Z}_+$,

$$\mu_{nk}^*(a_i) = P_\mu(X_1 = a_i \mid L_n^{\mathbf{X}} \in \Gamma), \quad \forall a_i \in \Sigma,$$

where Γ is defined in (3.3.13). Deduce that any limit point of μ_{nk}^* is a kth product of an element of $M_1(\Sigma')$. Hence, as $n \to \infty$ along integer multiples of k, the random variables $Y_i, i = 1, \ldots, k$ are asymptotically conditionally i.i.d.
(d) Prove that the preceding conclusion extends to n which need not be integer multiples of k.

3.4   The Hypothesis Testing Problem

Let $Y_1, \ldots, Y_n$ be a sequence of random variables. The hypothesis testing problem consists of deciding, based on the sequence $Y_1, \ldots, Y_n$, whether the law generating the sequence is P_{μ_0} or P_{μ_1}. We concentrate on the simplest situation, where both P_{μ_0} and P_{μ_1} are product measures, postponing the discussion of Markov chains to Exercise 3.4.18.

In mathematical terms, the problem is expressed as follows. Let Y_1, $\ldots, Y_n$ be distributed either according to the law μ_0^n (hypothesis H_0) or according to μ_1^n (hypothesis H_1), where μ_i^n denotes the product measure of $\mu_i \in M_1(\Sigma)$. The alphabet Σ may in general be quite arbitrary, provided that the probability measures μ_0^n and μ_1^n and the random variables used in the sequel are well-defined.

Definition 3.4.1 *A decision test $\mathcal{S}$ is a sequence of measurable (with respect to the product σ-field) maps $\mathcal{S}^n : \Sigma^n \rightarrow \{0,1\}$, with the interpretation that when $Y_1 = y_1, \ldots, Y_n = y_n$ is observed, then H_0 is accepted (H_1 rejected) if $\mathcal{S}^n(y_1, \ldots, y_n) = 0$, while H_1 is accepted (H_0 rejected) if $\mathcal{S}^n(y_1, \ldots, y_n) = 1$.*

The performance of a decision test $\mathcal{S}$ is determined by the error probabilities

$$\alpha_n \triangleq P_{\mu_0}(\mathcal{S}^n \text{ rejects } H_0), \quad \beta_n \triangleq P_{\mu_1}(\mathcal{S}^n \text{ rejects } H_1).$$

The aim is to minimize β_n. If no constraint is put on α_n, one may obtain $\beta_n = 0$ using the test $\mathcal{S}^n(y_1, \ldots, y_n) \equiv 1$ at the cost of $\alpha_n = 1$. Thus, a sensible criterion for optimality, originally suggested by Neyman and Pearson, is to seek a test that minimizes β_n subject to a constraint on α_n. Suppose now that the probability measures μ_0, μ_1 are known *a priori* and that they are *equivalent measures*, so the likelihood ratios $L_{0||1}(y) = d\mu_0/d\mu_1(y)$ and $L_{1||0}(y) = d\mu_1/d\mu_0(y)$ exist. (Some extensions for μ_0, μ_1 which are not mutually absolutely continuous, are given in Exercise 3.4.17.) This assumption is valid, for example, when μ_0, μ_1 are discrete measures of the same support, or when $\Sigma = \mathbb{R}^d$ and both μ_0 and μ_1 possess strictly positive densities. In order to avoid trivialities, it is further assumed that μ_0 and μ_1 are distinguishable, i.e., they differ on a set whose probability is positive.

Let $X_j \triangleq \log L_{1||0}(Y_j) = -\log L_{0||1}(Y_j)$ be the observed log-likelihood ratios. These are i.i.d. real valued random variables that are nonzero with positive probability. Moreover,

$$\overline{x}_0 \triangleq E_{\mu_0}[X_1] = E_{\mu_1}[X_1 e^{-X_1}]$$

exists (with possibly $\overline{x}_0 = -\infty$) as $xe^{-x} \leq 1$. Similarly,

$$\overline{x}_1 \triangleq E_{\mu_1}[X_1] = E_{\mu_0}[X_1 e^{X_1}] > E_{\mu_0}[X_1] = \overline{x}_0$$

exists (with possibly $\overline{x}_1 = \infty$), and the preceding inequality is strict, since X_1 is nonzero with positive probability. In Exercise 3.4.14, $\overline{x}_0$ and $\overline{x}_1$ are both characterized in terms of relative entropy.

Definition 3.4.2 *A Neyman–Pearson test is a test in which for any $n \in \mathbb{Z}_+$, the normalized observed log-likelihood ratio*

$$\hat{S}_n \triangleq \frac{1}{n} \sum_{j=1}^{n} X_j$$

is compared to a threshold γ_n and H_1 is accepted (rejected) when $\hat{S}_n > \gamma_n$ (respectively, $\hat{S}_n \leq \gamma_n$).

It is well-known that Neyman–Pearson tests are optimal in the sense that there are neither tests with the same value of α_n and a smaller value of β_n nor tests with the same value of β_n and a smaller value of α_n. (See, for example, [CT91] and [Leh59] for simple proofs of this claim.)

The exponential rates of α_n and β_n for Neyman–Pearson tests with constant thresholds $\gamma \in (\overline{x}_0, \overline{x}_1)$ are thus of particular interest. These may be cast in terms of the large deviations of $\hat{S}_n$. In particular, since X_j are i.i.d. real valued random variables, the following theorem is a direct application of Corollary 2.2.19.

Theorem 3.4.3 *The Neyman–Pearson test with the constant threshold $\gamma \in (\overline{x}_0, \overline{x}_1)$ satisfies*

$$\lim_{n \to \infty} \frac{1}{n} \log \alpha_n = -\Lambda_0^*(\gamma) < 0 \qquad (3.4.4)$$

and

$$\lim_{n \to \infty} \frac{1}{n} \log \beta_n = \gamma - \Lambda_0^*(\gamma) < 0 \,, \qquad (3.4.5)$$

where $\Lambda_0^*(\cdot)$ is the Fenchel–Legendre transform of $\Lambda_0(\lambda) \triangleq \log E_{\mu_0} [e^{\lambda X_1}]$.

Proof: Note that

$$\alpha_n = \mathrm{P}_{\mu_0}(\hat{S}_n \in (\gamma, \infty)) \,.$$

Moreover, by dominated and monotone convergence,

$$\overline{x}_0 = \lim_{\lambda \searrow 0} \Lambda_0'(\lambda), \quad \overline{x}_1 = \lim_{\lambda \nearrow 1} \Lambda_0'(\lambda) \,.$$

Hence, $\overline{x}_0 < \gamma = \Lambda_0'(\eta)$ for some $\eta \in (0, 1)$, and the limit (3.4.4) follows by part (b) of Exercise 2.2.25.

By the definition of X_j, the logarithmic moment generating function associated with μ_1 is $\Lambda_0(\lambda + 1)$. Hence, when H_1 holds, $\hat{S}_n$ satisfies the LDP with the rate function $\Lambda_1^*(x) = \Lambda_0^*(x) - x$. Since $\gamma \in (-\infty, \overline{x}_1)$, it follows by Corollary 2.2.19 and the monotonicity of $\Lambda_1^*(\cdot)$ on $(-\infty, \overline{x}_1)$ that

$$\lim_{n \to \infty} \frac{1}{n} \log \beta_n = \lim_{n \to \infty} \frac{1}{n} \log \mathrm{P}_{\mu_1}(\hat{S}_n \in (-\infty, \gamma]) = -\Lambda_1^*(\gamma) \,,$$

and consequently (3.4.5) results.   □

Remark: For a refinement of the above see Exercise 3.7.12.

A corollary of the preceding theorem is Chernoff's asymptotic bound on the best achievable Bayes probability of error,

$$P_n^{(e)} \triangleq \alpha_n \mathrm{P}(H_0) + \beta_n \mathrm{P}(H_1) \,.$$

Corollary 3.4.6 (Chernoff's bound) *If* $0 < P(H_0) < 1$, *then*

$$\inf_{S} \liminf_{n \to \infty} \{\frac{1}{n} \log P_n^{(e)}\} = -\Lambda_0^*(0) ,$$

where the infimum is over all decision tests.

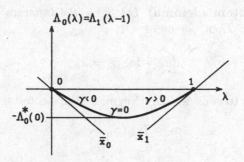

Figure 3.4.1: Geometrical interpretation of Λ_0 and Λ_0^*.

Remarks:
(a) Note that by Jensen's inequality, $\overline{x}_0 < \log E_{\mu_0}[e^{X_1}] = 0$ and $\overline{x}_1 > -\log E_{\mu_1}[e^{-X_1}] = 0$, and these inequalities are strict, since X_1 is nonzero with positive probability. Theorem 3.4.3 and Corollary 3.4.6 thus imply that the best Bayes exponential error rate is achieved by a Neyman–Pearson test with *zero threshold*.
(b) $\Lambda_0^*(0)$ is called Chernoff's information of the measures μ_0 and μ_1.

Proof: It suffices to consider only Neyman–Pearson tests. Let α_n^* and β_n^* be the error probabilities for the zero threshold Neyman–Pearson test. For any other Neyman–Pearson test, either $\alpha_n \geq \alpha_n^*$ (when $\gamma_n \leq 0$) or $\beta_n \geq \beta_n^*$ (when $\gamma_n \geq 0$). Thus, for any test,

$$\frac{1}{n} \log P_n^{(e)} \geq \frac{1}{n} \log [\min\{P(H_0), P(H_1)\}] + \min\{\frac{1}{n} \log \alpha_n^*, \frac{1}{n} \log \beta_n^*\} .$$

Hence, as $0 < P(H_0) < 1$,

$$\inf_{S} \liminf_{n \to \infty} \frac{1}{n} \log P_n^{(e)} \geq \liminf_{n \to \infty} \min\{\frac{1}{n} \log \alpha_n^*, \frac{1}{n} \log \beta_n^*\} .$$

By (3.4.4) and (3.4.5),

$$\lim_{n \to \infty} \frac{1}{n} \log \alpha_n^* = \lim_{n \to \infty} \frac{1}{n} \log \beta_n^* = -\Lambda_0^*(0) .$$

Consequently,

$$\liminf_{n\to\infty} \frac{1}{n} \log P_n^{(e)} \geq -\Lambda_0^*(0) \,,$$

with equality for the zero threshold Neyman–Pearson test.   □

Another related result is the following lemma, which determines the best exponential rate for β_n when α_n are bounded away from 1.

Lemma 3.4.7 (Stein's lemma) *Let β_n^ϵ be the infimum of β_n among all tests with $\alpha_n < \epsilon$. Then, for any $\epsilon < 1$,*

$$\lim_{n\to\infty} \frac{1}{n} \log \beta_n^\epsilon = \overline{x}_0 \,.$$

Proof: It suffices to consider only Neyman–Pearson tests. Then

$$\alpha_n = P_{\mu_0}(\hat{S}_n > \gamma_n)$$

and

$$\beta_n = P_{\mu_1}(\hat{S}_n \leq \gamma_n) = E_{\mu_1}[1_{\hat{S}_n \leq \gamma_n}] = E_{\mu_0}[1_{\hat{S}_n \leq \gamma_n} e^{n\hat{S}_n}] \,, \qquad (3.4.8)$$

where the last equality follows, since by definition X_j are the observed log-likelihood ratios. This identity yields the upper bound

$$\frac{1}{n} \log \beta_n = \frac{1}{n} \log E_{\mu_0}[1_{\hat{S}_n \leq \gamma_n} e^{n\hat{S}_n}] \leq \gamma_n \,. \qquad (3.4.9)$$

Suppose first that $\overline{x}_0 = -\infty$. Then, by Theorem 3.4.3, for any Neyman–Pearson test with a fixed threshold γ, eventually $\alpha_n < \epsilon$. Thus, by the preceding bound, $n^{-1} \log \beta_n^\epsilon \leq \gamma$ for all γ and all n large enough, and the proof of the lemma is complete.

Next, assume that $\overline{x}_0 > -\infty$. It may be assumed that

$$\liminf_{n\to\infty} \gamma_n \geq \overline{x}_0 \,,$$

for otherwise, by the weak law of large numbers, $\limsup_{n\to\infty} \alpha_n = 1$. Consequently, if $\alpha_n < \epsilon$, the weak law of large numbers implies

$$\liminf_{n\to\infty} P_{\mu_0}(\hat{S}_n \in [\overline{x}_0 - \eta, \gamma_n]) \geq 1 - \epsilon \quad \text{for all} \quad \eta > 0 \,. \qquad (3.4.10)$$

Hence, by the identity (3.4.8),

$$\frac{1}{n} \log \beta_n \geq \frac{1}{n} \log E_{\mu_0}[1_{\hat{S}_n \in [\overline{x}_0 - \eta, \gamma_n]} e^{n\hat{S}_n}]$$

$$\geq \overline{x}_0 - \eta + \frac{1}{n} \log P_{\mu_0}(\hat{S}_n \in [\overline{x}_0 - \eta, \gamma_n]) \,. \qquad (3.4.11)$$

Combining (3.4.10) and (3.4.11), the optimality of the Neyman–Pearson tests yields

$$\liminf_{n\to\infty} \frac{1}{n} \log \beta_n^\epsilon \geq \overline{x}_0 - \eta \quad \text{for all} \quad \eta > 0. \tag{3.4.12}$$

By Theorem 3.4.3, eventually $\alpha_n < \epsilon$ for any Neyman–Pearson test with a fixed threshold $\gamma > \overline{x}_0$. Hence, by (3.4.9),

$$\limsup_{n\to\infty} \frac{1}{n} \log \beta_n^\epsilon \leq \overline{x}_0 + \eta$$

for all $\eta > 0$ and all $\epsilon > 0$. The conclusion is a consequence of this bound coupled with the lower bound (3.4.12) and the arbitrariness of η. $\qquad\square$

Exercise 3.4.13 Prove that in Theorem 3.4.3,

$$\Lambda_0^*(\gamma) = \sup_{\lambda \in [0,1]} \{\lambda\gamma - \Lambda_0(\lambda)\}.$$

Exercise 3.4.14 Suppose that $Y_1, \ldots, Y_n$ are i.i.d. random variables taking values in the finite set $\Sigma \triangleq \{a_1, \ldots, a_{|\Sigma|}\}$, and μ_0, μ_1 are strictly positive on Σ.
(a) Prove that $\overline{x}_1 = H(\mu_1|\mu_0) < \infty$ and $\overline{x}_0 = -H(\mu_0|\mu_1) > -\infty$, where $H(\cdot|\cdot)$ is the relative entropy defined in Section 2.1.
(b) For $\eta \in [0,1]$, define the probability measures

$$\mu_\eta(a_j) \triangleq \frac{\mu_1(a_j)^\eta \mu_0(a_j)^{1-\eta}}{\sum_{k=1}^{|\Sigma|} \mu_1(a_k)^\eta \mu_0(a_k)^{1-\eta}} \quad j = 1, \ldots, |\Sigma|.$$

For $\gamma = H(\mu_\eta|\mu_0) - H(\mu_\eta|\mu_1)$, prove that $\Lambda_0^*(\gamma) = H(\mu_\eta|\mu_0)$.

Exercise 3.4.15 Consider the situation of Exercise 3.4.14.
(a) Define the conditional probability vectors

$$\mu_n^*(a_j) \triangleq P_{\mu_0}(Y_1 = a_j \mid H_0 \text{ rejected by } \mathcal{S}^n), \quad j = 1, \ldots, |\Sigma|, \tag{3.4.16}$$

where $\mathcal{S}$ is a Neyman–Pearson test with fixed threshold $\gamma = H(\mu_\eta|\mu_0) - H(\mu_\eta|\mu_1)$, and $\eta \in (0,1)$. Use Theorem 3.3.3 to deduce that $\mu_n^* \to \mu_\eta$.
(b) Consider now the kth joint conditional law

$$\mu_n^*(a_{j_1}, \ldots, a_{j_k}) \triangleq P_{\mu_0}(Y_1 = a_{j_1}, \ldots, Y_k = a_{j_k} \mid H_0 \text{ rejected by } \mathcal{S}^n),$$

$$a_{j_\ell} \in \Sigma, \quad \ell = 1, \ldots, k.$$

Apply Exercise 3.3.12 in order to deduce that for every fixed k,

$$\lim_{n\to\infty} \mu_n^*(a_{j_1}, \ldots, a_{j_k}) = \mu_\eta(a_{j_1})\mu_\eta(a_{j_2}) \cdots \mu_\eta(a_{j_k}).$$

Exercise 3.4.17 Suppose that $L_{1\|0}(y) = d\mu_1/d\mu_0(y)$ does not exist, while $L_{0\|1}(y) = d\mu_0/d\mu_1(y)$ does exist. Prove that Stein's lemma holds true whenever $\overline{x}_0 \triangleq - E_{\mu_0}[\log L_{0\|1}(Y_1)] > -\infty$.
Hint: Split μ_1 into its singular part with respect to μ_0 and its restriction to the support of the measure μ_0.

Exercise 3.4.18 Suppose that $Y_1, \ldots, Y_n$ is a realization of a Markov chain taking values in the finite set $\Sigma = \{1, 2, \ldots, |\Sigma|\}$, where the initial state of the chain Y_0 is known *a priori* to be some $\sigma \in \Sigma$. The transition matrix under hypothesis H_0 is Π_0, while under hypothesis II_1 it is Π_1, both of which are irreducible matrices with the same set of nonzero values. Here, the Neyman–Pearson tests are based upon $X_j \triangleq \log \frac{\pi_1(Y_{j-1}, Y_j)}{\pi_0(Y_{j-1}, Y_j)}$. Derive the analogs of Theorem 3.4.3 and of Stein's lemma by using the results of Section 3.1.3.

3.5 Generalized Likelihood Ratio Test for Finite Alphabets

This section is devoted to yet another version of the hypothesis testing problem presented in Section 3.4. The concept of decision test and the associated error probabilities are taken to be as defined there. While the law μ_0 is again assumed to be known *a priori*, here μ_1, the law of Y_j under the hypothesis H_1, is *unknown*. For that reason, neither the methods nor the results of Section 3.4 apply. Moreover, the error criterion has to be modified, since the requirement of uniformly small β_n over a large class of plausible laws μ_1 may be too strong and it may be that no test can satisfy such a condition. It is reasonable therefore to search for a criterion that involves asymptotic limits. The following criterion for finite alphabets $\Sigma = \{a_1, \ldots, a_{|\Sigma|}\}$ was suggested by Hoeffding.

Definition 3.5.1 *A test S is optimal (for a given $\eta > 0$) if, among all tests that satisfy*

$$\limsup_{n\to\infty} \frac{1}{n} \log \alpha_n \leq -\eta, \tag{3.5.2}$$

the test S has maximal exponential rate of error, i.e., uniformly over all possible laws μ_1, $-\limsup_{n\to\infty} n^{-1} \log \beta_n$ is maximal.

The following lemma states that it suffices to consider functions of the empirical measure when trying to construct an optimal test (i.e., the empirical measure is a sufficient statistic).

Lemma 3.5.3 *For every test $\mathcal{S}$ with error probabilities $\{\alpha_n, \beta_n\}_{n=1}^{\infty}$, there exists a test $\tilde{\mathcal{S}}$ of the form $\tilde{\mathcal{S}}^n(\mathbf{y}) = \tilde{\mathcal{S}}(L_n^{\mathbf{y}}, n)$ whose error probabilities $\{\tilde{\alpha}_n, \tilde{\beta}_n\}_{n=1}^{\infty}$ satisfy*

$$\limsup_{n \to \infty} \frac{1}{n} \log \tilde{\alpha}_n \leq \limsup_{n \to \infty} \frac{1}{n} \log \alpha_n \,,$$

$$\limsup_{n \to \infty} \frac{1}{n} \log \tilde{\beta}_n \leq \limsup_{n \to \infty} \frac{1}{n} \log \beta_n \,.$$

Proof: Let $\mathcal{S}_0^n \triangleq (\mathcal{S}^n)^{-1}(0)$ and $\mathcal{S}_1^n \triangleq (\mathcal{S}^n)^{-1}(1)$ denote the subsets of Σ^n that the maps $\mathcal{S}^n$ assign to H_0 and H_1, respectively. For $i = 0, 1$ and $\nu \in \mathcal{L}_n$, let $\mathcal{S}_i^{\nu,n} \triangleq \mathcal{S}_i^n \cap T_n(\nu)$, where $T_n(\nu)$ is the type class of ν from Definition 2.1.4. Define

$$\tilde{\mathcal{S}}(\nu, n) \triangleq \begin{cases} 0 & \text{if } |\mathcal{S}_0^{\nu,n}| \geq \frac{1}{2}|T_n(\nu)| \\ 1 & \text{otherwise} . \end{cases}$$

It will be shown that the maps $\tilde{\mathcal{S}}^n(\mathbf{y}) = \tilde{\mathcal{S}}(L_n^{\mathbf{y}}, n)$ play the role of the test $\tilde{\mathcal{S}}$ specified in the statement of the lemma.

Let $\mathbf{Y} = (Y_1, \ldots, Y_n)$. Recall that when Y_j are i.i.d. random variables, then for every $\mu \in M_1(\Sigma)$ and every $\nu \in \mathcal{L}_n$, the conditional measure $P_\mu(\mathbf{Y} \mid L_n^{\mathbf{Y}} = \nu)$ is a uniform measure on the type class $T_n(\nu)$. In particular, if $\tilde{\mathcal{S}}(\nu, n) = 0$, then

$$\frac{1}{2} P_{\mu_1}(L_n^{\mathbf{Y}} = \nu) \leq \frac{|\mathcal{S}_0^{\nu,n}|}{|T_n(\nu)|} P_{\mu_1}(L_n^{\mathbf{Y}} = \nu) = P_{\mu_1}(\mathbf{Y} \in \mathcal{S}_0^{\nu,n}) \,.$$

Therefore,

$$\begin{aligned} \tilde{\beta}_n &= \sum_{\{\nu : \tilde{\mathcal{S}}(\nu,n)=0\} \cap \mathcal{L}_n} P_{\mu_1}(L_n^{\mathbf{Y}} = \nu) \\ &\leq 2 \sum_{\{\nu : \tilde{\mathcal{S}}(\nu,n)=0\} \cap \mathcal{L}_n} P_{\mu_1}(\mathbf{Y} \in \mathcal{S}_0^{\nu,n}) \leq 2 P_{\mu_1}(\mathbf{Y} \in \mathcal{S}_0^n) = 2\beta_n \,. \end{aligned}$$

Consequently,

$$\limsup_{n \to \infty} \frac{1}{n} \log \tilde{\beta}_n \leq \limsup_{n \to \infty} \frac{1}{n} \log \beta_n \,.$$

A similar computation shows that $\tilde{\alpha}_n \leq 2\alpha_n$, completing the proof. $\square$

Considering hereafter tests that depend only on $L_n^{\mathbf{y}}$, the following theorem presents an optimal decision test.

Theorem 3.5.4 (Hoeffding) *Let the test $\mathcal{S}^*$ consist of the maps*

$$\mathcal{S}^{*n}(\mathbf{y}) = \begin{cases} 0 & \text{if } H(L_n^{\mathbf{y}}|\mu_0) < \eta \\ 1 & \text{otherwise.} \end{cases} \tag{3.5.5}$$

Then $\mathcal{S}^*$ is an optimal test for η.

Remark: The test (3.5.5) is referred to as the *Generalized Likelihood Ratio Test*. The reason is that one obtains (3.5.5) by taking the supremum, over all product measures μ_1^n, of the normalized observed log-likelihood ratio in the Neyman–Pearson test (Definition 3.4.2). Theorem 3.5.4 may thus be considered as a theoretical justification for this procedure.

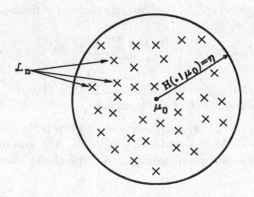

Figure 3.5.1: Optimal decision rule.

Proof: By the upper bound of Sanov's theorem (Theorem 2.1.10),

$$\limsup_{n \to \infty} \frac{1}{n} \log P_{\mu_0}(H_0 \text{ rejected by } \mathcal{S}^{*n})$$

$$= \limsup_{n \to \infty} \frac{1}{n} \log P_{\mu_0}(L_n^{\mathbf{Y}} \in \{\nu : H(\nu|\mu_0) \geq \eta\})$$

$$\leq - \inf_{\{\nu : H(\nu|\mu_0) \geq \eta\}} H(\nu|\mu_0) \leq -\eta .$$

Therefore, $\mathcal{S}^*$ satisfies the constraint (3.5.2) on α_n. Fix $\mu_1 \in M_1(\Sigma)$ and let β_n^* denote the β_n error probabilities associated with the test $\mathcal{S}^*$. Then, by the same upper bound,

$$\limsup_{n \to \infty} \frac{1}{n} \log \beta_n^* = \limsup_{n \to \infty} \frac{1}{n} \log P_{\mu_1}(L_n^{\mathbf{Y}} \in \{\nu : H(\nu|\mu_0) < \eta\})$$

$$\leq - \inf_{\{\nu : H(\nu|\mu_0) < \eta\}} H(\nu|\mu_1) \overset{\triangle}{=} -J(\eta) . \tag{3.5.6}$$

Let $\mathcal{S}$ be a test determined by the binary function $\mathcal{S}(L_n^{\mathbf{Y}}, n)$ on $M_1(\Sigma) \times \mathbb{Z}_+$. Suppose that for some $\delta > 0$ and for some n, there exists a $\nu \in \mathcal{L}_n$ such that $H(\nu|\mu_0) \leq (\eta - \delta)$ while $\mathcal{S}(\nu, n) = 1$. Then, by Lemma 2.1.9, for this test and this value of n,

$$\alpha_n \geq \mathrm{P}_{\mu_0}(L_n^{\mathbf{Y}} = \nu) \geq (n+1)^{-|\Sigma|}\, e^{-nH(\nu|\mu_0)} \geq (n+1)^{-|\Sigma|} e^{-n(\eta-\delta)} \,.$$

Thus, if the error probabilities α_n associated with $\mathcal{S}$ satisfy the constraint (3.5.2), then for every $\delta > 0$ and for all n large enough,

$$\mathcal{L}_n \cap \{\nu : H(\nu|\mu_0) \leq \eta - \delta\} \subset \mathcal{L}_n \cap \{\nu : \mathcal{S}(\nu, n) = 0\} \,.$$

Therefore, for every $\delta > 0$,

$$\limsup_{n \to \infty} \frac{1}{n} \log \beta_n \geq \liminf_{n \to \infty} \frac{1}{n} \log \beta_n$$

$$\geq \liminf_{n \to \infty} \frac{1}{n} \log \mathrm{P}_{\mu_1}(L_n^{\mathbf{Y}} \in \{\nu : H(\nu|\mu_0) < \eta - \delta\}) \,.$$

If $\Sigma_{\mu_0} = \Sigma$, then $\{\nu : H(\nu|\mu_0) < \eta - \delta\}$ is an open subset of $M_1(\Sigma)$. Hence, by the preceding inequality and the lower bound of Sanov's theorem (Theorem 2.1.10),

$$\limsup_{n \to \infty} \frac{1}{n} \log \beta_n \geq -\inf_{\delta > 0} \inf_{\{\nu: H(\nu|\mu_0) < \eta - \delta\}} H(\nu|\mu_1)$$

$$= -\inf_{\delta > 0} J(\eta - \delta) = -J(\eta) \,, \tag{3.5.7}$$

where the last equality follows from the strict inequality in the definition of $J(\cdot)$ in (3.5.6). When $\Sigma_{\mu_0} \neq \Sigma$, the sets $\{\nu : H(\nu|\mu_0) < \eta - \delta\}$ are not open. However, if $H(\nu|\mu_i) < \infty$ for $i = 0, 1$, then $\Sigma_\nu \subset \Sigma_{\mu_0} \cap \Sigma_{\mu_1}$, and consequently there exist $\nu_n \in \mathcal{L}_n$ such that $H(\nu_n|\mu_i) \to H(\nu|\mu_i)$ for $i = 0, 1$. Therefore, the lower bound (3.5.7) follows from (2.1.15) even when $\Sigma_{\mu_0} \neq \Sigma$.

The optimality of the test $\mathcal{S}^*$ for the law μ_1 results by comparing (3.5.6) and (3.5.7). Since μ_1 is arbitrary, the proof is complete. $\qquad\square$

Remarks:
(a) The finiteness of the alphabet is essential here, as (3.5.7) is obtained by applying the lower bounds of Lemma 2.1.9 for individual types instead of the large deviations lower bound for *open sets* of types. Indeed, for infinite alphabets a considerable weakening of the optimality criterion is necessary, as there are no non-trivial lower bounds for individual types. (See Section 7.1.) Note that the finiteness of Σ is also used in (3.5.6), where the upper bound for arbitrary (not necessarily closed) sets is used. However, this can be reproduced in a general situation by a careful approximation argument.

(b) As soon as an LDP exists, the results of this section may be extended to the hypothesis test problem for a known *joint law* $\mu_{0,n}$ versus a family of unknown *joint laws* $\mu_{1,n}$ provided that the random variables $Y_1, \ldots, Y_n$ are *finitely exchangeable* under $\mu_{0,n}$ and any possible $\mu_{1,n}$ so that the empirical measure is still a sufficient statistic. This extension is outlined in Exercises 3.5.10 and 3.5.11.

Exercise 3.5.8 (a) Let $\Delta = \{\Delta_r, r = 1, 2, \ldots\}$ be a deterministic increasing sequence of positive integers. Assume that $\{Y_j, j \notin \Delta\}$ are i.i.d. random variables taking values in the finite set $\Sigma = \Sigma_{\mu_0}$, while $\{Y_j, j \in \Delta\}$ are unknown deterministic points in Σ. Prove that if $\Delta_r / r \to \infty$, then the test $\mathcal{S}^*$ of (3.5.5) satisfies (3.5.2).
Hint: Let $L_n^{\mathbf{Y}^*}$ correspond to $\Delta = \emptyset$ and prove that almost surely $\limsup_{n \to \infty} d_V(L_n^{\mathbf{Y}}, L_n^{\mathbf{Y}^*}) = 0$, with some deterministic rate of convergence that depends only upon the sequence Δ. Conclude the proof by using the continuity of $H(\cdot | \mu_0)$ on $M_1(\Sigma)$.
(b) Construct a counterexample to the preceding claim when $\Sigma_{\mu_0} \neq \Sigma$.
Hint: Take $\Sigma = \{0, 1\}$, $\mu_0 = \delta_0$, and $Y_j = 1$ for some $j \in \Delta$.

Exercise 3.5.9 Prove that if in addition to the assumptions of part (a) of Exercise 3.5.8, $\Sigma_{\mu_1} = \Sigma$ for every possible choice of μ_1, then the test $\mathcal{S}^*$ of (3.5.5) is an optimal test.

Exercise 3.5.10 Suppose that for any $n \in \mathbb{Z}_+$, the random vector $\mathbf{Y}^{(n)} = (Y_1, \ldots, Y_n)$ taking values in the finite alphabet Σ^n possesses a known joint law $\mu_{0,n}$ under the null hypothesis H_0 and an unknown joint law $\mu_{1,n}$ under the alternative hypothesis H_1. Suppose that for all n, the coordinates of $\mathbf{Y}^{(n)}$ are exchangeable random variables under both $\mu_{0,n}$ and every possible $\mu_{1,n}$ (namely, the probability of any outcome $\mathbf{Y}^{(n)} = \mathbf{y}$ is invariant under permutations of indices in the vector $\mathbf{y}$). Let $L_n^{\mathbf{Y}}$ denote the empirical measure of $\mathbf{Y}^{(n)}$ and prove that Lemma 3.5.3 holds true in this case. (Note that $\{Y_k\}_{k=1}^n$ may well be dependent.)

Exercise 3.5.11 (a) Consider the setup of Exercise 3.5.10. Suppose that $I_n : M_1(\Sigma) \to [0, \infty]$ are such that

$$\lim_{n \to \infty} \sup_{\nu \in \mathcal{L}_n, I_n(\nu) < \infty} \left| \frac{1}{n} \log \mu_{0,n}(L_n^{\mathbf{Y}} = \nu) + I_n(\nu) \right| = 0 , \qquad (3.5.12)$$

and $\mu_{0,n}(L_n^{\mathbf{Y}} = \nu) = 0$ whenever $I_n(\nu) = \infty$. Prove that the test $\mathcal{S}^*$ of (3.5.5) with $H(L_n^{\mathbf{Y}} | \mu_0)$ replaced by $I_n(L_n^{\mathbf{Y}})$ is *weakly* optimal in the sense that for any possible law $\mu_{1,n}$, and any test for which $\limsup_{n \to \infty} \frac{1}{n} \log \alpha_n < -\eta$, $-\limsup_{n \to \infty} \{ \frac{1}{n} \log \beta_n^* \} \geq -\limsup_{n \to \infty} (\frac{1}{n} \log \beta_n)$.
(b) Apply part (a) to prove the weak optimality of $I(L_n^{\mathbf{Y}} | \frac{n}{m}, \mu_m) \gtrless \eta$, with $I(\cdot | \cdot, \cdot)$ of (2.1.32), when testing a given deterministic composition sequence μ_m in a sampling without replacement procedure against an unknown composition sequence for such a scheme.

3.6   Rate Distortion Theory

This section deals with one of the basic problems in information theory, namely, the problem of source coding. To understand the motivation behind this problem, think of a computer (the *source*) that creates long strings of 0 and 1. Suppose one wishes to store the output of the computer in memory. Of course, all the symbols could be stored in a long string. However, it is reasonable to assume that typical output sequences follow certain patterns; hence, by exploitation of these patterns the amount of digits required to store such sequences could be reduced. Note that the word *store* here could be replaced by the word *transmit*, and indeed it is the transmission problem that was initially emphasized.

This problem was tackled by Shannon in the late 1940s. His insight was that by looking at the source as a random source, one could hope to analyze this situation and find the fundamental performance limits (i.e., how much the data can be compressed), as well as methods to achieve these limits. Although Shannon's initial interest centered on schemes in which the compression involved no loss of data, later developments of the theory also allowed for some error in the reconstruction. This is considered here, and, as usual, we begin with a precise definition of the problem.

Let $\Omega = \Sigma^{\mathbb{Z}_+}$ be the space of semi-infinite sequences over Σ. By a stationary and ergodic source, with alphabet Σ, we mean a stationary ergodic probability measure $\mathcal{P}$ on Ω. Let $x_1, x_2, \ldots, x_n, \ldots$ denote an element of Ω, and note that since $\mathcal{P}$ is only ergodic, the corresponding random variables $X_1, X_2, \ldots, X_n, \ldots$ may well be dependent.

Next, let the measurable function $\rho(x, y) : \Sigma \times \Sigma \to [0, \rho_{\max}]$ be a *one symbol bounded distortion function* (where $\rho_{\max} < \infty$). Typically, $\rho(x, x) = 0$, and $\rho(x, y) \neq 0$ for $x \neq y$. Common examples are $\rho(x, y) = 0$ if $x = y$ and 1 otherwise (when Σ is a finite set) and $\rho(x, y) = |x - y|^2$ (when $\Sigma = [0, 1]$).

A *code* is a deterministic map $C_n : \Sigma^n \to \Sigma^n$, and

$$\rho_{C_n} \overset{\Delta}{=} \frac{1}{n} \sum_{i=1}^{n} E[\, \rho(X_i, (C_n(X_1, \ldots, X_n))_i)\,]$$

denotes the average distortion per symbol when the code C_n is used.

The basic problem of source coding is to find a sequence of codes $\{C_n\}_{n=1}^{\infty}$ having small *distortion*, defined to be $\limsup_{n \to \infty} \rho_{C_n}$. To this end, the range of C_n is to be a finite set, and the smaller this set, the fewer different "messages" exist. Therefore, the incoming information (of possible values in Σ^n) has been compressed to a smaller number of alternatives. The advantage of coding is that only the sequential number of the code word has to be stored. Thus, since there are only relatively few such code words, less information is retained per incoming string $X_1, \ldots, X_n$. This gain in transmission/storage requirement is referred to as the *coding gain*.

Clearly, by taking C_n to be the identity map, zero distortion is achieved but without coding gain. To get to a more meaningful situation, it would be desirable to have a reduction in the number of possible sequences when using C_n. Let $|C_n|$ denote the cardinality of the range of C_n. The *rate* of the code C_n is defined as

$$R_{C_n} = \frac{1}{n} \log |C_n|.$$

The smaller R_{C_n} is, the larger the coding gain when using C_n. (In the original application, the rate of the code measured how much information had to be transmitted per unit of time.) Shannon's source coding theorem asserts that it cannot be hoped to get R_{C_n} too small—i.e., under a bound on the distortion, R_{C_n} is generally bounded below by some positive quantity. Moreover, Shannon discovered, first for the case where distortion is not present and then for certain cases with distortion, that there are codes that are arbitrarily close to this bound.

Later proofs of Shannon's result, in its full generality, involve three distinct arguments: First, by information theory tools, it is shown that the rate cannot be too low. Next, by an ergodic theory argument, the problem is reduced to a token problem. Finally, this latter problem is analyzed by large deviations methods based on the Gärtner–Ellis theorem. Since this book deals with large deviations applications, we concentrate on the latter token problem, and only state and sketch the proof of Shannon's general result at the end of this section. In order to state the problem to be analyzed, another round of definitions is required.

The *distortion* associated with a probability measure Q on $\Sigma \times \Sigma$ is

$$\rho_Q \triangleq \int_{\Sigma \times \Sigma} \rho(x, y)\, Q(dx, dy).$$

Let Q_X and Q_Y be the marginals of Q. The *mutual information* associated with Q is

$$H(Q|Q_X \times Q_Y) \triangleq \int_{\Sigma \times \Sigma} \log \left(\frac{dQ}{dQ_X \times Q_Y} \right) dQ \qquad (3.6.1)$$

when the preceding integral is well-defined and finite and $H(Q|Q_X \times Q_Y) = \infty$ otherwise.[1]

The *single-symbol rate distortion function* is defined as

$$R_1(D) = \inf_{\{Q: \rho_Q \leq D,\, Q_X = \mathcal{P}_1\}} H(Q|Q_X \times Q_Y),$$

---

[1] In information theory, the mutual information is usually denoted by $I(X; Y)$. The notation $H(Q|Q_X \times Q_Y)$ is consistent with other notation in this book.

where $\mathcal{P}_1$ is the one-dimensional marginal distribution of the stationary measure $\mathcal{P}$ on $\Sigma^{\mathbb{Z}_+}$.

The following is the "token problem" alluded to earlier. It is proved via a sequence of lemmas based on the LDP of Section 2.3.

Theorem 3.6.2 (Shannon's weak source coding theorem) *For any $D \geq 0$ such that $R_1(D) < \infty$ and for any $\delta > 0$, there exists a sequence of codes $\{C_n\}_{n=1}^{\infty}$ with distortion at most D, and rates $R_{C_n} \leq R_1(D) + \delta$.*

The proof of this theorem is based on a random coding argument, where instead of explicitly constructing the codes C_n, the classes $\mathcal{C}_n$ of all codes of some fixed size are considered. A probability measure on $\mathcal{C}_n$ is constructed using a law that is independent of the sequence $\mathbf{X} \triangleq (X_1, \ldots, X_n, \ldots)$. Let $\bar{\rho}_n \triangleq E_{\mathcal{C}_n}[\rho_{C_n}]$ be the expectation of ρ_{C_n} over $\mathcal{C}_n$ according to this measure. Clearly, there exists at least one code in $\mathcal{C}_n$ for which $\rho_{C_n} \leq \bar{\rho}_n$. With this approach, Theorem 3.6.2 is a consequence of the upper bound on $\bar{\rho}_n$ to be derived in Lemma 3.6.5. This in turn is based on the following large deviations lower bound.

Lemma 3.6.3 *Suppose Q is a probability measure on $\Sigma \times \Sigma$ for which $H(Q|Q_X \times Q_Y) < \infty$ and $Q_X = \mathcal{P}_1$. Let $Z_n(\mathbf{x}) \triangleq \frac{1}{n} \sum_{j=1}^{n} \rho(x_j, Y_j)$, where Y_j are i.i.d. random variables, each distributed on Σ according to the law Q_Y, and all of which are independent of $\mathbf{x} \triangleq (x_1, \ldots, x_n, \ldots)$. Then, for every $\delta > 0$, and for $\mathcal{P}$ almost every semi-infinite sequence $\mathbf{x}$,*

$$\liminf_{n \to \infty} \frac{1}{n} \log \mathrm{P}(Z_n(\mathbf{x}) < \rho_Q + \delta) \geq -H(Q|Q_X \times Q_Y).$$

Proof: Let

$$\begin{aligned} \Lambda(\theta) &\triangleq \int_{\Sigma} \log \left(\int_{\Sigma} e^{\theta \rho(x,y)} Q_Y(dy) \right) \mathcal{P}_1(dx) \\ &= \int_{\Sigma} \log \left(\int_{\Sigma} e^{\theta \rho(x,y)} Q_Y(dy) \right) Q_X(dx) , \end{aligned}$$

where the second equality follows by our assumption that $Q_X = \mathcal{P}_1$. As $\rho(\cdot, \cdot)$ is uniformly bounded, $\Lambda(\cdot)$ is finite, and by dominated convergence it is also differentiable everywhere in $\mathbb{R}$.

Consider now the logarithmic moment generating functions $\Lambda_n(\theta) \triangleq \log E[e^{\theta Z_n(\mathbf{x})}]$. For $\mathcal{P}$ almost every sequence $\mathbf{x}$ emitted by the source, Birkhoff's ergodic theorem yields

$$\lim_{n \to \infty} \frac{1}{n} \Lambda_n(n\theta) = \lim_{n \to \infty} \frac{1}{n} \sum_{j=1}^{n} \log \int_{\Sigma} e^{\theta \rho(x_j, y)} Q_Y(dy) = \Lambda(\theta).$$

Fix $\mathbf{x} \in \Omega = \Sigma^{\mathbb{Z}_+}$ for which the preceding identity holds. Then, by the Gärtner–Ellis theorem (Theorem 2.3.6), the sequence of random variables $\{Z_n(\mathbf{x})\}$ satisfies the LDP with the rate function $\Lambda^*(\cdot)$. In particular, considering the open set $(-\infty, \rho_Q + \delta)$, this LDP yields

$$\liminf_{n \to \infty} \frac{1}{n} \log \mathrm{P}(Z_n(\mathbf{x}) < \rho_Q + \delta) \geq - \inf_{x < \rho_Q + \delta} \Lambda^*(x) \geq -\Lambda^*(\rho_Q). \quad (3.6.4)$$

It remains to be shown that $H(Q|Q_X \times Q_Y) \geq \Lambda^*(\rho_Q)$. To this end, associate with every $\lambda \in \mathbb{R}$ a probability measure Q_λ on $\Sigma \times \Sigma$ via

$$\frac{dQ_\lambda}{dQ_X \times Q_Y}(x, y) = \frac{e^{\lambda \rho(x,y)}}{\int_\Sigma e^{\lambda \rho(x,z)} Q_Y(dz)}.$$

Since $\rho(\cdot, \cdot)$ is bounded, the measures Q_λ and $Q_X \times Q_Y$ are mutually absolutely continuous. As $H(Q|Q_X \times Q_Y) < \infty$, the relative entropy $H(Q|Q_\lambda)$ is well-defined. Moreover,

$$
\begin{aligned}
0 \leq H(Q|Q_\lambda) &= \int_{\Sigma \times \Sigma} \log\left(\frac{dQ}{dQ_\lambda}(x, y)\right) Q(dx, dy) \\
&= \int_{\Sigma \times \Sigma} Q(dx, dy) \log\left\{\frac{dQ}{dQ_X \times Q_Y}(x, y) e^{-\lambda \rho(x,y)} \int_\Sigma e^{\lambda \rho(x,z)} Q_Y(dz)\right\} \\
&= H(Q|Q_X \times Q_Y) - \lambda \rho_Q + \Lambda(\lambda).
\end{aligned}
$$

Since this inequality holds for all λ, it follows that $H(Q|Q_X \times Q_Y) \geq \Lambda^*(\rho_Q)$, and the proof is complete in view of (3.6.4). □

Lemma 3.6.5 *Suppose Q is a probability measure on $\Sigma \times \Sigma$ for which $H(Q|Q_X \times Q_Y) < \infty$ and $Q_X = \mathcal{P}_1$. Fix $\delta > 0$ arbitrarily small and let $\mathcal{C}_n$ be the class of all codes C_n of size $k_n \triangleq \lfloor e^{n(H(Q|Q_X \times Q_Y) + \delta)} \rfloor$. Then there exist distributions on $\mathcal{C}_n$ for which $\limsup_{n \to \infty} \overline{\rho}_n \leq \rho_Q + \delta$.*

Proof: For any $\mathbf{x} = (x_1, \ldots, x_n, \ldots)$, define the set

$$S_n(\mathbf{x}) \triangleq \left\{ (y_1, \ldots, y_n) : \frac{1}{n} \sum_{j=1}^n \rho(x_j, y_j) < \rho_Q + \delta \right\},$$

and let $C_n(\mathbf{x})$ be any element of $C_n \cap S_n(\mathbf{x})$, where if this set is empty then $C_n(\mathbf{x})$ is arbitrarily chosen. For this mapping,

$$\frac{1}{n} \sum_{j=1}^n \rho(x_j, (C_n(\mathbf{x}))_j) \leq \rho_Q + \delta + \rho_{\max} 1_{\{C_n \cap S_n(\mathbf{x}) = \emptyset\}},$$

and consequently, for any measure on the class of codes $\mathcal{C}_n$,

$$\overline{\rho}_n \leq \rho_Q + \delta + \rho_{\max} P(\mathcal{C}_n \cap S_n(\mathbf{X}) = \emptyset) . \tag{3.6.6}$$

The probability distribution on the class of codes $\mathcal{C}_n$ is generated by considering the codes with code words $\mathbf{Y}^{(i)} \triangleq (Y_1^{(i)}, \ldots, Y_n^{(i)})$ for $i = 1, \ldots, k_n$, where $Y_j^{(i)}$, $j = 1, \ldots, n$, $i = 1, \ldots, k_n$, are i.i.d. random variables of law Q_Y, independent of the sequence emitted by the source. Hence,

$$
\begin{aligned}
P(\mathcal{C}_n \cap S_n(\mathbf{x}) = \emptyset) &= P(\mathbf{Y}^{(i)} \notin S_n(\mathbf{x}) \text{ for all } i) = [1 - P(\mathbf{Y}^{(1)} \in S_n(\mathbf{x}))]^{k_n} \\
&\leq \exp\left(-k_n P(\mathbf{Y}^{(1)} \in S_n(\mathbf{x}))\right) .
\end{aligned}
\tag{3.6.7}
$$

Note that the event $\{\mathbf{Y}^{(1)} \in S_n(\mathbf{x})\}$ has the same probability as the event $\{Z_n(\mathbf{x}) < \rho_Q + \delta\}$ considered in Lemma 3.6.3. Thus, by the definition of k_n and by the conclusion of Lemma 3.6.3, for $\mathcal{P}$ almost every semi-infinite sequence $\mathbf{x}$,

$$\lim_{n \to \infty} k_n P(\mathbf{Y}^{(1)} \in S_n(\mathbf{x})) = \infty .$$

Consequently, by the inequality (3.6.7),

$$\lim_{n \to \infty} P(\mathcal{C}_n \cap S_n(\mathbf{X}) = \emptyset) = 0 .$$

Substituting in (3.6.6) completes the proof. □

Proof of Theorem 3.6.2: Since $R_1(D) < \infty$, there exists a sequence of measures $\{Q^{(m)}\}_{m=1}^\infty$ such that $I_m \triangleq H(Q^{(m)}|Q_X^{(m)} \times Q_Y^{(m)}) \to R_1(D)$, while $\rho_{Q^{(m)}} \leq D$ and $Q_X^{(m)} = \mathcal{P}_1$. By applying Lemma 3.6.5 for $Q^{(m)}$ and $\delta_m = 1/m$, it follows that for all n there exists a distribution on the class of all codes of size $\lfloor e^{n(I_m + 1/m)} \rfloor$ such that $\limsup_{n \to \infty} \overline{\rho}_n \leq \rho_{Q^{(m)}} + 1/m$. Fix an arbitrary $\delta > 0$. For all m large enough and for all n, one can enlarge these codes to size $\lfloor e^{n(R_1(D) + \delta)} \rfloor$ with no increase in $\overline{\rho}_n$. Hence, for every n, there exists a distribution on the class of all codes of size $\lfloor e^{n(R_1(D) + \delta)} \rfloor$ such that

$$\limsup_{n \to \infty} \overline{\rho}_n \leq \limsup_{m \to \infty} \left(\rho_{Q^{(m)}} + \frac{1}{m}\right) \leq D .$$

The existence of a sequence of codes $\mathcal{C}_n$ of rates $R_{\mathcal{C}_n} \leq R_1(D) + \delta$ and of distortion $\limsup_{n \to \infty} \rho_{\mathcal{C}_n} \leq D$ is deduced by extracting the codes $\mathcal{C}_n$ of minimal $\rho_{\mathcal{C}_n}$ from these ensembles. □

We now return to the original problem of source coding discussed in the introduction of this section. Note that the one symbol distortion function $\rho(x, y)$ may be used to construct the corresponding J-symbol average distortion for $J = 2, 3, \ldots$,

$$\rho^{(J)}((x_1, \ldots, x_J), (y_1, \ldots, y_J)) \triangleq \frac{1}{J} \sum_{\ell=1}^{J} \rho(x_\ell, y_\ell) .$$

The J-symbol distortion of a probability measure Q on $\Sigma^J \times \Sigma^J$ is

$$\rho_Q^{(J)} = \int_{\Sigma^J \times \Sigma^J} \rho^{(J)}(\mathbf{x}, \mathbf{y})\, Q(d\mathbf{x}, d\mathbf{y}) \,.$$

The *mutual information* associated with the measure Q having marginals Q_X, Q_Y on Σ^J is defined via (3.6.1) with Σ^J instead of Σ. The J-symbol rate distortion function is defined as

$$R_J(D) \triangleq \inf_{\{Q:\, \rho_Q^{(J)} \le D,\, Q_X = \mathcal{P}_J\}} \frac{1}{J} H(Q|Q_X \times Q_Y),$$

where $\mathcal{P}_J$ is the J-dimensional marginal distribution of the stationary measure $\mathcal{P}$. Finally, the *rate distortion function* is

$$R(D) \triangleq \inf_{J \ge 1} R_J(D) \,.$$

Shannon's source coding theorem states that the rate distortion function is the optimal performance of any sequence of codes, and that it may be achieved to arbitrary accuracy.

Theorem 3.6.8 (Source coding theorem)

(a) Direct part: For any $D \ge 0$ such that $R(D) < \infty$ and any $\delta > 0$, there exists a sequence of codes $\{C_n\}_{n=1}^{\infty}$ with distortion at most D and rates $R_{C_n} \le R(D) + \delta$ for all sufficiently large n.

(b) Converse part: For any sequence of codes $\{C_n\}_{n=1}^{\infty}$ of distortion D and all $\delta > 0$, $\liminf_{n \to \infty} R_{C_n} \ge R(D + \delta)$.

Remarks:

(a) Note that $|\Sigma|$ may be infinite and there are no structural conditions on Σ besides the requirement that $\mathcal{P}$ be based on $\Sigma^{\mathbb{Z}_+}$, and conditioning and product measures are well-defined. (All the latter hold as soon as Σ is Polish.) On the other hand, whenever $R(D)$ is finite, the resulting codes always take values in some finite set and, in particular, may be represented by finite binary sequences.

(b) For i.i.d. source symbols, $R(D) = R_1(D)$ (see Exercise 3.6.11) and the direct part of the source coding theorem amounts to Theorem 3.6.2.

We sketch below the elements of the proof of Theorem 3.6.8. Beginning with the direct part, life is much easier when the source possesses strong ergodic properties. In this case, the proof of Lemma 3.6.9 is a repeat of the arguments used in the proof of Lemma 3.6.5 and is omitted.

Lemma 3.6.9 *Suppose $\mathcal{P}$ is ergodic with respect to the Jth shift operation (namely, it is ergodic in blocks of size J). Then, for any $\delta > 0$ and each* *probability measure Q on $\Sigma^J \times \Sigma^J$ with $Q_X = \mathcal{P}_J$ and $H(Q|Q_X \times Q_Y) < \infty$, there exists a sequence of codes C_n of rates $R_{C_n} \leq J^{-1} H(Q|Q_X \times Q_Y) + \delta$ and of distortion at most $\rho_Q^{(J)} + \delta$. In particular, if $\mathcal{P}$ is ergodic in blocks of size J for any $J \in \mathbb{Z}_+$, then the direct part of the source coding theorem holds.*

While in general an ergodic $\mathcal{P}$ might be non-ergodic in blocks, an ergodic theory argument shows that Lemma 3.6.9 holds true for any stationary and ergodic $\mathcal{P}$. A full proof of this fact may be found in [Ber71, pages 278–280], and [Ga68, pages 496–500]. This proof is based on showing that when considering blocks of size J, the emitted semi-infinite sequences of the source may almost surely be divided into J equally probable ergodic modes, $E_0, \ldots, E_{J-1}$, such that if the sequence $(x_1, x_2, \ldots, x_n, \ldots)$ belongs to mode E_i, then $(x_{1+k}, x_{2+k}, \ldots, x_{n+k}, \ldots)$ belongs to the mode $E_{(i+k) \bmod J}$.

Finally, we sketch the proof of the converse part of the source coding theorem. Clearly, it suffices to consider codes C_n of finite rates and of distortion D. Such a code C_n is a mapping from Σ^n to Σ^n. When its domain Σ^n is equipped with the probability measure $\mathcal{P}_n$ (the nth marginal of $\mathcal{P}$), C_n induces a (degenerate) joint measure $Q^{(n)}$ on $\Sigma^n \times \Sigma^n$. Note that $Q_X^{(n)} = \mathcal{P}_n$ and $\rho_{Q^{(n)}} = \rho_{C_n} \leq D + \delta$ for any $\delta > 0$ and all n large enough. Therefore, for such δ and n,

$$H(Q^{(n)}|Q_X^{(n)} \times Q_Y^{(n)}) \geq nR_n(D + \delta) \geq nR(D + \delta).$$

Since $Q_Y^{(n)}$ is supported on the finite range of C_n, the entropy

$$H(Q_Y^{(n)}) \triangleq -\sum_{i=1}^{|C_n|} Q_Y^{(n)}(y_i) \log Q_Y^{(n)}(y_i)$$

is a lower bound on $\log |C_n| = nR_{C_n}$. The Radon–Nikodym derivative $f_n(x, y_i) \triangleq dQ^{(n)}/dQ_X^{(n)} \times dQ_Y^{(n)}(x, y_i)$ is well-defined, and

$$\int_{\Sigma^n} f_n(x, y_i)\, Q_X^{(n)}(dx) = 1, \quad \sum_{i=1}^{|C_n|} f_n(x, y_i) Q_Y^{(n)}(y_i) = 1 \quad, Q_X^{(n)} - \text{a.e.}.$$

Thus, $Q_X^{(n)}$-a.e., $f_n(x, y_i) Q_Y^{(n)}(y_i) \leq 1$, and

$$H(Q_Y^{(n)}) - H(Q^{(n)}|Q_X^{(n)} \times Q_Y^{(n)})$$

$$= \sum_{i=1}^{|C_n|} Q_Y^{(n)}(y_i) \left[\log \frac{1}{Q_Y^{(n)}(y_i)} - \int_{\Sigma^n} f_n(x, y_i) \log f_n(x, y_i)\, Q_X^{(n)}(dx) \right]$$

$$= \int_{\Sigma^n} Q_X^{(n)}(dx) \left[\sum_{i=1}^{|C_n|} f_n(x, y_i) Q_Y^{(n)}(y_i) \log \frac{1}{f_n(x, y_i)\, Q_Y^{(n)}(y_i)} \right] \geq 0.$$

Therefore,

$$\liminf_{n\to\infty} R_{C_n} \geq \liminf_{n\to\infty} \frac{1}{n} H(Q_Y^{(n)})$$

$$\geq \liminf_{n\to\infty} \frac{1}{n} H(Q^{(n)}|Q_X^{(n)} \times Q_Y^{(n)}) \geq R(D+\delta).$$

Exercise 3.6.10 (a) Prove that if Σ is a finite set and $R_1(D) > 0$, then there exists a probability measure Q on $\Sigma \times \Sigma$ for which $Q_X = \mathcal{P}_1$, $\rho_Q = D$, and $H(Q|Q_X \times Q_Y) = R_1(D)$.
(b) Prove that for this measure also $\Lambda^*(\rho_Q) = H(Q|Q_X \times Q_Y)$.

Exercise 3.6.11 Prove that when $\mathcal{P}$ is a product measure (namely, the emitted symbols $X_1, X_2, \ldots, X_n$ are i.i.d.), then $R_J(D) = R_1(D)$ for all $J \in \mathbb{Z}_+$.
Hint: Apply Jensen's inequality to show, for any measure Q on $\Sigma^J \times \Sigma^J$ such that $Q_X = \mathcal{P}_J$, that

$$H(Q|Q_X \times Q_Y) \geq \sum_{i=1}^{|J|} H(Q^i|\mathcal{P}_1 \times Q_Y^i),$$

where Q^i (Q_Y^i) denote the ith coordinate marginal of Q $(Q_Y$, respectively). Then, use the convexity of $R_1(\cdot)$.

Exercise 3.6.12 (a) Show that for all integers m, n,

$$(m+n)R_{m+n}(D) \leq mR_m(D) + nR_n(D).$$

(b) Conclude that

$$\limsup_{J\to\infty} R_J(D) < \infty \Rightarrow R(D) = \lim_{J\to\infty} R_J(D).$$

This is a particular instance of the sub-additivity lemma (Lemma 6.1.11).

3.7 Moderate Deviations and Exact Asymptotics in $\mathbb{R}^d$

Cramér's theorem deals with the tails of the empirical mean $\hat{S}_n$ of i.i.d. random variables. On a finer scale, the random variables $\sqrt{n}\hat{S}_n$ possess a limiting Normal distribution by the central limit theorem. In this situation, for $\beta \in (0, 1/2)$, the renormalized empirical mean $n^\beta \hat{S}_n$ satisfies an LDP but always with a quadratic (Normal-like) rate function. This statement is made precise in the following theorem. (Choose $a_n = n^{(2\beta-1)}$ in the theorem to obtain $Z_n = n^\beta \hat{S}_n$.)

Theorem 3.7.1 (Moderate Deviations) *Let $X_1, \ldots, X_n$ be a sequence of $\mathbb{R}^d$-valued i.i.d. random vectors such that $\Lambda_X(\lambda) \triangleq \log E[e^{\langle \lambda, X_i \rangle}] < \infty$ in some ball around the origin, $E(X_i) = 0$, and $\mathbf{C}$, the covariance matrix of X_1, is invertible. Fix $a_n \to 0$ such that $n a_n \to \infty$ as $n \to \infty$, and let $Z_n \triangleq \sqrt{a_n/n} \sum_{i=1}^n X_i = \sqrt{n a_n} \hat{S}_n$. Then, for every measurable set Γ,*

$$-\frac{1}{2} \inf_{x \in \Gamma^o} \langle x, \mathbf{C}^{-1} x \rangle \leq \liminf_{n \to \infty} a_n \log P(Z_n \in \Gamma)$$

$$\leq \limsup_{n \to \infty} a_n \log P(Z_n \in \Gamma)$$

$$\leq -\frac{1}{2} \inf_{x \in \bar{\Gamma}} \langle x, \mathbf{C}^{-1} x \rangle. \tag{3.7.2}$$

Proof: Let $\Lambda(\lambda) \triangleq E[\langle \lambda, X_1 \rangle^2]/2 = \langle \lambda, \mathbf{C}\lambda \rangle/2$. It follows that the Fenchel–Legendre transform of $\Lambda(\cdot)$ is

$$\Lambda^*(x) = \sup_{\lambda \in \mathbb{R}^d} \{\langle \lambda, x \rangle - \Lambda(\lambda)\}$$

$$= \sup_{\lambda \in \mathbb{R}^d} \{\langle \lambda, x \rangle - \frac{1}{2} \langle \lambda, \mathbf{C}\lambda \rangle\} = \frac{1}{2} \langle x, \mathbf{C}^{-1} x \rangle.$$

Let $\Lambda_n(\cdot)$ be the logarithmic moment generating function of Z_n. As will be shown, for every $\lambda \in \mathbb{R}^d$,

$$\Lambda(\lambda) = \lim_{n \to \infty} a_n \Lambda_n(a_n^{-1} \lambda). \tag{3.7.3}$$

Consequently, the theorem is an application of the Gärtner–Ellis theorem, where a_n replaces $1/n$ throughout. Indeed, $\Lambda(\cdot)$ is finite and differentiable everywhere.

Turning now to establish the limit (3.7.3), observe that

$$\Lambda_n(a_n^{-1} \lambda) = \log E \left(e^{a_n^{-1} \langle \lambda, Z_n \rangle} \right)$$

$$= \sum_{i=1}^n \log E \left(e^{(n a_n)^{-1/2} \langle \lambda, X_i \rangle} \right) = n \log E \left(e^{(n a_n)^{-1/2} \langle \lambda, X_1 \rangle} \right).$$

Since $\Lambda_X(\lambda) < \infty$ in a ball around the origin and $n a_n \to \infty$, it follows that $E\left(\exp((n a_n)^{-1/2} \langle \lambda, X_1 \rangle)\right) < \infty$ for each $\lambda \in \mathbb{R}^d$, and all n large enough. By dominated convergence,

$$E \left(\exp((n a_n)^{-1/2} \langle \lambda, X_1 \rangle) \right)$$

$$= 1 + (n a_n)^{-1/2} E[\langle \lambda, X_1 \rangle] + \frac{1}{2} (n a_n)^{-1} E[\langle \lambda, X_1 \rangle^2] + \mathbf{O}\left((n a_n)^{-3/2}\right),$$

where $g(n) = \mathbf{O}\left((na_n)^{-3/2}\right)$ means that $\limsup_{n\to\infty}(na_n)^{3/2}g(n) < \infty$. Hence, since $E[\langle\lambda, X_1\rangle] = 0$,

$$a_n\Lambda_n(a_n^{-1}\lambda) = na_n\log\left\{1 + \frac{1}{2}(na_n)^{-1}E[\langle\lambda, X_1\rangle^2] + \mathbf{O}((na_n)^{-3/2})\right\}.$$

Consequently,

$$\lim_{n\to\infty} a_n\Lambda_n(a_n^{-1}\lambda) = \frac{1}{2}E[\langle\lambda, X_1\rangle^2] = \Lambda(\lambda).\qquad\square$$

Remarks:

(a) Note that the preceding theorem is nothing more than what is obtainable from a naive Taylor expansion applied on the ersatz $\mathrm{P}(\hat{S}_n = x) \approx e^{-nI(x)}$, where $I(\cdot)$ is the rate function of Cramér's theorem.

(b) A similar result may be obtained in the context of the Markov additive process discussed in Section 3.1.1.

(c) Theorem 3.7.1 is representative of the so-called *Moderate Deviation Principle* (MDP), in which for some $\gamma(\cdot)$ and a whole range of $a_n \to 0$, the sequences $\{\gamma(a_n)Y_n\}$ satisfy the LDP with the same rate function. Here, $Y_n = \sqrt{n}\hat{S}_n$ and $\gamma(a) = a^{1/2}$ (as in other situations in which Y_n obeys the central limit theorem).

Another refinement of Cramér's theorem involves a more accurate estimate of the law μ_n of $\hat{S}_n$. Specifically, for a "nice" set A, one seeks an estimate J_n^{-1} of $\mu_n(A)$ in the sense that $\lim_{n\to\infty} J_n\mu_n(A) = 1$. Such an estimate is an improvement over the normalized logarithmic limit implied by the LDP. The following theorem, a representative of the so-called *exact asymptotics*, deals with the estimate J_n for certain half intervals $A = [q, \infty) \subset \mathbb{R}$.

Theorem 3.7.4 (Bahadur and Rao) *Let μ_n denote the law of $\hat{S}_n = \frac{1}{n}\sum_{i=1}^n X_i$, where X_i are i.i.d. real valued random variables with logarithmic moment generating function $\Lambda(\lambda) = \log E[e^{\lambda X_1}]$. Consider the set $A = [q, \infty)$, where $q = \Lambda'(\eta)$ for some positive $\eta \in \mathcal{D}_\Lambda^o$.*
(a) If the law of X_1 is non-lattice, then

$$\lim_{n\to\infty} J_n\mu_n(A) = 1,\qquad(3.7.5)$$

where $J_n = \eta\sqrt{\Lambda''(\eta)\,2\pi n}\,e^{n\Lambda^*(q)}$.

(b) Suppose X_1 has a lattice law, i.e., for some x_0, d, the random variable $d^{-1}(X_1 - x_0)$ is (a.s.) an integer number, and d is the largest number with this property. Assume further that $1 > \mathrm{P}(X_1 = q) > 0$. (In particular, this implies that $d^{-1}(q - x_0)$ is an integer and that $\Lambda''(\eta) > 0$.) Then

$$\lim_{n\to\infty} J_n\mu_n(A) = \frac{\eta d}{1 - e^{-\eta d}}.\qquad(3.7.6)$$

Remarks:
(a) Recall that by part (c) of Lemma 2.2.5 and Exercise 2.2.24, $\Lambda^*(q) = \eta q - \Lambda(\eta)$, $\Lambda(\cdot)$ is C^∞ in some open neighborhood of η, $\eta = \Lambda^{*\prime}(q)$ and $\Lambda^{*\prime\prime}(q) = 1/\Lambda''(\eta)$. Hence, $J_n = \Lambda^{*\prime}(q)\sqrt{2\pi n/\Lambda^{*\prime\prime}(q)}e^{n\Lambda^*(q)}$.
(b) Theorem 3.7.4 holds even when A is a small interval of size of order $\mathbf{O}(\log n/n)$. (See Exercise 3.7.10.)
(c) The proof of this theorem is based on an exponential translation of a local CLT. This approach is applicable for the dependent case of Section 2.3 and to a certain extent applies also in $\mathbb{R}^d$, $d > 1$.

Proof: Consider the probability measure $\tilde{\mu}$ defined by $d\tilde{\mu}/d\mu(x) = e^{\eta x - \Lambda(\eta)}$, and let $Y_i \triangleq (X_i - q)/\sqrt{\Lambda''(\eta)}$, for $i = 1, 2, \ldots, n$. Note that $Y_1, \ldots, Y_n$ are i.i.d. random variables, with $E_{\tilde{\mu}}[Y_1] = 0$ and $E_{\tilde{\mu}}[Y_1^2] = 1$. Let $\psi_n \triangleq \eta\sqrt{n\Lambda''(\eta)}$ and let $F_n(\cdot)$ denote the distribution function of $W_n \triangleq n^{-1/2}\sum_{i=1}^n Y_i$ when X_i are i.i.d. with marginal law $\tilde{\mu}$. Since $\hat{S}_n = q + \sqrt{\Lambda''(\eta)/n}\, W_n$, it follows that

$$\mu_n(A) = \mu_n([q, \infty)) = E_{\tilde{\mu}}[e^{-n[\eta\hat{S}_n - \Lambda(\eta)]}1_{\{\hat{S}_n \geq q\}}]$$

$$= e^{-n\Lambda^*(q)}E_{\tilde{\mu}}\left[e^{-\psi_n W_n}1_{\{W_n \geq 0\}}\right] = e^{-n\Lambda^*(q)}\int_0^\infty e^{-\psi_n x}dF_n.$$

Hence, by an integration by parts,

$$J_n\mu_n(A) = \sqrt{2\pi}\int_0^\infty \psi_n^2 e^{-\psi_n x}[F_n(x) - F_n(0)]\,dx$$

$$= \sqrt{2\pi}\int_0^\infty \psi_n e^{-t}\left[F_n\left(\frac{t}{\psi_n}\right) - F_n(0)\right]dt. \qquad (3.7.7)$$

(a) When X_i are non-lattice, the Berry–Esséen expansion of $F_n(x)$ (see [Fel71, page 538]) results in

$$\lim_{n\to\infty}\left\{\sqrt{n}\,\sup_x\left|F_n(x) - \Phi(x) - \frac{m_3}{6\sqrt{n}}(1 - x^2)\phi(x)\right|\right\} = 0, \qquad (3.7.8)$$

where $m_3 \triangleq E_{\tilde{\mu}}[Y_1^3] < \infty$, $\phi(x) = 1/\sqrt{2\pi}\exp(-x^2/2)$ is the standard Normal density, and $\Phi(x) = \int_{-\infty}^x \phi(\theta)d\theta$. Define

$$c_n \triangleq \sqrt{2\pi}\int_0^\infty \psi_n e^{-t}\left(\Phi\left(\frac{t}{\psi_n}\right) + \frac{m_3}{6\sqrt{n}}\left[1 - \left(\frac{t}{\psi_n}\right)^2\right]\right.$$

$$\left.\cdot\phi\left(\frac{t}{\psi_n}\right) - \Phi(0) - \frac{m_3}{6\sqrt{n}}\phi(0)\right)dt. \qquad (3.7.9)$$

Comparing (3.7.7) and (3.7.9), observe that the Berry–Esséen expansion (3.7.8) yields the relation $\lim_{n\to\infty}|J_n\mu_n(A) - c_n| = 0$. Moreover, since

$$\sup_{x\geq 0}|\phi'(x)| < \infty, \quad \lim_{x\to 0}|\phi'(x)| = 0,$$

it follows by a Taylor expansion of $\Phi(t/\psi_n)$ and the dominated convergence theorem that

$$\lim_{n\to\infty} c_n = \lim_{n\to\infty} \sqrt{2\pi} \int_0^\infty \psi_n e^{-t} \left[\Phi\left(\frac{t}{\psi_n}\right) - \Phi(0) \right] dt$$

$$= \sqrt{2\pi}\,\phi(0) \int_0^\infty t e^{-t}\, dt = 1\,.$$

(b) In the lattice case, the range of Y_i is $\{md/\sqrt{\Lambda''(\eta)} : m \in \mathbb{Z}\}$. Consequently, here the Berry–Esséen expansion (see [Fel71, page 540] or [Pet75, page 171], Theorem 6), is

$$\lim_{n\to\infty} \left(\sqrt{n} \sup_x \left| F_n(x) - \Phi(x) - \frac{m_3}{6\sqrt{n}}\,(1 - x^2)\,\phi(x) \right.\right.$$

$$\left.\left. -\phi(x)g\left(x, \frac{d}{\sqrt{\Lambda''(\eta)n}}\right) \right| \right) = 0\,,$$

where $g(x,h) = h/2 - (x \bmod h)$ if $(x \bmod h) \neq 0$ and $g(x,h) = -h/2$ if $(x \bmod h) = 0$.

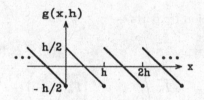

Figure 3.7.1: The function $g(x, h)$.

Thus, paraphrasing the preceding argument for the lattice case,

$$\lim_{n\to\infty} J_n \mu_n(A) = 1 + \lim_{n\to\infty}$$

$$\sqrt{2\pi} \int_0^\infty \psi_n e^{-t} \left[\phi\left(\frac{t}{\psi_n}\right) g\left(\frac{t}{\psi_n}, \frac{\eta d}{\psi_n}\right) - \phi(0) g\left(0, \frac{\eta d}{\psi_n}\right) \right] dt\,.$$

Since $\psi_n g(t/\psi_n, \eta d/\psi_n) = g(t, \eta d)$, it follows that

$$\lim_{n\to\infty} J_n \mu_n(A)$$

$$= 1 + \lim_{n\to\infty} \sqrt{2\pi} \int_0^\infty e^{-t} \left(\phi\left(\frac{t}{\psi_n}\right) g(t, \eta d) - \phi(0) g(0, \eta d) \right) dt$$

$$= 1 + \sqrt{2\pi}\,\phi(0) \int_0^\infty e^{-t}\,[g(t, \eta d) - g(0, \eta d)]\,dt\,.$$

The proof is completed by combining the preceding limit with

$$\int_0^\infty e^{-t}[g(t, \eta d) - g(0, \eta d)]\, dt$$

$$= \left(\sum_{n=0}^\infty e^{-n\eta d}\right) \int_0^{\eta d} e^{-t}(\eta d - t)\, dt = \frac{\eta d}{1 - e^{-\eta d}} - 1. \qquad \square$$

Exercise 3.7.10 (a) Let $A = [q, q + a/n)$, where in the lattice case a/d is restricted to being an integer. Prove that for any $a \in (0, \infty)$, both (3.7.5) and (3.7.6) hold with $J_n = \eta\sqrt{\Lambda''(\eta)\, 2\pi n}\; e^{n\Lambda^*(q)}/(1 - e^{-\eta a})$.
(b) As a consequence of part (a), conclude that for any set $A = [q, q + b_n)$, both (3.7.5) and (3.7.6) hold for the J_n given in Theorem 3.7.4 as long as $\lim_{n\to\infty} nb_n = \infty$.

Exercise 3.7.11 Let $\eta > 0$ denote the minimizer of $\Lambda(\lambda)$ and suppose that $\Lambda(\lambda) < \infty$ in some open interval around η.
(a) Based on Exercise 3.7.10, prove that when X_1 has a non-lattice distribution, the limiting distribution of $S_n \triangleq \sum_{i=1}^n X_i$ conditioned on $\{S_n \geq 0\}$ is Exponential(η) .
(b) Suppose now that X_1 has a lattice distribution of span d and $1 > P(X_1 = 0) > 0$. Prove that the limiting distribution of S_n/d conditioned on $\{S_n \geq 0\}$ is Geometric(p), with $p = 1 - e^{-\eta d}$ (i.e., $P(S_n = kd|S_n \geq 0) \to pq^k$ for $k = 0, 1, 2, \ldots$).

Exercise 3.7.12 Returning to the situation discussed in Section 3.4, consider a Neyman–Pearson test with constant threshold $\gamma \in (\overline{x}_0, \overline{x}_1)$. Suppose that $X_1 = \log(d\mu_1/d\mu_0(Y_1))$ has a non-lattice distribution. Let $\lambda_\gamma \in (0, 1)$ be the unique solution of $\Lambda_0'(\lambda) = \gamma$. Deduce from (3.7.5) that

$$\lim_{n\to\infty} \left(\alpha_n e^{n\Lambda_0^*(\gamma)}\lambda_\gamma\sqrt{\Lambda_0''(\lambda_\gamma)\, 2\pi n}\right) = 1 \qquad (3.7.13)$$

and

$$\lim_{n\to\infty} \left(\frac{e^{n\gamma}\alpha_n}{\beta_n}\right) = \frac{1 - \lambda_\gamma}{\lambda_\gamma}. \qquad (3.7.14)$$

3.8 Historical Notes and References

The large deviations statements for Markov chains have a long history, and several approaches exist to derive them. To avoid repetition, they are partially described in the historical notes of Chapter 6. In the finite state setup, an early reference is Miller [Mil61]. The approach taken here is based, in part, on the ideas of Ellis [Ell84] and [Ell88]. The method of types and estimates as in Exercise 3.1.21 are applied in [AM92] to provide tight large deviation bounds for certain sets with empty interior.

Theorem 3.2.1 is motivated by the results in [ArGW90]. Exercise 3.2.7 is taken from [DKZ94a], where part (b) is strengthened to a full convergence. See also [DKZ94b] for limit laws and references to related works. For similar results in the context of $\mathbb{Z}^d$-indexed Gibbs fields, see [Com94].

Gibbs's conditioning principle has served as a driving force behind Ruelle and Lanford's treatment of large deviations (without calling it by that name) [Rue65, Rue67, Lan73]. The form of the Gibbs's principle here was proved using large deviations methods in the discrete, uniform case in [Vas80], via the method of types by Campenhout and Cover [CaC81] and more generally by Csiszár [Cs84], and Stroock and Zeitouni [StZ91]. For other references to the statistical mechanics literature, see the historical notes of Chapter 7.

One of the early reasons for deriving the LDP was their use in evaluating the performance of estimators and decision rules, and in comparing different estimators (see, e.g., [Che52]). A good summary of this early work may be found in [Bah67, Bah71]. More recent accounts are [BaZG80, Rad83, KK86] (in the i.i.d. case) and [Bah83] (for Markov chains). The material of Section 3.4 may be traced back to Chernoff [Che52, Che56], who derived Theorem 3.4.3. Lemma 3.4.7 appears first in [Che56], which attributes it to an unpublished manuscript of C. Stein. Other statistical applications, some of which are based on abstract LDPs, may be found in the collection [As79].

The generalized likelihood ratio test of Section 3.5 was considered by Hoeffding [Hoe65], whose approach we basically follow here. See the historical notes of Chapter 7 for extensions to other situations.

The source coding theorem is attributed to Shannon [Sha59]. See also [Sha48] and [Ber71]. Our treatment of it in Section 3.6 is a combination of the method of this book and the particular case treated by Bucklew in [Buc90].

Theorem 3.7.1 follows [Pet75], although some of the methods are much older and may be found in Feller's book [Fel71]. Theorem 3.7.4 was proved in the discrete case by [BH59] and in the form here by [BaR60]. (The saddle point approximation of [Dan54] is also closely related.) A review of related asymptotic expansions may be found in [Hea67, Pet75, BhR76]. Other refinements and applications in statistics may be found in [Barn78]. An early $\mathbb{R}^d$ local CLT type expansion is contained in [BoR65]. For related results, see [Boo75]. The extension of Theorem 3.7.4 to $\mathbb{R}^d$-valued random variables is provided in [Ilt95], which builds upon the results of [Ney83]. For a full asymptotic expansion, even when the exponential moments condition is not satisfied, see [Roz86] and references therein.

Chapter 4

General Principles

In this chapter, we initiate the investigation of large deviation principles (LDPs) for families of measures on general spaces. As will be obvious in subsequent chapters, the objects on which the LDP is sought may vary considerably. Hence, it is necessary to undertake a study of the LDP in an abstract setting. We shall focus our attention on the abstract statement of the LDP as presented in Section 1.2 and give conditions for the existence of such a principle and various approaches for the identification of the resulting rate function.

Since this chapter deals with different approaches to the LDP, some of its sections are independent of the others. A rough structure of it is as follows. In Section 4.1, extensions of the basic properties of the LDP are provided. In particular, relations between the topological structure of the space, the existence of certain limits, and the existence and uniqueness of the LDP are explored. Section 4.2 describes how to move around the LDP from one space to another. Thus, under appropriate conditions, the LDP can be proved in a simple situation and then effortlessly transferred to a more complex one. Of course, one should not be misled by the word *effortlessly*: It often occurs in applications that much of the technical work to be done is checking that the conditions for such a transformation are satisfied! Sections 4.3 and 4.4 investigate the relation between the LDP and the computation of exponential integrals. Although in some applications the computation of the exponential integrals is a goal in itself, it is more often the case that such computations are an intermediate step in deriving the LDP. Such a situation has already been described, though implicitly, in the treatment of the Chebycheff upper bound in Section 2.2. This line of thought is tackled again in Section 4.5.1, in the case where $\mathcal{X}$ is a topological vector space, such that the Fenchel–Legendre transform is well-defined

A. Dembo, O. Zeitouni, *Large Deviations Techniques and Applications*, 115
Stochastic Modelling and Applied Probability 38,
DOI 10.1007/978-3-642-03311-7_4,
© Springer-Verlag Berlin Heidelberg 1998, corrected printing 2010

and an upper bound may be derived based on it. Section 4.5.2 complements this approach by providing the tools that will enable us to exploit convexity, again in the case of topological vector spaces, to derive the lower bound. The attentive reader may have already suspected that such an attack on the LDP is possible when he followed the arguments of Section 2.3. Section 4.6 is somewhat independent of the rest of the chapter. Its goal is to show that the LDP is preserved under projective limits. Although at first sight this may not look useful for applications, it will become clear to the patient reader that this approach is quite general and may lead from finite dimensional computations to the LDP in abstract spaces. Finally, Section 4.7 draws attention to the similarity between the LDP and weak convergence in metric spaces.

Since this chapter deals with the LDP in abstract spaces, some topological and analytical preliminaries are in order. The reader may find Appendices B and C helpful reminders of a particular definition or theorem.

The convention that $\mathcal{B}$ contains the Borel σ-field $\mathcal{B}_{\mathcal{X}}$ is used throughout this chapter, except in Lemma 4.1.5, Theorem 4.2.1, Exercise 4.2.9, Exercise 4.2.32, and Section 4.6.

4.1   Existence of an LDP and Related Properties

If a set $\mathcal{X}$ is given the coarse topology $\{\emptyset, \mathcal{X}\}$, the only information implied by the LDP is that $\inf_{x \in \mathcal{X}} I(x) = 0$, and many rate functions satisfy this requirement. To avoid such trivialities, we must put some constraint on the topology of the set $\mathcal{X}$. Recall that a topological space is Hausdorff if, for every pair of distinct points x and y, there exist disjoint neighborhoods of x and y. The natural condition that prevails throughout this book is that, in addition to being Hausdorff, $\mathcal{X}$ is a *regular space* as defined next.

Definition 4.1.1 *A Hausdorff topological space $\mathcal{X}$ is regular if, for any closed set $F \subset \mathcal{X}$ and any point $x \notin F$, there exist disjoint open subsets G_1 and G_2 such that $F \subset G_1$ and $x \in G_2$.*

In the rest of the book, the term *regular* will mean Hausdorff and regular. The following observations regarding regular spaces are of crucial importance here:
(a) For any neighborhood G of $x \in \mathcal{X}$, there exists a neighborhood A of x such that $\overline{A} \subset G$.
(b) Every metric space is regular. Moreover, if a real topological vector space is Hausdorff, then it is regular. All examples of an LDP considered

in this book are either for metric spaces, or for Hausdorff real topological vector spaces.

(c) A lower semicontinuous function f satisfies, at every point x,

$$f(x) = \sup_{\{G \text{ neighborhood of } x\}} \inf_{y \in G} f(y). \tag{4.1.2}$$

Therefore, for any $x \in \mathcal{X}$ and any $\delta > 0$, one may find a neighborhood $G = G(x, \delta)$ of x, such that $\inf_{y \in G} f(y) \geq (f(x) - \delta) \wedge 1/\delta$. Let $A = A(x, \delta)$ be a neighborhood of x such that $\overline{A} \subset G$. (Such a set exists by property (a).) One then has

$$\inf_{y \in \overline{A}} f(y) \geq \inf_{y \in G} f(y) \geq (f(x) - \delta) \wedge \frac{1}{\delta}. \tag{4.1.3}$$

The sets $G = G(x, \delta)$ frequently appear in the proofs of large deviations statements and properties. Observe that in a metric space, $G(x, \delta)$ may be taken as a ball centered at x and having a small enough radius.

4.1.1 Properties of the LDP

The first desirable consequence of the assumption that $\mathcal{X}$ is a regular topological space is the uniqueness of the rate function associated with the LDP.

Lemma 4.1.4 *A family of probability measures $\{\mu_\epsilon\}$ on a regular topological space can have at most one rate function associated with its LDP.*

Proof: Suppose there exist two rate functions $I_1(\cdot)$ and $I_2(\cdot)$, both associated with the LDP for $\{\mu_\epsilon\}$. Without loss of generality, assume that for some $x_0 \in \mathcal{X}$, $I_1(x_0) > I_2(x_0)$. Fix $\delta > 0$ and consider the open set A for which $x_0 \in A$, while $\inf_{y \in \overline{A}} I_1(y) \geq (I_1(x_0) - \delta) \wedge 1/\delta$. Such a set exists by (4.1.3). It follows by the LDP for $\{\mu_\epsilon\}$ that

$$- \inf_{y \in \overline{A}} I_1(y) \geq \limsup_{\epsilon \to 0} \epsilon \log \mu_\epsilon(A) \geq \liminf_{\epsilon \to 0} \epsilon \log \mu_\epsilon(A) \geq - \inf_{y \in A} I_2(y).$$

Therefore,

$$I_2(x_0) \geq \inf_{y \in A} I_2(y) \geq \inf_{y \in \overline{A}} I_1(y) \geq (I_1(x_0) - \delta) \wedge \frac{1}{\delta}.$$

Since δ is arbitrary, this contradicts the assumption that $I_1(x_0) > I_2(x_0)$. $\square$

Remarks:

(a) It is evident from the proof that if $\mathcal{X}$ is a locally compact space (e.g.,

$\mathcal{X} = \mathbb{R}^d$), the rate function is unique as soon as a weak LDP holds. As shown in Exercise 4.1.30, if $\mathcal{X}$ is a Polish space, then also the rate function is unique as soon as a weak LDP holds.

(b) The uniqueness of the rate function does not depend on the Hausdorff part of the definition of regular spaces. However, the rate function assigns the same value to any two points of $\mathcal{X}$ that are not separated. (See Exercise 4.1.9.) Thus, in terms of the LDP, such points are indistinguishable.

As shown in the next lemma, the LDP is preserved under suitable inclusions. Hence, in applications, one may first prove an LDP in a space that possesses additional structure (for example, a topological vector space), and then use this lemma to deduce the LDP in the subspace of interest.

Lemma 4.1.5 *Let $\mathcal{E}$ be a measurable subset of $\mathcal{X}$ such that $\mu_\epsilon(\mathcal{E}) = 1$ for all $\epsilon > 0$. Suppose that $\mathcal{E}$ is equipped with the topology induced by $\mathcal{X}$.*

(a) If $\mathcal{E}$ is a closed subset of $\mathcal{X}$ and $\{\mu_\epsilon\}$ satisfies the LDP in $\mathcal{E}$ with rate function I, then $\{\mu_\epsilon\}$ satisfies the LDP in $\mathcal{X}$ with rate function I' such that $I' = I$ on $\mathcal{E}$ and $I' = \infty$ on $\mathcal{E}^c$.

(b) If $\{\mu_\epsilon\}$ satisfies the LDP in $\mathcal{X}$ with rate function I and $\mathcal{D}_I \subset \mathcal{E}$, then the same LDP holds in $\mathcal{E}$. In particular, if $\mathcal{E}$ is a closed subset of $\mathcal{X}$, then $\mathcal{D}_I \subset \mathcal{E}$ and hence the LDP holds in $\mathcal{E}$.

Proof: In the topology induced on $\mathcal{E}$ by $\mathcal{X}$, the open sets are the sets of the form $G \cap \mathcal{E}$ with $G \subseteq \mathcal{X}$ open. Similarly, the closed sets in this topology are the sets of the form $F \cap \mathcal{E}$ with $F \subseteq \mathcal{X}$ closed. Furthermore, $\mu_\epsilon(\Gamma) = \mu_\epsilon(\Gamma \cap \mathcal{E})$ for any $\Gamma \in \mathcal{B}$.

(a) Suppose that an LDP holds in $\mathcal{E}$, which is a closed subset of $\mathcal{X}$. Extend the rate function I to be a lower semicontinuous function on $\mathcal{X}$ by setting $I(x) = \infty$ for any $x \in \mathcal{E}^c$. Thus, $\inf_{x \in \Gamma} I(x) = \inf_{x \in \Gamma \cap \mathcal{E}} I(x)$ for any $\Gamma \subset \mathcal{X}$ and the large deviations lower (upper) bound holds.

(b) Suppose that an LDP holds in $\mathcal{X}$. If $\mathcal{E}$ is closed, then $\mathcal{D}_I \subset \mathcal{E}$ by the large deviations lower bound (since $\mu_\epsilon(\mathcal{E}^c) = 0$ for all $\epsilon > 0$ and $\mathcal{E}^c$ is open). Now, $\mathcal{D}_I \subset \mathcal{E}$ implies that $\inf_{x \in \Gamma} I(x) = \inf_{x \in \Gamma \cap \mathcal{E}} I(x)$ for any $\Gamma \subset \mathcal{X}$ and the large deviations lower (upper) bound holds for all measurable subsets of $\mathcal{E}$. Further, since the level sets $\Psi_I(\alpha)$ are closed subsets of $\mathcal{E}$, the rate function I remains lower semicontinuous when restricted to $\mathcal{E}$. $\square$

Remarks:

(a) The preceding lemma also holds for the weak LDP, since compact subsets of $\mathcal{E}$ are just the compact subsets of $\mathcal{X}$ contained in $\mathcal{E}$. Similarly, under the assumptions of the lemma, I is a good rate function on $\mathcal{X}$ iff it is a good rate function when restricted to $\mathcal{E}$.

(b) Lemma 4.1.5 holds without any change in the proof even when $\mathcal{B}_\mathcal{X} \not\subseteq \mathcal{B}$.

The following is an important property of good rate functions.

Lemma 4.1.6 *Let I be a good rate function.*
(a) Let $\{F_\delta\}_{\delta>0}$ be a nested family of closed sets, i.e., $F_\delta \subseteq F_{\delta'}$ if $\delta < \delta'$.
Define $F_0 = \cap_{\delta>0} F_\delta$. Then

$$\inf_{y \in F_0} I(y) = \lim_{\delta \to 0} \inf_{y \in F_\delta} I(y).$$

(b) Suppose $(\mathcal{X}, d)$ is a metric space. Then, for any set A,

$$\inf_{y \in \overline{A}} I(y) = \lim_{\delta \to 0} \inf_{y \in A^\delta} I(y), \tag{4.1.7}$$

where

$$A^\delta \stackrel{\triangle}{=} \{y : d(y, A) = \inf_{z \in A} d(y, z) \le \delta\} \tag{4.1.8}$$

denotes the closed blowup of A.

Proof: (a) Since $F_0 \subseteq F_\delta$ for all $\delta > 0$, it suffices to prove that for all $\eta > 0$,

$$\gamma \stackrel{\triangle}{=} \lim_{\delta \to 0} \inf_{y \in F_\delta} I(y) \ge \inf_{y \in F_0} I(y) - \eta.$$

This inequality holds trivially when $\gamma = \infty$. If $\gamma < \infty$, fix $\eta > 0$ and let $\alpha = \gamma + \eta$. The sets $F_\delta \cap \Psi_I(\alpha)$, $\delta > 0$, are non-empty, nested, and compact. Consequently,

$$F_0 \cap \Psi_I(\alpha) = \bigcap_{\delta > 0} F_\delta \cap \Psi_I(\alpha)$$

is also non-empty, and the proof of part (a) is thus completed.
(b) Note that $d(\cdot, A)$ is a continuous function and hence $\{A^\delta\}_{\delta>0}$ are nested, closed sets. Moreover,

$$\bigcap_{\delta > 0} A^\delta = \{y : d(y, A) = 0\} = \overline{A}. \qquad \square$$

Exercise 4.1.9 Suppose that for any closed subset F of $\mathcal{X}$ and any point $x \notin F$, there exist two disjoint open sets G_1 and G_2 such that $F \subset G_1$ and $\dot{x} \in G_2$. Prove that if $I(x) \ne I(y)$ for some lower semicontinuous function I, then there exist disjoint neighborhoods of x and y.

Exercise 4.1.10 [[LyS87], Lemma 2.6. See also [Puk91], Theorem (P).]
Let $\{\mu_n\}$ be a sequence of probability measures on a Polish space $\mathcal{X}$.
(a) Show that $\{\mu_n\}$ is exponentially tight if for every $\alpha < \infty$ and every $\eta > 0$, there exist $m \in \mathbb{Z}_+$ and $x_1, \ldots, x_m \in \mathcal{X}$ such that for all n,

$$\mu_n\left(\left[\bigcup_{i=1}^m B_{x_i, \eta}\right]^c\right) \le e^{-\alpha n}.$$

Hint: Observe that for every sequence $\{m_k\}$ and any $x_i^{(k)} \in \mathcal{X}$, the set $\cap_{k=1}^{\infty} \cup_{i=1}^{m_k} B_{x_i^{(k)},1/k}$ is pre-compact.

(b) Suppose that $\{\mu_n\}$ satisfies the large deviations upper bound with a good rate function. Show that for every countable dense subset of $\mathcal{X}$, e.g., $\{x_i\}$, every $\eta > 0$, every $\alpha < \infty$, and every m large enough,

$$\limsup_{n\to\infty} \frac{1}{n} \log \mu_n \left(\left[\bigcup_{i=1}^{m} B_{x_i,\eta} \right]^c \right) < -\alpha .$$

Hint: Use Lemma 4.1.6.

(c) Deduce that if $\{\mu_n\}$ satisfies the large deviations upper bound with a good rate function, then $\{\mu_n\}$ is exponentially tight.

Remark: When a non-countable family of measures $\{\mu_\epsilon, \epsilon > 0\}$ satisfies the large deviations upper bound in a Polish space with a good rate function, the preceding yields the exponential tightness of every sequence $\{\mu_{\epsilon_n}\}$, where $\epsilon_n \to 0$ as $n \to \infty$. As far as large deviations results are concerned, this is indistinguishable from exponential tightness of the whole family.

4.1.2 The Existence of an LDP

The following theorem introduces a general, indirect approach for establishing the *existence* of a *weak LDP*.

Theorem 4.1.11 *Let $\mathcal{A}$ be a base of the topology of $\mathcal{X}$. For every $A \in \mathcal{A}$, define*

$$\mathcal{L}_A \triangleq -\liminf_{\epsilon\to 0} \epsilon \log \mu_\epsilon(A) \qquad (4.1.12)$$

and

$$I(x) \triangleq \sup_{\{A\in\mathcal{A}: x\in A\}} \mathcal{L}_A . \qquad (4.1.13)$$

Suppose that for all $x \in \mathcal{X}$,

$$I(x) = \sup_{\{A\in\mathcal{A}: x\in A\}} \left[-\limsup_{\epsilon\to 0} \epsilon \log \mu_\epsilon(A) \right] . \qquad (4.1.14)$$

Then μ_ϵ satisfies the weak LDP with the rate function $I(x)$.

Remarks:

(a) Observe that condition (4.1.14) holds when the limits $\lim_{\epsilon\to 0} \epsilon \log \mu_\epsilon(A)$ exist for all $A \in \mathcal{A}$ (with $-\infty$ as a possible value).

(b) When $\mathcal{X}$ is a locally convex, Hausdorff topological vector space, the base $\mathcal{A}$ is often chosen to be the collection of open, convex sets. For concrete examples, see Sections 6.1 and 6.3.

Proof: Since $\mathcal{A}$ is a base for the topology of $\mathcal{X}$, for any open set G and any point $x \in G$ there exists an $A \in \mathcal{A}$ such that $x \in A \subset G$. Therefore, by definition,

$$\liminf_{\epsilon \to 0} \epsilon \log \mu_\epsilon(G) \geq \liminf_{\epsilon \to 0} \epsilon \log \mu_\epsilon(A) = -\mathcal{L}_A \geq -I(x).$$

As seen in Section 1.2, this is just one of the alternative statements of the large deviations lower bound.

Clearly, $I(x)$ is a nonnegative function. Moreover, if $I(x) > \alpha$, then $\mathcal{L}_A > \alpha$ for some $A \in \mathcal{A}$ such that $x \in A$. Therefore, $I(y) \geq \mathcal{L}_A > \alpha$ for every $y \in A$. Hence, the sets $\{x : I(x) > \alpha\}$ are open, and consequently I is a rate function.

Note that the lower bound and the fact that I is a rate function do not depend on (4.1.14). This condition is used in the proof of the upper bound. Fix $\delta > 0$ and a compact $F \subset \mathcal{X}$. Let I^δ be the δ-rate function, i.e., $I^\delta(x) \triangleq \min\{I(x) - \delta, 1/\delta\}$. Then, (4.1.14) implies that for every $x \in F$, there exists a set $A_x \in \mathcal{A}$ (which may depend on δ) such that $x \in A_x$ and

$$-\limsup_{\epsilon \to 0} \epsilon \log \mu_\epsilon(A_x) \geq I^\delta(x).$$

Since F is compact, one can extract from the open cover $\cup_{x \in F} A_x$ of F a finite cover of F by the sets $A_{x_1}, \ldots, A_{x_m}$. Thus,

$$\mu_\epsilon(F) \leq \sum_{i=1}^{m} \mu_\epsilon(A_{x_i}),$$

and consequently,

$$\begin{aligned}
\limsup_{\epsilon \to 0} \epsilon \log \mu_\epsilon(F) &\leq \max_{i=1,\ldots,m} \limsup_{\epsilon \to 0} \epsilon \log \mu_\epsilon(A_{x_i}) \\
&\leq -\min_{i=1,\ldots,m} I^\delta(x_i) \leq -\inf_{x \in F} I^\delta(x).
\end{aligned}$$

The proof of the upper bound for compact sets is completed by considering the limit as $\delta \to 0$. $\qquad\square$

Theorem 4.1.11 is extended in the following lemma, which concerns the LDP of a family of probability measures $\{\mu_{\epsilon,\sigma}\}$ that is indexed by an additional parameter σ. For a concrete application, see Section 6.3, where σ is the initial state of a Markov chain.

Lemma 4.1.15 *Let $\mu_{\epsilon,\sigma}$ be a family of probability measures on $\mathcal{X}$, indexed by σ, whose range is the set Σ. Let $\mathcal{A}$ be a base for the topology of $\mathcal{X}$. For each $A \in \mathcal{A}$, define*

$$\mathcal{L}_A \overset{\triangle}{=} -\liminf_{\epsilon \to 0} \epsilon \log \Big[\inf_{\sigma \in \Sigma} \mu_{\epsilon,\sigma}(A) \Big]. \tag{4.1.16}$$

Let

$$I(x) = \sup_{\{A \in \mathcal{A} : x \in A\}} \mathcal{L}_A .$$

If for every $x \in \mathcal{X}$,

$$I(x) = \sup_{\{A \in \mathcal{A} : x \in A\}} \left\{ - \limsup_{\epsilon \to 0} \epsilon \log \left[\sup_{\sigma \in \Sigma} \mu_{\epsilon,\sigma}(A) \right] \right\} , \qquad (4.1.17)$$

then, for each $\sigma \in \Sigma$, *the measures* $\mu_{\epsilon,\sigma}$ *satisfy a weak LDP with the (same) rate function* $I(\cdot)$.

Proof: The proof parallels that of Theorem 4.1.11. (See Exercise 4.1.29.) $\square$

It is aesthetically pleasing to know that the following partial converse of Theorem 4.1.11 holds.

Theorem 4.1.18 *Suppose that* $\{\mu_\epsilon\}$ *satisfies the LDP in a regular topological space* $\mathcal{X}$ *with rate function* I. *Then, for any base* $\mathcal{A}$ *of the topology of* $\mathcal{X}$, *and for any* $x \in \mathcal{X}$,

$$\begin{aligned} I(x) &= \sup_{\{A \in \mathcal{A} : x \in A\}} \left\{ - \liminf_{\epsilon \to 0} \epsilon \log \mu_\epsilon(A) \right\} \\ &= \sup_{\{A \in \mathcal{A} : x \in A\}} \left\{ - \limsup_{\epsilon \to 0} \epsilon \log \mu_\epsilon(A) \right\} . \qquad (4.1.19) \end{aligned}$$

Remark: As shown in Exercise 4.1.30, for a Polish space $\mathcal{X}$ suffices to assume in Theorem 4.1.18 that $\{\mu_\epsilon\}$ satisfies the weak LDP. Consequently, by Theorem 4.1.11, in this context (4.1.19) is *equivalent* to the weak LDP.

Proof: Fix $x \in \mathcal{X}$ and let

$$\ell(x) = \sup_{\{A \in \mathcal{A} : x \in A\}} \inf_{y \in \overline{A}} I(y) . \qquad (4.1.20)$$

Suppose that $I(x) > \ell(x)$. Then, in particular, $\ell(x) < \infty$ and $x \in \Psi_I(\alpha)^c$ for some $\alpha > \ell(x)$. Since $\Psi_I(\alpha)^c$ is an open set and $\mathcal{A}$ is a base for the topology of the regular space $\mathcal{X}$, there exists a set $A \in \mathcal{A}$ such that $x \in A$ and $\overline{A} \subseteq \Psi_I(\alpha)^c$. Therefore, $\inf_{y \in \overline{A}} I(y) \geq \alpha$, which contradicts (4.1.20). We conclude that $\ell(x) \geq I(x)$. The large deviations lower bound implies

$$I(x) \geq \sup_{\{A \in \mathcal{A} : x \in A\}} \left\{ - \liminf_{\epsilon \to 0} \epsilon \log \mu_\epsilon(A) \right\} ,$$

while the large deviations upper bound implies that for all $A \in \mathcal{A}$,

$$- \liminf_{\epsilon \to 0} \epsilon \log \mu_\epsilon(A) \geq - \limsup_{\epsilon \to 0} \epsilon \log \mu_\epsilon(\overline{A}) \geq \inf_{y \in \overline{A}} I(y) .$$

These two inequalities yield (4.1.19), since $\ell(x) \geq I(x)$. $\qquad\qquad\qquad\square$

The characterization of the rate function in Theorem 4.1.11 and Lemma 4.1.15 involves the supremum over a large collection of sets. Hence, it does not yield a convenient explicit formula. As shown in Section 4.5.2, if $\mathcal{X}$ is a Hausdorff topological vector space, this rate function can sometimes be identified with the Fenchel–Legendre transform of a limiting logarithmic moment generating function. This approach requires an *a priori* proof that the rate function is *convex*. The following lemma improves on Theorem 4.1.11 by giving a sufficient condition for the convexity of the rate function. Throughout, for any sets $A_1, A_2 \in \mathcal{X}$,

$$\frac{A_1 + A_2}{2} \triangleq \{x : \; x = (x_1 + x_2)/2, \; x_1 \in A_1, x_2 \in A_2\}.$$

Lemma 4.1.21 *Let $\mathcal{A}$ be a base for a Hausdorff topological vector space $\mathcal{X}$, such that in addition to condition (4.1.14), for every $A_1, A_2 \in \mathcal{A}$,*

$$\limsup_{\epsilon \to 0} \epsilon \log \mu_\epsilon \left(\frac{A_1 + A_2}{2}\right) \geq -\frac{1}{2} \left(\mathcal{L}_{A_1} + \mathcal{L}_{A_2}\right). \qquad (4.1.22)$$

Then the rate function I of (4.1.13), which governs the weak LDP associated with $\{\mu_\epsilon\}$, is convex.

Proof: It suffices to show that the condition (4.1.22) yields the convexity of the rate function I of (4.1.13). To this end, fix $x_1, x_2 \in \mathcal{X}$ and $\delta > 0$. Let $x = (x_1 + x_2)/2$ and let I^δ denote the δ-rate function. Then, by (4.1.14), there exists an $A \in \mathcal{A}$ such that $x \in A$ and $-\limsup_{\epsilon \to 0} \epsilon \log \mu_\epsilon(A) \geq I^\delta(x)$. The pair (x_1, x_2) belongs to the set $\{(y_1, y_2) : (y_1 + y_2)/2 \in A\}$, which is an open subset of $\mathcal{X} \times \mathcal{X}$. Therefore, there exist open sets $A_1 \subseteq \mathcal{X}$ and $A_2 \subseteq \mathcal{X}$ with $x_1 \in A_1$ and $x_2 \in A_2$ such that $(A_1 + A_2)/2 \subseteq A$. Furthermore, since $\mathcal{A}$ is a base for the topology of $\mathcal{X}$, one may take A_1 and A_2 in $\mathcal{A}$. Thus, our assumptions imply that

$$
\begin{aligned}
-I^\delta(x) &\geq \limsup_{\epsilon \to 0} \epsilon \log \mu_\epsilon(A) \\
&\geq \limsup_{\epsilon \to 0} \epsilon \log \mu_\epsilon \left(\frac{A_1 + A_2}{2}\right) \geq -\frac{1}{2} \left(\mathcal{L}_{A_1} + \mathcal{L}_{A_2}\right).
\end{aligned}
$$

Since $x_1 \in A_1$ and $x_2 \in A_2$, it follows that

$$\frac{1}{2} I(x_1) + \frac{1}{2} I(x_2) \geq \frac{1}{2} \mathcal{L}_{A_1} + \frac{1}{2} \mathcal{L}_{A_2} \geq I^\delta(x) = I^\delta \left(\frac{1}{2} x_1 + \frac{1}{2} x_2\right).$$

Considering the limit $\delta \searrow 0$, one obtains

$$\frac{1}{2} I(x_1) + \frac{1}{2} I(x_2) \geq I \left(\frac{1}{2} x_1 + \frac{1}{2} x_2\right).$$

By iterating, this inequality can be extended to any x of the form $(k/2^n)x_1 + (1-k/2^n)x_2$ with $k, n \in \mathbb{Z}_+$. The definition of a topological vector space and the lower semicontinuity of I imply that $I(\beta x_1 + (1-\beta)x_2) : [0,1] \to [0,\infty]$ is a lower semicontinuous function of β. Hence, the preceding inequality holds for all convex combinations of x_1, x_2 and the proof of the lemma is complete. $\square$

When combined with exponential tightness, Theorem 4.1.11 implies the following large deviations analog of Prohorov's theorem (Theorem D.9).

Lemma 4.1.23 *Suppose the topological space $\mathcal{X}$ has a countable base. For any family of probability measures $\{\mu_\epsilon\}$, there exists a sequence $\epsilon_k \to 0$ such that $\{\mu_{\epsilon_k}\}$ satisfies the weak LDP in $\mathcal{X}$. If $\{\mu_\epsilon\}$ is an exponentially tight family of probability measures, then $\{\mu_{\epsilon_k}\}$ also satisfies the LDP with a good rate function.*

Proof: Fix a countable base $\mathcal{A}$ for the topology of $\mathcal{X}$ and a sequence $\epsilon_n \to 0$. By Tychonoff's theorem (Theorem B.3), the product topology makes $\mathcal{Y} = [0,1]^{\mathcal{A}}$ into a compact metrizable space. Since $\mathcal{Y}$ is sequentially compact (Theorem B.2) and $\mu_\epsilon(\cdot)^\epsilon : \mathcal{A} \to [0,1]$ is in $\mathcal{Y}$ for each $\epsilon > 0$, the sequence $\mu_{\epsilon_n}(\cdot)^{\epsilon_n}$ has a convergent subsequence in $\mathcal{Y}$. Hence, passing to the latter subsequence, denoted ϵ_k, the limits $\lim_{k\to\infty} \epsilon_k \log \mu_{\epsilon_k}(A)$ exist for all $A \in \mathcal{A}$ (with $-\infty$ as a possible value). In particular, condition (4.1.14) holds and by Theorem 4.1.11, $\{\mu_{\epsilon_k} : k \in \mathbb{Z}_+\}$ satisfies the weak LDP. Applying Lemma 1.2.18, the LDP with a good rate function follows when $\{\mu_\epsilon\}$ is an exponentially tight family of probability measures. $\square$

The next lemma applies for tight Borel probability measures μ_ϵ on metric spaces. In this context, it allows replacement of the assumed LDP in either Lemma 4.1.4 or Theorem 4.1.18 by a weak LDP (see Exercise 4.1.30).

Lemma 4.1.24 *Suppose $\{\mu_\epsilon\}$ is a family of tight (Borel) probability measures on a metric space $(\mathcal{X}, d)$, such that the upper bound (1.2.12) holds for all compact sets and some rate function $I(\cdot)$. Then, for any base $\mathcal{A}$ of the topology of $\mathcal{X}$, and for any $x \in \mathcal{X}$,*

$$I(x) \leq \sup_{\{A \in \mathcal{A} : x \in A\}} \left\{ -\limsup_{\epsilon \to 0} \epsilon \log \mu_\epsilon(A) \right\} . \qquad (4.1.25)$$

Proof: We argue by contradiction, fixing a base $\mathcal{A}$ of the metric topology and $x \in \mathcal{X}$ for which (4.1.25) fails. For any $m \in \mathbb{Z}_+$, there exists some $A \in \mathcal{A}$ such that $x \in A \subset B_{x,m^{-1}}$. Hence, for some $\delta > 0$ and any $m \in \mathbb{Z}_+$,

$$\limsup_{\epsilon \to 0} \epsilon \log \mu_\epsilon(B_{x,m^{-1}}) > -I^\delta(x) = -\min\{I(x) - \delta, 1/\delta\} ,$$

implying that for some $\epsilon_m \to 0$,

$$\mu_{\epsilon_m}(B_{x,m^{-1}}) > e^{-I^\delta(x)/\epsilon_m} \qquad \forall m \in \mathbb{Z}_+ . \tag{4.1.26}$$

Recall that the probability measures μ_{ϵ_m} are regular (by Theorem C.5), hence in (4.1.26) we may replace each open set $B_{x,m^{-1}}$ by some closed subset F_m. With each μ_{ϵ_m} assumed tight, we may further replace the closed sets F_m by compact subsets $K_m \subset F_m \subset B_{x,m^{-1}}$ such that

$$\mu_{\epsilon_m}(K_m) > e^{-I^\delta(x)/\epsilon_m} \qquad \forall m \in \mathbb{Z}_+ . \tag{4.1.27}$$

Note that the sets $K_r^* = \{x\} \cup_{m \geq r} K_m$ are also compact. Indeed, in any open covering of K_r^* there is an open set G_o such that $x \in G_o$ and hence $\cup_{m>m_0} K_m \subset B_{x,m_0^{-1}} \subset G_o$ for some $m_o \in \mathbb{Z}_+$, whereas the compact set $\cup_{m=r}^{m_o} K_m$ is contained in the union of some G_i, $i = 1, \ldots, M$, from this cover. In view of (4.1.27), the upper bound (1.2.12) yields for $K_r^* \subset B_{x,r^{-1}}$ that,

$$-\inf_{y \in B_{x,r^{-1}}} I(y) \geq -\inf_{y \in K_r^*} I(y) \geq \limsup_{\epsilon \to 0} \epsilon \log \mu_\epsilon(K_r^*) \tag{4.1.28}$$

$$\geq \limsup_{m \to \infty} \epsilon_m \log \mu_{\epsilon_m}(K_m) \geq -I^\delta(x) .$$

By lower semicontinuity, $\lim_{r \to \infty} \inf_{y \in B_{x,r^{-1}}} I(y) = I(x) > I^\delta(x)$, in contradiction with (4.1.28). Necessarily, (4.1.25) holds for any $x \in \mathcal{X}$ and any base $\mathcal{A}$. $\qquad\square$

Exercise 4.1.29 Prove Lemma 4.1.15 using the following steps.
(a) Check that the large deviations lower bound (for each $\sigma \in \Sigma$) and the lower semicontinuity of I may be proved exactly as done in Theorem 4.1.11.
(b) Fix $\sigma \in \Sigma$ and prove the large deviations upper bound for compact sets.

Exercise 4.1.30 Suppose a family of tight (Borel) probability measures $\{\mu_\epsilon\}$ satisfies the weak LDP in a metric space $(\mathcal{X}, d)$ with rate function $I(\cdot)$.
(a) Combine Lemma 4.1.24 with the large deviations lower bound to conclude that (4.1.19) holds for any base $\mathcal{A}$ of the topology of $\mathcal{X}$ and any $x \in \mathcal{X}$.
(b) Conclude that in this context the rate function $I(\cdot)$ associated with the weak LDP is unique.

Exercise 4.1.31 Suppose $X_i \in \mathbb{R}^{d-1}$, $d \geq 2$, with $|X_i| \leq C$ and $Y_i \in [m, M]$ for some $0 < m < M, C < \infty$, are such that $n^{-1} \sum_{i=1}^n (X_i, Y_i)$ satisfy the LDP in $\mathbb{R}^d$ with a good rate function $J(x, y)$. Let $\tau_\epsilon = \inf\{n : \sum_{i=1}^n Y_i > \epsilon^{-1}\}$. Show that $(\epsilon \sum_{i=1}^{\tau_\epsilon} X_i, (\epsilon \tau_\epsilon)^{-1})$ satisfies the LDP in $\mathbb{R}^d$ with good rate function $y^{-1} J(xy, y)$.
Hint: A convenient way to handle the move from the random variables

$n^{-1} \sum_{i=1}^{n} (X_i, Y_i)$ to $(\epsilon \sum_{i=1}^{\tau_\epsilon} X_i, (\epsilon \tau_\epsilon)^{-1})$ is in looking at small balls in $\mathbb{R}^d$ and applying the characterization of the weak LDP as in Theorem 4.1.11.

Remark: Such transformations appear, for example, in the study of regenerative (or renewal) processes [KucC91, Jia94, PuW97], and of multifractal formalism [Rei95, Zoh96].

Exercise 4.1.32 Suppose the topological space $\mathcal{X}$ has a countable base. Show that for any rate function $I(\cdot)$ such that $\inf_x I(x) = 0$, the LDP with rate function $I(\cdot)$ holds for some family of probability measures $\{\mu_\epsilon\}$ on $\mathcal{X}$.
Hint: For $\mathcal{A}$ a countable base for the topology of $\mathcal{X}$ and each $A \in \mathcal{A}$, let $x_{A,m} \in A$ be such that $I(x_{A,m}) \to \inf_{x \in A} I(x)$ as $m \to \infty$. Let $\mathcal{Y} = \{y_k : k \in \mathbb{Z}_+\}$ denote the countable set $\cup_{A \in \mathcal{A}} \cup_m x_{A,m}$. Check that $\inf_{x \in G} I(x) = \inf_{x \in \mathcal{Y} \cap G} I(x)$ for any open set $G \subset \mathcal{X}$, and try the probability measures μ_ϵ such that $\mu_\epsilon(\{y_k\}) = c_\epsilon^{-1} \exp(-k - I(y_k)/\epsilon)$ for $y_k \in \mathcal{Y}$ and $c_\epsilon = \sum_k \exp(-k - I(y_k)/\epsilon)$.

4.2 Transformations of LDPs

This section is devoted to transformations that preserve the LDP, although, possibly, changing the rate function. Once the LDP with a good rate function is established for μ_ϵ, the basic *contraction principle* yields the LDP for $\mu_\epsilon \circ f^{-1}$, where f is any continuous map. The *inverse contraction principle* deals with f which is the inverse of a continuous bijection, and this is a useful tool for strengthening the topology under which the LDP holds. These two transformations are presented in Section 4.2.1. Section 4.2.2 is devoted to exponentially good approximations and their implications; for example, it is shown that when two families of measures defined on the same probability space are exponentially equivalent, then one can infer the LDP for one family from the other. A direct consequence is Theorem 4.2.23, which extends the contraction principle to "approximately continuous" maps.

4.2.1 Contraction Principles

The LDP is preserved under continuous mappings, as the following elementary theorem shows.

Theorem 4.2.1 (Contraction principle) *Let $\mathcal{X}$ and $\mathcal{Y}$ be Hausdorff topological spaces and $f : \mathcal{X} \to \mathcal{Y}$ a continuous function. Consider a good rate function $I : \mathcal{X} \to [0, \infty]$.*
(a) For each $y \in \mathcal{Y}$, define

$$I'(y) \overset{\triangle}{=} \inf\{I(x) : x \in \mathcal{X}, \quad y = f(x)\}. \tag{4.2.2}$$

Then I' is a good rate function on $\mathcal{Y}$, where as usual the infimum over the empty set is taken as ∞.
(b) If I controls the LDP associated with a family of probability measures $\{\mu_\epsilon\}$ on $\mathcal{X}$, then I' controls the LDP associated with the family of probability measures $\{\mu_\epsilon \circ f^{-1}\}$ on $\mathcal{Y}$.

Proof: (a) Clearly, I' is nonnegative. Since I is a good rate function, for all $y \in f(\mathcal{X})$ the infimum in the definition of I' is obtained at some point of $\mathcal{X}$. Thus, the level sets of I', $\Psi_{I'}(\alpha) \triangleq \{y : I'(y) \le \alpha\}$, are

$$\Psi_{I'}(\alpha) = \{f(x) : I(x) \le \alpha\} = f(\Psi_I(\alpha)),$$

where $\Psi_I(\alpha)$ are the corresponding level sets of I. As $\Psi_I(\alpha) \subset \mathcal{X}$ are compact, so are the sets $\Psi_{I'}(\alpha) \subset \mathcal{Y}$.
(b) The definition of I' implies that for any $A \subset \mathcal{Y}$,

$$\inf_{y \in A} I'(y) = \inf_{x \in f^{-1}(A)} I(x). \tag{4.2.3}$$

Since f is continuous, the set $f^{-1}(A)$ is an open (closed) subset of $\mathcal{X}$ for any open (closed) $A \subset \mathcal{Y}$. Therefore, the LDP for $\mu_\epsilon \circ f^{-1}$ follows as a consequence of the LDP for μ_ϵ and (4.2.3). □

Remarks:
(a) This theorem holds even when $\mathcal{B}_\mathcal{X} \not\subseteq \mathcal{B}$, since for any (measurable) set $A \subset \mathcal{Y}$, both $\overline{f^{-1}(A)} \subset f^{-1}(\overline{A})$ and $f^{-1}(A^o) \subset (f^{-1}(A))^o$.
(b) Note that the upper and lower bounds implied by part (b) of Theorem 4.2.1 hold even when I is not a good rate function. However, if I is *not* a good rate function, it may happen that I' is not a rate function, as the example $\mathcal{X} = \mathcal{Y} = \mathbb{R}$, $I(x) = 0$, and $f(x) = e^x$ demonstrates.
(c) Theorem 4.2.1 holds as long as f is continuous at every $x \in \mathcal{D}_I$; namely, for every $x \in \mathcal{D}_I$ and every neighborhood G of $f(x) \in \mathcal{Y}$, there exists a neighborhood A of x such that $A \subseteq f^{-1}(G)$. This suggests that the contraction principle may be further extended to cover a certain class of "approximately continuous" maps. Such an extension will be pursued in Theorem 4.2.23.

We remind the reader that in what follows, it is always assumed that $\mathcal{B}_\mathcal{X} \subseteq \mathcal{B}$, and therefore open sets are always measurable. The following theorem shows that in the presence of exponential tightness, the contraction principle can be made to work in the reverse direction. This property is extremely useful for strengthening large deviations results from a coarse topology to a finer one.

Theorem 4.2.4 (Inverse contraction principle) *Let $\mathcal{X}$ and $\mathcal{Y}$ be Hausdorff topological spaces. Suppose that $g : \mathcal{Y} \to \mathcal{X}$ is a continuous bijection,*

and that $\{\nu_\epsilon\}$ is an exponentially tight family of probability measures on $\mathcal{Y}$.
If $\{\nu_\epsilon \circ g^{-1}\}$ satisfies the LDP with the rate function $I : \mathcal{X} \to [0, \infty]$, then
$\{\nu_\epsilon\}$ satisfies the LDP with the good rate function $I'(\cdot) \triangleq I(g(\cdot))$.

Remarks:
(a) In view of Lemma 4.1.5, it suffices for g to be a continuous injection, for
then the exponential tightness of $\{\nu_\epsilon\}$ implies that $\mathcal{D}_I \subseteq g(\mathcal{Y})$ even if the
latter is not a closed subset of $\mathcal{X}$.
(b) The requirement that $\mathcal{B}_\mathcal{Y} \subseteq \mathcal{B}$ is relaxed in Exercise 4.2.9.

Proof: Note first that for every $\alpha < \infty$, by the continuity of g, the level
set $\{y : I'(y) \le \alpha\} = g^{-1}(\Psi_I(\alpha))$ is closed. Moreover, $I' \ge 0$, and hence
I' is a rate function. Next, because $\{\nu_\epsilon\}$ is an exponentially tight family,
it suffices to prove a weak LDP with the rate function $I'(\cdot)$. Starting with
the upper bound, fix an arbitrary compact set $K \subset \mathcal{Y}$ and apply the large
deviations upper bound for $\nu_\epsilon \circ g^{-1}$ on the compact set $g(K)$ to obtain

$$\limsup_{\epsilon \to 0} \epsilon \log \nu_\epsilon(K) = \limsup_{\epsilon \to 0} \epsilon \log[\nu_\epsilon \circ g^{-1}(g(K))]$$
$$\le - \inf_{x \in g(K)} I(x) = - \inf_{y \in K} I'(y) ,$$

which is the specified upper bound for ν_ϵ.

To prove the large deviations lower bound, fix $y \in \mathcal{Y}$ with $I'(y) = I(g(y)) = \alpha < \infty$, and a neighborhood G of y. Since $\{\nu_\epsilon\}$ is exponentially
tight, there exists a compact set $K_\alpha \subset \mathcal{Y}$ such that

$$\limsup_{\epsilon \to 0} \epsilon \log \nu_\epsilon(K_\alpha^c) < -\alpha. \qquad (4.2.5)$$

Because g is a bijection, $K_\alpha^c = g^{-1} \circ g(K_\alpha^c)$ and $g(K_\alpha^c) = g(K_\alpha)^c$. By the
continuity of g, the set $g(K_\alpha)$ is compact, and consequently $g(K_\alpha)^c$ is an
open set. Thus, the large deviations lower bound for the measures $\{\nu_\epsilon \circ g^{-1}\}$
results in

$$- \inf_{x \in g(K_\alpha^c)} I(x) \le \liminf_{\epsilon \to 0} \epsilon \log \nu_\epsilon(K_\alpha^c) < -\alpha.$$

Recall that $I(g(y)) = \alpha$, and thus by the preceding inequality it must be
that $y \in K_\alpha$. Since g is a continuous bijection, it is a homeomorphism
between the compact sets K_α and $g(K_\alpha)$. Therefore, the set $g(G \cap K_\alpha)$ is a
neighborhood of $g(y)$ in the induced topology on $g(K_\alpha) \subset \mathcal{X}$. Hence, there
exists a neighborhood G' of $g(y)$ in $\mathcal{X}$ such that

$$G' \subset g(G \cap K_\alpha) \cup g(K_\alpha)^c = g(G \cup K_\alpha^c) ,$$

where the last equality holds because g is a bijection. Consequently, for
every $\epsilon > 0$,

$$\nu_\epsilon(G) + \nu_\epsilon(K_\alpha^c) \ge \nu_\epsilon \circ g^{-1}(G') ,$$

and by the large deviations lower bound for $\{\nu_\epsilon \circ g^{-1}\}$,

$$\max\left\{\liminf_{\epsilon\to 0}\ \epsilon\log\nu_\epsilon(G),\ \limsup_{\epsilon\to 0}\ \epsilon\log\nu_\epsilon(K_\alpha^c)\right\}$$
$$\geq \liminf_{\epsilon\to 0}\ \epsilon\log\{\nu_\epsilon\circ g^{-1}(G')\}$$
$$\geq -I(g(y)) = -I'(y).$$

Since $I'(y) = \alpha$, it follows by combining this inequality and (4.2.5) that

$$\liminf_{\epsilon\to 0}\ \epsilon\log\nu_\epsilon(G) \geq -I'(y).$$

The proof is complete, since the preceding holds for every $y \in \mathcal{Y}$ and every neighborhood G of y. $\square$

Corollary 4.2.6 *Let $\{\mu_\epsilon\}$ be an exponentially tight family of probability measures on $\mathcal{X}$ equipped with the topology τ_1. If $\{\mu_\epsilon\}$ satisfies an LDP with respect to a Hausdorff topology τ_2 on $\mathcal{X}$ that is coarser than τ_1, then the same LDP holds with respect to the topology τ_1.*

Proof: The proof follows from Theorem 4.2.4 by using as g the natural embedding of $(\mathcal{X}, \tau_1)$ onto $(\mathcal{X}, \tau_2)$, which is continuous because τ_1 is finer than τ_2. Note that, since g is continuous, the measures μ_ϵ are well-defined as Borel measures on $(\mathcal{X}, \tau_2)$. $\square$

Exercise 4.2.7 Suppose that $\mathcal{X}$ is a separable regular space, and that for all $\epsilon > 0$, (X_ϵ, Y_ϵ) is distributed according to the product measure $\mu_\epsilon \times \nu_\epsilon$ on $\mathcal{B}_\mathcal{X} \times \mathcal{B}_\mathcal{X}$ (namely, X_ϵ is independent of Y_ϵ). Assume that $\{\mu_\epsilon\}$ satisfies the LDP with the good rate function $I_X(\cdot)$, while ν_ϵ satisfies the LDP with the good rate function $I_Y(\cdot)$, and both $\{\mu_\epsilon\}$ and $\{\nu_\epsilon\}$ are exponentially tight. Prove that for any continuous $F : \mathcal{X} \times \mathcal{X} \to \mathcal{Y}$, the family of laws induced on $\mathcal{Y}$ by $Z_\epsilon = F(X_\epsilon, Y_\epsilon)$ satisfies the LDP with the good rate function

$$I_Z(z) = \inf_{\{(x,y):z=F(x,y)\}} I_X(x) + I_Y(y). \tag{4.2.8}$$

Hint: Recall that $\mathcal{B}_\mathcal{X} \times \mathcal{B}_\mathcal{X} = \mathcal{B}_{\mathcal{X}\times\mathcal{X}}$ by Theorem D.4. To establish the LDP for $\mu_\epsilon \times \nu_\epsilon$, apply Theorems 4.1.11 and 4.1.18.

Exercise 4.2.9 (a) Prove that Theorem 4.2.4 holds even when the exponentially tight $\{\nu_\epsilon : \epsilon > 0\}$ are not Borel measures on $\mathcal{Y}$, provided $\{\nu_\epsilon \circ g^{-1} : \epsilon > 0\}$ are Borel probability measures on $\mathcal{X}$.
(b) Show that in particular, Corollary 4.2.6 holds as soon as $\mathcal{B}$ contains the Borel σ-field of $(\mathcal{X}, \tau_2)$ and all compact subsets of $(\mathcal{X}, \tau_1)$.

4.2.2   Exponential Approximations

In order to extend the contraction principle beyond the continuous case, it
is obvious that one should consider approximations by continuous functions.
It is beneficial to consider a somewhat wider question, namely, when the
LDP for a family of laws $\{\tilde{\mu}_\epsilon\}$ can be deduced from the LDP for a family $\{\mu_\epsilon\}$. The application to approximate contractions follows from these
general results.

Definition 4.2.10 *Let $(\mathcal{Y}, d)$ be a metric space. The probability measures
$\{\mu_\epsilon\}$ and $\{\tilde{\mu}_\epsilon\}$ on $\mathcal{Y}$ are called exponentially equivalent if there exist probability spaces $\{(\Omega, \mathcal{B}_\epsilon, P_\epsilon)\}$ and two families of $\mathcal{Y}$-valued random variables $\{Z_\epsilon\}$
and $\{\tilde{Z}_\epsilon\}$ with joint laws $\{P_\epsilon\}$ and marginals $\{\mu_\epsilon\}$ and $\{\tilde{\mu}_\epsilon\}$, respectively,
such that the following condition is satisfied:*

For each $\delta > 0$, the set $\{\omega : (\tilde{Z}_\epsilon, Z_\epsilon) \in \Gamma_\delta\}$ is $\mathcal{B}_\epsilon$ measurable, and

$$\limsup_{\epsilon \to 0} \epsilon \log P_\epsilon(\Gamma_\delta) = -\infty, \tag{4.2.11}$$

where

$$\Gamma_\delta \overset{\triangle}{=} \{(\tilde{y}, y) : d(\tilde{y}, y) > \delta\} \subset \mathcal{Y} \times \mathcal{Y}. \tag{4.2.12}$$

Remarks:
(a) The random variables $\{Z_\epsilon\}$ and $\{\tilde{Z}_\epsilon\}$ in Definition 4.2.10 are called *exponentially equivalent*.
(b) It is relatively easy to check that the measurability requirement is satisfied whenever $\mathcal{Y}$ is a separable space, or whenever the laws $\{P_\epsilon\}$ are induced
by separable real-valued stochastic processes and d is the supremum norm.

   As far as the LDP is concerned, exponentially equivalent measures are
indistinguishable, as the following theorem attests.

Theorem 4.2.13 *If an LDP with a good rate function $I(\cdot)$ holds for the
probability measures $\{\mu_\epsilon\}$, which are exponentially equivalent to $\{\tilde{\mu}_\epsilon\}$, then
the same LDP holds for $\{\tilde{\mu}_\epsilon\}$.*

Proof: This theorem is a consequence of the forthcoming Theorem 4.2.16.
To avoid repetitions, a direct proof is omitted.   □

   As pointed out in the beginning of this section, an important goal in
considering exponential equivalence is the treatment of approximations. To
this end, the notion of exponential equivalence is replaced by the notion of
exponential approximation, as follows.

Definition 4.2.14 *Let $\mathcal{Y}$ and Γ_δ be as in Definition 4.2.10. For each $\epsilon > 0$ and all $m \in \mathbb{Z}_+$, let $(\Omega, \mathcal{B}_\epsilon, P_{\epsilon,m})$ be a probability space, and let the $\mathcal{Y}$-valued random variables $\tilde{Z}_\epsilon$ and $Z_{\epsilon,m}$ be distributed according to the joint law $P_{\epsilon,m}$, with marginals $\tilde{\mu}_\epsilon$ and $\mu_{\epsilon,m}$, respectively. $\{Z_{\epsilon,m}\}$ are called exponentially good approximations of $\{\tilde{Z}_\epsilon\}$ if, for every $\delta > 0$, the set $\{\omega : (\tilde{Z}_\epsilon, Z_{\epsilon,m}) \in \Gamma_\delta\}$ is $\mathcal{B}_\epsilon$ measurable and*

$$\lim_{m \to \infty} \limsup_{\epsilon \to 0} \epsilon \log P_{\epsilon,m}(\Gamma_\delta) = -\infty . \qquad (4.2.15)$$

Similarly, the measures $\{\mu_{\epsilon,m}\}$ are exponentially good approximations of $\{\tilde{\mu}_\epsilon\}$ if one can construct probability spaces $\{(\Omega, \mathcal{B}_\epsilon, P_{\epsilon,m})\}$ as above.

It should be obvious that Definition 4.2.14 reduces to Definition 4.2.10 if the laws $P_{\epsilon,m}$ do not depend on m.

The main (highly technical) result of this section is the following relation between the LDPs of exponentially good approximations.

Theorem 4.2.16 *Suppose that for every m, the family of measures $\{\mu_{\epsilon,m}\}$ satisfies the LDP with rate function $I_m(\cdot)$ and that $\{\mu_{\epsilon,m}\}$ are exponentially good approximations of $\{\tilde{\mu}_\epsilon\}$. Then*
(a) $\{\tilde{\mu}_\epsilon\}$ satisfies a weak LDP with the rate function

$$I(y) \overset{\triangle}{=} \sup_{\delta > 0} \liminf_{m \to \infty} \inf_{z \in B_{y,\delta}} I_m(z) , \qquad (4.2.17)$$

where $B_{y,\delta}$ denotes the ball $\{z : d(y, z) < \delta\}$.
(b) If $I(\cdot)$ is a good rate function and for every closed set F,

$$\inf_{y \in F} I(y) \leq \limsup_{m \to \infty} \inf_{y \in F} I_m(y) , \qquad (4.2.18)$$

then the full LDP holds for $\{\tilde{\mu}_\epsilon\}$ with rate function I.

Remarks:
(a) The sets Γ_δ may be replaced by sets $\tilde{\Gamma}_{\delta,m}$ such that the sets $\{\omega : (\tilde{Z}_\epsilon, Z_{\epsilon,m}) \in \tilde{\Gamma}_{\delta,m}\}$ differ from $\mathcal{B}_\epsilon$ measurable sets by $P_{\epsilon,m}$ null sets, and $\tilde{\Gamma}_{\delta,m}$ satisfy both (4.2.15) and $\Gamma_\delta \subset \tilde{\Gamma}_{\delta,m}$.
(b) If the rate functions $I_m(\cdot)$ are independent of m, and are good rate functions, then by Theorem 4.2.16, $\{\tilde{\mu}_\epsilon\}$ satisfies the LDP with $I(\cdot) = I_m(\cdot)$. In particular, Theorem 4.2.13 is a direct consequence of Theorem 4.2.16.

Proof: (a) Throughout, let $\{Z_{\epsilon,m}\}$ be the exponentially good approximations of $\{\tilde{Z}_\epsilon\}$, having the joint laws $\{P_{\epsilon,m}\}$ with marginals $\{\mu_{\epsilon,m}\}$ and $\{\tilde{\mu}_\epsilon\}$, respectively, and let Γ_δ be as defined in (4.2.12). The weak LDP is obtained by applying Theorem 4.1.11 for the base $\{B_{y,\delta}\}_{y\in\mathcal{Y},\delta>0}$ of $(\mathcal{Y}, d)$. Specifically, it suffices to show that

$$I(y) = -\inf_{\delta>0} \limsup_{\epsilon\to 0} \epsilon \log \tilde{\mu}_\epsilon(B_{y,\delta}) = -\inf_{\delta>0} \liminf_{\epsilon\to 0} \epsilon \log \tilde{\mu}_\epsilon(B_{y,\delta}). \quad (4.2.19)$$

To this end, fix $\delta > 0$, $y \in \mathcal{Y}$. Note that for every $m \in \mathbb{Z}_+$ and every $\epsilon > 0$,

$$\{Z_{\epsilon,m} \in B_{y,\delta}\} \subseteq \{\tilde{Z}_\epsilon \in B_{y,2\delta}\} \cup \{(\tilde{Z}_\epsilon, Z_{\epsilon,m}) \in \Gamma_\delta\}.$$

Hence, by the union of events bound,

$$\mu_{\epsilon,m}(B_{y,\delta}) \le \tilde{\mu}_\epsilon(B_{y,2\delta}) + P_{\epsilon,m}(\Gamma_\delta).$$

By the large deviations lower bounds for $\{\mu_{\epsilon,m}\}$,

$$\begin{aligned}
- \inf_{z\in B_{y,\delta}} I_m(z) &\le \liminf_{\epsilon\to 0} \epsilon \log \mu_{\epsilon,m}(B_{y,\delta}) \\
&\le \liminf_{\epsilon\to 0} \epsilon \log [\tilde{\mu}_\epsilon(B_{y,2\delta}) + P_{\epsilon,m}(\Gamma_\delta)] \qquad (4.2.20) \\
&\le \liminf_{\epsilon\to 0} \epsilon \log \tilde{\mu}_\epsilon(B_{y,2\delta}) \vee \limsup_{\epsilon\to 0} \epsilon \log P_{\epsilon,m}(\Gamma_\delta).
\end{aligned}$$

Because $\{\mu_{\epsilon,m}\}$ are exponentially good approximations of $\{\tilde{\mu}_\epsilon\}$,

$$\liminf_{\epsilon\to 0} \epsilon \log \tilde{\mu}_\epsilon(B_{y,2\delta}) \ge \limsup_{m\to\infty} \left\{ - \inf_{z\in B_{y,\delta}} I_m(z) \right\}.$$

Repeating the derivation leading to (4.2.20) with the roles of $Z_{\epsilon,m}$ and $\tilde{Z}_\epsilon$ reversed yields

$$\limsup_{\epsilon\to 0} \epsilon \log \tilde{\mu}_\epsilon(B_{y,\delta}) \le \liminf_{m\to\infty} \left\{ - \inf_{z\in \overline{B}_{y,2\delta}} I_m(z) \right\}.$$

Since $\overline{B}_{y,2\delta} \subset B_{y,3\delta}$, (4.2.19) follows by considering the infimum over $\delta > 0$ in the preceding two inequalities (recall the definition (4.2.17) of $I(\cdot)$). Moreover, this argument also implies that

$$I(y) = \sup_{\delta>0} \limsup_{m\to\infty} \inf_{z\in\overline{B}_{y,\delta}} I_m(z) = \sup_{\delta>0} \limsup_{m\to\infty} \inf_{z\in B_{y,\delta}} I_m(z).$$

(b) Fix $\delta > 0$ and a closed set $F \subseteq \mathcal{Y}$. Observe that for $m = 1, 2, \ldots$, and for all $\epsilon > 0$,

$$\{\tilde{Z}_\epsilon \in F\} \subseteq \{Z_{\epsilon,m} \in F^\delta\} \cup \{(\tilde{Z}_\epsilon, Z_{\epsilon,m}) \in \Gamma_\delta\},$$

where $F^\delta = \{z : d(z, F) \le \delta\}$ is the closed blowup of F. Thus, the large deviations upper bounds for $\{\mu_{\epsilon,m}\}$ imply that for every m,

$$\begin{aligned}
\limsup_{\epsilon\to 0} \epsilon \log \tilde{\mu}_\epsilon(F) &\le \limsup_{\epsilon\to 0} \epsilon \log \mu_{\epsilon,m}(F^\delta) \vee \limsup_{\epsilon\to 0} \epsilon \log P_{\epsilon,m}(\Gamma_\delta) \\
&\le [- \inf_{y\in F^\delta} I_m(y)] \vee \limsup_{\epsilon\to 0} \epsilon \log P_{\epsilon,m}(\Gamma_\delta).
\end{aligned}$$

Hence, as $\{Z_{\epsilon,m}\}$ are exponentially good approximations of $\{\tilde{Z}_\epsilon\}$, considering $m \to \infty$, it follows that

$$\limsup_{\epsilon \to 0} \epsilon \log \tilde{\mu}_\epsilon(F) \le -\limsup_{m \to \infty} \inf_{y \in F^\delta} I_m(y) \le -\inf_{y \in F^\delta} I(y) ,$$

where the second inequality is just our condition (4.2.18) for the closed set F^δ. Taking $\delta \to 0$, Lemma 4.1.6 yields the large deviations upper bound and completes the proof of the full LDP .   □

It should be obvious that the results on exponential approximations imply results on approximate contractions. We now present two such results. The first is related to Theorem 4.2.13 and considers approximations that are ϵ dependent. The second allows one to consider approximations that depend on an auxiliary parameter.

Corollary 4.2.21 *Suppose $f : \mathcal{X} \to \mathcal{Y}$ is a continuous map from a Hausdorff topological space $\mathcal{X}$ to the metric space $(\mathcal{Y}, d)$ and that $\{\mu_\epsilon\}$ satisfy the LDP with the good rate function $I : \mathcal{X} \to [0, \infty]$. Suppose further that for all $\epsilon > 0$, $f_\epsilon : \mathcal{X} \to \mathcal{Y}$ are measurable maps such that for all $\delta > 0$, the set $\Gamma_{\epsilon,\delta} \triangleq \{x \in \mathcal{X} : d(f(x), f_\epsilon(x)) > \delta\}$ is measurable, and*

$$\limsup_{\epsilon \to 0} \epsilon \log \mu_\epsilon(\Gamma_{\epsilon,\delta}) = -\infty . \tag{4.2.22}$$

Then the LDP with the good rate function $I'(\cdot)$ of (4.2.2) holds for the measures $\mu_\epsilon \circ f_\epsilon^{-1}$ on $\mathcal{Y}$.

Proof: The contraction principle (Theorem 4.2.1) yields the desired LDP for $\{\mu_\epsilon \circ f^{-1}\}$. By (4.2.22), these measures are exponentially equivalent to $\{\mu_\epsilon \circ f_\epsilon^{-1}\}$, and the corollary follows from Theorem 4.2.13.   □

A special case of Theorem 4.2.16 is the following extension of the contraction principle to maps that are not continuous, but that can be approximated well by continuous maps.

Theorem 4.2.23 *Let $\{\mu_\epsilon\}$ be a family of probability measures that satisfies the LDP with a good rate function I on a Hausdorff topological space $\mathcal{X}$, and for $m = 1, 2, \ldots$, let $f_m : \mathcal{X} \to \mathcal{Y}$ be continuous functions, with $(\mathcal{Y}, d)$ a metric space. Assume there exists a measurable map $f : \mathcal{X} \to \mathcal{Y}$ such that for every $\alpha < \infty$,*

$$\limsup_{m \to \infty} \sup_{\{x : I(x) \le \alpha\}} d(f_m(x), f(x)) = 0 . \tag{4.2.24}$$

Then any family of probability measures $\{\tilde{\mu}_\epsilon\}$ for which $\{\mu_\epsilon \circ f_m^{-1}\}$ are exponentially good approximations satisfies the LDP in $\mathcal{Y}$ with the good rate function $I'(y) = \inf\{I(x) : y = f(x)\}$.

Remarks:

(a) The condition (4.2.24) implies that for every $\alpha < \infty$, the function f is continuous on the level set $\Psi_I(\alpha) = \{x : I(x) \leq \alpha\}$. Suppose that in addition,

$$\lim_{m\to\infty} \inf_{x\in\Psi_I(m)^c} I(x) = \infty . \qquad (4.2.25)$$

Then the LDP for $\mu_\epsilon \circ f^{-1}$ follows as a direct consequence of Theorem 4.2.23 by considering a sequence f_m of continuous extensions of f from $\Psi_I(m)$ to $\mathcal{X}$. (Such a sequence exists whenever $\mathcal{X}$ is a completely regular space.) That (4.2.25) need not hold true, even when $\mathcal{X} = \mathbb{R}$, may be seen by considering the following example. It is easy to check that $\mu_\epsilon = (\delta_{\{0\}} + \delta_{\{\epsilon\}})/2$ satisfies the LDP on $\mathbb{R}$ with the good rate function $I(0) = 0$ and $I(x) = \infty, x \neq 0$. On the other hand, the closure of the complement of any level set is the whole real line. If one now considers the function $f : \mathbb{R} \to \mathbb{R}$ such that $f(0) = 0$ and $f(x) = 1, x \neq 0$, then $\mu_\epsilon \circ f^{-1}$ does not satisfy the LDP with the rate function $I'(y) = \inf\{I(x) : x \in \mathbb{R}, y = f(x)\}$, i.e., $I'(0) = 0$ and $I'(y) = \infty, y \neq 0$.

(b) Suppose for each $m \in \mathbb{Z}_+$, the family of measures $\{\mu_{\epsilon,m}\}$ satisfies the LDP on $\mathcal{Y}$ with the good rate function $I_m(\cdot)$ of (4.2.26), where the continuous functions $f_m : \mathcal{D}_I \to \mathcal{Y}$ and the measurable function $f : \mathcal{D}_I \to \mathcal{Y}$ satisfy condition (4.2.24). Then any $\{\tilde{\mu}_\epsilon\}$ for which $\{\mu_{\epsilon,m}\}$ are exponentially good approximations satisfies the LDP in $\mathcal{Y}$ with good rate function $I'(\cdot)$. This easy adaptation of the proof of Theorem 4.2.23 is left for the reader.

Proof: By assumption, the functions $f_m : \mathcal{X} \to \mathcal{Y}$ are continuous. Hence, by the contraction principle (Theorem 4.2.1), for each $m \in \mathbb{Z}_+$, the family of measures $\{\mu_\epsilon \circ f_m^{-1}\}$ satisfies the LDP on $\mathcal{Y}$ with the good rate function

$$I_m(y) = \inf\{I(x) : x \in \mathcal{X}, \quad y = f_m(x)\} . \qquad (4.2.26)$$

Recall that the condition (4.2.24) implies that f is continuous on each level set $\Psi_I(\alpha)$. Therefore, I' is a good rate function on $\mathcal{Y}$ with level sets $f(\Psi_I(\alpha))$ (while the corresponding level set of I_m is $f_m(\Psi_I(\alpha))$).

Fix a closed set F and for any $m \in \mathbb{Z}_+$, let

$$\gamma_m \overset{\triangle}{=} \inf_{y\in F} I_m(y) = \inf_{x\in f_m^{-1}(F)} I(x) .$$

Assume first that $\gamma \overset{\triangle}{=} \liminf_{m\to\infty} \gamma_m < \infty$, and pass to a subsequence of m's such that $\gamma_m \to \gamma$ and $\sup_m \gamma_m = \alpha < \infty$. Since $I(\cdot)$ is a good rate function and $f_m^{-1}(F)$ are closed sets, there exist $x_m \in \mathcal{X}$ such that $f_m(x_m) \in F$ and $I(x_m) = \gamma_m \leq \alpha$. Now, the uniform convergence assumption of (4.2.24) implies that $f(x_m) \in F^\delta$ for every $\delta > 0$ and all m large enough. Therefore, $\inf_{y\in F^\delta} I'(y) \leq I'(f(x_m)) \leq I(x_m) = \gamma_m$ for all m large enough. Hence, for all $\delta > 0$,

$$\inf_{y \in F^\delta} I'(y) \leq \liminf_{m \to \infty} \inf_{y \in F} I_m(y).$$

(Note that this inequality trivially holds when $\gamma = \infty$.) Taking $\delta \to 0$, it follows from Lemma 4.1.6 that for every closed set F,

$$\inf_{y \in F} I'(y) \leq \liminf_{m \to \infty} \inf_{y \in F} I_m(y). \qquad (4.2.27)$$

In particular, this inequality implies that (4.2.18) holds for the good rate function $I'(\cdot)$. Moreover, considering $F = \overline{B}_{y,\delta}$, and taking $\delta \to 0$, it follows from Lemma 4.1.6 that

$$I'(y) = \sup_{\delta > 0} \inf_{z \in \overline{B}_{y,\delta}} I'(z) \leq \sup_{\delta > 0} \liminf_{m \to \infty} \inf_{z \in B_{y,\delta}} I_m(z) \overset{\triangle}{=} \bar{I}(y).$$

Note that $\bar{I}(\cdot)$ is the rate function defined in Theorem 4.2.16, and consequently the proof is complete as soon as we show that $\bar{I}(y) \leq I'(y)$ for all $y \in \mathcal{Y}$. To this end, assume with no loss of generality that $I'(y) = \alpha < \infty$. Then, $y \in f(\Psi_I(\alpha))$, i.e., there exists $x \in \Psi_I(\alpha)$ such that $f(x) = y$. Note that $y_m = f_m(x) \in f_m(\Psi_I(\alpha))$, and consequently $I_m(y_m) \leq \alpha$ for all $m \in \mathbb{Z}_+$. The condition (4.2.24) then implies that $d(y, y_m) \to 0$, and hence $\bar{I}(y) \leq \liminf_{m \to \infty} I_m(y_m) \leq \alpha$, as required. $\square$

Exercise 4.2.28 [Based on [DV75a]] Let $\Sigma = \{1, \cdots, r\}$, and let Y_t be a Σ-valued *continuous time* Markov process with irreducible generator $A = \{a(i,j)\}$. In this exercise, you derive the LDP for the empirical measures

$$L_\epsilon^y(i) = \epsilon \int_0^{1/\epsilon} 1_i(Y_t) dt, \quad i = 1, \ldots, r.$$

(a) Define

$$L_{\epsilon,m}^y(i) = \frac{\epsilon}{m} \sum_{j=1}^{\lfloor \frac{m}{\epsilon} \rfloor} 1_i(Y_{\frac{j}{m}}), \quad i = 1, \ldots, r.$$

Show that $\{L_{\epsilon,m}^y\}$ are exponentially good approximations of $\{L_\epsilon^y\}$.
Hint: Note that

$$|L_\epsilon^y(i) - L_{\epsilon,m}^y(i)| \leq \frac{\epsilon}{m} \left\{ \begin{array}{c} \text{total number of jumps in} \\ \text{the path } Y_t, \ t \in [0, 1/\epsilon] \end{array} \right\} \overset{\triangle}{=} \frac{\epsilon}{m} N_\epsilon,$$

and N_ϵ is stochastically dominated by a Poisson(c/ϵ) random variable for some constant $c < \infty$.
(b) Note that $L_{\epsilon,m}^y$ is the empirical measure of a Σ-valued, discrete time Markov process with irreducible transition probability matrix $e^{A/m}$. Using Theorem

3.1.6 and Exercise 3.1.11, show that for every m, $L^{\mathcal{Y}}_{\epsilon,m}$ satisfies the LDP with the good rate function

$$I_m(q) = m \sup_{u \gg 0} \sum_{j=1}^{r} q_j \log \left[\frac{u_j}{(e^{A/m} u)_j} \right] ,$$

where $q \in M_1(\Sigma)$.

(c) Applying Theorem 4.2.16, prove that $\{L^{\mathcal{Y}}_\epsilon\}$ satisfies the LDP with the good rate function

$$I(q) = \sup_{u \gg 0} \left\{ - \sum_{j=1}^{r} q_j \frac{(Au)_j}{u_j} \right\} .$$

Hint: Check that for all $q \in M_1(\Sigma)$, $I(q) \geq I_m(q)$, and that for each fixed $u \gg 0$, $I_m(q) \geq - \sum_j q_j \frac{(Au)_j}{u_j} - \frac{c(u)}{m}$ for some $c(u) < \infty$.

(d) Assume that A is symmetric and check that then

$$I(q) = - \sum_{i,j=1}^{r} \sqrt{q_i}\, a(i,j) \sqrt{q_j} .$$

Exercise 4.2.29 Suppose that for every m, the family of measures $\{\mu_{\epsilon,m}\}$ satisfies the LDP with good rate function $I_m(\cdot)$ and that $\{\mu_{\epsilon,m}\}$ are exponentially good approximations of $\{\tilde{\mu}_\epsilon\}$.

(a) Show that if $(\mathcal{Y}, d)$ is a Polish space, then $\{\tilde{\mu}_{\epsilon_n}\}$ is exponentially tight for any $\epsilon_n \to 0$. Hence, by part (a) of Theorem 4.2.16, $\{\tilde{\mu}_\epsilon\}$ satisfies the LDP with the good rate function $I(\cdot)$ of (4.2.17).

Hint: See Exercise 4.1.10.

(b) Let $\mathcal{Y} = \{1/m, m \in \mathbb{Z}_+\}$ with the metric $d(\cdot, \cdot)$ induced on $\mathcal{Y}$ by $\mathbb{R}$ and $\mathcal{Y}$-valued random variables Y_m such that $P(Y_m = 1$ for every $m) = 1/2$, and $P(Y_m = 1/m$ for every $m) = 1/2$. Check that $Z_{\epsilon,m} \triangleq Y_m$ are exponentially good approximations of $\tilde{Z}_\epsilon \triangleq Y_{[1/\epsilon]}$ ($\epsilon \leq 1$), which for any fixed $m \in \mathbb{Z}_+$ satisfy the LDP in $\mathcal{Y}$ with the good rate function $I_m(y) = 0$ for $y = 1$, $y = 1/m$, and $I_m(y) = \infty$ otherwise. Check that in this case, the good rate function $I(\cdot)$ of (4.2.17) is such that $I(y) = \infty$ for every $y \neq 1$ and in particular, the large deviations upper bound fails for $\{\tilde{Z}_\epsilon \neq 1\}$ and this rate function.

Remark: This example shows that when $(\mathcal{Y}, d)$ is not a Polish space one can not dispense of condition (4.2.18) in Theorem 4.2.16.

Exercise 4.2.30 For any $\delta > 0$ and probability measures ν, μ on the metric space $(\mathcal{Y}, d)$ let

$$\rho_\delta(\nu, \mu) \triangleq \sup\{\nu(A) - \mu(A^\delta) : A \in \mathcal{B}_{\mathcal{Y}} \} .$$

(a) Show that if $\{\mu_{\epsilon,m}\}$ are exponentially good approximations of $\{\tilde{\mu}_\epsilon\}$ then

$$\lim_{m \to \infty} \limsup_{\epsilon \to 0} \epsilon \log \rho_\delta(\mu_{\epsilon,m}, \tilde{\mu}_\epsilon) = -\infty . \tag{4.2.31}$$

(b) Show that if $(\mathcal{Y}, d)$ is a Polish space and (4.2.31) holds for any $\delta > 0$, then $\{\mu_{\epsilon,m}\}$ are exponentially good approximations of $\{\tilde{\mu}_\epsilon\}$.

Hint: Recall the following consequence of [Str65, Theorem 11]. For any open set $\Gamma \subset \mathcal{Y}^2$ and any Borel probability measures ν, μ on the Polish space $(\mathcal{Y}, d)$ there exists a Borel probability measure P on $\mathcal{Y}^2$ with marginals μ, ν such that

$$P(\Gamma) = \sup\{\nu(G) - \mu(\{\tilde{y} : \exists y \in G, \text{ such that } (\tilde{y}, y) \in \Gamma^c\}) : G \subset \mathcal{Y} \text{ open}\}.$$

Conclude that $P_{\epsilon,m}(\Gamma_{\delta'}) \leq \rho_\delta(\mu_{\epsilon,m}, \tilde{\mu}_\epsilon)$ for any $m, \epsilon > 0$, and $\delta' > \delta > 0$.

Exercise 4.2.32 Prove Theorem 4.2.13, assuming that $\{\mu_\epsilon\}$ are Borel probability measures, but $\{\tilde{\mu}_\epsilon\}$ are not necessarily such.

4.3 Varadhan's Integral Lemma

Throughout this section, $\{Z_\epsilon\}$ is a family of random variables taking values in the regular topological space $\mathcal{X}$, and $\{\mu_\epsilon\}$ denotes the probability measures associated with $\{Z_\epsilon\}$. The next theorem could actually be used as a starting point for developing the large deviations paradigm. It is a very useful tool in many applications of large deviations. For example, the asymptotics of the partition function in statistical mechanics can be derived using this theorem.

Theorem 4.3.1 (Varadhan) *Suppose that* $\{\mu_\epsilon\}$ *satisfies the LDP with a good rate function* $I : \mathcal{X} \to [0, \infty]$, *and let* $\phi : \mathcal{X} \to \mathbb{R}$ *be any continuous function. Assume further either the tail condition*

$$\lim_{M \to \infty} \limsup_{\epsilon \to 0} \epsilon \log E\left[e^{\phi(Z_\epsilon)/\epsilon} 1_{\{\phi(Z_\epsilon) \geq M\}}\right] = -\infty, \qquad (4.3.2)$$

or the following moment condition for some $\gamma > 1$,

$$\limsup_{\epsilon \to 0} \epsilon \log E\left[e^{\gamma\phi(Z_\epsilon)/\epsilon}\right] < \infty. \qquad (4.3.3)$$

Then

$$\lim_{\epsilon \to 0} \epsilon \log E\left[e^{\phi(Z_\epsilon)/\epsilon}\right] = \sup_{x \in \mathcal{X}} \{\phi(x) - I(x)\}.$$

Remark: This theorem is the natural extension of Laplace's method to infinite dimensional spaces. Indeed, let $\mathcal{X} = \mathbb{R}$ and assume for the moment that the density of μ_ϵ with respect to Lebesgue's measure is such that $d\mu_\epsilon/dx \approx e^{-I(x)/\epsilon}$. Then

$$\int_{\mathbb{R}} e^{\phi(x)/\epsilon} \mu_\epsilon(dx) \approx \int_{\mathbb{R}} e^{(\phi(x) - I(x))/\epsilon} dx.$$

Assume that $I(\cdot)$ and $\phi(\cdot)$ are twice differentiable, with $(\phi(x)-I(x))$ concave and possessing a unique global maximum at some $\bar{x}$. Then

$$\phi(x) - I(x) = \phi(\bar{x}) - I(\bar{x}) + \frac{(x - \bar{x})^2}{2}(\phi(x) - I(x))''|_{x=\xi} \, ,$$

where $\xi \in [\bar{x}, x]$. Therefore,

$$\int_{\mathbb{R}} e^{\phi(x)/\epsilon} \mu_\epsilon(dx) \approx e^{(\phi(\bar{x})-I(\bar{x}))/\epsilon} \int_{\mathbb{R}} e^{-B(x)(x-\bar{x})^2/2\epsilon} dx \, ,$$

where $B(\cdot) \geq 0$. The content of Laplace's method (and of Theorem 4.3.1) is that on a logarithmic scale the rightmost integral may be ignored.

Theorem 4.3.1 is a direct consequence of the following three lemmas.

Lemma 4.3.4 *If $\phi : \mathcal{X} \to \mathbb{R}$ is lower semicontinuous and the large deviations lower bound holds with $I : \mathcal{X} \to [0, \infty]$, then*

$$\liminf_{\epsilon \to 0} \epsilon \log E\left[e^{\phi(Z_\epsilon)/\epsilon}\right] \geq \sup_{x \in \mathcal{X}} \{\phi(x) - I(x)\} \, . \qquad (4.3.5)$$

Lemma 4.3.6 *If $\phi : \mathcal{X} \to \mathbb{R}$ is an upper semicontinuous function for which the tail condition (4.3.2) holds, and the large deviations upper bound holds with the good rate function $I : \mathcal{X} \to [0, \infty]$, then*

$$\limsup_{\epsilon \to 0} \epsilon \log E\left[e^{\phi(Z_\epsilon)/\epsilon}\right] \leq \sup_{x \in \mathcal{X}} \{\phi(x) - I(x)\} \, . \qquad (4.3.7)$$

Lemma 4.3.8 *Condition (4.3.3) implies the tail condition (4.3.2).*

Proof of Lemma 4.3.4: Fix $x \in \mathcal{X}$ and $\delta > 0$. Since $\phi(\cdot)$ is lower semicontinuous, it follows that there exists a neighborhood G of x such that $\inf_{y \in G} \phi(y) \geq \phi(x) - \delta$. Hence,

$$\begin{aligned}
\liminf_{\epsilon \to 0} \epsilon \log E\left[e^{\phi(Z_\epsilon)/\epsilon}\right] &\geq \liminf_{\epsilon \to 0} \epsilon \log E\left[e^{\phi(Z_\epsilon)/\epsilon} 1_{\{Z_\epsilon \in G\}}\right] \\
&\geq \inf_{y \in G} \phi(y) + \liminf_{\epsilon \to 0} \epsilon \log \mu_\epsilon(G) \, .
\end{aligned}$$

By the large deviations lower bound and the choice of G,

$$\inf_{y \in G} \phi(y) + \liminf_{\epsilon \to 0} \epsilon \log \mu_\epsilon(G) \geq \inf_{y \in G} \phi(y) - \inf_{y \in G} I(y) \geq \phi(x) - I(x) - \delta \, .$$

The inequality (4.3.5) now follows, since $\delta > 0$ and $x \in \mathcal{X}$ are arbitrary. $\square$

Proof of Lemma 4.3.6: Consider first a function ϕ bounded above, i.e., $\sup_{x \in \mathcal{X}} \phi(x) \leq M < \infty$. For such functions, the tail condition (4.3.2) holds trivially. Fix $\alpha < \infty$ and $\delta > 0$, and let $\Psi_I(\alpha) = \{x : I(x) \leq \alpha\}$ denote the compact level set of the good rate function I. Since $I(\cdot)$ is lower semicontinuous, $\phi(\cdot)$ is upper semicontinuous, and $\mathcal{X}$ is a regular topological space, for every $x \in \Psi_I(\alpha)$, there exists a neighborhood A_x of x such that

$$\inf_{y \in \overline{A_x}} I(y) \geq I(x) - \delta \quad , \quad \sup_{y \in \overline{A_x}} \phi(y) \leq \phi(x) + \delta . \qquad (4.3.9)$$

From the open cover $\cup_{x \in \Psi_I(\alpha)} A_x$ of the compact set $\Psi_I(\alpha)$, one can extract a finite cover of $\Psi_I(\alpha)$, e.g., $\cup_{i=1}^N A_{x_i}$. Therefore,

$$
\begin{aligned}
E\left[e^{\phi(Z_\epsilon)/\epsilon}\right] &\leq \sum_{i=1}^N E\left[e^{\phi(Z_\epsilon)/\epsilon} 1_{\{Z_\epsilon \in A_{x_i}\}}\right] + e^{M/\epsilon} \mu_\epsilon\left(\left(\bigcup_{i=1}^N A_{x_i}\right)^c\right) \\
&\leq \sum_{i=1}^N e^{(\phi(x_i)+\delta)/\epsilon} \mu_\epsilon(\overline{A_{x_i}}) + e^{M/\epsilon} \mu_\epsilon\left(\left(\bigcup_{i=1}^N A_{x_i}\right)^c\right)
\end{aligned}
$$

where the last inequality follows by (4.3.9). Applying the large deviations upper bound to the sets $\overline{A_{x_i}}$, $i = 1, \ldots, N$ and $(\cup_{i=1}^N A_{x_i})^c \subseteq \Psi_I(\alpha)^c$, one obtains (again, in view of (4.3.9)),

$$
\begin{aligned}
\limsup_{\epsilon \to 0} \epsilon \log E &\left[e^{\phi(Z_\epsilon)/\epsilon}\right] \\
&\leq \max\left\{\max_{i=1}^N\{\phi(x_i) + \delta - \inf_{y \in \overline{A_{x_i}}} I(y)\}, M - \inf_{y \in (\cup_{i=1}^N A_{x_i})^c} I(y)\right\} \\
&\leq \max\left\{\max_{i=1}^N\{\phi(x_i) - I(x_i) + 2\delta\}, M - \alpha\right\} \\
&\leq \max\left\{\sup_{x \in \mathcal{X}}\{\phi(x) - I(x)\}, M - \alpha\right\} + 2\delta .
\end{aligned}
$$

Thus, for any $\phi(\cdot)$ bounded above, the lemma follows by taking the limits $\delta \to 0$ and $\alpha \to \infty$.

To treat the general case, set $\phi_M(x) = \phi(x) \wedge M \leq \phi(x)$, and use the preceding to show that for every $M < \infty$,

$$
\begin{aligned}
\limsup_{\epsilon \to 0} \epsilon \log E &\left[e^{\phi(Z_\epsilon)/\epsilon}\right] \\
&\leq \sup_{x \in \mathcal{X}}\{\phi(x) - I(x)\} \vee \limsup_{\epsilon \to 0} \epsilon \log E \left[e^{\phi(Z_\epsilon)/\epsilon} 1_{\{\phi(Z_\epsilon) \geq M\}}\right] .
\end{aligned}
$$

The tail condition (4.3.2) completes the proof of the lemma by taking the limit $M \to \infty$. $\qquad \square$

Proof of Lemma 4.3.8: For $\epsilon > 0$, define $X_\epsilon \triangleq \exp((\phi(Z_\epsilon) - M)/\epsilon)$, and let $\gamma > 1$ be the constant given in the moment condition (4.3.3). Then

$$
\begin{aligned}
e^{-M/\epsilon} E\left[e^{\phi(Z_\epsilon)/\epsilon} 1_{\{\phi(Z_\epsilon)\geq M\}}\right] &= E\left[X_\epsilon 1_{\{X_\epsilon\geq 1\}}\right] \\
&\leq E\left[(X_\epsilon)^\gamma\right] = e^{-\gamma M/\epsilon} E\left[e^{\gamma\phi(Z_\epsilon)/\epsilon}\right].
\end{aligned}
$$

Therefore,

$$
\limsup_{\epsilon\to 0} \epsilon \log E\left[e^{\phi(Z_\epsilon)/\epsilon} 1_{\{\phi(Z_\epsilon)\geq M\}}\right]
$$
$$
\leq -(\gamma-1)M + \limsup_{\epsilon\to 0} \epsilon \log E\left[e^{\gamma\phi(Z_\epsilon)/\epsilon}\right].
$$

The right side of this inequality is finite by the moment condition (4.3.3). In the limit $M \to \infty$, it yields the tail condition (4.3.2). $\qquad\square$

Exercise 4.3.10 Let $\phi : \mathcal{X} \to [-\infty, \infty]$ be an upper semicontinuous function, and let $I(\cdot)$ be a good rate function. Prove that in any closed set $F \subset \mathcal{X}$ on which ϕ is bounded above, there exists a point x_0 such that

$$
\phi(x_0) - I(x_0) = \sup_{x\in F} \{\phi(x) - I(x)\}.
$$

Exercise 4.3.11 [From [DeuS89b], Exercise 2.1.24]. Assume that $\{\mu_\epsilon\}$ satisfies the LDP with good rate function $I(\cdot)$ and that the tail condition (4.3.2) holds for the continuous function $\phi : \mathcal{X} \to \mathbb{R}$. Show that

$$
\liminf_{\epsilon\to 0} \epsilon \log\left(\int_G e^{\phi(x)/\epsilon} d\mu_\epsilon\right) \geq \sup_{x\in G} \{\phi(x) - I(x)\}, \quad \forall G \quad \text{open},
$$

$$
\limsup_{\epsilon\to 0} \epsilon \log\left(\int_F e^{\phi(x)/\epsilon} d\mu_\epsilon\right) \leq \sup_{x\in F} \{\phi(x) - I(x)\}, \quad \forall F \quad \text{closed}.
$$

Exercise 4.3.12 The purpose of this exercise is to demonstrate that some tail condition like (4.3.2) is necessary for Lemma 4.3.6 to hold. In particular, this lemma may not hold for linear functions.

Consider a family of real valued random variables $\{Z_\epsilon\}$, where $P(Z_\epsilon = 0) = 1 - 2p_\epsilon$, $P(Z_\epsilon = -m_\epsilon) = p_\epsilon$, and $P(Z_\epsilon = m_\epsilon) = p_\epsilon$.
(a) Prove that if

$$
\lim_{\epsilon\to 0} \epsilon \log p_\epsilon = -\infty,
$$

then the laws of $\{Z_\epsilon\}$ are exponentially tight, and moreover they satisfy the LDP with the convex, good rate function

$$
I(x) = \begin{cases} 0 & x = 0 \\ \infty & \text{otherwise}. \end{cases}
$$

(b) Let $m_\epsilon = -\epsilon \log p_\epsilon$ and define

$$\Lambda(\lambda) = \lim_{\epsilon \to 0} \epsilon \log E\left(e^{\lambda Z_\epsilon / \epsilon}\right).$$

Prove that

$$\Lambda(\lambda) = \begin{cases} 0 & |\lambda| \leq 1 \\ \infty & \text{otherwise}, \end{cases}$$

and its Fenchel–Legendre transform is $\Lambda^*(x) = |x|$.

(c) Observe that $\Lambda(\lambda) \neq \sup_{x \in \mathbb{R}} \{\lambda x - I(x)\}$, and $\Lambda^*(\cdot) \neq I(\cdot)$.

4.4 Bryc's Inverse Varadhan Lemma

As will be seen in Section 4.5, in the setting of topological vector spaces, linear functionals play an important role in establishing the LDP, particularly when convexity is involved. Note, however, that Varadhan's lemma applies to nonlinear functions as well. It is the goal of this section to derive the *inverse* of Varadhan's lemma. Specifically, let $\{\mu_\epsilon\}$ be a family of probability measures on a topological space $\mathcal{X}$. For each Borel measurable function $f : \mathcal{X} \to \mathbb{R}$, define

$$\Lambda_f \stackrel{\triangle}{=} \lim_{\epsilon \to 0} \epsilon \log \int_{\mathcal{X}} e^{f(x)/\epsilon} \mu_\epsilon(dx), \qquad (4.4.1)$$

provided the limit exists. For example, when $\mathcal{X}$ is a vector space, then the $\{\Lambda_f\}$ for continuous linear functionals (i.e., for $f \in \mathcal{X}^*$) are just the values of the logarithmic moment generating function defined in Section 4.5. The main result of this section is that the LDP is a consequence of exponential tightness and the existence of the limits (4.4.1) for every $f \in \mathcal{G}$, for appropriate families of functions $\mathcal{G}$. This result is used in Section 6.4, where the smoothness assumptions of the Gärtner–Ellis theorem (Theorem 2.3.6) are replaced by mixing assumptions en route to the LDP for the empirical measures of Markov chains.

Throughout this section, it is assumed that $\mathcal{X}$ is a completely regular topological space, i.e., $\mathcal{X}$ is Hausdorff, and for any closed set $F \subset \mathcal{X}$ and any point $x \notin F$, there exists a continuous function $f : \mathcal{X} \to [0, 1]$ such that $f(x) = 1$ and $f(y) = 0$ for all $y \in F$. It is also not hard to verify that such a space is regular and that both metric spaces and Hausdorff topological vector spaces are completely regular.

The class of all bounded, real valued continuous functions on $\mathcal{X}$ is denoted throughout by $C_b(\mathcal{X})$.

Theorem 4.4.2 (Bryc) *Suppose that the family $\{\mu_\epsilon\}$ is exponentially tight and that the limit Λ_f in (4.4.1) exists for every $f \in C_b(\mathcal{X})$. Then $\{\mu_\epsilon\}$ satisfies the LDP with the good rate function*

$$I(x) = \sup_{f \in C_b(\mathcal{X})} \{f(x) - \Lambda_f\} . \tag{4.4.3}$$

Furthermore, for every $f \in C_b(\mathcal{X})$,

$$\Lambda_f = \sup_{x \subset \mathcal{X}} \{f(x) - I(x)\} . \tag{4.4.4}$$

Remark: In the case where $\mathcal{X}$ is a topological vector space, it is tempting to compare (4.4.3) and (4.4.4) with the Fenchel–Legendre transform pair $\Lambda(\cdot)$ and $\Lambda^*(\cdot)$ of Section 4.5. Note, however, that here the rate function $I(x)$ need not be convex.

Proof: Since $\Lambda_0 = 0$, it follows that $I(\cdot) \geq 0$. Moreover, $I(x)$ is lower semicontinuous, since it is the supremum of continuous functions. Due to the exponential tightness of $\{\mu_\epsilon\}$, the LDP asserted follows once the weak LDP (with rate function $I(\cdot)$) is proved. Moreover, by an application of Varadhan's lemma (Theorem 4.3.1), the identity (4.4.4) then holds. It remains, therefore, only to prove the weak LDP, which is a consequence of the following two lemmas.

Lemma 4.4.5 (Upper bound) *If Λ_f exists for each $f \in C_b(\mathcal{X})$, then, for every compact $\Gamma \subset \mathcal{X}$,*

$$\limsup_{\epsilon \to 0} \epsilon \log \mu_\epsilon(\Gamma) \leq -\inf_{x \in \Gamma} I(x) .$$

Lemma 4.4.6 (Lower bound) *If Λ_f exists for each $f \in C_b(\mathcal{X})$, then, for every open $G \subset \mathcal{X}$ and each $x \in G$,*

$$\liminf_{\epsilon \to 0} \epsilon \log \mu_\epsilon(G) \geq -I(x) .$$

Proof of Lemma 4.4.5: The proof is almost identical to the proof of part (b) of Theorem 4.5.3, substituting $f(x)$ for $\langle \lambda, x \rangle$. To avoid repetition, the details are omitted. $\square$

Proof of Lemma 4.4.6: Fix $x \in \mathcal{X}$ and a neighborhood G of x. Since $\mathcal{X}$ is a completely regular topological space, there exists a continuous function $f : \mathcal{X} \to [0,1]$, such that $f(x) = 1$ and $f(y) = 0$ for all $y \in G^c$. For $m > 0$, define $f_m(\cdot) \stackrel{\triangle}{=} m(f(\cdot) - 1)$. Then

$$\int_{\mathcal{X}} e^{f_m(x)/\epsilon} \mu_\epsilon(dx) \leq e^{-m/\epsilon} \mu_\epsilon(G^c) + \mu_\epsilon(G) \leq e^{-m/\epsilon} + \mu_\epsilon(G) .$$

Since $f_m \in C_b(\mathcal{X})$ and $f_m(x) = 0$, it now follows that

$$\max\{ \liminf_{\epsilon \to 0} \epsilon \log \mu_\epsilon(G), \, -m \}$$

$$\geq \liminf_{\epsilon \to 0} \epsilon \log \int_\mathcal{X} e^{f_m(x)/\epsilon} \mu_\epsilon(dx) = \Lambda_{f_m}$$

$$= -[f_m(x) - \Lambda_{f_m}] \geq - \sup_{f \in C_b(\mathcal{X})} \{f(x) - \Lambda_f\} = -I(x) \,,$$

and the lower bound follows by letting $m \to \infty$. $\square$

This proof works because indicators on open sets are approximated well enough by bounded continuous functions. It is clear, however, that not all of $C_b(\mathcal{X})$ is needed for that purpose. The following definition is the tool for relaxing the assumptions of Theorem 4.4.2.

Definition 4.4.7 *A class $\mathcal{G}$ of continuous, real valued functions on a topological space $\mathcal{X}$ is said to be well-separating if:*
(1) $\mathcal{G}$ contains the constant functions.
(2) $\mathcal{G}$ is closed under finite pointwise minima, i.e., $g_1, g_2 \in \mathcal{G} \Rightarrow g_1 \wedge g_2 \in \mathcal{G}$.
(3) $\mathcal{G}$ separates points of $\mathcal{X}$, i.e., given two points $x, y \in \mathcal{X}$ with $x \neq y$, and $a, b \in \mathbb{R}$, there exists a function $g \in \mathcal{G}$ such that $g(x) = a$ and $g(y) = b$.

Remark: It is easy to check that if $\mathcal{G}$ is well-separating, so is $\mathcal{G}^+$, the class of all bounded above functions in $\mathcal{G}$.

When $\mathcal{X}$ is a vector space, a particularly useful class of well-separating functions exists.

Lemma 4.4.8 *Let $\mathcal{X}$ be a locally convex, Hausdorff topological vector space. Then the class $\mathcal{G}$ of all continuous, bounded above, concave functions on $\mathcal{X}$ is well-separating.*

Proof: Let $\mathcal{X}^*$ denote the topological dual of $\mathcal{X}$, and let $\mathcal{G}_0 \triangleq \{\lambda(x) + c : \lambda \in \mathcal{X}^*, c \in \mathbb{R}\}$. Note that $\mathcal{G}_0$ contains the constant functions, and by the Hahn–Banach theorem, $\mathcal{G}_0$ separates points of $\mathcal{X}$. Since $\mathcal{G}_0$ consists of continuous, concave functions, it follows that the class of all continuous, concave functions separates points. Moreover, as the pointwise minimum of concave, continuous functions is concave and continuous, this class of functions is well-separating. Finally, by the earlier remark, it suffices to consider only the bounded above, continuous, concave functions. $\square$

The following lemma, whose proof is deferred to the end of the section, states the specific approximation property of well-separating classes of functions that allows their use instead of $C_b(\mathcal{X})$. It will be used in the proof of Theorem 4.4.10.

Lemma 4.4.9 *Let $\mathcal{G}$ be a well-separating class of functions on $\mathcal{X}$. Then for any compact set $\Gamma \subset \mathcal{X}$, any $f \in C_b(\Gamma)$, and any $\delta > 0$, there exists an integer $d < \infty$ and functions $g_1, \ldots, g_d \in \mathcal{G}$ such that*

$$\sup_{x \in \Gamma} |f(x) - \max_{i=1}^{d} g_i(x)| \leq \delta$$

and

$$\sup_{x \in \mathcal{X}} g_i(x) \leq \sup_{x \in \Gamma} f(x) < \infty \ .$$

Theorem 4.4.10 *Let $\{\mu_\epsilon\}$ be an exponentially tight family of probability measures on a completely regular topological space $\mathcal{X}$, and suppose $\mathcal{G}$ is a well-separating class of functions on $\mathcal{X}$. If Λ_g exists for each $g \in \mathcal{G}$, then Λ_f exists for each $f \in C_b(\mathcal{X})$. Consequently, all the conclusions of Theorem 4.4.2 hold.*

Proof: Fix a bounded continuous function $f(x)$ with $|f(x)| \leq M$. Since the family $\{\mu_\epsilon\}$ is exponentially tight, there exists a compact set Γ such that for all ϵ small enough,

$$\mu_\epsilon(\Gamma^c) \leq e^{-3M/\epsilon} \ .$$

Fix $\delta > 0$ and let $g_1, \ldots, g_d \in \mathcal{G}$, $d < \infty$ be as in Lemma 4.4.9, with $h(x) \triangleq \max_{i=1}^{d} g_i(x)$. Then, for every $\epsilon > 0$,

$$\max_{i=1}^{d} \left\{ \int_{\mathcal{X}} e^{g_i(x)/\epsilon} \mu_\epsilon(dx) \right\} \leq \int_{\mathcal{X}} e^{h(x)/\epsilon} \mu_\epsilon(dx) \leq \sum_{i=1}^{d} \int_{\mathcal{X}} e^{g_i(x)/\epsilon} \mu_\epsilon(dx) \ .$$

Hence, by the assumption of the theorem, the limit

$$\Lambda_h = \lim_{\epsilon \to 0} \epsilon \log \int_{\mathcal{X}} e^{h(x)/\epsilon} \mu_\epsilon(dx)$$

exists, and $\Lambda_h = \max_{i=1}^{d} \Lambda_{g_i}$. Moreover, by Lemma 4.4.9, $h(x) \leq M$ for all $x \in \mathcal{X}$, and $h(x) \geq (f(x) - \delta) \geq -(M + \delta)$ for all $x \in \Gamma$. Consequently, for all ϵ small enough,

$$\int_{\Gamma^c} e^{h(x)/\epsilon} \mu_\epsilon(dx) \leq e^{-2M/\epsilon}$$

and

$$\int_{\Gamma} e^{h(x)/\epsilon} \mu_\epsilon(dx) \geq e^{-(M+\delta)/\epsilon} \mu_\epsilon(\Gamma) \geq \frac{1}{2} e^{-(M+\delta)/\epsilon}.$$

Hence, for any $\delta < M$,

$$\Lambda_h = \lim_{\epsilon \to 0} \epsilon \log \int_{\Gamma} e^{h(x)/\epsilon} \mu_\epsilon(dx) \ .$$

Since $\sup_{x \in \Gamma} |f(x) - h(x)| \le \delta$,

$$
\begin{aligned}
\limsup_{\epsilon \to 0} \epsilon \log \int_\Gamma e^{f(x)/\epsilon} \mu_\epsilon(dx) & \le \quad \delta + \limsup_{\epsilon \to 0} \epsilon \log \int_\Gamma e^{h(x)/\epsilon} \mu_\epsilon(dx) \\
& = \delta + \Lambda_h = \quad \delta + \liminf_{\epsilon \to 0} \epsilon \log \int_\Gamma e^{h(x)/\epsilon} \mu_\epsilon(dx) \\
& \le \quad 2\delta + \liminf_{\epsilon \to 0} \epsilon \log \int_\Gamma e^{f(x)/\epsilon} \mu_\epsilon(dx).
\end{aligned}
$$

Thus, taking $\delta \to 0$, it follows that

$$
\lim_{\epsilon \to 0} \epsilon \log \int_\Gamma e^{f(x)/\epsilon} \mu_\epsilon(dx)
$$

exists. This limit equals Λ_f, since, for all ϵ small enough,

$$
\int_{\Gamma^c} e^{f(x)/\epsilon} \mu_\epsilon(dx) \le e^{-2M/\epsilon}, \qquad \int_\Gamma e^{f(x)/\epsilon} \mu_\epsilon(dx) \ge \frac{1}{2} e^{-M/\epsilon}. \qquad \square
$$

Proof of Lemma 4.4.9: Fix $\Gamma \subset \mathcal{X}$ compact, $f \in C_b(\Gamma)$ and $\delta > 0$. Let $x, y \in \Gamma$ with $x \ne y$. Since $\mathcal{G}$ separates points in Γ, there is a function $g_{x,y}(\cdot) \in \mathcal{G}$ such that $g_{x,y}(x) = f(x)$ and $g_{x,y}(y) = f(y)$. Because each of the functions $f(\cdot) - g_{x,y}(\cdot)$ is continuous, one may find for each $y \in \Gamma$ a neighborhood U_y of y such that

$$
\inf_{u \in U_y} \{f(u) - g_{x,y}(u)\} \ge -\delta.
$$

The neighborhoods $\{U_y\}$ form a cover of Γ; hence, Γ may be covered by a finite collection $U_{y_1}, \ldots, U_{y_m}$ of such neighborhoods. For every $x \in \Gamma$, define

$$
g_x(\cdot) = g_{x,y_1}(\cdot) \wedge g_{x,y_2}(\cdot) \wedge \cdots \wedge g_{x,y_m}(\cdot) \in \mathcal{G}.
$$

Then

$$
\inf_{u \in \Gamma} \{f(u) - g_x(u)\} \ge -\delta. \tag{4.4.11}
$$

Recall now that, for all i, $g_{x,y_i}(x) = f(x)$ and hence $g_x(x) = f(x)$. Since each of the functions $f(\cdot) - g_x(\cdot)$ is continuous, one may find a finite cover $V_1, \ldots, V_d$ of Γ and functions $g_{x_1}, \ldots, g_{x_d} \in \mathcal{G}$ such that

$$
\sup_{v \in V_i} \{f(v) - g_{x_i}(v)\} \le \delta. \tag{4.4.12}
$$

By the two preceding inequalities,

$$
\sup_{v \in \Gamma} |f(v) - \max_{i=1}^{d} g_{x_i}(v)| \le \delta.
$$

To complete the proof, observe that the constant $M \triangleq \sup_{x \in \Gamma} f(x)$ belongs to $\mathcal{G}$, and hence so does $g_i(\cdot) = g_{x_i}(\cdot) \wedge M$, while for all $v \in \Gamma$,

$$|f(v) - \max_{i=1}^{d} g_i(v)| \leq |f(v) - \max_{i=1}^{d} g_{x_i}(v)| . \qquad \square$$

The following variant of Theorem 4.4.2 dispenses with the exponential tightness of $\{\mu_\epsilon\}$, assuming instead that (4.4.4) holds for some good rate function $I(\cdot)$. See Section 6.6 for an application of this result.

Theorem 4.4.13 *Let $I(\cdot)$ be a good rate function. A family of probability measures $\{\mu_\epsilon\}$ satisfies the LDP in $\mathcal{X}$ with the rate function $I(\cdot)$ if and only if the limit Λ_f in (4.4.1) exists for every $f \in C_b(\mathcal{X})$ and satisfies (4.4.4).*

Proof: Suppose first that $\{\mu_\epsilon\}$ satisfies the LDP in $\mathcal{X}$ with the good rate function $I(\cdot)$. Then, by Varadhan's Lemma (Theorem 4.3.1), the limit Λ_f in (4.4.1) exists for every $f \in C_b(\mathcal{X})$ and satisfies (4.4.4).

Conversely, suppose that the limit Λ_f in (4.4.1) exists for every $f \in C_b(\mathcal{X})$ and satisfies (4.4.4) for some good rate function $I(\cdot)$. The relation (4.4.4) implies that $\Lambda_f - f(x) \geq -I(x)$ for any $x \in \mathcal{X}$ and any $f \in C_b(\mathcal{X})$. Therefore, by Lemma 4.4.6, the existence of Λ_f implies that $\{\mu_\epsilon\}$ satisfies the large deviations lower bound, with the good rate function $I(\cdot)$. Turning to prove the complementary upper bound, it suffices to consider closed sets $F \subset \mathcal{X}$ for which $\inf_{x \in F} I(x) > 0$. Fix such a set and $\delta > 0$ small enough so that $\alpha \triangleq \inf_{x \in F} I^\delta(x) \in (0, \infty)$ for the δ-rate function $I^\delta(\cdot) = \min\{I(\cdot) - \delta, \frac{1}{\delta}\}$. With $\Lambda_0 = 0$, the relation (4.4.4) implies that $\Psi_I(\alpha)$ is non-empty. Since F and $\Psi_I(\alpha)$ are disjoint subsets of the completely regular topological space $\mathcal{X}$, for any $y \in \Psi_I(\alpha)$ there exists a continuous function $f_y : \mathcal{X} \to [0, 1]$ such that $f_y(y) = 1$ and $f_y(x) = 0$ for all $x \in F$. The neighborhoods $U_y \triangleq \{z : f_y(z) > 1/2\}$ form a cover of $\Psi_I(\alpha)$; hence, the compact set $\Psi_I(\alpha)$ may be covered by a finite collection $U_{y_1}, \ldots, U_{y_n}$ of such neighborhoods. For any $m \in \mathbb{Z}_+$, the non-negative function $h_m(\cdot) \triangleq 2m \max_{i=1}^{n} f_{y_i}(\cdot)$ is continuous and bounded, with $h_m(x) = 0$ for all $x \in F$ and $h_m(y) \geq m$ for all $y \in \Psi_I(\alpha)$. Therefore, by (4.4.4),

$$\begin{aligned} \limsup_{\epsilon \to 0} \epsilon \log \mu_\epsilon(F) &\leq \limsup_{\epsilon \to 0} \epsilon \log \int_{\mathcal{X}} e^{-h_m(x)/\epsilon} \mu_\epsilon(dx) \\ &= \Lambda_{-h_m} = -\inf_{x \in \mathcal{X}} \{h_m(x) + I(x)\} . \end{aligned}$$

Note that $h_m(x) + I(x) \geq m$ for any $x \in \Psi_I(\alpha)$, whereas $h_m(x) + I(x) \geq \alpha$ for any $x \notin \Psi_I(\alpha)$. Consequently, taking $m \geq \alpha$,

$$\limsup_{\epsilon \to 0} \epsilon \log \mu_\epsilon(F) \leq -\alpha .$$

Since $\delta > 0$ is arbitrarily small, the large deviations upper bound holds (see (1.2.11)).   □

Exercise 4.4.14 Let $\{\mu_\epsilon\}$ be an exponentially tight family of probability measures on a completely regular topological space $\mathcal{X}$. Let $\mathcal{G}$ be a well-separating class of real valued, continuous functions on $\mathcal{X}$, and let $\mathcal{G}^+$ denote the functions in $\mathcal{G}$ that are bounded above.
(a) Suppose that Λ_g exists for all $g \in \mathcal{G}^+$. For $g \notin \mathcal{G}^+$, define

$$\Lambda_g = \liminf_{\epsilon \to 0} \epsilon \log \int_{\mathcal{X}} e^{g(x)/\epsilon} \mu_\epsilon(dx) .$$

Let $\hat{I}(x) = \sup_{g \in \mathcal{G}^+}\{g(x) - \Lambda_g\}$ and show that

$$\hat{I}(x) = \sup_{g \in \mathcal{G}}\{g(x) - \Lambda_g\} .$$

Hint: Observe that for every $g \in \mathcal{G}$ and every constant $M < \infty$, both $g(x) \wedge M \in \mathcal{G}^+$ and $\Lambda_{g \wedge M} \leq \Lambda_g$.
(b) Note that $\mathcal{G}^+$ is well-separating, and hence $\{\mu_\epsilon\}$ satisfies the LDP with the good rate function

$$I(x) = \sup_{f \in C_b(\mathcal{X})} \{f(x) - \Lambda_f\} .$$

Prove that $I(\cdot) = \hat{I}(\cdot)$.
Hint: Varadhan's lemma applies to every $g \in \mathcal{G}^+$. Consequently, $I(x) \geq \hat{I}(x)$. Fix $x \in \mathcal{X}$ and $f \in C_b(\mathcal{X})$. Following the proof of Theorem 4.4.10 with the compact set Γ enlarged to ensure that $x \in \Gamma$, show that

$$f(x) - \Lambda_f \leq \sup_{d < \infty} \sup_{g_i \in \mathcal{G}^+} \left\{ \max_{i=1}^{d} g_i(x) - \max_{i=1}^{d} \Lambda_{g_i} \right\} = \hat{I}(x) .$$

(c) To derive the converse of Theorem 4.4.10, suppose now that $\{\mu_\epsilon\}$ satisfies the LDP with rate function $I(\cdot)$. Use Varadhan's lemma to deduce that Λ_g exists for all $g \in \mathcal{G}^+$, and consequently by parts (a) and (b) of this exercise,

$$I(x) = \sup_{g \in \mathcal{G}}\{g(x) - \Lambda_g\} .$$

Exercise 4.4.15 Suppose the topological space $\mathcal{X}$ has a countable base. Let $\mathcal{G}$ be a class of continuous, bounded above, real valued functions on $\mathcal{X}$ such that for any good rate function $J(\cdot)$,

$$J(y) \leq \sup_{g \in \mathcal{G}} \inf_{x \in \mathcal{X}} \{ g(y) - g(x) + J(x) \} . \qquad (4.4.16)$$

(a) Suppose the family of probability measures $\{\mu_\epsilon\}$ satisfies the LDP in $\mathcal{X}$ with a good rate function $I(\cdot)$. Then, by Varadhan's Lemma, Λ_g exists for

$g \in \mathcal{G}$ and is given by (4.4.4). Show that $I(\cdot) = \hat{I}(\cdot) \triangleq \sup_{g \in \mathcal{G}} \{g(\cdot) - \Lambda_g\}$.

(b) Suppose $\{\mu_\epsilon\}$ is an exponentially tight family of probability measures, such that Λ_g exists for any $g \in \mathcal{G}$. Show that $\{\mu_\epsilon\}$ satisfies the LDP in $\mathcal{X}$ with the good rate function $\hat{I}(\cdot)$.

Hint: By Lemma 4.1.23 for any sequence $\epsilon_n \to 0$, there exists a subsequence $n(k) \to \infty$ such that $\{\mu_{\epsilon_{n(k)}}\}$ satisfies the LDP with a good rate function. Use part (a) to show that this good rate function is independent of $\epsilon_n \to 0$.

(c) Show that (4.4.16) holds if for any compact set $K \subset \mathcal{X}$, $y \notin K$ and $\alpha, \delta > 0$, there exists $g \in \mathcal{G}$ such that $\sup_{x \in \mathcal{X}} g(x) \le g(y) + \delta$ and $\sup_{x \in K} g(x) \le g(y) - \alpha$.

Hint: Consider $g \in \mathcal{G}$ corresponding to $K = \Psi_J(\alpha)$, $\alpha \nearrow J(y)$ and $\delta \to 0$.

(d) Use part (c) to verify that (4.4.16) holds for $\mathcal{G} = C_b(\mathcal{X})$ and $\mathcal{X}$ a completely regular topological space, thus providing an alternative proof of Theorem 4.4.2 under somewhat stronger conditions.

Hint: See the construction of $h_m(\cdot)$ in Theorem 4.4.13.

Exercise 4.4.17 Complete the proof of Lemma 4.4.5.

4.5 LDP in Topological Vector Spaces

In Section 2.3, it was shown that when a limiting logarithmic moment generating function exists for a family of $\mathbb{R}^d$-valued random variables, then its Fenchel–Legendre transform is the natural candidate rate function for the LDP associated with these variables. The goal of this section is to extend this result to topological vector spaces. As will be seen, convexity plays a major role as soon as the linear structure is introduced. For this reason, after the upper bound is established for all compact sets in Section 4.5.1, Section 4.5.2 turns to the study of some generalities involving the convex duality of Λ and Λ^*. These convexity considerations play an essential role in applications. Finally, Section 4.5.3 is devoted to a direct derivation of a weak version of the Gärtner–Ellis theorem in an abstract setup (Theorem 4.5.20), and to a Banach space variant of it.

Throughout this section, $\mathcal{X}$ is a Hausdorff (real) topological *vector* space. Recall that such spaces are regular, so the results of Sections 4.1 and 4.3 apply. The dual space of $\mathcal{X}$, namely, the space of all *continuous linear functionals* on $\mathcal{X}$, is denoted throughout by $\mathcal{X}^*$. Let Z_ϵ be a family of random variables taking values in $\mathcal{X}$, and let $\mu_\epsilon \in M_1(\mathcal{X})$ denote the probability measure associated with Z_ϵ. By analogy with the $\mathbb{R}^d$ case presented in Section 2.3, the logarithmic moment generating function $\Lambda_{\mu_\epsilon} : \mathcal{X}^* \to (-\infty, \infty]$ is defined to be

$$\Lambda_{\mu_\epsilon}(\lambda) = \log E\left[e^{\langle \lambda, Z_\epsilon \rangle}\right] = \log \int_{\mathcal{X}} e^{\lambda(x)} \mu_\epsilon(dx), \quad \lambda \in \mathcal{X}^*,$$

where for $x \in \mathcal{X}$ and $\lambda \in \mathcal{X}^*$, $\langle \lambda, x \rangle$ denotes the value of $\lambda(x) \in \mathbb{R}$.

Let

$$\bar{\Lambda}(\lambda) \stackrel{\triangle}{=} \limsup_{\epsilon \to 0} \epsilon \Lambda_{\mu_\epsilon} \left(\frac{\lambda}{\epsilon} \right), \tag{4.5.1}$$

using the notation $\Lambda(\lambda)$ whenever the *limit exists*. In most of the examples considered in Chapter 2, when $\epsilon \Lambda_{\mu_\epsilon}(\cdot/\epsilon)$ converges pointwise to $\Lambda(\cdot)$ for $\mathcal{X} = \mathbb{R}^d$ and an LDP holds for $\{\mu_\epsilon\}$, the rate function associated with this LDP is the Fenchel–Legendre transform of $\Lambda(\cdot)$. In the current setup, the Fenchel–Legendre transform of a function $f : \mathcal{X}^* \to [-\infty, \infty]$ is defined as

$$f^*(x) \stackrel{\triangle}{=} \sup_{\lambda \in \mathcal{X}^*} \{\langle \lambda, x \rangle - f(\lambda)\}, \quad x \in \mathcal{X}. \tag{4.5.2}$$

Thus, $\bar{\Lambda}^*$ denotes the Fenchel–Legendre transform of $\bar{\Lambda}$, and Λ^* denotes that of Λ when the latter exists for all $\lambda \in \mathcal{X}^*$.

4.5.1   A General Upper Bound

As in the $\mathbb{R}^d$ case, $\bar{\Lambda}^*$ plays a prominent role in the LDP bounds.

Theorem 4.5.3
(a) $\bar{\Lambda}(\cdot)$ of (4.5.1) is convex on $\mathcal{X}^*$ and $\bar{\Lambda}^*(\cdot)$ is a convex rate function.
(b) For any compact set $\Gamma \subset \mathcal{X}$,

$$\limsup_{\epsilon \to 0} \epsilon \log \mu_\epsilon(\Gamma) \leq - \inf_{x \in \Gamma} \bar{\Lambda}^*(x). \tag{4.5.4}$$

Remarks:
(a) In Theorem 2.3.6, which corresponds to $\mathcal{X} = \mathbb{R}^d$, it was assumed, for the purpose of establishing exponential tightness, that $0 \in \mathcal{D}_\Lambda^o$. In the abstract setup considered here, the exponential tightness does not follow from this assumption, and therefore must be handled on a case-by-case basis. (See, however, [deA85a] for a criterion for exponential tightness which is applicable in a variety of situations.)
(b) Note that any bound of the form $\bar{\Lambda}(\lambda) \leq K(\lambda)$ for all $\lambda \in \mathcal{X}^*$ implies that the Fenchel–Legendre transform $K^*(\cdot)$ may be substituted for $\bar{\Lambda}^*(\cdot)$ in (4.5.4). This is useful in situations in which $\bar{\Lambda}(\lambda)$ is easy to bound but hard to compute.
(c) The inequality (4.5.4) may serve as the upper bound related to a weak LDP. Thus, when $\{\mu_\epsilon\}$ is an exponentially tight family of measures, (4.5.4) extends to all closed sets. If in addition, the large deviations lower bound is also satisfied with $\bar{\Lambda}^*(\cdot)$, then this is a good rate function that controls the large deviations of the family $\{\mu_\epsilon\}$.

Proof: (a) The proof is similar to the proof of these properties in the special case $\mathcal{X} = \mathbb{R}^d$, which is presented in the context of the Gärtner–Ellis theorem.

Using the linearity of (λ/ϵ) and applying Hölder's inequality, one shows that the functions $\Lambda_{\mu_\epsilon}(\lambda/\epsilon)$ are convex. Thus, $\bar{\Lambda}(\cdot) = \limsup_{\epsilon \to 0} \epsilon\Lambda_{\mu_\epsilon}(\cdot/\epsilon)$, is also a convex function. Since $\Lambda_{\mu_\epsilon}(0) = 0$ for all $\epsilon > 0$, it follows that $\bar{\Lambda}(0) = 0$. Consequently, $\bar{\Lambda}^*(\cdot)$ is a nonnegative function. Since the supremum of a family of continuous functions is lower semicontinuous, the lower semicontinuity of $\bar{\Lambda}^*(\cdot)$ follows from the continuity of $g_\lambda(x) = \langle \lambda, x \rangle - \bar{\Lambda}(\lambda)$ for every $\lambda \in \mathcal{X}^*$. The convexity of $\bar{\Lambda}^*(\cdot)$ is a direct consequence of its definition via (4.5.2).

(b) The proof of the upper bound (4.5.4) is a repeat of the relevant part of the proof of Theorem 2.2.30. In particular, fix a compact set $\Gamma \subset \mathcal{X}$ and a $\delta > 0$. Let I^δ be the δ-rate function associated with $\bar{\Lambda}^*$, i.e., $I^\delta(x) \triangleq \min\{\bar{\Lambda}^*(x) - \delta, 1/\delta\}$. Then, for any $x \in \Gamma$, there exists a $\lambda_x \in \mathcal{X}^*$ such that

$$\langle \lambda_x, x \rangle - \bar{\Lambda}(\lambda_x) \geq I^\delta(x).$$

Since λ_x is a continuous functional, there exists a neighborhood of x, denoted A_x, such that

$$\inf_{y \in A_x} \{\langle \lambda_x, y \rangle - \langle \lambda_x, x \rangle\} \geq -\delta.$$

For any $\theta \in \mathcal{X}^*$, by Chebycheff's inequality,

$$\mu_\epsilon(A_x) \leq E\left[e^{\langle \theta, Z_\epsilon \rangle - \langle \theta, x \rangle}\right] \exp\left(-\inf_{y \in A_x} \{\langle \theta, y \rangle - \langle \theta, x \rangle\}\right).$$

Substituting $\theta = \lambda_x/\epsilon$ yields

$$\epsilon \log \mu_\epsilon(A_x) \leq \delta - \left\{\langle \lambda_x, x \rangle - \epsilon\Lambda_{\mu_\epsilon}\left(\frac{\lambda_x}{\epsilon}\right)\right\}.$$

A finite cover, $\cup_{i=1}^N A_{x_i}$, can be extracted from the open cover $\cup_{x \in \Gamma} A_x$ of the compact set Γ. Therefore, by the union of events bound,

$$\epsilon \log \mu_\epsilon(\Gamma) \leq \epsilon \log N + \delta - \min_{i=1,\ldots,N} \left\{\langle \lambda_{x_i}, x_i \rangle - \epsilon\Lambda_{\mu_\epsilon}\left(\frac{\lambda_{x_i}}{\epsilon}\right)\right\}.$$

Thus, by (4.5.1) and the choice of λ_x,

$$
\begin{aligned}
\limsup_{\epsilon \to 0} \epsilon \log \mu_\epsilon(\Gamma) &\leq \delta - \min_{i=1,\ldots,N} \{\langle \lambda_{x_i}, x_i \rangle - \bar{\Lambda}(\lambda_{x_i})\} \\
&\leq \delta - \min_{i=1,\ldots,N} I^\delta(x_i).
\end{aligned}
$$

Moreover, $x_i \in \Gamma$ for each i, yielding the inequality

$$\limsup_{\epsilon \to 0} \epsilon \log \mu_\epsilon(\Gamma) \leq \delta - \inf_{x \in \Gamma} I^\delta(x).$$

The proof of the theorem is complete by taking $\delta \to 0$. $\qquad\qquad\qquad$ $\square$

Exercise 4.5.5 An upper bound, valid for all ϵ, is developed in this exercise. This bound may be made specific in various situations (*c.f.* Exercise 6.2.19).
(a) Let $\mathcal{X}$ be a Hausdorff topological vector space and $V \subset \mathcal{X}$ a compact, convex set. Prove that for any $\epsilon > 0$,

$$\mu_\epsilon(V) \leq \exp\left(-\frac{1}{\epsilon} \inf_{x \in V} \Lambda_\epsilon^*(x) \right), \qquad (4.5.6)$$

where

$$\Lambda_\epsilon^*(x) = \sup_{\lambda \in \mathcal{X}^*} \left\{ \langle \lambda, x \rangle - \epsilon \Lambda_{\mu_\epsilon}\left(\frac{\lambda}{\epsilon} \right) \right\}.$$

Hint: Recall the following version of the *min–max theorem* ([Sio58], Theorem 4.2'). Let $f(x, \lambda)$ be concave in λ and convex and lower semicontinuous in x. Then

$$\sup_{\lambda \in \mathcal{X}^*} \inf_{x \in V} f(x, \lambda) = \inf_{x \in V} \sup_{\lambda \in \mathcal{X}^*} f(x, \lambda).$$

To prove (4.5.6), first use Chebycheff's inequality and then apply the min–max theorem to the function

$$f(x, \lambda) = [\langle \lambda, x \rangle - \epsilon \Lambda_{\mu_\epsilon}(\lambda/\epsilon)].$$

(b) Suppose that $\mathcal{E}$ is a convex metric subspace of $\mathcal{X}$ (in a metric compatible with the induced topology). Assume that all balls in $\mathcal{E}$ are convex, pre-compact subsets of $\mathcal{X}$. Show that for every measurable set $A \in \mathcal{E}$,

$$\mu_\epsilon(A) \leq \inf_{\delta > 0} \left\{ m(A, \delta) \exp\left(-\frac{1}{\epsilon} \inf_{x \in A^\delta} \Lambda_\epsilon^*(x) \right) \right\}, \qquad (4.5.7)$$

where A^δ is the closed δ blowup of A, and $m(A, \delta)$ denotes the *metric entropy* of A, i.e., the minimal number of balls of radius δ needed to cover A.

4.5.2 Convexity Considerations

The implications of the existence of an LDP with a convex rate function to the structure of Λ and Λ^* are explored here. Building on Varadhan's lemma and Theorem 4.5.3, it is first shown that when the quantities $\epsilon \Lambda_{\mu_\epsilon}(\lambda/\epsilon)$ are uniformly bounded (in ϵ) and an LDP holds with a good convex rate function, then $\epsilon \Lambda_{\mu_\epsilon}(\cdot/\epsilon)$ converges pointwise to $\Lambda(\cdot)$ and the rate function equals $\Lambda^*(\cdot)$. Consequently, the assumptions of Lemma 4.1.21 together with the exponential tightness of $\{\mu_\epsilon\}$ and the uniform boundedness mentioned earlier, suffice to establish the LDP with rate function $\Lambda^*(\cdot)$. Alternatively, if the relation (4.5.15) between I and Λ^* holds, then $\Lambda^*(\cdot)$ controls a weak LDP even when $\Lambda(\lambda) = \infty$ for some λ and $\{\mu_\epsilon\}$ are not exponentially tight. This statement is the key to Cramér's theorem at its most general.

Before proceeding with the attempt to identify the rate function of the LDP as $\Lambda^*(\cdot)$, note that while $\Lambda^*(\cdot)$ is always convex by Theorem 4.5.3, the rate function may well be non-convex. For example, such a situation may occur when contractions using non-convex functions are considered. However, it may be expected that $I(\cdot)$ is identical to $\Lambda^*(\cdot)$ when $I(\cdot)$ is convex.

An instrumental tool in the identification of I as Λ^* is the following duality property of the Fenchel–Legendre transform, whose proof is deferred to the end of this section.

Lemma 4.5.8 (Duality lemma) *Let $\mathcal{X}$ be a locally convex Hausdorff topological vector space. Let $f : \mathcal{X} \to (-\infty, \infty]$ be a lower semicontinuous, convex function, and define*

$$g(\lambda) = \sup_{x \in \mathcal{X}} \{\langle \lambda, x \rangle - f(x)\}.$$

Then $f(\cdot)$ is the Fenchel–Legendre transform of $g(\cdot)$, i.e.,

$$f(x) = \sup_{\lambda \in \mathcal{X}^*} \{\langle \lambda, x \rangle - g(\lambda)\}. \tag{4.5.9}$$

Remark: This lemma has the following geometric interpretation. For every hyperplane defined by λ, $g(\lambda)$ is the largest amount one may push up the tangent before it hits $f(\cdot)$ and becomes a tangent hyperplane. The duality lemma states the "obvious result" that to reconstruct $f(\cdot)$, one only needs to find the tangent at x and "push it down" by $g(\lambda)$. (See Fig. 4.5.2.)

The first application of the duality lemma is in the following theorem, where convex rate functions are identified as $\Lambda^*(\cdot)$.

Theorem 4.5.10 *Let $\mathcal{X}$ be a locally convex Hausdorff topological vector space. Assume that μ_ϵ satisfies the LDP with a good rate function I. Suppose in addition that*

$$\bar{\Lambda}(\lambda) \stackrel{\triangle}{=} \limsup_{\epsilon \to 0} \epsilon \Lambda_{\mu_\epsilon}(\lambda/\epsilon) < \infty, \quad \forall \lambda \in \mathcal{X}^*. \tag{4.5.11}$$

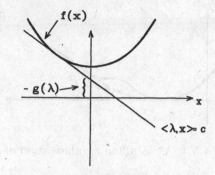

Figure 4.5.1: Duality lemma.

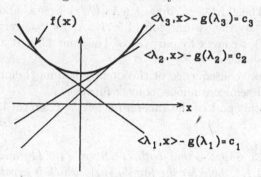

Figure 4.5.2: Duality reconstruction. $c_i = f(x_i)$ and x_i is the point of tangency of the line with slope λ_i to the graph of $f(\cdot)$.

(a) For each $\lambda \in \mathcal{X}^*$, the limit $\Lambda(\lambda) = \lim_{\epsilon \to 0} \epsilon \Lambda_{\mu_\epsilon}(\lambda/\epsilon)$ exists, is finite, and satisfies

$$\Lambda(\lambda) = \sup_{x \in \mathcal{X}} \{\langle \lambda, x \rangle - I(x)\} . \qquad (4.5.12)$$

(b) If I is convex, then it is the Fenchel–Legendre transform of Λ, namely,

$$I(x) = \Lambda^*(x) \overset{\triangle}{=} \sup_{\lambda \in \mathcal{X}^*} \{\langle \lambda, x \rangle - \Lambda(\lambda)\} .$$

(c) If I is not convex, then Λ^* is the affine regularization of I, i.e., $\Lambda^*(\cdot) \leq I(\cdot)$, and for any convex rate function f, $f(\cdot) \leq I(\cdot)$ implies $f(\cdot) \leq \Lambda^*(\cdot)$. (See Fig. 4.5.3.)

Remark: The weak* topology on $\mathcal{X}^*$ makes the functions $\langle \lambda, x \rangle - I(x)$ continuous in λ for all $x \in \mathcal{X}$. By part (a), $\Lambda(\cdot)$ is lower semicontinuous with respect to this topology, which explains why lower semicontinuity of $\Lambda(\cdot)$ is necessary in Rockafellar's lemma (Lemma 2.3.12).

Proof: (a) Fix $\lambda \in \mathcal{X}^*$ and $\gamma > 1$. By assumption, $\bar{\Lambda}(\gamma \lambda) < \infty$, and Varadhan's lemma (Theorem 4.3.1) applies for the continuous function

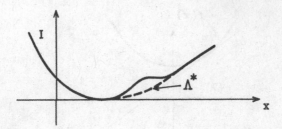

Figure 4.5.3: Λ^* as affine regularization of I.

$\lambda : \mathcal{X} \to \mathbb{R}$. Thus, $\Lambda(\lambda) = \lim_{\epsilon \to 0} \epsilon \Lambda_{\mu_\epsilon}(\lambda/\epsilon)$ exists, and satisfies the identity (4.5.12). By the assumption (4.5.11), $\Lambda(\cdot) < \infty$ everywhere. Since $\Lambda(0) = 0$ and $\Lambda(\cdot)$ is convex by part (a) of Theorem 4.5.3, it also holds that $\Lambda(\lambda) > -\infty$ everywhere.
(b) This is a direct consequence of the duality lemma (Lemma 4.5.8), applied to the lower semicontinuous, convex function I.
(c) The proof of this part of the theorem is left as Exercise 4.5.18. $\square$

Corollary 4.5.13 *Suppose that both condition (4.5.11) and the assumptions of Lemma 4.1.21 hold for the family $\{\mu_\epsilon\}$, which is exponentially tight. Then $\{\mu_\epsilon\}$ satisfies in $\mathcal{X}$ the LDP with the good, convex rate function Λ^*.*

Proof: By Lemma 4.1.21, $\{\mu_\epsilon\}$ satisfies a weak LDP with a convex rate function. As $\{\mu_\epsilon\}$ is exponentially tight, it is deduced that it satisfies the *full* LDP with a convex, good rate function. The corollary then follows from parts (a) and (b) of Theorem 4.5.10. $\square$

Theorem 4.5.10 is not applicable when $\Lambda(\cdot)$ exists but is infinite at some $\lambda \in \mathcal{X}^*$, and moreover, it requires the *full* LDP with a convex, *good* rate function. As seen in the case of Cramér's theorem in $\mathbb{R}$, these conditions are not necessary. The following theorem replaces the finiteness conditions on Λ by an appropriate inequality on *open* half-spaces. Of course, there is a price to pay: The resulting Λ^* may not be a good rate function and only the weak LDP is proved.

Theorem 4.5.14 *Suppose that $\{\mu_\epsilon\}$ satisfies a weak LDP with a convex rate function $I(\cdot)$, and that $\mathcal{X}$ is a locally convex, Hausdorff topological vector space. Assume that for each $\lambda \in \mathcal{X}^*$, the limits $\Lambda_\lambda(t) = \lim_{\epsilon \to 0} \epsilon \Lambda_{\mu_\epsilon}(t\lambda/\epsilon)$ exist as extended real numbers, and that $\Lambda_\lambda(t)$ is a lower semicontinuous function of $t \in \mathbb{R}$. Let $\Lambda_\lambda^*(\cdot)$ be the Fenchel–Legendre* *transform of* $\Lambda_\lambda(\cdot)$, *i.e.,*

$$\Lambda_\lambda^*(z) \overset{\triangle}{=} \sup_{\theta \in \mathbb{R}} \left\{ \theta z - \Lambda_\lambda(\theta) \right\}.$$

If for every $\lambda \in \mathcal{X}^*$ *and every* $a \in \mathbb{R}$,

$$\inf_{\{x : (\langle \lambda, x \rangle - a) > 0\}} I(x) \leq \inf_{z > a} \Lambda_\lambda^*(z), \qquad (4.5.15)$$

then $I(\cdot) = \Lambda^*(\cdot)$, *and consequently,* Λ^* *controls a weak LDP associated with* $\{\mu_\epsilon\}$.

Proof: Fix $\lambda \in \mathcal{X}^*$. By the inequality (4.5.15),

$$\begin{aligned}
\sup_{x \in \mathcal{X}} \{\langle \lambda, x \rangle - I(x)\} &= \sup_{a \in \mathbb{R}} \sup_{\{x : (\langle \lambda, x \rangle - a) > 0\}} \{\langle \lambda, x \rangle - I(x)\} \\
&\geq \sup_{a \in \mathbb{R}} \left\{ a - \inf_{\{x : (\langle \lambda, x \rangle - a) > 0\}} I(x) \right\} \qquad (4.5.16) \\
&\geq \sup_{a \in \mathbb{R}} \left\{ a - \inf_{z > a} \Lambda_\lambda^*(z) \right\} = \sup_{z \in \mathbb{R}} \left\{ z - \Lambda_\lambda^*(z) \right\}.
\end{aligned}$$

Note that $\Lambda_\lambda(\cdot)$ is convex with $\Lambda_\lambda(0) = 0$ and is assumed lower semicontinuous. Therefore, it can not attain the value $-\infty$. Hence, by applying the duality lemma (Lemma 4.5.8) to $\Lambda_\lambda : \mathbb{R} \to (-\infty, \infty]$, it follows that

$$\Lambda_\lambda(1) = \sup_{z \in \mathbb{R}} \{z - \Lambda_\lambda^*(z)\}.$$

Combining this identity with (4.5.16) yields

$$\sup_{x \in \mathcal{X}} \{\langle \lambda, x \rangle - I(x)\} \geq \Lambda_\lambda(1) = \Lambda(\lambda).$$

The opposite inequality follows by applying Lemma 4.3.4 to the continuous linear functional $\lambda \in \mathcal{X}^*$. Thus, the identity (4.5.12) holds for all $\lambda \in \mathcal{X}^*$, and the proof of the theorem is completed by applying the duality lemma (Lemma 4.5.8) to the convex rate function I. $\qquad\qquad\square$

Proof of Lemma 4.5.8: Consider the sets $\mathcal{X} \times \mathbb{R}$ and $\mathcal{X}^* \times \mathbb{R}$. Each of these can be made into a locally convex, Hausdorff topological vector space in the obvious way. If f is identically ∞, then g is identically $-\infty$ and the lemma trivially holds. Assume otherwise and define

$$\begin{aligned}
\mathcal{E} &= \{(x, \alpha) \in \mathcal{X} \times \mathbb{R} : f(x) \leq \alpha\}, \\
\mathcal{E}^* &= \{(\lambda, \beta) \in \mathcal{X}^* \times \mathbb{R} : g(\lambda) \leq \beta\}.
\end{aligned}$$

Note that for any $(\lambda, \beta) \in \mathcal{E}^*$ and any $x \in \mathcal{X}$,

$$f(x) \geq \langle \lambda, x \rangle - \beta.$$

Therefore, it also holds that

$$f(x) \geq \sup_{(\lambda,\beta)\in\mathcal{E}^*} \{\langle\lambda,x\rangle - \beta\} = \sup_{\lambda\in\mathcal{X}^*} \{\langle\lambda,x\rangle - g(\lambda)\}\,.$$

It thus suffices to show that for any $(x,\alpha) \notin \mathcal{E}$ (i.e., $f(x) > \alpha$), there exists a $(\lambda,\beta) \in \mathcal{E}^*$ such that

$$\langle\lambda,x\rangle - \beta > \alpha\,, \tag{4.5.17}$$

in order to complete the proof of the lemma.

Since f is a lower semicontinuous function, the set $\mathcal{E}$ is closed (alternatively, the set $\mathcal{E}^c$ is open). Indeed, whenever $f(x) > \gamma$, there exists a neighborhood V of x such that $\inf_{y\in V} f(y) > \gamma$, and thus $\mathcal{E}^c$ contains a neighborhood of (x,γ). Moreover, since $f(\cdot)$ is convex and not identically ∞, the set $\mathcal{E}$ is a non-empty convex subset of $\mathcal{X} \times \mathbb{R}$.

Fix $(x,\alpha) \notin \mathcal{E}$. The product space $\mathcal{X} \times \mathbb{R}$ is locally convex and therefore, by the Hahn–Banach theorem (Theorem B.6), there exists a hyperplane in $\mathcal{X} \times \mathbb{R}$ that strictly separates the non-empty, closed, and convex set $\mathcal{E}$ and the point (x,α) in its complement. Hence, as the topological dual of $\mathcal{X} \times \mathbb{R}$ is $\mathcal{X}^* \times \mathbb{R}$, for some $\mu \in \mathcal{X}^*$, $\rho \in \mathbb{R}$, and $\gamma \in \mathbb{R}$,

$$\sup_{(y,\xi)\in\mathcal{E}} \{\langle\mu,y\rangle - \rho\xi\} \leq \gamma < \langle\mu,x\rangle - \rho\alpha\,.$$

In particular, since f is not identically ∞, it follows that $\rho \geq 0$, for otherwise a contradiction results when $\xi \to \infty$. Moreover, by considering $(y,\xi) = (x,f(x))$, the preceding inequality implies that $\rho > 0$ whenever $f(x) < \infty$.

Suppose first that $\rho > 0$. Then, (4.5.17) holds for the point $(\mu/\rho,\,\gamma/\rho)$. This point must be in $\mathcal{E}^*$, for otherwise there exists a $y_0 \in \mathcal{X}$ such that $\langle\mu,y_0\rangle - \rho f(y_0) > \gamma$, contradicting the previous construction of the separating hyperplane (since $(y_0,f(y_0)) \in \mathcal{E}$). In particular, since $f(x) < \infty$ for some $x \in \mathcal{X}$ it follows that $\mathcal{E}^*$ is non-empty.

Now suppose that $\rho = 0$ so that

$$\sup_{\{y:f(y)<\infty\}} \{\langle\mu,y\rangle - \gamma\} \leq 0\,,$$

while $\langle\mu,x\rangle - \gamma > 0$. Consider the points

$$(\lambda_\delta,\beta_\delta) \triangleq \left(\frac{\mu}{\delta} + \lambda_0, \frac{\gamma}{\delta} + \beta_0\right)\,, \qquad \forall\delta > 0\,,$$

where (λ_0,β_0) is an arbitrary point in $\mathcal{E}^*$. Then, for all $y \in \mathcal{X}$,

$$\langle\lambda_\delta,y\rangle - \beta_\delta = \frac{1}{\delta}(\langle\mu,y\rangle - \gamma) + (\langle\lambda_0,y\rangle - \beta_0) \leq f(y)\,.$$

Therefore, $(\lambda_\delta, \beta_\delta) \in \mathcal{E}^*$ for any $\delta > 0$. Moreover,

$$\lim_{\delta \to 0} (\langle \lambda_\delta, x \rangle - \beta_\delta) = \lim_{\delta \to 0} \left\{ \frac{1}{\delta} (\langle \mu, x \rangle - \gamma) + (\langle \lambda_0, x \rangle - \beta_0) \right\} = \infty .$$

Thus, for any $\alpha < \infty$, there exists $\delta > 0$ small enough so that $\langle \lambda_\delta, x \rangle - \beta_\delta > \alpha$. This completes the proof of (4.5.17) and of Lemma 4.5.8.   □

Exercise 4.5.18 Prove part (c) of Theorem 4.5.10.

Exercise 4.5.19 Consider the setup of Exercise 4.2.7, except that now $\mathcal{X} = \mathcal{Y}$ is a locally convex, separable, Hausdorff topological vector space. Let $Z_\epsilon = X_\epsilon + Y_\epsilon$.
(a) Prove that if I_X and I_Y are convex, then so is I_Z.
(b) Deduce that if in addition, the condition (4.5.11) holds for both μ_ϵ—the laws of X_ϵ and ν_ϵ—the laws of Y_ϵ, then I_Z is the Fenchel–Legendre transform of $\Lambda_X(\cdot) + \Lambda_Y(\cdot)$.

4.5.3 Abstract Gärtner–Ellis Theorem

Having seen a general upper bound in Section 4.5.1, we turn next to sufficient conditions for the existence of a complementary lower bound. To this end, recall that a point $x \in \mathcal{X}$ is called an *exposed* point of $\bar{\Lambda}^*$ if there exists an *exposing hyperplane* $\lambda \in \mathcal{X}^*$ such that

$$\langle \lambda, x \rangle - \bar{\Lambda}^*(x) > \langle \lambda, z \rangle - \bar{\Lambda}^*(z) , \quad \forall z \neq x .$$

An exposed point of $\bar{\Lambda}^*$ is, in convex analysis parlance, an exposed point of the epigraph of $\bar{\Lambda}^*$. For a geometrical interpretation, see Fig. 2.3.2.

Theorem 4.5.20 (Baldi) *Suppose that $\{\mu_\epsilon\}$ are exponentially tight probability measures on $\mathcal{X}$.*
(a) For every closed set $F \subset \mathcal{X}$,

$$\limsup_{\epsilon \to 0} \epsilon \log \mu_\epsilon(F) \leq - \inf_{x \in F} \bar{\Lambda}^*(x) .$$

(b) Let $\mathcal{F}$ be the set of exposed points of $\bar{\Lambda}^*$ with an exposing hyperplane λ for which

$$\Lambda(\lambda) = \lim_{\epsilon \to 0} \epsilon \Lambda_{\mu_\epsilon} \left(\frac{\lambda}{\epsilon} \right) \quad \text{exists and } \bar{\Lambda}(\gamma \lambda) < \infty \quad \text{for some} \quad \gamma > 1 . \quad (4.5.21)$$

Then, for every open set $G \subset \mathcal{X}$,

$$\liminf_{\epsilon \to 0} \epsilon \log \mu_\epsilon(G) \geq - \inf_{x \in G \cap \mathcal{F}} \bar{\Lambda}^*(x) .$$

(c) If for every open set G,

$$\inf_{x \in G \cap \mathcal{F}} \bar{\Lambda}^*(x) = \inf_{x \in G} \bar{\Lambda}^*(x) , \qquad (4.5.22)$$

then $\{\mu_\epsilon\}$ satisfies the LDP with the good rate function $\bar{\Lambda}^*$.

Proof: (a) The upper bound is a consequence of Theorem 4.5.3 and the assumed exponential tightness.

(b) If $\bar{\Lambda}(\lambda) = -\infty$ for some $\lambda \in \mathcal{X}^*$, then $\bar{\Lambda}^*(\cdot) \equiv \infty$ and the large deviations lower bound trivially holds. So, without loss of generality, it is assumed throughout that $\bar{\Lambda} : \mathcal{X}^* \to (-\infty, \infty]$. Fix an open set G, an exposed point $y \in G \cap \mathcal{F}$, and $\delta > 0$ arbitrarily small. Let η be an exposing hyperplane for $\bar{\Lambda}^*$ at y such that (4.5.21) holds. The proof is now a repeat of the proof of (2.3.13). Indeed, by the continuity of η, there exists an open subset of G, denoted B_δ, such that $y \in B_\delta$ and

$$\sup_{z \in B_\delta} \{\langle \eta, z - y \rangle\} < \delta .$$

Observe that $\Lambda(\eta) < \infty$ in view of (4.5.21). Hence, by (4.5.1), $\Lambda_{\mu_\epsilon}(\eta/\epsilon) < \infty$ for all ϵ small enough. Thus, for all $\epsilon > 0$ small enough, define the probability measures $\tilde{\mu}_\epsilon$ via

$$\frac{d\tilde{\mu}_\epsilon}{d\mu_\epsilon}(z) = \exp\left[\left\langle \frac{\eta}{\epsilon}, z \right\rangle - \Lambda_{\mu_\epsilon}\left(\frac{\eta}{\epsilon}\right)\right] . \qquad (4.5.23)$$

Using this definition,

$$\epsilon \log \mu_\epsilon(B_\delta) = \epsilon \Lambda_{\mu_\epsilon}\left(\frac{\eta}{\epsilon}\right) - \langle \eta, y \rangle + \epsilon \log \int_{z \in B_\delta} \exp\left(\left\langle \frac{\eta}{\epsilon}, y - z \right\rangle\right) \tilde{\mu}_\epsilon(dz)$$

$$\geq \epsilon \Lambda_{\mu_\epsilon}\left(\frac{\eta}{\epsilon}\right) - \langle \eta, y \rangle - \delta + \epsilon \log \tilde{\mu}_\epsilon(B_\delta) .$$

Therefore, by (4.5.21),

$$\liminf_{\epsilon \to 0} \epsilon \log \mu_\epsilon(G) \geq \lim_{\delta \to 0} \liminf_{\epsilon \to 0} \epsilon \log \mu_\epsilon(B_\delta) \qquad (4.5.24)$$

$$\geq \Lambda(\eta) - \langle \eta, y \rangle + \lim_{\delta \to 0} \liminf_{\epsilon \to 0} \epsilon \log \tilde{\mu}_\epsilon(B_\delta)$$

$$\geq -\bar{\Lambda}^*(y) + \lim_{\delta \to 0} \liminf_{\epsilon \to 0} \epsilon \log \tilde{\mu}_\epsilon(B_\delta) .$$

Recall that $\{\mu_\epsilon\}$ are exponentially tight, so for each $\alpha < \infty$, there exists a compact set K_α such that

$$\limsup_{\epsilon \to 0} \epsilon \log \mu_\epsilon(K_\alpha^c) < -\alpha .$$

If for all $\delta > 0$ and all $\alpha < \infty$,

$$\limsup_{\epsilon \to 0} \epsilon \log \tilde{\mu}_\epsilon(B_\delta^c \cap K_\alpha) < 0 \,, \qquad (4.5.25)$$

and for all α large enough,

$$\limsup_{\epsilon \to 0} \epsilon \log \tilde{\mu}_\epsilon(K_\alpha^c) < 0 \,, \qquad (4.5.26)$$

then $\tilde{\mu}_\epsilon(B_\delta) \to 1$ when $\epsilon \to 0$ and part (b) of the theorem follows by (4.5.24), since $y \in G \cap \mathcal{F}$ is arbitrary.

To establish (4.5.25), let $\Lambda_{\tilde{\mu}_\epsilon}(\cdot)$ denote the logarithmic moment generating function associated with the law $\tilde{\mu}_\epsilon$. By the definition (4.5.23), for every $\theta \in \mathcal{X}^*$,

$$\epsilon \Lambda_{\tilde{\mu}_\epsilon}\left(\frac{\theta}{\epsilon}\right) = \epsilon \Lambda_{\mu_\epsilon}\left(\frac{\theta + \eta}{\epsilon}\right) - \epsilon \Lambda_{\mu_\epsilon}\left(\frac{\eta}{\epsilon}\right) \,.$$

Hence, by (4.5.1) and (4.5.21),

$$\tilde{\Lambda}(\theta) \stackrel{\triangle}{=} \limsup_{\epsilon \to 0} \epsilon \Lambda_{\tilde{\mu}_\epsilon}\left(\frac{\theta}{\epsilon}\right) = \bar{\Lambda}(\theta + \eta) - \Lambda(\eta) \,.$$

Let $\tilde{\Lambda}^*$ denote the Fenchel–Legendre transform of $\tilde{\Lambda}$. It follows that for all $z \in \mathcal{X}$,

$$\tilde{\Lambda}^*(z) = \bar{\Lambda}^*(z) + \Lambda(\eta) - \langle \eta, z \rangle \geq \bar{\Lambda}^*(z) - \bar{\Lambda}^*(y) - \langle \eta, z - y \rangle \,.$$

Since η is an exposing hyperplane for $\bar{\Lambda}^*$ at y, this inequality implies that $\tilde{\Lambda}^*(z) > 0$ for all $z \neq y$. Theorem 4.5.3, applied to the measures $\tilde{\mu}_\epsilon$ and the compact sets $B_\delta^c \cap K_\alpha$, now yields

$$\limsup_{\epsilon \to 0} \epsilon \log \tilde{\mu}_\epsilon(B_\delta^c \cap K_\alpha) \leq - \inf_{z \in B_\delta^c \cap K_\alpha} \tilde{\Lambda}^*(z) < 0 \,,$$

where the strict inequality follows because $\tilde{\Lambda}^*(\cdot)$ is a lower semicontinuous function and $y \in B_\delta$.

Turning now to establish (4.5.26), consider the open half-spaces

$$H_\rho = \{z \in \mathcal{X} : \langle \eta, z \rangle - \rho < 0\} \,.$$

By Chebycheff's inequality, for any $\beta > 0$,

$$\begin{aligned}
\epsilon \log \tilde{\mu}_\epsilon(H_\rho^c) &= \epsilon \log \int_{\{z:\langle \eta, z \rangle \geq \rho\}} \tilde{\mu}_\epsilon(dz) \\
&\leq \epsilon \log \left[\int_{\mathcal{X}} \exp\left(\frac{\beta \langle \eta, z \rangle}{\epsilon}\right) \tilde{\mu}_\epsilon(dz) \right] - \beta \rho \\
&= \epsilon \Lambda_{\tilde{\mu}_\epsilon}\left(\frac{\beta \eta}{\epsilon}\right) - \beta \rho \,.
\end{aligned}$$

Hence,

$$\limsup_{\epsilon \to 0} \epsilon \log \tilde{\mu}_\epsilon(H_\rho^c) \le \inf_{\beta > 0} \{\tilde{\Lambda}(\beta\eta) - \beta\rho\} .$$

Due to condition (4.5.21), $\tilde{\Lambda}(\beta\eta) < \infty$ for some $\beta > 0$, implying that for large enough ρ,

$$\limsup_{\epsilon \to 0} \epsilon \log \tilde{\mu}_\epsilon(H_\rho^c) < 0 .$$

Now, for every α and every $\rho > 0$,

$$\limsup_{\epsilon \to 0} \epsilon \log \tilde{\mu}_\epsilon(K_\alpha^c \cap II_\rho)$$

$$= \limsup_{\epsilon \to 0} \epsilon \log \int_{K_\alpha^c \cap H_\rho} \exp\left[\left\langle \frac{\eta}{\epsilon}, z \right\rangle - \Lambda_{\mu_\epsilon}\left(\frac{\eta}{\epsilon}\right)\right] \mu_\epsilon(dz)$$

$$< \rho - \Lambda(\eta) - \alpha .$$

Finally, (4.5.26) follows by combining the two preceding inequalities.
(c) Starting with (4.5.22), the LDP is established by combining parts (a) and (b). □

In the following corollary, the smoothness of $\Lambda(\cdot)$ yields the identity (4.5.22) for exponentially tight probability measures on a Banach space, resulting in the LDP. Its proof is based on a theorem of Brønsted and Rockafellar whose proof is not reproduced here. Recall that a function $f : \mathcal{X}^* \to \mathbb{R}$ is *Gateaux differentiable* if, for every $\lambda, \theta \in \mathcal{X}^*$, the function $f(\lambda + t\theta)$ is differentiable with respect to t at $t = 0$.

Corollary 4.5.27 *Let $\{\mu_\epsilon\}$ be exponentially tight probability measures on the Banach space $\mathcal{X}$. Suppose that $\Lambda(\cdot) = \lim_{\epsilon \to 0} \epsilon \Lambda_{\mu_\epsilon}(\cdot/\epsilon)$ is finite valued, Gateaux differentiable, and lower semicontinuous in $\mathcal{X}^*$ with respect to the weak* topology. Then $\{\mu_\epsilon\}$ satisfies the LDP with the good rate function Λ^*.*

Remark: For a somewhat stronger version, see Corollary 4.6.14.

Proof: By Baldi's theorem (Theorem 4.5.20), it suffices to show that for any $x \in \mathcal{D}_{\Lambda^*}$, there exists a sequence of exposed points x_k such that $x_k \to x$ and $\Lambda^*(x_k) \to \Lambda^*(x)$. Let $\lambda \in \partial\Lambda^*(x)$ iff

$$\langle \lambda, x \rangle - \Lambda^*(x) = \sup_{z \in \mathcal{X}} \{\langle \lambda, z \rangle - \Lambda^*(z)\} ,$$

and define

$$\operatorname{dom} \partial\Lambda^* \stackrel{\triangle}{=} \{x : \exists \lambda \in \partial\Lambda^*(x)\} .$$

Note that it may be assumed that the convex, lower semicontinuous function $\Lambda^* : \mathcal{X} \to [0, \infty]$ is proper (i.e., $\mathcal{D}_{\Lambda^*}$ is not empty). Therefore, by the Brønsted–Rockafellar theorem (see [BrR65], Theorem 2), for every $x \in \mathcal{D}_{\Lambda^*}$, there exists a sequence $x_k \to x$ such that $x_k \in \text{dom } \partial\Lambda^*$ and $\Lambda^*(x_k) \to \Lambda^*(x)$.

It is therefore enough to prove that when Λ is Gateaux differentiable and weak* lower semicontinuous, any point in dom $\partial\Lambda^*$ is also an exposed point. To this end, fix $x \in \text{dom}\partial\Lambda^*$ and $\lambda \in \partial\Lambda^*(x)$. Observe that $\mathcal{X}^*$ when equipped with the weak* topology is a locally convex, Hausdorff topological vector space with $\mathcal{X}$ being its topological dual. Hence, it follows by applying the duality lemma (Lemma 4.5.8) for the convex, lower semicontinuous function $\Lambda : \mathcal{X}^* \to \mathbb{R}$ that

$$\Lambda(\lambda) = \sup_{z \in \mathcal{X}} \{\langle \lambda, z \rangle - \Lambda^*(z)\} = \langle \lambda, x \rangle - \Lambda^*(x) \, .$$

Therefore, for any $t > 0$, and any $\theta \in \mathcal{X}^*$,

$$\langle \theta, x \rangle \leq \frac{1}{t} [\Lambda(\lambda + t\theta) - \Lambda(\lambda)] \, .$$

Thus, by the Gateaux differentiability of Λ, it follows that

$$\langle \theta, x \rangle \leq \lim_{t \searrow 0} \frac{1}{t} [\Lambda(\lambda + t\theta) - \Lambda(\lambda)] \stackrel{\triangle}{=} D\Lambda(\theta) \, .$$

Moreover, $D\Lambda(\theta) = -D\Lambda(-\theta)$, and consequently $\langle \theta, x \rangle = D\Lambda(\theta)$ for all $\theta \in \mathcal{X}^*$. Similarly, if there exists $y \in \mathcal{X}$, $y \neq x$, such that

$$\langle \lambda, x \rangle - \Lambda^*(x) = \langle \lambda, y \rangle - \Lambda^*(y) \, ,$$

then, by exactly the same argument, $\langle \theta, y \rangle = D\Lambda(\theta)$ for all $\theta \in \mathcal{X}^*$. Since $\langle \theta, x - y \rangle = 0$ for all $\theta \in \mathcal{X}^*$, it follows that $x = y$. Hence, x is an exposed point and the proof is complete. $\qquad\square$

4.6 Large Deviations for Projective Limits

In this section, we develop a method of lifting a collection of LDPs in "small" spaces into the LDP in the "large" space $\mathcal{X}$, which is their projective limit. (See definition below.) The motivation for such an approach is as follows. Suppose we are interested in proving the LDP associated with a sequence of random variables $X_1, X_2, \ldots$ in some abstract space $\mathcal{X}$. The identification of $\mathcal{X}^*$ (if $\mathcal{X}$ is a topological vector space) and the computation of the Fenchel–Legendre transform of the moment generating function may involve the solution of variational problems in an infinite dimensional setting. Moreover, proving exponential tightness in $\mathcal{X}$, the main tool of getting at the upper bound, may be a difficult task. On the other hand, the evaluation of the limiting logarithmic moment generating function involves probabilistic computations at the level of real-valued random variables, albeit an infinite number of such computations. It is often relatively easy to derive the LDP for every finite collection of these real-valued random variables. Hence, it is reasonable to inquire if this implies that the laws of the original, $\mathcal{X}$-valued random variables satisfy the LDP.

An affirmative result is derived shortly in a somewhat abstract setting that will serve us well in diverse situations. The idea is to identify $\mathcal{X}$ with the projective limit of a family of spaces $\{\mathcal{Y}_j\}_{j \in J}$ with the hope that the LDP for any given family $\{\mu_\epsilon\}$ of probability measures on $\mathcal{X}$ follows as the consequence of the fact that the LDP holds for any of the projections of μ_ϵ to $\{\mathcal{Y}_j\}_{j \in J}$.

To make the program described precise, we first review a few standard topological definitions. Let $(J, \leq)$ be a partially ordered, right-filtering set. (The latter notion means that for any i, j in J, there exists $k \in J$ such that both $i \leq k$ and $j \leq k$.) Note that J need not be countable. A projective system $(\mathcal{Y}_j, p_{ij})_{i \leq j \in J}$ consists of Hausdorff topological spaces $\{\mathcal{Y}_j\}_{j \in J}$ and continuous maps $p_{ij} : \mathcal{Y}_j \rightarrow \mathcal{Y}_i$ such that $p_{ik} = p_{ij} \circ p_{jk}$ whenever $i \leq j \leq k$ ($\{p_{jj}\}_{j \in J}$ are the appropriate identity maps). The *projective limit* of this system, denoted by $\mathcal{X} = \varprojlim \mathcal{Y}_j$, is the subset of the topological product space $\mathcal{Y} = \prod_{j \in J} \mathcal{Y}_j$, consisting of all the elements $\mathbf{x} = (y_j)_{j \in J}$ for which $y_i = p_{ij}(y_j)$ whenever $i \leq j$, equipped with the topology induced by $\mathcal{Y}$. Projective limits of closed subsets $F_j \subseteq \mathcal{Y}_j$ are defined analogously and denoted $F = \varprojlim F_j$. The canonical projections of $\mathcal{X}$, which are the restrictions $p_j : \mathcal{X} \rightarrow \mathcal{Y}_j$ of the coordinate maps from $\mathcal{Y}$ to $\mathcal{Y}_j$, are continuous. Some properties of projective limits are recalled in Appendix B.

The following theorem yields the LDP in $\mathcal{X}$ as a consequence of the LDPs associated with $\{\mu_\epsilon \circ p_j^{-1}, \epsilon > 0\}$. In order to have a specific example in mind, think of $\mathcal{X}$ as the space of all maps $f : [0,1] \rightarrow \mathbb{R}$ such that $f(0) = 0$, equipped with the topology of pointwise convergence. Then $p_j : \mathcal{X} \rightarrow \mathbb{R}^d$ is the projection of functions onto their values at the time instances $0 \leq t_1 < t_2 < \cdots < t_d \leq 1$, with the partial ordering induced on the set $J = \cup_{d=1}^\infty \{(t_1, \ldots, t_d) : 0 \leq t_1 < t_2 < \cdots < t_d \leq 1\}$ by inclusions. For details of this construction, see Section 5.1.

Theorem 4.6.1 (Dawson–Gärtner) *Let* $\{\mu_\epsilon\}$ *be a family of probability measures on* $\mathcal{X}$, *such that for any* $j \in J$ *the Borel probability measures* $\mu_\epsilon \circ p_j^{-1}$ *on* $\mathcal{Y}_j$ *satisfy the LDP with the good rate function* $I_j(\cdot)$. *Then* $\{\mu_\epsilon\}$ *satisfies the LDP with the good rate function*

$$I(\mathbf{x}) = \sup_{j \in J} \{ I_j(p_j(\mathbf{x})) \}, \qquad \mathbf{x} \in \mathcal{X}. \tag{4.6.2}$$

Remark: Throughout this section, we drop the blanket assumption that $\mathcal{B}_{\mathcal{X}} \subseteq \mathcal{B}$. This is natural in view of the fact that the set J need not be countable. It is worthwhile to note that $\mathcal{B}$ is required to contain all sets $p_j^{-1}(B_j)$, where $B_j \in \mathcal{B}_{\mathcal{Y}_j}$.

Proof: Clearly, $I(\mathbf{x})$ is nonnegative. For any $\alpha \in [0,\infty)$ and $j \in J$, let $\Psi_{I_j}(\alpha)$ denote the *compact* level set of I_j, i.e., $\Psi_{I_j}(\alpha) \triangleq \{y_j : I_j(y_j) \leq \alpha\}$. Recall that for any $i \leq j \in J$, $p_{ij} : \mathcal{Y}_j \to \mathcal{Y}_i$ is a continuous map and $\mu_\epsilon \circ p_i^{-1} = (\mu_\epsilon \circ p_j^{-1}) \circ p_{ij}^{-1}$. Hence, by the contraction principle (Theorem 4.2.1), $I_i(y_i) = \inf_{y_j \in p_{ij}^{-1}(y_i)} I_j(y_j)$, or alternatively, $\Psi_{I_i}(\alpha) = p_{ij}(\Psi_{I_j}(\alpha))$. Therefore,

$$\Psi_I(\alpha) = \mathcal{X} \cap \prod_{j \in J} \Psi_{I_j}(\alpha) = \varprojlim \Psi_{I_j}(\alpha) , \qquad (4.6.3)$$

and $I(\mathbf{x})$ is a good rate function, since by Tychonoff's theorem (Theorem B.3), the projective limit of compact subsets of $\mathcal{Y}_j$, $j \in J$, is a compact subset of $\mathcal{X}$.

In order to prove the large deviations lower bound, it suffices to show that for every measurable set $A \subset \mathcal{X}$ and each $\mathbf{x} \in A^o$, there exists a $j \in J$ such that

$$\liminf_{\epsilon \to 0} \epsilon \log \mu_\epsilon(A) \geq -I_j(p_j(\mathbf{x})) .$$

Since the collection $\{p_j^{-1}(U_j) : U_j \subset \mathcal{Y}_j \text{ is open}\}$ is a base of the topology of $\mathcal{X}$, there exists some $j \in J$ and an open set $U_j \subset \mathcal{Y}_j$ such that $\mathbf{x} \in p_j^{-1}(U_j) \subset A^o$. Thus, by the large deviations lower bound for $\{\mu_\epsilon \circ p_j^{-1}\}$,

$$\liminf_{\epsilon \to 0} \epsilon \log \mu_\epsilon(A) \geq \liminf_{\epsilon \to 0} \epsilon \log(\mu_\epsilon \circ p_j^{-1}(U_j))$$
$$\geq -\inf_{y \in U_j} I_j(y) \geq -I_j(p_j(\mathbf{x})) ,$$

as desired.

Considering the large deviations upper bound, fix a measurable set $A \subset \mathcal{X}$ and let $A_j \triangleq p_j(\overline{A})$. Then, $A_i = p_{ij}(A_j)$ for any $i \leq j$, implying that $p_{ij}(\overline{A_j}) \subseteq \overline{A_i}$ (since p_{ij} are continuous). Hence, $\overline{A} \subseteq \varprojlim \overline{A_j}$. To prove the converse inclusion, fix $\mathbf{x} \in (\overline{A})^c$. Since $(\overline{A})^c$ is an open subset of $\mathcal{X}$, there exists some $j \in J$ and an open set $U_j \subseteq \mathcal{Y}_j$ such that $\mathbf{x} \in p_j^{-1}(U_j) \subseteq (\overline{A})^c$. Consequently, for this value of j, $p_j(\mathbf{x}) \in U_j \subseteq A_j^c$, implying that $p_j(\mathbf{x}) \notin \overline{A_j}$. Hence,

$$\overline{A} = \varprojlim \overline{A_j} . \qquad (4.6.4)$$

Combining this identity with (4.6.3), it follows that for every $\alpha < \infty$,

$$\overline{A} \cap \Psi_I(\alpha) = \varprojlim \left(\overline{A_j} \cap \Psi_{I_j}(\alpha)\right) .$$

Fix $\alpha < \inf_{x \in \overline{A}} I(x)$, for which $\overline{A} \cap \Psi_I(\alpha) = \emptyset$. Then, by Theorem B.4, $\overline{A_j} \cap \Psi_{I_j}(\alpha) = \emptyset$ for some $j \in J$. Therefore, as $A \subseteq p_j^{-1}(\overline{A_j})$, by the LDP upper bound associated with the Borel measures $\{\mu_\epsilon \circ p_j^{-1}\}$,

$$\limsup_{\epsilon \to 0} \epsilon \log \mu_\epsilon(A) \le \limsup_{\epsilon \to 0} \epsilon \log \mu_\epsilon \circ p_j^{-1}(\overline{A_j}) \le -\alpha.$$

This inequality holds for every measurable A and $\alpha < \infty$ such that $\overline{A} \cap \Psi_I(\alpha) = \emptyset$. Consequently, it yields the LDP upper bound for $\{\mu_\epsilon\}$. □

The following lemma is often useful for simplifying the formula (4.6.2) of the Dawson–Gärtner rate function.

Lemma 4.6.5 *If $I(\cdot)$ is a good rate function on $\mathcal{X}$ such that*

$$I_j(y) = \inf\{I(\mathbf{x}) : \mathbf{x} \in \mathcal{X}, \quad y = p_j(\mathbf{x})\}, \tag{4.6.6}$$

for any $y \in \mathcal{Y}_j$, $j \in J$, then the identity (4.6.2) holds.

Proof: Fix $\alpha \in [0, \infty)$ and let A denote the compact level set $\Psi_I(\alpha)$. Since $p_j : \mathcal{X} \to \mathcal{Y}_j$ is continuous for any $j \in J$, by (4.6.6) $A_j \triangleq \Psi_{I_j}(\alpha) = p_j(A)$ is a compact subset of $\mathcal{Y}_j$. With $A_i = p_{ij}(A_j)$ for any $i \le j$, the set $\{\mathbf{x} : \sup_{j \in J} I_j(p_j(\mathbf{x})) \le \alpha\}$ is the projective limit of the closed sets A_j, and as such it is merely the closed set $A = \Psi_I(\alpha)$ (see (4.6.4)). The identity (4.6.2) follows since $\alpha \in [0, \infty)$ is arbitrary. □

The preceding theorem is particularly suitable for situations involving topological vector spaces that satisfy the following assumptions.

Assumption 4.6.7 *Let $\mathcal{W}$ be an infinite dimensional real vector space, and $\mathcal{W}'$ its algebraic dual, i.e., the space of all linear functionals $\lambda \mapsto \langle \lambda, x \rangle : \mathcal{W} \to \mathbb{R}$. The topological (vector) space $\mathcal{X}$ consists of $\mathcal{W}'$ equipped with the $\mathcal{W}$-topology, i.e., the weakest topology such that for each $\lambda \in \mathcal{W}$, the linear functional $x \mapsto \langle \lambda, x \rangle : \mathcal{X} \to \mathbb{R}$ is continuous.*

Remark: The $\mathcal{W}$-topology of $\mathcal{W}'$ makes $\mathcal{W}$ into the topological dual of $\mathcal{X}$, i.e., $\mathcal{W} = \mathcal{X}^*$.

For any $d \in \mathbb{Z}_+$ and $\lambda_1, \ldots, \lambda_d \in \mathcal{W}$, define the projection $p_{\lambda_1, \ldots, \lambda_d} : \mathcal{X} \to \mathbb{R}^d$ by $p_{\lambda_1, \ldots, \lambda_d}(x) = (\langle \lambda_1, x \rangle, \langle \lambda_2, x \rangle, \ldots, \langle \lambda_d, x \rangle)$.

Assumption 4.6.8 *Let $(\mathcal{X}, \mathcal{B}, \mu_\epsilon)$ be probability spaces such that:*
(a) $\mathcal{X}$ satisfies Assumption 4.6.7.
(b) For any $\lambda \in \mathcal{W}$ and any Borel set B in $\mathbb{R}$, $p_\lambda^{-1}(B) \in \mathcal{B}$.

Remark: Note that if $\{\mu_\epsilon\}$ are Borel measures, then Assumption 4.6.8 reduces to Assumption 4.6.7.

Theorem 4.6.9 *Let Assumption 4.6.8 hold. Further assume that for every* $d \in \mathbb{Z}_+$ *and every* $\lambda_1, \ldots, \lambda_d \in \mathcal{W}$, *the measures* $\{\mu_\epsilon \circ p_{\lambda_1, \ldots, \lambda_d}^{-1}, \epsilon > 0\}$ *satisfy the LDP with the good rate function* $I_{\lambda_1, \ldots, \lambda_d}(\cdot)$. *Then* $\{\mu_\epsilon\}$ *satisfies the LDP in* $\mathcal{X}$, *with the good rate function*

$$I(x) = \sup_{d \in \mathbb{Z}_+} \sup_{\lambda_1, \ldots, \lambda_d \in \mathcal{W}} I_{\lambda_1, \ldots, \lambda_d}((\langle \lambda_1, x \rangle, \langle \lambda_2, x \rangle, \ldots, \langle \lambda_d, x \rangle)). \quad (4.6.10)$$

Remark: In most applications, one is interested in obtaining an LDP on $\mathcal{E}$ that is a non-closed subset of $\mathcal{X}$. Hence, the relatively effortless projective limit approach is then followed by an application specific check that $\mathcal{D}_I \subset \mathcal{E}$, as needed for Lemma 4.1.5. For example, in the study of empirical measures on a Polish space Σ, it is known *a priori* that $\mu_\epsilon(M_1(\Sigma)) = 1$ for all $\epsilon > 0$, where $M_1(\Sigma)$ is the space of Borel probability measures on Σ, equipped with the $B(\Sigma)$-topology, and $B(\Sigma) = \{f : \Sigma \to \mathbb{R}, f \text{ bounded}, \text{Borel measurable}\}$. Identifying each $\nu \in M_1(\Sigma)$ with the linear functional $f \mapsto \int_\Sigma f d\nu$, $\forall f \in B(\Sigma)$, it follows that $M_1(\Sigma)$ is homeomorphic to $\mathcal{E} \subset \mathcal{X}$, where here $\mathcal{X}$ denotes the algebraic dual of $B(\Sigma)$ equipped with the $B(\Sigma)$-topology. Thus, $\mathcal{X}$ satisfies Assumption 4.6.7, and $\mathcal{E}$ is not a closed subset of $\mathcal{X}$. It is worthwhile to note that in this setup, μ_ϵ is not necessarily a Borel probability measure.

Proof: Let $\mathcal{V}$ be the system of all finite dimensional linear subspaces of $\mathcal{W}$, equipped with the partial ordering defined by inclusion. To each $V \in \mathcal{V}$, attach its (finite dimensional) algebraic dual V' equipped with the V-topology. The latter are clearly Hausdorff topological spaces. For any $V \subseteq U$ and any linear functional $f : U \to \mathbb{R}$, let $p_{V,U}(f) : V \to \mathbb{R}$ be the restriction of f on the subspace V. The projections $p_{V,U} : U' \to V'$ thus defined are continuous, and compatible with the inclusion ordering of $\mathcal{V}$. Let $\tilde{\mathcal{X}}$ be the projective limit of the system $(V', p_{V,U})$. Consider the map $x \mapsto \tilde{x} = (p_V(x)) \in \tilde{\mathcal{X}}$, where for each $V \in \mathcal{V}$, $p_V(x) \in V'$ is the linear functional $\lambda \mapsto \langle \lambda, x \rangle$, $\forall \lambda \in V$. This map is a bijection between $\mathcal{W}'$ and $\tilde{\mathcal{X}}$, since the consistency conditions in the definition of $\tilde{\mathcal{X}}$ imply that any $\tilde{x} \in \tilde{\mathcal{X}}$ is determined by its values on the *one-dimensional* linear subspaces of $\mathcal{W}$, and any such collection of values determines a point in $\tilde{\mathcal{X}}$. By Assumption 4.6.7, $\mathcal{X}$ consists of the vector space $\mathcal{W}'$ equipped with the $\mathcal{W}$-topology that is generated by the sets $\{x : |\langle \lambda, x \rangle - \rho| < \delta\}$ for $\lambda \in \mathcal{W}, \rho \in \mathbb{R}, \delta > 0$. It is not hard to check that the image of these sets under the map $x \mapsto \tilde{x}$ generates the projective topology of $\tilde{\mathcal{X}}$. Consequently, this map is a homeomorphism between $\mathcal{X}$ and $\tilde{\mathcal{X}}$. Hence, if for every $V \in \mathcal{V}$, $\{\mu_\epsilon \circ p_V^{-1}, \epsilon > 0\}$ satisfies the LDP in V' with the good rate function $I_V(\cdot)$, then by Theorem 4.6.1, $\{\mu_\epsilon\}$ satisfies the LDP in $\mathcal{X}$ with the good rate function $\sup_{V \in \mathcal{V}} I_V(p_V(\cdot))$.

Fix $d \in \mathbb{Z}_+$ and $V \in \mathcal{V}$, a d-dimensional linear subspace of $\mathcal{W}$. Let $\lambda_1, \ldots, \lambda_d$ be any algebraic base of V. Observe that the map $f \mapsto (f(\lambda_1), \ldots, f(\lambda_d))$ is a homeomorphism between V' and $\mathbb{R}^d$ under which the image of $p_V(x) \in V'$ is $p_{\lambda_1, \ldots, \lambda_d}(x) = (\langle \lambda_1, x \rangle, \ldots, \langle \lambda_d, x \rangle) \in \mathbb{R}^d$. Consequently, by our assumptions, the family of Borel probability measures $\{\mu_\epsilon \circ p_V^{-1}, \epsilon > 0\}$ satisfies the LDP in V', and moreover, $I_V(p_V(x)) = I_{\lambda_1, \ldots, \lambda_d}((\langle \lambda_1, x \rangle, \ldots, \langle \lambda_d, x \rangle))$. The proof is complete, as the preceding holds for every $V \in \mathcal{V}$, while because of the contraction principle (Theorem 4.2.1), there is no need to consider only linearly independent $\lambda_1, \ldots, \lambda_d$ in (4.6.10). $\square$

When using Theorem 4.6.9, either the convexity of $I_{\lambda_1, \ldots, \lambda_d}(\cdot)$ or the existence and smoothness of the limiting logarithmic moment generating function $\Lambda(\cdot)$ are relied upon in order to identify the good rate function of (4.6.10) with $\Lambda^*(\cdot)$, in a manner similar to that encountered in Section 4.5.2. This is spelled out in the following corollary.

Corollary 4.6.11 *Let Assumption 4.6.8 hold.*
(a) Suppose that for each $\lambda \in \mathcal{W}$, the limit

$$\Lambda(\lambda) = \lim_{\epsilon \to 0} \epsilon \log \int_{\mathcal{X}} e^{\epsilon^{-1} \langle \lambda, x \rangle} \mu_\epsilon(dx) \qquad (4.6.12)$$

exists as an extended real number, and moreover that for any $d \in \mathbb{Z}_+$ and any $\lambda_1, \ldots, \lambda_d \in \mathcal{W}$, the function

$$g((t_1, \ldots, t_d)) \triangleq \Lambda(\sum_{i=1}^{d} t_i \lambda_i) : \mathbb{R}^d \to (-\infty, \infty]$$

is essentially smooth, lower semicontinuous, and finite in some neighborhood of 0.
Then $\{\mu_\epsilon\}$ satisfies the LDP in $(\mathcal{X}, \mathcal{B})$ with the convex, good rate function

$$\Lambda^*(x) = \sup_{\lambda \in \mathcal{W}} \{\langle \lambda, x \rangle - \Lambda(\lambda)\}. \qquad (4.6.13)$$

(b) Alternatively, if for any $\lambda_1, \ldots, \lambda_d \in \mathcal{W}$, there exists a compact set $K \subset \mathbb{R}^d$ such that $\mu_\epsilon \circ p_{\lambda_1, \ldots, \lambda_d}^{-1}(K) = 1$, and moreover $\{\mu_\epsilon \circ p_{\lambda_1, \ldots, \lambda_d}^{-1}, \epsilon > 0\}$ satisfies the LDP with a convex rate function, then $\Lambda : \mathcal{W} \to \mathbb{R}$ exists, is finite everywhere, and $\{\mu_\epsilon\}$ satisfies the LDP in $(\mathcal{X}, \mathcal{B})$ with the convex, good rate function $\Lambda^*(\cdot)$ as defined in (4.6.13).

Remark: Since $\mathcal{X}$ satisfies Assumption 4.6.7, the only continuous linear functionals on $\mathcal{X}$ are of the form $x \mapsto \langle \lambda, x \rangle$, where $\lambda \in \mathcal{W}$. Consequently, $\mathcal{X}^*$ may be identified with $\mathcal{W}$, and $\Lambda^*(\cdot)$ is the Fenchel–Legendre transform of $\Lambda(\cdot)$ as defined in Section 4.5.

Proof: (a) Fix $d \in \mathbb{Z}_+$ and $\lambda_1, \ldots, \lambda_d \in \mathcal{W}$. Note that the limiting logarithmic moment generating function associated with $\{\mu_\epsilon \circ p_{\lambda_1, \ldots, \lambda_d}^{-1}, \epsilon > 0\}$ is $g((t_1, \ldots, t_d))$. Hence, by our assumptions, the Gärtner–Ellis theorem (Theorem 2.3.6) implies that these measures satisfy the LDP in $\mathbb{R}^d$ with the good rate function $I_{\lambda_1, \ldots, \lambda_d} = g^* : \mathbb{R}^d \to [0, \infty]$, where

$$I_{\lambda_1, \ldots, \lambda_d}(((\langle \lambda_1, x \rangle, \langle \lambda_2, x \rangle, \ldots, \langle \lambda_d, x \rangle)))$$

$$= \sup_{t_1, \ldots, t_d \in \mathbb{R}} \left\{ \sum_{i=1}^d t_i \langle \lambda_i, x \rangle - \Lambda\left(\sum_{i=1}^d t_i \lambda_i \right) \right\}.$$

Consequently, for every $x \in \mathcal{X}$,

$$I_{\lambda_1, \ldots, \lambda_d}(((\langle \lambda_1, x \rangle, \langle \lambda_2, x \rangle, \ldots, \langle \lambda_d, x \rangle))) \leq \Lambda^*(x) = \sup_{\lambda \in \mathcal{W}} I_\lambda((\langle \lambda, x \rangle)).$$

Since the preceding holds for every $\lambda_1, \ldots, \lambda_d \in \mathcal{W}$, the LDP of $\{\mu_\epsilon\}$ with the good rate function $\Lambda^*(\cdot)$ is a direct consequence of Theorem 4.6.9.

(b) Fix $d \in \mathbb{Z}_+$ and $\lambda_1, \ldots, \lambda_d \in \mathcal{W}$. Since $\mu_\epsilon \circ p_{\lambda_1, \ldots, \lambda_d}^{-1}$ are supported on a compact set K, they satisfy the boundedness condition (4.5.11). Hence, by our assumptions, Theorem 4.5.10 applies. It then follows that the limiting moment generating function $g(\cdot)$ associated with $\{\mu_\epsilon \circ p_{\lambda_1, \ldots, \lambda_d}^{-1}, \epsilon > 0\}$ exists, and the LDP for these probability measures is controlled by $g^*(\cdot)$. With $I_{\lambda_1, \ldots, \lambda_d} = g^*$ for any $\lambda_1, \ldots, \lambda_d \in \mathcal{W}$, the proof is completed as in part (a). $\qquad \square$

The following corollary of the projective limit approach is a somewhat stronger version of Corollary 4.5.27.

Corollary 4.6.14 *Let $\{\mu_\epsilon\}$ be an exponentially tight family of Borel probability measures on the locally convex Hausdorff topological vector space $\mathcal{E}$. Suppose $\Lambda(\cdot) = \lim_{\epsilon \to 0} \epsilon \Lambda_{\mu_\epsilon}(\cdot/\epsilon)$ is finite valued and Gateaux differentiable. Then $\{\mu_\epsilon\}$ satisfies the LDP in $\mathcal{E}$ with the convex, good rate function Λ^*.*

Proof: Let $\mathcal{W}$ be the topological dual of $\mathcal{E}$. Suppose first that $\mathcal{W}$ is an infinite dimensional vector space, and define $\mathcal{X}$ according to Assumption 4.6.7. Let $i : \mathcal{E} \to \mathcal{X}$ denote the map $x \mapsto i(x)$, where $i(x)$ is the linear functional $\lambda \mapsto \langle \lambda, x \rangle$, $\forall \lambda \in \mathcal{W}$. Since $\mathcal{E}$ is a locally convex topological vector space, by the Hahn–Banach theorem, $\mathcal{W}$ is separating. Therefore, $\mathcal{E}$ when equipped with the *weak* topology is Hausdorff, and i is a homeomorphism between this topological space and $i(\mathcal{E}) \subset \mathcal{X}$. Consequently, $\{\mu_\epsilon \circ i^{-1}\}$ are Borel probability measures on $\mathcal{X}$ such that $\mu_\epsilon \circ i^{-1}(i(\mathcal{E})) = 1$ for all $\epsilon > 0$. All the conditions in part (a) of Corollary 4.6.11 hold for $\{\mu_\epsilon \circ i^{-1}\}$, since we assumed that $\Lambda : \mathcal{W} \to \mathbb{R}$ exists, and is a finite valued, Gateaux differentiable function. Hence, $\{\mu_\epsilon \circ i^{-1}\}$ satisfies the LDP in $\mathcal{X}$ with the convex, good rate function $\Lambda^*(\cdot)$. Recall that $i : \mathcal{E} \to \mathcal{X}$ is a continuous injection with respect to the weak topology on $\mathcal{E}$, and hence it is also continuous with respect to the original topology on $\mathcal{E}$. Now, the exponential tightness of $\{\mu_\epsilon\}$, Theorem 4.2.4, and the remark following it, imply that $\{\mu_\epsilon\}$ satisfies the LDP in $\mathcal{E}$ with the good rate function $\Lambda^*(\cdot)$.

We now turn to settle the (trivial) case where $\mathcal{W}$ is a d-dimensional vector space for some $d < \infty$. Observe that then $\mathcal{X}$ is of the same dimension as $\mathcal{W}$. The finite dimensional topological vector space $\mathcal{X}$ can be represented as $\mathbb{R}^d$. Hence, our assumptions about the function $\Lambda(\cdot)$ imply the LDP in $\mathcal{X}$ associated with $\{\mu_\epsilon \circ i^{-1}\}$ by a direct application of the Gärtner–Ellis theorem (Theorem 2.3.6). The LDP (in $\mathcal{E}$) associated with $\{\mu_\epsilon\}$ follows exactly as in the infinite dimensional case. $\square$

Exercise 4.6.15 Suppose that all the conditions of Corollary 4.6.14 hold except for the exponential tightness of $\{\mu_\epsilon\}$. Prove that $\{\mu_\epsilon\}$ satisfies a weak LDP with respect to the *weak* topology on $\mathcal{E}$, with the rate function $\Lambda^*(\cdot)$ defined in (4.6.13).
Hint: Follow the proof of the corollary and observe that the LDP of $\{\mu_\epsilon \circ i^{-1}\}$ in $\mathcal{X}$ still holds. Note that if $K \subset \mathcal{E}$ is weakly compact, then $i(K) \subset i(\mathcal{E})$ is a compact subset of $\mathcal{X}$.

4.7 The LDP and Weak Convergence in Metric Spaces

Throughout this section $(\mathcal{X}, d)$ is a metric space and all probability measures are Borel. For $\delta > 0$, let

$$A^{\delta,o} \triangleq \{y : \; d(y, A) \triangleq \inf_{z \in A} d(y, z) < \delta\} \tag{4.7.1}$$

denote the open blowups of A (compare with (4.1.8)), with $A^{-\delta} = ((A^c)^{\delta,o})^c$ a closed set (possibly empty). The proof of the next lemma which summarizes immediate relations between these sets is left as Exercise 4.7.18.

Lemma 4.7.2 *For any $\delta > 0$, $\eta > 0$ and $\Gamma \subset \mathcal{X}$*
(a) $(\Gamma^{-\delta})^{\delta,o} \subset \Gamma \subset (\Gamma^{\delta,o})^{-\delta}$.
(b) $\Gamma^{-(\delta+\eta)} \subset (\Gamma^{-\delta})^{-\eta}$ and $(\Gamma^{\delta,o})^{\eta,o} \subset \Gamma^{(\delta+\eta),o}$.
(c) $G^{-\delta}$ increases to G for any open set G and $F^{\delta,o}$ decreases to F for any closed set F.

Let $\mathcal{Q}(\mathcal{X})$ denote the collection of set functions $\nu : \mathcal{B}_\mathcal{X} \to [0,1]$ such that:

(a) $\nu(\emptyset) = 0$.

(b) $\nu(\Gamma) = \inf\{\nu(G) : \Gamma \subset G \text{ open}\}$ for any $\Gamma \in \mathcal{B}_\mathcal{X}$.

(c) $\nu(\cup_{i=1}^\infty \Gamma_i) \leq \sum_{i=1}^\infty \nu(\Gamma_i)$ for any. $\Gamma_i \in \mathcal{B}_\mathcal{X}$.

(d) $\nu(G) = \lim_{\delta \to 0} \nu(G^{-\delta})$ for any open set $G \subset \mathcal{X}$.

Condition (b) implies the monotonicity property $\nu(A) \leq \nu(B)$ whenever $A \subset B$.

The following important subset of $\mathcal{Q}(\mathcal{X})$ represents the rate functions.

Definition 4.7.3 *A set function $\nu : \mathcal{B}_\mathcal{X} \to [0,1]$ is called a sup-measure if $\nu(\Gamma) = \sup_{y \in \Gamma} \nu(\{y\})$ for any $\Gamma \in \mathcal{B}_\mathcal{X}$ and $\nu(\{y\})$ is an upper semicontinuous function of $y \in \mathcal{X}$. With a sup-measure ν uniquely characterized by the rate function $I(y) = -\log \nu(\{y\})$, we adopt the notation $\nu = e^{-I}$.*

The next lemma explains why $\mathcal{Q}(\mathcal{X})$ is useful for exploring similarities between the LDP and the well known theory of weak convergence of probability measures.

Lemma 4.7.4 $\mathcal{Q}(\mathcal{X})$ *contains all sup-measures and all set functions of the form μ^ϵ for μ a probability measure on $\mathcal{X}$ and $\epsilon \in (0,1]$.*

Proof: Conditions (a) and (c) trivially hold for any sup-measure. Since any point y in an open set G is also in $G^{-\delta}$ for some $\delta = \delta(y) > 0$, all sup-measures satisfy condition (d). For (b), let $\nu(\{y\}) = e^{-I(y)}$. Fix $\Gamma \in \mathcal{B}_\mathcal{X}$ and $G(x, \delta)$ as in (4.1.3), such that

$$e^\delta \nu(\{x\}) = e^{-(I(x)-\delta)} \geq e^{-\inf_{y \in G(x,\delta)} I(y)} = \sup_{y \in G(x,\delta)} \nu(\{y\}) .$$

It follows that for the open set $G_\delta = \cup_{x \in \Gamma} G(x, \delta)$,

$$e^\delta \nu(\Gamma) = e^\delta \sup_{x \in \Gamma} \nu(\{x\}) \geq \sup_{y \in G_\delta} \nu(\{y\}) = \nu(G_\delta) .$$

Taking $\delta \to 0$, we have condition (b) holding for an arbitrary $\Gamma \in \mathcal{B}_\mathcal{X}$.

Turning to the second part of the lemma, note that conditions (a)–(d) hold when ν is a probability measure. Suppose next that $\nu(\cdot) = f(\mu(\cdot))$ for a probability measure μ and $f \in C_b([0,1])$ non-decreasing such that $f(0) = 0$ and $f(p+q) \leq f(p) + f(q)$ for $0 \leq p \leq 1 - q \leq 1$. By induction, $f(\sum_{i=1}^k p_i) \leq \sum_{i=1}^k f(p_i)$ for all $k \in \mathbb{Z}_+$ and non-negative p_i such that $\sum_{i=1}^k p_i \leq 1$. The continuity of $f(\cdot)$ at 0 extends this property to $k =$

∞. Therefore, condition (c) holds for ν by the subadditivity of μ and monotonicity of $f(\cdot)$. The set function ν inherits conditions (b) and (d) from μ by the continuity and monotonicity of $f(\cdot)$. Similarly, it inherits condition (a) because $f(0) = 0$. In particular, this applies to $f(p) = p^\epsilon$ for all $\epsilon \in (0, 1]$. $\square$

The next definition of convergence in $\mathcal{Q}(\mathcal{X})$ coincides by the Portmanteau theorem with weak convergence when restricted to probability measures ν_ϵ, ν_0 (see Theorem D.10 for $\mathcal{X}$ Polish).

Definition 4.7.5 $\nu_\epsilon \to \nu_0$ in $\mathcal{Q}(\mathcal{X})$ *if for any closed set* $F \subset \mathcal{X}$

$$\limsup_{\epsilon \to 0} \nu_\epsilon(F) \le \nu_0(F), \qquad (4.7.6)$$

and for any open set $G \subset \mathcal{X}$,

$$\liminf_{\epsilon \to 0} \nu_\epsilon(G) \ge \nu_0(G). \qquad (4.7.7)$$

For probability measures ν_ϵ, ν_0 the two conditions (4.7.6) and (4.7.7) are equivalent. However, this is not the case in general. For example, if $\nu_0(\cdot) \equiv 0$ (an element of $\mathcal{Q}(\mathcal{X})$), then (4.7.7) holds for any ν_ϵ but (4.7.6) fails unless $\nu_\epsilon(\mathcal{X}) \to 0$.

For a family of probability measures $\{\mu_\epsilon\}$, the convergence of $\nu_\epsilon = \mu_\epsilon^\epsilon$ to a sup-measure $\nu_0 = e^{-I}$ is exactly the LDP statement (compare (4.7.6) and (4.7.7) with (1.2.12) and (1.2.13), respectively).

With this in mind, we next extend the definition of tightness and uniform tightness from $M_1(\mathcal{X})$ to $\mathcal{Q}(\mathcal{X})$ in such a way that a sup-measure $\nu = e^{-I}$ is tight if and only if the corresponding rate function is good, and exponential tightness of $\{\mu_\epsilon\}$ is essentially the same as uniform tightness of the set functions $\{\mu_\epsilon^\epsilon\}$.

Definition 4.7.8 *A set function* $\nu \in \mathcal{Q}(\mathcal{X})$ *is tight if for each* $\eta > 0$, *there exists a compact set* $K_\eta \subset \mathcal{X}$ *such that* $\nu(K_\eta^c) < \eta$. *A collection* $\{\nu_\epsilon\} \subset \mathcal{Q}(\mathcal{X})$ *is uniformly tight if the set* K_η *may be chosen independently of* ϵ.

The following lemma provides a useful consequence of tightness in $\mathcal{Q}(\mathcal{X})$.

Lemma 4.7.9 *If* $\nu \in \mathcal{Q}(\mathcal{X})$ *is tight, then for any* $\Gamma \in \mathcal{B}_{\mathcal{X}}$,

$$\nu(\overline{\Gamma}) = \lim_{\delta \to 0} \nu(\Gamma^{\delta, o}). \qquad (4.7.10)$$

Remark: For sup-measures this is merely part (b) of Lemma 4.1.6.

Proof: Fix a non-empty set $\Gamma \in \mathcal{B}_\mathcal{X}$, $\eta > 0$ and a compact set $K = K_\eta$ for which $\nu(K_\eta^c) < \eta$. For any open set $G \subset \mathcal{X}$ such that $\overline{\Gamma} \subset G$, either $K \subset G$ or else the non-empty compact set $K \cap G^c$ and the closed set $\overline{\Gamma}$ are disjoint, with $\inf_{x \in K \cap G^c} d(x, \overline{\Gamma}) > 0$. In both cases, $\Gamma^{\delta,o} \cap K = \overline{\Gamma}^{\delta,o} \cap K \subset G$ for some $\delta > 0$, and by properties (b), (c) and monotonicity of set functions in $\mathcal{Q}(\mathcal{X})$,

$$\nu(\overline{\Gamma}) = \inf\{\nu(G) : \overline{\Gamma} \subset G \text{ open }\} \geq \lim_{\delta \to 0} \nu(\Gamma^{\delta,o} \cap K)$$
$$\geq \lim_{\delta \to 0} \nu(\Gamma^{\delta,o}) - \eta \geq \nu(\overline{\Gamma}) - \eta .$$

The limit as $\eta \to 0$ yields (4.7.10). $\qquad\square$

For $\tilde{\nu}, \nu \in \mathcal{Q}(\mathcal{X})$, let

$$\rho(\tilde{\nu}, \nu) \overset{\triangle}{=} \inf\{\delta > 0 : \tilde{\nu}(F) \leq \nu(F^{\delta,o}) + \delta \ \forall F \subset \mathcal{X} \text{ closed,}$$
$$\tilde{\nu}(G) \geq \nu(G^{-\delta}) - \delta \ \forall G \subset \mathcal{X} \text{ open }\} \qquad (4.7.11)$$

When $\rho(\cdot, \cdot)$ is restricted to $M_1(\mathcal{X}) \times M_1(\mathcal{X})$, it coincides with the Lévy metric (see Theorem D.8). Indeed, in this special case, if $\delta > 0$ is such that $\tilde{\nu}(F) \leq \nu(F^{\delta,o}) + \delta$ for a closed set $F \subset \mathcal{X}$, then $\tilde{\nu}(G) \geq \nu((F^{\delta,o})^c) - \delta = \nu(G^{-\delta}) - \delta$ for the open set $G = F^c$.

The next theorem shows that in analogy with the theory of weak convergence, $(\mathcal{Q}(\mathcal{X}), \rho)$ is a metric space for which convergence to a tight limit point is characterized by Definition 4.7.5.

Theorem 4.7.12
(a) $\rho(\cdot, \cdot)$ is a metric on $\mathcal{Q}(\mathcal{X})$.
(b) For ν_0 tight, $\rho(\nu_\epsilon, \nu_0) \to 0$ if and only if $\nu_\epsilon \to \nu_0$ in $\mathcal{Q}(\mathcal{X})$.

Remarks:
(a) By Theorem 4.7.12, the Borel probability measures $\{\mu_\epsilon\}$ satisfy the LDP in $(\mathcal{X}, d)$ with good rate function $I(\cdot)$ if and only if $\rho(\mu_\epsilon, e^{-I}) \to 0$.
(b) In general, one can not dispense of tightness of $\nu_0 = e^{-I}$ when relating the $\rho(\nu_\epsilon, \nu_0)$ convergence to the LDP. Indeed, with μ_1 a probability measure on $\mathbb{R}$ such that $d\mu_1/dx = C/(1 + |x|^2)$ it is easy to check that $\mu_\epsilon(\cdot) \overset{\triangle}{=} \mu_1(\cdot/\epsilon)$ satisfies the LDP in $\mathbb{R}$ with rate function $I(\cdot) \equiv 0$ while considering the open sets $G_x = (x, \infty)$ for $x \to \infty$ we see that $\rho(\mu_\epsilon, e^{-I}) = 1$ for all $\epsilon > 0$.
(c) By part (a) of Lemma 4.7.2, $F \subset G^{-\delta}$ for the open set $G = F^{\delta,o}$ and $F^{\delta,o} \subset G$ for the closed set $F = G^{-\delta}$. Therefore, the monotonicity of the set functions $\tilde{\nu}, \nu \in \mathcal{Q}(\mathcal{X})$, results with

$$\rho(\tilde{\nu}, \nu) = \inf\{\delta > 0 : \quad \tilde{\nu}(F) \leq \nu(F^{\delta,o}) + \delta \quad \text{and} \qquad (4.7.13)$$
$$\nu(F) \leq \tilde{\nu}(F^{\delta,o}) + \delta \ \forall F \subset \mathcal{X} \text{ closed }\}.$$

Proof: (a) The alternative definition (4.7.13) of ρ shows that it is a non-negative, symmetric function, such that $\rho(\nu, \nu) = 0$ (by the monotonicity of set functions in $\mathcal{Q}(\mathcal{X})$). If $\rho(\tilde{\nu}, \nu) = 0$, then by (4.7.11), for any open set $G \subset \mathcal{X}$,

$$\tilde{\nu}(G) \geq \limsup_{\delta \to 0}[\nu(G^{-\delta}) - \delta] = \nu(G)$$

(see property (d) of set functions in $\mathcal{Q}(\mathcal{X})$). Since ρ is symmetric, by same reasoning also $\nu(G) \geq \tilde{\nu}(G)$, so that $\tilde{\nu}(G) = \nu(G)$ for every open set $G \subset \mathcal{X}$. Thus, by property (b) of set functions in $\mathcal{Q}(\mathcal{X})$ we conclude that $\tilde{\nu} = \nu$.

Fix $\tilde{\nu}, \nu, \omega \in \mathcal{Q}(\mathcal{X})$ and $\delta > \rho(\tilde{\nu}, \omega)$, $\eta > \rho(\omega, \nu)$. Then, by (4.7.11) and part (b) of Lemma 4.7.2, for any closed set $F \subset \mathcal{X}$,

$$\tilde{\nu}(F) \leq \omega(F^{\delta,o}) + \delta \leq \nu((F^{\delta,o})^{\eta,o}) + \delta + \eta \leq \nu(F^{(\delta+\eta),o}) + \delta + \eta \,.$$

By symmetry of ρ we can reverse the roles of $\tilde{\nu}$ and ν, hence concluding by (4.7.13) that $\rho(\tilde{\nu}, \nu) \leq \delta + \eta$. Taking $\delta \to \rho(\tilde{\nu}, \omega)$ and $\eta \to \rho(\omega, \nu)$ we have the triangle inequality $\rho(\tilde{\nu}, \nu) \leq \rho(\tilde{\nu}, \omega) + \rho(\omega, \nu)$.

(b) Suppose $\rho(\nu_\epsilon, \nu_0) \to 0$ for tight $\nu_0 \in \mathcal{Q}(\mathcal{X})$. By (4.7.11), for any open set $G \subset \mathcal{X}$,

$$\liminf_{\epsilon \to 0} \nu_\epsilon(G) \geq \lim_{\delta \to 0}(\nu_0(G^{-\delta}) - \delta) = \nu_0(G) \,,$$

yielding the lower bound (4.7.7). Similarly, by (4.7.11) and Lemma 4.7.9, for any closed set $F \subset \mathcal{X}$

$$\limsup_{\epsilon \to 0} \nu_\epsilon(F) \leq \lim_{\delta \to 0} \nu_0(F^{\delta,o}) = \nu_0(F) \,.$$

Thus, the upper bound (4.7.6) holds for any closed set $F \subset \mathcal{X}$ and so $\nu_\epsilon \to \nu_0$.

Suppose now that $\nu_\epsilon \to \nu_0$ for tight $\nu_0 \in \mathcal{Q}(\mathcal{X})$. Fix $\eta > 0$ and a compact set $K = K_\eta$ such that $\nu_0(K^c) < \eta$. Extract a finite cover of K by open balls of radius $\eta/2$, each centered in K. Let $\{\Gamma_i; i = 0, \ldots, M\}$ be the *finite* collection of all unions of elements of this cover, with $\Gamma_0 \supset K$ denoting the union of all the elements of the cover. Since $\nu_0(\Gamma_0^c) < \eta$, by (4.7.6) also $\nu_\epsilon(\Gamma_0^c) \leq \eta$ for some $\epsilon_0 > 0$ and all $\epsilon \leq \epsilon_0$. For any closed set $F \subset \mathcal{X}$ there exists an $i \in \{0, \ldots, M\}$ such that

$$(F \cap \Gamma_0) \subset \overline{\Gamma}_i \subset F^{2\eta,o} \tag{4.7.14}$$

(take for Γ_i the union of those elements of the cover that intersect $F \cap \Gamma_0$). Thus, for $\epsilon \leq \epsilon_0$, by monotonicity and subadditivity of ν_ϵ, ν_0 and by the choice of K,

$$\nu_\epsilon(F) \leq \nu_\epsilon(F \cap \Gamma_0) + \nu_\epsilon(\Gamma_0^c) \leq \max_{0 \leq i \leq M}\{\nu_\epsilon(\overline{\Gamma}_i) - \nu_0(\overline{\Gamma}_i)\} + \nu_0(F^{2\eta,o}) + \eta.$$

With ϵ_0, M, and $\{\Gamma_i\}$ independent of F, since $\nu_\epsilon \to \nu_0$, it thus follows that

$$\limsup_{\epsilon \to 0} \sup_{F \text{ closed}} (\nu_\epsilon(F) - \nu_0(F^{2\eta,o})) \leq \eta. \qquad (4.7.15)$$

For an open set $G \subset \mathcal{X}$, let $F = G^{-2\eta}$ and note that (4.7.14) still holds with Γ_i replacing $\overline{\Gamma}_i$. Hence, reverse the roles of ν_0 and ν_ϵ to get for all $\epsilon \leq \epsilon_0$,

$$\nu_0(G^{-2\eta}) \leq \max_{0 \leq i \leq M} \{\nu_0(\Gamma_i) - \nu_\epsilon(\Gamma_i)\} + \nu_\epsilon((G^{-2\eta})^{2\eta,o}) + \eta. \qquad (4.7.16)$$

Recall that $(G^{-2\eta})^{2\eta,o} \subset G$ by Lemma 4.7.2. Hence, by (4.7.7), (4.7.16), and monotonicity of ν_ϵ

$$\limsup_{\epsilon \to 0} \sup_{G \text{ open}} (\nu_0(G^{-2\eta}) - \nu_\epsilon(G)) \leq \eta. \qquad (4.7.17)$$

Combining (4.7.15) and (4.7.17), we see that $\rho(\nu_\epsilon, \nu_0) \leq 2\eta$ for all ϵ small enough. Taking $\eta \to 0$, we conclude that $\rho(\nu_\epsilon, \nu_0) \to 0$. $\qquad \square$

Exercise 4.7.18 Prove Lemma 4.7.2.

4.8 Historical Notes and References

A statement of the LDP in a general setup appears in various places, *c.f.* [Var66, FW84, St84, Var84]. As mentioned in the historical notes referring to Chapter 2, various forms of this principle in specific applications have appeared earlier. The motivation for Theorem 4.1.11 and Lemma 4.1.21 comes from the analysis of [Rue67] and [Lan73].

Exercise 4.1.10 is taken from [LyS87]. Its converse, Lemma 4.1.23, is proved in [Puk91]. In that paper and in its follow-up [Puk94a], Pukhalskii derives many other parallels between exponential convergence in the form of large deviations and weak convergence. Our exposition of Lemma 4.1.23 follows that of [deA97a]. Other useful criteria for exponential tightness exist; see, for example, Theorem 3.1 in [deA85a].

The contraction principle was used by Donsker and Varadhan [DV76] in their treatment of Markov chains empirical measures. Statements of approximate contraction principles play a predominant role in Azencott's study of the large deviations for sample paths of diffusion processes [Aze80]. A general approximate contraction principle appears also in [DeuS89b]. The concept of exponentially good approximation is closely related to the comparison principle of [BxJ88, BxJ96]. In particular, the latter motivates

Exercises 4.2.29 and 4.2.30. For the extension of most of the results of Section 4.2.2 to $\mathcal{Y}$ a completely regular topological space, see [EicS96]. Finally, the inverse contraction principle in the form of Theorem 4.2.4 and Corollary 4.2.6 is taken from [Io91a].

The original version of Varadhan's lemma appears in [Var66]. As mentioned in the text, this lemma is related to *Laplace's method* in an abstract setting. See [Mal82] for a simple application in $\mathbb{R}^1$. For more on this method and its refinements, see the historical notes of Chapters 5 and 6. The inverse to Varadhan's lemma stated here is a modification of [Bry90], which also proves a version of Theorem 4.4.10.

The form of the upper bound presented in Section 4.5.1 dates back (for the empirical mean of real valued i.i.d. random variables) to Cramér and Chernoff. The bound of Theorem 4.5.3 appears in [Gär77] under additional restrictions, which are removed by Stroock [St84] and de Acosta [deA85a]. A general procedure for extending the upper bound from compact sets to closed sets without an exponential tightness condition is described in [DeuS89b], Chapter 5.1. For another version geared towards weak topologies see [deA90]. Exercise 4.5.5 and the specific computation in Exercise 6.2.19 are motivated by the derivation in [ZK95].

Convex analysis played a prominent role in the derivation of the LDP. As seen in Chapter 2, convex analysis methods had already made their entrance in $\mathbb{R}^d$. They were systematically used by Lanford and Ruelle in their treatment of thermodynamical limits via sub-additivity, and later applied in the derivation of Sanov's theorem (*c.f.* the historical notes of Chapter 6). Indeed, the statements here build on [DeuS89b] with an eye to the weak LDP presented by Bahadur and Zabell [BaZ79]. The extension of the Gärtner–Ellis theorem to the general setup of Section 4.5.3 borrows mainly from [Bal88] (who proved implicitly Theorem 4.5.20) and [Io91b]. For other variants of Corollaries 4.5.27 and 4.6.14, see also [Kif90a, deA94c, OBS96].

The projective limits approach to large deviations was formalized by Dawson and Gärtner in [DaG87], and was used in the context of obtaining the LDP for the empirical process by Ellis [Ell88] and by Deuschel and Stroock [DeuS89b]. It is a powerful tool for proving large deviations statements, as demonstrated in Section 5.1 (when combined with the inverse contraction principle) and in Section 6.4. The identification Lemma 4.6.5 is taken from [deA97a], where certain variants and generalizations of Theorem 4.6.1 are also provided. See also [deA94c] for their applications.

Our exposition of Section 4.7 is taken from [Jia95] as is Exercise 4.1.32. In [OBV91, OBV95, OBr96], O'Brien and Vervaat provide a comprehensive abstract unified treatment of weak convergence and of large deviation theory, a small part of which inspired Lemma 4.1.24 and its consequences.

Chapter 5

Sample Path Large Deviations

The finite dimensional LDPs considered in Chapter 2 allow computations of the tail behavior of rare events associated with various sorts of empirical means. In many problems, the interest is actually in rare events that depend on a collection of random variables, or, more generally, on a random process. Whereas some of these questions may be cast in terms of empirical measures, this is not always the most fruitful approach. Interest often lies in the probability that a *path* of a random process hits a particular set. Questions of this nature are addressed in this chapter. In Section 5.1, the case of a random walk, the simplest example of all, is analyzed. The Brownian motion counterpart is then an easy application of exponential equivalence, and the diffusion case follows by suitable approximate contractions. The range of applications presented in this chapter is also representative: stochastic dynamical systems (Sections 5.4, 5.7, and 5.8), DNA matching problems and statistical change point questions (Section 5.5).

In this chapter, all probability measures are Borel with the appropriate completion. Since all processes involved here are separable, all measurability issues are obvious, and we shall not bother to make them precise. A word of caution is that these issues have to be considered when more complex processes are involved, particularly in the case of general continuous time Markov processes.

A. Dembo, O. Zeitouni, *Large Deviations Techniques and Applications*, 175
Stochastic Modelling and Applied Probability 38,
DOI 10.1007/978-3-642-03311-7_5,
© Springer-Verlag Berlin Heidelberg 1998, corrected printing 2010

5.1 Sample Path Large Deviations for Random Walks

Let $X_1, X_2, \ldots$ be a sequence of i.i.d. random vectors taking values in $\mathbb{R}^d$, with $\Lambda(\lambda) \triangleq \log E(e^{\langle \lambda, X_1 \rangle}) < \infty$ for all $\lambda \in \mathbb{R}^d$. Cramér's theorem (Theorem 2.2.30) allows the analysis of the large deviations of $\frac{1}{n} \sum_{i=1}^n X_i$. Similarly, the large deviations behavior of the pair of random variables $\frac{1}{n} \sum_{i=1}^n X_i$ and $\frac{1}{n} \sum_{i=1}^{[n/2]} X_i$ can be obtained, where $[c]$ as usual denotes the integer part of c. In this section, the large deviations joint behavior of a family of random variables indexed by t is considered.

Define

$$Z_n(t) = \frac{1}{n} \sum_{i=1}^{[nt]} X_i, \quad 0 \le t \le 1, \tag{5.1.1}$$

and let μ_n be the law of $Z_n(\cdot)$ in $L_\infty([0,1])$. Throughout, $|x| \triangleq \sqrt{\langle x, x \rangle}$ denotes the Euclidean norm on $\mathbb{R}^d$, $\| f \|$ denotes the supremum norm on $L_\infty([0,1])$, and $\Lambda^*(x) \triangleq \sup_{\lambda \in \mathbb{R}^d} [\langle \lambda, x \rangle - \Lambda(\lambda)]$ denotes the Fenchel–Legendre transform of $\Lambda(\cdot)$.

The following theorem is the main result of this section.

Theorem 5.1.2 (Mogulskii) *The measures μ_n satisfy in $L_\infty([0,1])$ the LDP with the good rate function*

$$I(\phi) = \begin{cases} \int_0^1 \Lambda^*(\dot{\phi}(t)) \, dt, & \text{if } \phi \in \mathcal{AC}, \phi(0) = 0 \\ \\ \infty & \text{otherwise}, \end{cases} \tag{5.1.3}$$

where $\mathcal{AC}$ denotes the space of absolutely continuous functions, i.e.,

$$\mathcal{AC} \triangleq \Big\{ \phi \in C([0,1]) :$$

$$\sum_{\ell=1}^k |t_\ell - s_\ell| \to 0, s_\ell < t_\ell \le s_{\ell+1} < t_{\ell+1} \implies \sum_{\ell=1}^k |\phi(t_\ell) - \phi(s_\ell)| \to 0 \Big\}.$$

Remarks:
(a) Recall that $\phi : [0,1] \to \mathbb{R}^d$ absolutely continuous implies that ϕ is differentiable almost everywhere; in particular, that it is the integral of an $L_1([0,1])$ function.
(b) Since $\{\mu_n\}$ are supported on the space of functions continuous from the right and having left limits, of which $\mathcal{D}_I$ is a subset, the preceding LDP holds in this space when equipped with the supremum norm topology. In

fact, all steps of the proof would have been the same had we been working in that space, instead of $L_\infty([0,1])$, throughout.

(c) Theorem 5.1.2 possesses extensions to stochastic processes with jumps at random times; To avoid measurability problems, one usually works in the space of functions continuous from the right and having left limits, equipped with a topology which renders the latter Polish (the Skorohod topology). Results may then be strengthened to the supremum norm topology by using Exercise 4.2.9.

The proof of Theorem 5.1.2 is based on the following three lemmas, whose proofs follow the proof of the theorem. For an alternative proof, see Section 7.2.

Lemma 5.1.4 *Let $\tilde{\mu}_n$ denote the law of $\tilde{Z}_n(\cdot)$ in $L_\infty([0,1])$, where*

$$\tilde{Z}_n(t) \stackrel{\triangle}{=} Z_n(t) + \left(t - \frac{[nt]}{n}\right) X_{[nt]+1} \tag{5.1.5}$$

is the polygonal approximation of $Z_n(t)$. Then the probability measures μ_n and $\tilde{\mu}_n$ are exponentially equivalent in $L_\infty([0,1])$.

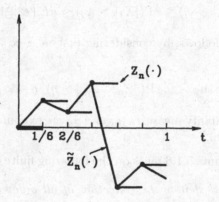

Figure 5.1.1: Z_n and $\tilde{Z}_n$ for $n = 6$.

Lemma 5.1.6 *Let $\mathcal{X}$ consist of all the maps from $[0,1]$ to $\mathbb{R}^d$ such that $t = 0$ is mapped to the origin, and equip $\mathcal{X}$ with the topology of pointwise convergence on $[0,1]$. Then the probability measures $\tilde{\mu}_n$ of Lemma 5.1.4 (defined on $\mathcal{X}$ by the natural embedding) satisfy the LDP in this Hausdorff topological space with the good rate function $I(\cdot)$ of (5.1.3).*

Lemma 5.1.7 *The probability measures $\tilde{\mu}_n$ are exponentially tight in the space $C_0([0,1])$ of all continuous functions $f : [0,1] \to \mathbb{R}^d$ such that $f(0) = 0$, equipped with the supremum norm topology.*

Proof of Theorem 5.1.2: By Lemma 5.1.6, $\{\tilde{\mu}_n\}$ satisfies the LDP in $\mathcal{X}$. Note that $\mathcal{D}_I \subset C_0([0,1])$, and by (5.1.1) and (5.1.5), $\tilde{\mu}_n(C_0([0,1])) = 1$ for all n. Thus, by Lemma 4.1.5, the LDP for $\{\tilde{\mu}_n\}$ also holds in the space $C_0([0,1])$ when equipped with the relative (Hausdorff) topology induced by $\mathcal{X}$. The latter is the pointwise convergence topology, which is generated by the sets $V_{t,x,\delta} \triangleq \{g \in C_0([0,1]) : |g(t) - x| < \delta\}$ with $t \in (0,1]$, $x \in \mathbb{R}^d$ and $\delta > 0$. Since each $V_{t,x,\delta}$ is an open set under the supremum norm, the latter topology is finer (stronger) than the pointwise convergence topology. Hence, the exponential tightness of $\{\tilde{\mu}_n\}$ as established in Lemma 5.1.7 allows, by Corollary 4.2.6, for the strengthening of the LDP to the supremum norm topology on $C_0([0,1])$. Since $C_0([0,1])$ is a closed subset of $L_\infty([0,1])$, the same LDP holds in $L_\infty([0,1])$ by again using Lemma 4.1.5, now in the opposite direction. Finally, in view of Lemma 5.1.4, the LDP of $\{\mu_n\}$ in the metric space $L_\infty([0,1])$ follows from that of $\{\tilde{\mu}_n\}$ by an application of Theorem 4.2.13. $\qquad\square$

Proof of Lemma 5.1.4: The sets $\{\omega : \|\tilde{Z}_n - Z_n\| > \eta\}$ are obviously measurable. Note that $|\tilde{Z}_n(t) - Z_n(t)| \leq |X_{[nt]+1}|/n$. Thus, for any $\eta > 0$ and any $\lambda > 0$,

$$\mathrm{P}(\|\tilde{Z}_n - Z_n\| > \eta) \leq n\mathrm{P}(|X_1| > n\eta) \leq nE\left(e^{\lambda|X_1|}\right)e^{-\lambda n\eta}.$$

Since $\mathcal{D}_\Lambda = \mathbb{R}^d$, it follows, by considering first $n \to \infty$ and then $\lambda \to \infty$, that for any $\eta > 0$,

$$\limsup_{n\to\infty} \frac{1}{n} \log\mathrm{P}(\|\tilde{Z}_n - Z_n\| > \eta) = -\infty.$$

Therefore, the probability measures μ_n and $\tilde{\mu}_n$ are exponentially equivalent. (See Definition 4.2.10.) $\qquad\square$

The proof of Lemma 5.1.6 relies on the following finite dimensional LDP.

Lemma 5.1.8 *Let J denote the collection of all ordered finite subsets of $(0,1]$. For any $j = \{0 < t_1 < t_2 < \cdots < t_{|j|} \leq 1\} \in J$ and any $f : [0,1] \to \mathbb{R}^d$, let $p_j(f)$ denote the vector $(f(t_1), f(t_2), \ldots, f(t_{|j|})) \in (\mathbb{R}^d)^{|j|}$. Then the sequence of laws $\{\mu_n \circ p_j^{-1}\}$ satisfies the LDP in $(\mathbb{R}^d)^{|j|}$ with the good rate function*

$$I_j(\mathbf{z}) = \sum_{\ell=1}^{|j|}(t_\ell - t_{\ell-1})\Lambda^*\left(\frac{z_\ell - z_{\ell-1}}{t_\ell - t_{\ell-1}}\right), \qquad (5.1.9)$$

where $\mathbf{z} = (z_1, \ldots, z_{|j|})$ and $t_0 = 0$, $z_0 = 0$.

Proof: Fix $j \in J$ and observe that $\mu_n \circ p_j^{-1}$ is the law of the random vector

$$Z_n^j \triangleq (Z_n(t_1), Z_n(t_2), \ldots, Z_n(t_{|j|})).$$

Let
$$Y_n^j \stackrel{\triangle}{=} (Z_n(t_1), Z_n(t_2) - Z_n(t_1), \ldots, Z_n(t_{|j|}) - Z_n(t_{|j|-1})) .$$

Since the map $Y_n^j \mapsto Z_n^j$ of $(\mathbb{R}^d)^{|j|}$ onto itself is continuous and one to one, the specified LDP for Z_n^j follows by the contraction principle (Theorem 4.2.1) from an LDP for Y_n^j, with rate function

$$\Lambda_j^*(\mathbf{y}) \stackrel{\triangle}{=} \sum_{\ell=1}^{|j|} (t_\ell - t_{\ell-1}) \Lambda^* \left(\frac{y_\ell}{t_\ell - t_{\ell-1}} \right) ,$$

where $\mathbf{y} = (y_1, \ldots, y_{|j|}) \in (\mathbb{R}^d)^{|j|}$. Note that

$$
\begin{aligned}
\Lambda_j^*(\mathbf{y}) &= \sum_{\ell=1}^{|j|} (t_\ell - t_{\ell-1}) \sup_{\lambda_\ell \in \mathbb{R}^d} \{\langle \lambda_\ell, y_\ell/(t_\ell - t_{\ell-1}) \rangle - \Lambda(\lambda_\ell)\} \\
&= \sup_{\underline{\lambda} \in (\mathbb{R}^d)^{|j|}} \left\{ \sum_{\ell=1}^{|j|} \langle \lambda_\ell, y_\ell \rangle - (t_\ell - t_{\ell-1}) \Lambda(\lambda_\ell) \right\} \\
&= \sup_{\underline{\lambda} \in (\mathbb{R}^d)^{|j|}} \{\langle \underline{\lambda}, \mathbf{y} \rangle - \Lambda_j(\underline{\lambda})\} ,
\end{aligned}
$$

where $\underline{\lambda} \stackrel{\triangle}{=} (\lambda_1, \ldots, \lambda_{|j|}) \in (\mathbb{R}^d)^{|j|}$ and

$$\Lambda_j(\underline{\lambda}) = \sum_{\ell=1}^{|j|} (t_\ell - t_{\ell-1}) \Lambda(\lambda_\ell) .$$

Thus, $\Lambda_j^*(\mathbf{y})$ is the Fenchel–Legendre transform of the finite and differentiable function $\Lambda_j(\underline{\lambda})$. The LDP for Y_n^j now follows from the Gärtner–Ellis theorem (Theorem 2.3.6), since by the independence of X_i,

$$\lim_{n \to \infty} \frac{1}{n} \log E[e^{n\langle \underline{\lambda}, Y_n^j \rangle}] = \lim_{n \to \infty} \sum_{\ell=1}^{|j|} \frac{1}{n} ([n\,t_\ell] - [n\,t_{\ell-1}]) \Lambda(\lambda_\ell) = \Lambda_j(\underline{\lambda}) . \qquad \square$$

The probability measures $\{\mu_n \circ p_j^{-1}\}$ and $\{\tilde{\mu}_n \circ p_j^{-1}\}$ are exponentially equivalent in $(\mathbb{R}^d)^{|j|}$ as a consequence of Lemma 5.1.4. Thus, the following is an immediate corollary of Lemma 5.1.8.

Corollary 5.1.10 *For any $j \in J$, $\{\tilde{\mu}_n \circ p_j^{-1}\}$ satisfies the LDP in $(\mathbb{R}^d)^{|j|}$ with the good rate function I_j of (5.1.9).*

Proof of Lemma 5.1.6: A partial order by inclusions is defined on J as follows. For $i, j \in J$, $i = \{s_1, \ldots, s_{|i|}\} \le j = \{t_1, \ldots, t_{|j|}\}$ iff for any ℓ, $s_\ell = t_{q(\ell)}$ for some $q(\ell)$. Then, for $i \le j \in J$, the projection

$$p_{ij} : (\mathbb{R}^d)^{|j|} \to (\mathbb{R}^d)^{|i|}$$

is defined in the natural way. Let $\tilde{\mathcal{X}}$ denote the projective limit of $\{\mathcal{Y}_j = (\mathbb{R}^d)^{|j|}\}_{j \in J}$ with respect to the projections p_{ij}, i.e., $\tilde{\mathcal{X}} = \varprojlim \mathcal{Y}_j$. Actually, $\tilde{\mathcal{X}}$ may be identified with the space $\mathcal{X}$. Indeed, each $f \in \mathcal{X}$ corresponds to $(p_j(f))_{j \in J}$, which belongs to $\tilde{\mathcal{X}}$ since $p_i(f) = p_{ij}(p_j(f))$ for $i \leq j \in J$. In the reverse direction, each point $\mathbf{x} = (x_j)_{j \in J}$ of $\tilde{\mathcal{X}}$ may be identified with the map $f : [0,1] \to \mathbb{R}^d$, where $f(t) = x_{\{t\}}$ for $t > 0$ and $f(0) = 0$. Further, with this identification, the projective topology on $\tilde{\mathcal{X}}$ coincides with the pointwise convergence topology of $\mathcal{X}$, and p_j as defined in the statement of Lemma 5.1.8 are the canonical projections for $\tilde{\mathcal{X}}$. The LDP for $\{\tilde{\mu}_n\}$ in the Hausdorff topological space $\mathcal{X}$ thus follows by applying the Dawson–Gärtner theorem (Theorem 4.6.1) in conjunction with Corollary 5.1.10. (Note that $(\mathbb{R}^d)^{|j|}$ are Hausdorff spaces and I_j are good rate functions.) The rate function governing this LDP is

$$I_{\mathcal{X}}(f) = \sup_{0 = t_0 < t_1 < t_2 < \ldots < t_k \leq 1} \sum_{\ell=1}^{k} (t_\ell - t_{\ell-1}) \Lambda^* \left(\frac{f(t_\ell) - f(t_{\ell-1})}{t_\ell - t_{\ell-1}} \right). \quad (5.1.11)$$

Since Λ^* is nonnegative, without loss of generality, assume hereafter that $t_k = 1$. It remains to be shown that $I_{\mathcal{X}}(\cdot) = I(\cdot)$. The convexity of Λ^* implies by Jensen's inequality that $I(\phi) \geq I_{\mathcal{X}}(\phi)$. As for the opposite inequality, first consider $\phi \in \mathcal{AC}$. Let $g(t) \triangleq d\phi(t)/dt \in L_1([0,1])$ and, for $k \geq 1$, define

$$g^k(t) \triangleq k \int_{[kt]/k}^{([kt]+1)/k} g(s) ds \ \ t \in [0,1), \quad g^k(1) = k \int_{1-1/k}^{1} g(s) ds.$$

With these notations, observe that

$$I_{\mathcal{X}}(\phi) \geq \liminf_{k \to \infty} \sum_{\ell=1}^{k} \frac{1}{k} \Lambda^* \left(k \left[\phi\left(\frac{\ell}{k}\right) - \phi\left(\frac{\ell-1}{k}\right) \right] \right)$$

$$= \liminf_{k \to \infty} \int_{0}^{1} \Lambda^*(g^k(t)) dt. \quad (5.1.12)$$

By Lebesgue's theorem (Theorem C.13), $\lim_{k \to \infty} g^k(t) = g(t)$ almost everywhere in $[0,1]$. Hence, by Fatou's lemma and the lower semicontinuity of $\Lambda^*(\cdot)$,

$$\liminf_{k \to \infty} \int_{0}^{1} \Lambda^*(g^k(t)) dt \geq \int_{0}^{1} \liminf_{k \to \infty} \Lambda^*(g^k(t)) dt$$

$$\geq \int_{0}^{1} \Lambda^*(g(t)) dt = I(\phi). \quad (5.1.13)$$

The inequality $I_{\mathcal{X}}(\phi) \geq I(\phi)$ results by combining (5.1.12) and (5.1.13).

Finally, suppose that $\phi \in \mathcal{X}$ and $\phi \notin \mathcal{AC}$. Then there exist $\delta > 0$ and $\{s_1^n < t_1^n \leq \cdots \leq s_{k_n}^n < t_{k_n}^n\}$ such that $\sum_{\ell=1}^{k_n}(t_\ell^n - s_\ell^n) \to 0$, while $\sum_{\ell=1}^{k_n} |\phi(t_\ell^n) - \phi(s_\ell^n)| \geq \delta$. Note that, since Λ^* is nonnegative,

$$I_{\mathcal{X}}(\phi) = \sup_{\substack{0 < t_1 < t_2 < \ldots < t_k \\ \lambda_1, \ldots, \lambda_k \in \mathbb{R}^d}} \sum_{\ell=1}^{k} [\langle \lambda_\ell, \phi(t_\ell) - \phi(t_{\ell-1}) \rangle - (t_\ell - t_{\ell-1}) \Lambda(\lambda_\ell)]$$

$$\geq \sup_{\substack{0 \leq s_1 < t_1 \leq s_2 < t_2 \leq \ldots \leq s_k < t_k \\ \lambda_1, \ldots, \lambda_k \in \mathbb{R}^d}} \sum_{\ell=1}^{k} [\langle \lambda_\ell, \phi(t_\ell) - \phi(s_\ell) \rangle - (t_\ell - s_\ell) \Lambda(\lambda_\ell)].$$

Hence, for $t_\ell = t_\ell^n$, $s_\ell = s_\ell^n$, and λ_ℓ proportional to $\phi(t_\ell) - \phi(s_\ell)$ and with $|\lambda_\ell| = \rho$, the following bound is obtained:

$$I_{\mathcal{X}}(\phi) \geq \limsup_{n \to \infty} \left\{ \rho \sum_{\ell=1}^{k_n} |\phi(t_\ell^n) - \phi(s_\ell^n)| - [\sup_{|\lambda|=\rho} \Lambda(\lambda)] \sum_{\ell=1}^{k_n} (t_\ell^n - s_\ell^n) \right\} \geq \rho \delta.$$

(Recall that $\Lambda(\cdot)$ is continuous everywhere.) The arbitrariness of ρ implies that $I_{\mathcal{X}}(\phi) = \infty$, completing the proof of the lemma. $\qquad\square$

The proof of Lemma 5.1.7 relies on the following one-dimensional result.

Lemma 5.1.14 *Let X be a real valued random variable distributed according to the law ν. Then $E\left[e^{\delta \Lambda_\nu^*(X)}\right] < \infty$ for all $\delta < 1$.*

Proof: Let Λ_ν denotes the logarithmic moment generating function of X. If $\Lambda_\nu(\lambda) = \infty$ for all $\lambda \neq 0$, then Λ_ν^* is identically zero and the lemma trivially holds. Assume otherwise and recall that then $\overline{x} = E_\nu[X]$ exists, possibly as an extended real number. Observe that for any $x \in \mathbb{R}$,

$$e^{\Lambda_\nu(\lambda)} \geq \begin{cases} e^{\lambda x} \nu([x, \infty)) & \text{if } \lambda \geq 0 \\ \\ e^{\lambda x} \nu((-\infty, x]) & \text{if } \lambda \leq 0. \end{cases}$$

Hence, by (2.2.6) and (2.2.7),

$$\Lambda_\nu^*(x) \leq \begin{cases} -\log \nu([x, \infty)) & \text{if } x \geq \overline{x} \\ \\ -\log \nu((-\infty, x]) & \text{if } x \leq \overline{x}. \end{cases}$$

Let $\delta < 1$. Then

$$E\left[e^{\delta \Lambda_\nu^*(X)}\right] = \int_{-\infty}^{\overline{x}} \nu(dx) e^{\delta \Lambda_\nu^*(x)} + \int_{\overline{x}}^{\infty} \nu(dx) e^{\delta \Lambda_\nu^*(x)}$$

$$\leq \int_{-\infty}^{\overline{x}} \frac{\nu(dx)}{\nu((-\infty, x])^\delta} + \int_{\overline{x}}^{\infty} \frac{\nu(dx)}{\nu([x, \infty))^\delta}. \quad (5.1.15)$$

For any $M < \bar{x}$ such that $\nu((-\infty, M]) > 0$, integration by parts yields

$$\int_M^{\bar{x}} \frac{\nu(dx)}{\nu((-\infty, x])^\delta} = \nu((-\infty, \bar{x}])^{1-\delta} - \nu((-\infty, M])^{1-\delta} + \delta \int_M^{\bar{x}} \frac{\nu(dx)}{\nu((-\infty, x])^\delta} .$$

Hence,

$$\int_M^{\bar{x}} \frac{\nu(dx)}{\nu((-\infty, x])^\delta} = \frac{1}{1-\delta} \left\{ \nu((-\infty, \bar{x}])^{1-\delta} - \nu((-\infty, M])^{1-\delta} \right\} \leq \frac{1}{1-\delta} .$$

By monotone convergence, one may set $M = -\infty$. Then, substituting the last inequality into (5.1.15) and repeating this procedure for the integral on $[\bar{x}, \infty)$ yields

$$E\left[e^{\delta \Lambda_\nu^*(X)} \right] \leq \frac{2}{1-\delta} < \infty. \tag{5.1.16}$$

$\square$

Proof of Lemma 5.1.7: To see the exponential tightness of $\tilde{\mu}_n$ in $C_0([0,1])$ when equipped with the supremum norm topology, denote by X_1^j the jth component of X_1, define

$$\Lambda_j(\lambda) \triangleq \log \left(E[\exp(\lambda X_1^j)] \right) ,$$

with $\Lambda_j^*(\cdot)$ being the Fenchel–Legendre transform of $\Lambda_j(\cdot)$. Fix $\alpha > 0$ and

$$K_\alpha^j \triangleq \{ f \in \mathcal{AC} \ : \ f(0) = 0, \ \int_0^1 \Lambda_j^*(\dot{f}_j(\theta))d\theta \leq \alpha \} ,$$

where $f_j(\cdot)$ is the jth component of $f : [0,1] \to \mathbb{R}^d$. Now let $K_\alpha \triangleq \cap_{j=1}^d K_\alpha^j$. Note that $d\tilde{Z}_n(t)/dt = X_{[nt]+1}$ for almost all $t \in [0,1)$. Thus,

$$\tilde{\mu}_n(K_\alpha^c) \leq d \max_{j=1}^d P\left(\frac{1}{n} \sum_{i=1}^n \Lambda_j^*(X_i^j) > \alpha \right).$$

Since $\{X_i\}_{i=1}^n$ are independent, it now follows by Chebycheff's inequality that for any $\delta > 0$,

$$\frac{1}{n} \log \tilde{\mu}_n(K_\alpha^c) \leq -\delta\alpha + \frac{1}{n} \log d + \max_{j=1}^d \log E\left[e^{\delta \Lambda_j^*(X_1^j)} \right] .$$

In view of Lemma 5.1.14, it follows by considering $\delta = \frac{1}{2}$ and $\alpha \to \infty$ that $\lim_{\alpha\to\infty} \limsup_{n\to\infty} \frac{1}{n} \log \tilde{\mu}_n(K_\alpha^c) = -\infty$.

By the Arzelà–Ascoli theorem (Theorem C.8), the proof of the lemma is complete as soon as we show that K_α is a bounded set of equicontinuous functions. To see the equicontinuity, note that if $f \in K_\alpha$, then the continuous function f is differentiable almost everywhere, and for all $0 \leq s < t \leq 1$ and $j = 1, 2, \ldots, d$,

$$\Lambda_j^* \left(\frac{f_j(t) - f_j(s)}{t - s} \right) \leq \frac{1}{t - s} \int_s^t \Lambda_j^*(\dot{f}_j(\theta)) d\theta \leq \frac{\alpha}{t - s} \ .$$

Since $\Lambda_j^*(x) \geq M|x| - \{\Lambda_j(M) \vee \Lambda_j(-M)\}$ for all $M > 0$, it follows that for all $(t - s) \leq \delta$,

$$|f_j(t) - f_j(s)| \leq \frac{1}{M} (\alpha + \delta\{\Lambda_j(M) \vee \Lambda_j(-M)\}) \ . \tag{5.1.17}$$

Since $\Lambda_j(\cdot)$ is continuous on $\mathbb{R}$, there exist $M_j = M_j(\delta)$ such that $\Lambda_j(M_j) \leq 1/\delta$, $\Lambda_j(-M_j) \leq 1/\delta$, and $\lim_{\delta \to 0} M_j(\delta) = \infty$. Hence, $\epsilon(\delta) \triangleq \max_{j=1,\cdots,d}(\alpha + 1)/M_j(\delta)$ is a uniform modulus of continuity for the set K_α. Finally, K_α is bounded by (5.1.17) (for $s = 0, \delta = 1$). $\qquad\square$

Theorem 5.1.2 can be extended to the laws ν_ϵ of

$$Y_\epsilon(t) = \epsilon \sum_{i=1}^{[\frac{t}{\epsilon}]} X_i, \quad 0 \leq t \leq 1, \tag{5.1.18}$$

where μ_n (and $Z_n(t)$) correspond to the special case of $\epsilon = n^{-1}$. The precise statement is given in the following theorem.

Theorem 5.1.19 *The probability measures ν_ϵ induced on $L_\infty([0,1])$ by $Y_\epsilon(\cdot)$ satisfy the LDP with the good rate function $I(\cdot)$ of (5.1.3).*

Proof: For any sequence $\epsilon_m \to 0$ such that ϵ_m^{-1} are integers, Theorem 5.1.19 is a consequence of Theorem 5.1.2. Consider now an arbitrary sequence $\epsilon_m \to 0$ and let $n_m \triangleq [\epsilon_m^{-1}]$. By Theorem 5.1.2, $\{\mu_{n_m}\}_{m=1}^\infty$ satisfies an LDP with the rate function $I(\cdot)$ of (5.1.3) and rate $1/n_m$. Since $n_m\epsilon_m \to 1$, the proof of the theorem is completed by applying Theorem 4.2.13, provided that for any $\delta > 0$,

$$\limsup_{m \to \infty} \frac{1}{n_m} \log \mathrm{P}(\| Y_{\epsilon_m} - Z_{n_m} \| \geq \delta) = -\infty \ . \tag{5.1.20}$$

To this end, observe that $\epsilon_m n_m \in [1 - \epsilon_m, 1]$ and $\left[\frac{t}{\epsilon_m}\right] \in \{[n_m t], [n_m t] + 1\}$. Hence, by (5.1.1) and (5.1.18),

$$
\begin{aligned}
|Y_{\epsilon_m}(t) - Z_{n_m}(t)| &\leq (1 - \epsilon_m n_m)|Z_{n_m}(t)| + \epsilon_m |X_{[\frac{t}{\epsilon_m}]}| \\
&\leq 2\epsilon_m \max_{i=1,\ldots,n_m} |X_i| \ .
\end{aligned}
$$

Now, by the union of events bound,

$$\frac{1}{n_m} \log P(\| Y_{\epsilon_m} - Z_{n_m} \| \geq \delta) \leq \frac{1}{n_m} \log n_m + \frac{1}{n_m} \log P(|X_1| \geq \frac{\delta}{2\epsilon_m}),$$

and the limit (5.1.20) follows, since, by Exercise 5.1.24,

$$\lim_{\epsilon \to 0} \epsilon \log P(|X_1| \geq \frac{1}{\epsilon}) = -\infty. \qquad (5.1.21)$$

$\square$

Exercise 5.1.22 Establish the LDP associated with $Y_\epsilon(\cdot)$ over the time interval $[0, T]$, where T is arbitrary (but finite), i.e., prove that $\{Y_\epsilon(\cdot)\}$ satisfies the LDP in $L_\infty([0, T])$ with the good rate function

$$I_T(\phi) = \begin{cases} \int_0^T \Lambda^*(\dot{\phi}(t)) \, dt, & \text{if } \phi \in \mathcal{AC}^T, \phi(0) = 0 \\ \\ \infty & \text{otherwise}, \end{cases} \qquad (5.1.23)$$

where $\mathcal{AC}^T$ is defined in the obvious way as the space of absolutely continuous functions on $[0, T]$.

Exercise 5.1.24 Prove (5.1.21).
Hint: Include the event $\{|X_1| \geq 1/\epsilon\}$ within the union of $2d$ simpler one-dimensional events.

Remark: Observe that (5.1.21) is false when $\Lambda(\lambda) = \infty$ for some $\lambda \in \mathbb{R}^d$. For example, check that (5.1.21) is false for $d = 1$ and $X_1 \sim$ Exponential(1).

Exercise 5.1.25 Let $Z_n(t) \triangleq \frac{1}{n} S_{[nt]}$, where

$$S_k = S_{k-1} + g(\frac{1}{n} S_{k-1}) + X_k, \quad k \geq 1, \quad S_0 = 0,$$

X_k are as in Theorem 5.1.2, and $g : \mathbb{R}^d \to \mathbb{R}^d$ is a bounded, deterministic Lipschitz continuous function. Prove that $Z_n(\cdot)$ satisfy the LDP in $L_\infty([0, 1])$ with the good rate function

$$I(\phi) = \begin{cases} \int_0^1 \Lambda^*(\dot{\phi}(t) - g(\phi(t))) \, dt, & \text{if } \phi \in \mathcal{AC}, \phi(0) = 0 \\ \\ \infty & \text{otherwise}. \end{cases} \qquad (5.1.26)$$

Note that Theorem 5.1.2 corresponds to $g = 0$.
Hint: You may want to take a look at the proof of Theorem 5.8.14.

Exercise 5.1.27 Prove Theorem 5.1.2 for $X_i = f(Y_i)$ where f is a deterministic function, $\{Y_i\}$ is the realization of a finite state, irreducible Markov chain (c.f. Section 3.1), and Λ^* is replaced by $I(\cdot)$ of Theorem 3.1.2.

5.2 Brownian Motion Sample Path Large Deviations

Let w_t, $t \in [0,1]$ denote a standard Brownian motion in $\mathbb{R}^d$. Consider the process

$$w_\epsilon(t) = \sqrt{\epsilon} w_t,$$

and let ν_ϵ be the probability measure induced by $w_\epsilon(\cdot)$ on $C_0([0,1])$, the space of all continuous functions $\phi : [0,1] \to \mathbb{R}^d$ such that $\phi(0) = 0$, equipped with the supremum norm topology. The process $w_\epsilon(\cdot)$ is a candidate for an LDP similar to the one developed for $Y_\epsilon(\cdot)$ in Section 5.1. Indeed, $\| w_\epsilon \| \xrightarrow[\epsilon \to 0]{} 0$ in probability (actually, almost surely) and exponentially fast in $1/\epsilon$ as implied by the following useful (though elementary) lemma whose proof is deferred to the end of this section.

Lemma 5.2.1 *For any integer d and any $\tau, \epsilon, \delta > 0$,*

$$\mathrm{P}\left(\sup_{0 \le t \le \tau} |w_\epsilon(t)| \ge \delta \right) \le 4d e^{-\delta^2/2d\tau\epsilon} . \tag{5.2.2}$$

The LDP for $w_\epsilon(\cdot)$ is stated in the following theorem. Let $H_1 \stackrel{\triangle}{=} \{ \int_0^t f(s)ds : f \in L_2([0,1]) \}$ denote the space of all absolutely continuous functions with value 0 at 0 that possess a square integrable derivative, equipped with the norm $\|g\|_{H_1} = [\int_0^1 |\dot{g}(t)|^2 \, dt]^{\frac{1}{2}}$.

Theorem 5.2.3 (Schilder) *$\{\nu_\epsilon\}$ satisfies, in $C_0([0,1])$, an LDP with good rate function*

$$I_w(\phi) = \begin{cases} \frac{1}{2} \int_0^1 |\dot{\phi}(t)|^2 \, dt, & \phi \in H_1 \\ \infty & \text{otherwise}. \end{cases}$$

Proof: Observe that the process

$$\hat{w}_\epsilon(t) \stackrel{\triangle}{=} w_\epsilon\left(\epsilon \left[\frac{t}{\epsilon} \right] \right)$$

is merely the process $Y_\epsilon(\cdot)$ of Section 5.1, for the particular choice of X_i, which are standard Normal random variables in $\mathbb{R}^d$ (namely, of zero mean and of the identity covariance matrix).

Thus, by Theorem 5.1.19, the probability laws of $\hat{w}_\epsilon(\cdot)$ satisfy the LDP in $L_\infty([0,1])$ with the good rate function $I(\cdot)$ of (5.1.3). For the standard Normal variables considered here,

$$\Lambda(\lambda) = \log E\left[e^{\langle \lambda, X_1 \rangle} \right] = \frac{1}{2} |\lambda|^2 ,$$

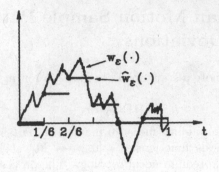

Figure 5.2.1: Typical w_ϵ and $\hat{w}_\epsilon$ for $\epsilon = 1/6$.

implying that

$$\Lambda^*(x) = \sup_{\lambda \in \mathbb{R}^d} \left\{ \langle \lambda, x \rangle - \frac{1}{2} |\lambda|^2 \right\} = \frac{1}{2} |x|^2 . \qquad (5.2.4)$$

Hence, for these variables $\mathcal{D}_I = H_1$, and the rate function $I(\cdot)$ specializes to $I_w(\cdot)$.

Observe that for any $\delta > 0$,

$$P(\| w_\epsilon - \hat{w}_\epsilon \| \geq \delta) \;\; \leq \;\; ([1/\epsilon] + 1) P \left(\sup_{0 \leq t \leq \epsilon} |w_\epsilon(t)| \geq \delta \right)$$

$$\leq \;\; 4d\epsilon^{-1}(1 + \epsilon) e^{-\delta^2/(2d\epsilon^2)},$$

where the first inequality follows by the time-homogeneity of increments of the Brownian motion, and the second by (5.2.2). Consequently,

$$\limsup_{\epsilon \to 0} \epsilon \log P(\| w_\epsilon - \hat{w}_\epsilon \| \geq \delta) = -\infty ,$$

and by Theorem 4.2.13, it follows that $\{\nu_\epsilon\}$ satisfies the LDP in $L_\infty([0,1])$ with the good rate function $I_w(\cdot)$. The restriction to $C_0([0,1])$ follows from Lemma 4.1.5, since $w_\epsilon(\cdot) \in C_0([0,1])$ with probability one. $\square$

Proof of Lemma 5.2.1: First note that

$$P \left(\sup_{0 \leq t \leq \tau} |w_\epsilon(t)| \geq \delta \right) \;\; = \;\; P \left(\sup_{0 \leq t \leq \tau} |w_t|^2 \geq \epsilon^{-1}\delta^2 \right)$$

$$\leq \;\; d \, P \left(\sup_{0 \leq t \leq \tau} (w_t)_1^2 \geq \frac{\delta^2}{d\epsilon} \right) , \qquad (5.2.5)$$

where $(w_t)_1$ is a Brownian motion in $\mathbb{R}$ and the last inequality is a consequence of the set inclusion

$$\{x \in \mathbb{R}^d : |x|^2 \geq \alpha\} \subset \bigcup_{i=1}^{d} \left\{x \in \mathbb{R}^d : |x_i|^2 \geq \frac{\alpha}{d}\right\},$$

where x_i is the ith coordinate of $x \in \mathbb{R}^d$. Since the laws of w_t and $\sqrt{\tau} w_{t/\tau}$ are identical, one obtains from (5.2.5) by time rescaling

$$\mathrm{P}\left(\sup_{0 < t \leq \tau} |w_\epsilon(t)| \geq \delta\right) \leq d\, \mathrm{P}\left(\| (w_t)_1 \| \geq \frac{\delta}{\sqrt{d\tau\epsilon}}\right). \tag{5.2.6}$$

Let $\overline{w}_t \overset{\triangle}{=} (w_t)_1$, where $\overline{w}_t$ is a one-dimensional Brownian motion. Since $\overline{w}_t$ and $-\overline{w}_t$ possess the same law in $C_0([0,1])$,

$$\mathrm{P}(\| \overline{w}_t \| \geq \eta) \leq 2\mathrm{P}(\sup_{0 \leq t \leq 1} \overline{w}_t \geq \eta) = 4\mathrm{P}(\overline{w}_1 \geq \eta) \leq 4e^{-\eta^2/2}, \tag{5.2.7}$$

where the equality is Désiré André's reflection principle (Theorem E.4), and the last inequality follows from Chebycheff's bound. Substituting (5.2.7) into (5.2.6) yields the lemma. $\qquad\qquad\square$

Exercise 5.2.8 Establish, for any $T < \infty$, Schilder's theorem in $C_0([0,T])$, the space of all continuous functions $\phi : [0,T] \to \mathbb{R}^d$ such that $\phi(0) = 0$, equipped with the supremum norm topology. Here, the rate function is

$$I_w(\phi) = \begin{cases} \frac{1}{2} \int_0^T |\dot\phi(t)|^2 \, dt, & \phi \in H_1([0,T]) \\ \infty & \text{otherwise}, \end{cases} \tag{5.2.9}$$

where $H_1([0,T]) \overset{\triangle}{=} \{\int_0^t f(s)ds : f \in L_2([0,T])\}$ denotes the space of absolutely continuous functions with value 0 at 0 that possess a square integrable derivative.

Exercise 5.2.10 Note that, as a by-product of the proof of Theorem 5.2.3, it is known that $I_w(\cdot)$ is a good rate function. Prove this fact directly.

Exercise 5.2.11 Obviously, Theorem 5.2.3 may be proved directly.
(a) Prove the upper bound by considering a discretized version of w_t and estimating the distance between w_t and its discretized version.
(b) Let $\phi \in H_1$ be given and let μ_ϵ be the measure induced on $C_0([0,1])$ by the process $X_t = -\phi(t) + \sqrt{\epsilon} w_t$. Compute the Radon–Nikodym derivative $d\mu_\epsilon/d\nu_\epsilon$ and prove the lower bound by mimicking the arguments in the proof of Theorem 2.2.30.

Exercise 5.2.12 Prove the analog of Schilder's theorem for Poisson process. Specifically, let μ_ϵ be the probability measures induced on $L_\infty([0,1])$

by $\epsilon \hat{N}(t/\epsilon)$, where $\hat{N}(\cdot)$ is a Poisson process on $[0, \infty)$ of intensity one. Prove that $\{\mu_\epsilon\}$ satisfies the LDP with the good rate function

$$I_{\hat{N}}(\phi) = \begin{cases} \int_0^1 [\dot{\phi}(t) \log \dot{\phi}(t) - \dot{\phi}(t) + 1] \, dt \\ \qquad \text{if } \phi \in \mathcal{AC}, \phi \text{ is increasing}, \phi(0) = 0 \\ \infty \qquad \text{otherwise}. \end{cases}$$

Hint: The process $\epsilon \hat{N}([\frac{t}{\epsilon}])$ is a particular instance of $Y_\epsilon(\cdot)$ of Section 5.1 for $d = 1$ and X_i which are Poisson(1) random variables. Use part (a) of Exercise 2.2.23 to determine the rate function. Complete the proof by establishing the analog of Theorem 5.2.3 and using Exercise 4.2.32.

Exercise 5.2.13 Show that the results of Exercise 5.2.12 hold for any Poisson process $N(\cdot)$ of intensity $\psi(t)$ over $[0, \infty)$ provided that $\lim_{t \to \infty} \psi(t) = 1$.
Hint: Let $\Psi(t) \triangleq \int_0^t \psi(s) ds$, and represent $N(t) = \hat{N}(\Psi(t))$. Then show that $\{\epsilon \hat{N}(t/\epsilon), \epsilon > 0\}$ and $\{\epsilon N(t/\epsilon), \epsilon > 0\}$ are exponentially equivalent.

Exercise 5.2.14 For $\alpha < 1/2$, let $\mathrm{Lip}_\alpha([0,1])$ denote the space of α Hölder continuous functions in $C_0([0,1])$, equipped with the norm $\|f\|_\alpha \triangleq \sup_{t \neq s} |f(t) - f(s)|/|t - s|^\alpha$. Establish Schilder's theorem in $\mathrm{Lip}_\alpha([0,1])$.
Hint: Since $\nu_\epsilon(\mathrm{Lip}_\alpha([0,1])) = 1$, it is enough to prove the exponential tightness of $\{\nu_\epsilon\}$ in $\mathrm{Lip}_\alpha([0,1])$. Fix $\alpha' \in (\alpha, 1/2)$ and consider the following pre-compact subsets of $\mathrm{Lip}_\alpha([0,1])$ (*c.f.* [Joh70], Corollary 3.3)

$$K_\beta \triangleq \{\phi \in C_0([0,1]) : \|\phi\|_{\alpha'} \leq \beta\}.$$

Recall Borell's inequality [Bore75], which states that any centered Gaussian process $X_{t,s}$, with a.s. bounded sample paths, satisfies

$$P\left(\sup_{0 \leq t, s \leq 1} |X_{t,s}| \geq \delta\right) \leq 2e^{-(\delta - E)^2/2V} \qquad (5.2.15)$$

for all $\delta > E$, where $E \triangleq E\left(\sup_{0 \leq t, s \leq 1} |X_{t,s}|\right) < \infty$, $V = \sup_{0 \leq s, t \leq 1} E|X_{t,s}|^2$. Apply this inequality to $X_{t,s} = \frac{1}{(t-s)^{\alpha'}}(w_t - w_s)$ (where $X_{t,t}$ is defined to be zero).

5.3 Multivariate Random Walk and Brownian Sheet

The results of Sections 5.1 and 5.2 may be extended to the situation where more than one time index is present. To avoid cumbersome notations, real valued random variables are considered. The case of $\mathbb{R}^k$-valued random variables, being similar, is presented in Exercise 5.3.5.

Let d be a given integer. Throughout, $i = (i_1, \ldots, i_d) \in \mathbb{Z}_+^d$ is a multi-index. Let $\{X_i\}$ denote a family of i.i.d. random variables with

$$\Lambda(\lambda) \triangleq \log E\left[e^{\lambda X_i}\right] < \infty \quad \text{for all } \lambda \in \mathbb{R}.$$

As usual, $\Lambda^*(x) = \sup_{\lambda \in \mathbb{R}}\{\lambda x - \Lambda(\lambda)\}$ denotes the Fenchel–Legendre transform of $\Lambda(\cdot)$.

Let $t = (t_1, \ldots, t_d) \in [0, 1]^d$. Define the multi-index random process

$$Z_n(t) = \frac{1}{n} \sum_{i_1=1}^{[n^{1/d}t_1]} \cdots \sum_{i_d=1}^{[n^{1/d}t_d]} X_i \,,$$

and let μ_n be the law of $Z_n(\cdot)$ on $L_\infty([0,1]^d)$.

Motivated by the results of Section 5.1, it is natural to look for the large deviations behavior of $\{\mu_n\}$. As a preliminary step, the notion of absolute continuity is defined for functions on $[0,1]^d$, in order to describe the resulting rate function. Let ϕ be a map from $[0,1]^d$ to $\mathbb{R}$, and let $\epsilon > 0$ be a given constant. The ϵ approximate derivative of ϕ in the jth direction, $j = 1, \ldots, d$, is defined as

$$\Delta_j^\epsilon \phi(t) = \frac{1}{\epsilon}\left[\phi(t_1, \ldots, t_j + \epsilon, \ldots, t_d) - \phi(t_1, \ldots, t_j, \ldots, t_d)\right].$$

Similarly, the $\epsilon = (\epsilon_1, \ldots, \epsilon_d)$ mixed derivative of ϕ at t is defined as

$$\Delta^\epsilon \phi(t) = (\Delta_d^{\epsilon_d} \cdots (\Delta_2^{\epsilon_2}(\Delta_1^{\epsilon_1}\phi)))(t).$$

Let $\mathcal{Q}$ denote the following collection of cubes in $[0,1]^d$; $q \in \mathcal{Q}$ if either q is the empty set, or $q = [a_1, b_1) \times [a_2, b_2) \times \cdots \times [a_d, b_d)$ for some $0 \le a_j < b_j \le 1$, $j = 1, \ldots, d$. For any $\phi : [0,1]^d \to \mathbb{R}$, define

$$\phi(q) = \Delta^\epsilon \phi(a_1, \ldots, a_d) \prod_{k=1}^d (b_k - a_k),$$

where $\epsilon = (b_1 - a_1, b_2 - a_2, \ldots, b_d - a_d)$, and $\phi(\emptyset) = 0$. The function ϕ is of *bounded variation* if

$$\sup_k \sup_{\substack{q_1,\ldots,q_k \in \mathcal{Q} \\ q_1,\ldots,q_k \ disjoint}} \sum_{\ell=1}^k |\phi(q_\ell)| < \infty.$$

Hence, each $\phi : [0,1]^d \to \mathbb{R}$ defines an additive set function on $\mathcal{Q}$. This set function is extended in the obvious manner to an additive set function on the field generated by $\mathcal{Q}$. If ϕ is of bounded variation, then this extension is also bounded and countably additive. In this case, it possesses a unique extension, denoted μ_ϕ, to a bounded (signed) measure on the σ-field generated by $\mathcal{Q}$. The latter is exactly the Borel σ-field of $[0,1]^d$ (since every open set in $[0,1]^d$ may be expressed as a countable union of elements of $\mathcal{Q}$). The function ϕ is called *absolutely continuous* if it is of bounded variation and μ_ϕ is absolutely continuous with respect to the Lebesgue measure on $[0,1]^d$. By the Radon–Nikodym theorem (Theorem C.9), when $\partial^d \phi / \partial t_1 \cdots \partial t_d \in L_1(m)$, where m denotes the Lebesgue measure on $[0,1]^d$, then $d\mu_\phi/dm = \partial^d \phi / \partial t_1 \cdots \partial t_d \in L_1(m)$ (m a.e.). Let

$$\mathcal{AC}_0 \;\triangleq\; \{\phi: \phi \text{ absolutely continuous},$$
$$\phi(0, t_2, \ldots, t_d) = \phi(t_1, 0, \ldots, t_d) = \cdots = \phi(t_1, \ldots, t_{d-1}, 0) = 0\}.$$

The following is the analog of Mogulskii's theorem (Theorem 5.1.2).

Theorem 5.3.1 *The sequence $\{\mu_n\}$ satisfies in $L_\infty([0,1]^d)$ the LDP with the good rate function*

$$I(\phi) = \begin{cases} \int_{[0,1]^d} \Lambda^* \left(\frac{d\mu_\phi}{dm} \right) dm & \text{if } \phi \in \mathcal{AC}_0 \\ \infty & \text{otherwise}. \end{cases} \tag{5.3.2}$$

Remark: Note that $\mathcal{AC}_0$ is a subset of

$$C_0([0,1]^d) \triangleq \{\phi \in C([0,1]^d) :$$
$$\phi(0, t_2, \ldots, t_d) = \phi(t_1, 0, \ldots, t_d) = \cdots = \phi(t_1, \ldots, t_{d-1}, 0) = 0\}.$$

Proof: The proof of the theorem follows closely the proof of Theorem 5.1.2. Let $\tilde{\mu}_n$ denote the law on $L_\infty([0,1]^d)$ induced by the natural polygonal interpolation of $Z_n(t)$. E.g., for $d = 2$, it is induced by the random variables

$$\tilde{Z}_n(t) = Z_n\left(\frac{[\sqrt{n}t_1]}{\sqrt{n}}, \frac{[\sqrt{n}t_2]}{\sqrt{n}}\right)$$
$$+ \sqrt{n}\left(t_1 - \frac{[\sqrt{n}t_1]}{\sqrt{n}}\right) \left\{ Z_n\left(\frac{[\sqrt{n}t_1]+1}{\sqrt{n}}, \frac{[\sqrt{n}t_2]}{\sqrt{n}}\right) - Z_n\left(\frac{[\sqrt{n}t_1]}{\sqrt{n}}, \frac{[\sqrt{n}t_2]}{\sqrt{n}}\right) \right\}$$
$$+ \sqrt{n}\left(t_2 - \frac{[\sqrt{n}t_2]}{\sqrt{n}}\right) \left\{ Z_n\left(\frac{[\sqrt{n}t_1]}{\sqrt{n}}, \frac{[\sqrt{n}t_2]+1}{\sqrt{n}}\right) - Z_n\left(\frac{[\sqrt{n}t_1]}{\sqrt{n}}, \frac{[\sqrt{n}t_2]}{\sqrt{n}}\right) \right\}$$
$$+ \left(t_1 - \frac{[\sqrt{n}t_1]}{\sqrt{n}}\right)\left(t_2 - \frac{[\sqrt{n}t_2]}{\sqrt{n}}\right) X_{([\sqrt{n}t_1]+1, [\sqrt{n}t_2]+1)} \,.$$

By the same proof as in Lemma 5.1.4, $\{\mu_n\}$ and $\{\tilde{\mu}_n\}$ are exponentially equivalent on $L_\infty([0,1]^d)$. Moreover, mimicking the proof of Lemma

5.1.7 (*c.f.* Exercise 5.3.6), it follows that $\{\tilde{\mu}_n\}$ are exponentially tight on $C_0([0,1]^d)$ when the latter is equipped with the supremum norm topology. Thus, Theorem 5.3.1 is a consequence of the following lemma in the same way that Theorem 5.1.2 is a consequence of Lemma 5.1.6.

Lemma 5.3.3 *Let $\mathcal{X}$ consist of all the maps from $[0,1]^d$ to $\mathbb{R}$ such that the axis $(0, t_2, \ldots, t_d)$, $(t_1, 0, \ldots, t_d)$, ..., $(t_1, \ldots, t_{d-1}, 0)$ are mapped to zero, and equip $\mathcal{X}$ with the topology of pointwise convergence on $[0,1]^d$. Then the probability measures $\tilde{\mu}_n$ satisfy the LDP in $\mathcal{X}$ with the good rate function $I(\cdot)$ of (5.3.2).* $\qquad\square$

Proof of Lemma 5.3.3: Applying the projective limit argument as in the proof of Lemma 5.1.6, one concludes that $\tilde{\mu}_n$ satisfies the LDP in $\mathcal{X}$ with the good rate function

$$I_{\mathcal{X}}(\phi) \;=\; \sup_{k<\infty} \;\; \sup_{\substack{q_1,\ldots,q_k \in \mathcal{Q} \\ q_1,\ldots,q_k \; disjoint \\ [0,1)^d = \bigcup_{\ell=1}^{k} q_\ell}} \;\; \sum_{\ell=1}^{k} m(q_\ell)\Lambda^*\left(\frac{\phi(q_\ell)}{m(q_\ell)}\right).$$

By the convexity of Λ^*, $I(\phi) \geq I_{\mathcal{X}}(\phi)$. As for the opposite inequality, first consider ϕ absolutely continuous such that $d\mu_\phi/dm \in L_1([0,1]^d)$. Let $\{\tilde{q}_k(\ell)\}_{\ell=1}^{k^d}$ denote the cover of $[0,1)^d$ by disjoint cubes of volume k^{-d} and equal side length. Then

$$I_{\mathcal{X}}(\phi) \;\geq\; \liminf_{k\to\infty} \sum_{\ell=1}^{k^d} \frac{1}{k^d}\Lambda^*\left(k^d\mu_\phi(\tilde{q}_k(\ell))\right)$$

$$= \; \liminf_{k\to\infty} \int_{[0,1]^d} \Lambda^*(g_k(t))dt\,,$$

where $g_k(t)$ is constant on each of the cubes $\tilde{q}_k(\ell)$, $\ell = 1, \ldots, k^d$. Moreover, for each $t \in [0,1)^d$, the sequence of cubes $q_k \in \{\tilde{q}_k(\cdot)\}$, chosen such that $t \in q_k$ for all k, shrinks nicely to t in the sense of Theorem C.13. Hence, by this theorem, $g_n(t) \xrightarrow[n\to\infty]{} \frac{d\mu_\phi}{dm}(t)$, for m-a.e. values of t. Therefore, by Fatou's lemma and the lower semicontinuity of $\Lambda^*(\cdot)$,

$$I_{\mathcal{X}}(\phi) \;\geq\; \liminf_{k\to\infty} \int_{[0,1]^d} \Lambda^*(g_k(t))\,dt$$

$$\geq \; \int_{[0,1]^d} \liminf_{k\to\infty} \Lambda^*(g_k(t))\,dt$$

$$\geq \; \int_{[0,1]^d} \Lambda^*\left(\frac{d\mu_\phi}{dm}(t)\right)dt = I(\phi).$$

The remaining step is to check that $I_{\mathcal{X}}(\phi) = \infty$ whenever μ_ϕ is not absolutely continuous with respect to Lebesgue's measure. Since for every disjoint collection of cubes $\{q_\ell\}_{\ell=1}^k$,

$$
\begin{aligned}
I_{\mathcal{X}}(\phi) &\geq \sup_{\lambda_1,\dots,\lambda_k \in \mathbb{R}} \sum_{\ell=1}^k \left[\lambda_\ell \phi(q_\ell) - m(q_\ell)\Lambda(\lambda_\ell)\right] \\
&\geq \sum_{\ell=1}^k |\phi(q_\ell)| - \max\{\Lambda(-1), \Lambda(1)\},
\end{aligned}
$$

it follows that $I_{\mathcal{X}}(\phi) = \infty$ whenever ϕ is of unbounded variation. Assume now that ϕ is of bounded variation, but μ_ϕ is not absolutely continuous with respect to Lebesgue measure. Then there exist a $\delta > 0$ and a sequence of measurable sets A_k, with $m(A_k) \xrightarrow[k\to\infty]{} 0$ and $|\mu_\phi(A_k)| \geq \delta$. Using the regularity and boundedness (over $[0,1]^d$) of both m and μ_ϕ, it is enough to consider A_k open. Hence, without loss of generality, assume that each A_k is a disjoint union of countably many cubes. Fix k, and let $A_k = \bigcup_\ell q_\ell$. Then for every $\rho > 0$,

$$
\begin{aligned}
I_{\mathcal{X}}(\phi) &\geq \sup_{\lambda_\ell \in \mathbb{R}} \sum_\ell \left[\lambda_\ell \mu_\phi(q_\ell) - m(q_\ell)\Lambda(\lambda_\ell)\right] \\
&\geq \rho \sum_\ell |\mu_\phi(q_\ell)| - m(A_k)\max\{\Lambda(-\rho), \Lambda(\rho)\} \\
&\geq \rho|\mu_\phi(A_k)| - m(A_k)\max\{\Lambda(-\rho), \Lambda(\rho)\}.
\end{aligned}
$$

Hence, considering $k \to \infty$, we obtain

$$
I_{\mathcal{X}}(\phi) \geq \limsup_{k\to\infty} \{\rho|\mu_\phi(A_k)| - m(A_k)\max\{\Lambda(-\rho), \Lambda(\rho)\}\} \geq \rho\delta.
$$

Taking $\rho \to \infty$ yields $I_{\mathcal{X}}(\phi) = \infty$ and completes the proof.   □

The following corollary is obtained in the same way as Theorem 5.2.3 is obtained from Theorem 5.1.2, and yields the LDP for sample paths of the Brownian sheet.

Corollary 5.3.4 *Let Z_t, $t \in [0,1]^2$ be the Brownian sheet, i.e., the Gaussian process on $[0,1]^2$ with zero mean and covariance*

$$
E(Z_t Z_s) = (s_1 \wedge t_1)(s_2 \wedge t_2).
$$

Let μ_ϵ denote the law of $\sqrt{\epsilon}Z_t$. Then $\{\mu_\epsilon\}$ satisfies in $C_0([0,1]^2)$ the LDP with the good rate function

$$
I(\phi) = \begin{cases} \frac{1}{2}\int_{[0,1]^2}\left(\frac{d\mu_\phi}{dm}\right)^2 dm & \text{if } \frac{d\mu_\phi}{dm} \in L_2(m) \\ \infty & \text{otherwise}. \end{cases}
$$

Exercise 5.3.5 Prove that Theorem 5.3.1 remains valid if $\{X_i\}$ take values in $\mathbb{R}^k$, with the rate function now being

$$I(\phi) = \begin{cases} \int_{[0,1]^d} \Lambda^* \left(\frac{d\mu_{\phi_1}}{dm}, \ldots, \frac{d\mu_{\phi_k}}{dm} \right) dm & \text{if, for all } j, \frac{d\mu_{\phi_j}}{dm} \in L_1([0,1]^d) \\ \infty & \text{otherwise}, \end{cases}$$

where $\phi = (\phi_1, \ldots, \phi_k)$ and $\phi_j \in C_0([0,1]^d)$ for $j = 1, \ldots, k$.

Exercise 5.3.6 Prove the exponential tightness of $\{\tilde{\mu}_n\}$ in $C_0([0,1]^d)$.
Hint: Define the sets

$$K_\alpha = \left\{ \phi \text{ of bounded variation} : \int_{[0,1]^d} \Lambda^* \left(\frac{d\mu_\phi}{dm} \right) dm \leq \alpha \right\}.$$

Show that for every $\alpha < \infty$, the set K_α is bounded, and equicontinuous. Then follow the proof of Lemma 5.1.7.

Exercise 5.3.7 Complete the proof of Corollary 5.3.4.
Hint: As in Exercise 5.2.14, use Borell's inequality [Bore75].

5.4 Performance Analysis of DMPSK Modulation

Large deviations techniques are useful in the asymptotic performance analysis of various communication systems. An example of such an analysis for Differential Multiple Phase Shift Key (DMPSK) modulation, which is extensively used in digital optical communication systems, is presented here. For references to the literature on such systems, see the historical notes at the end of this chapter.

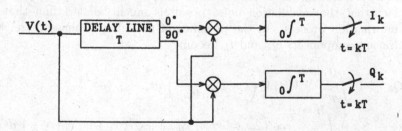

Figure 5.4.1: A DMPSK receiver.

A typical DMPSK system is depicted in Fig. 5.4.1. Let γ_k, $k = 1, 2, \ldots$ be a sequence of angles (phases) such that

$$\gamma_k \in \left\{ 0, \frac{2\pi}{M}, \frac{4\pi}{M}, \ldots, \frac{(M-1)\,2\pi}{M} \right\}.$$

The phase difference $(\gamma_{k-1} - \gamma_k)$ represents the information to be transmitted (with $M \geq 2$ possible values). Ideally, the modulator transmits the signal

$$V(t) - \cos(\omega_o t + \gamma_{[\frac{t}{T}]}) = \cos(\gamma_{[\frac{t}{T}]}) \cos \omega_o t - \sin(\gamma_{[\frac{t}{T}]}) \sin \omega_o t , \quad (5.4.1)$$

where ω_o is the carrier frequency and T is the time duration devoted to each information symbol. Typically, $\omega_o \gg \frac{1}{T}$, and, without loss of generality, assume that $\omega_o T (\mathrm{mod}\, 2\pi) = 0$. The (synchronous) DMPSK detection scheme creates the quadrature signals

$$Q_k = \int_{kT}^{(k+1)T} V(t) V\left(t - T + \frac{\pi}{2\omega_o}\right) dt$$

$$I_k = \int_{kT}^{(k+1)T} V(t) V(t - T)\, dt .$$

Note that $I_k = T\cos(\gamma_k - \gamma_{k-1} + \omega_o T)/2 = T\cos(\gamma_k - \gamma_{k-1})/2$, and since $\omega_o T \gg 1$, $Q_k \simeq T\sin(\gamma_k - \gamma_{k-1} + \omega_o T)/2 = T\sin(\gamma_k - \gamma_{k-1})/2$. Therefore, $\Delta\hat{\gamma}_k$, the phase of the complex number $I_k + iQ_k$, is a very good estimate of the phase difference $(\gamma_k - \gamma_{k-1})$.

In the preceding situation, M may be taken arbitrarily large and yet almost no error is made in the detection scheme. Unfortunately, in practice, a phase noise always exists. This is modeled by the noisy modulated signal

$$V(t) = \cos(\omega_o t + \gamma_{[\frac{t}{T}]} + \sqrt{\epsilon}\, w_t) , \quad (5.4.2)$$

where w_t is a standard Brownian motion and the coefficient $\sqrt{\epsilon}$ emphasizes the fact that the modulation noise power is much smaller than that of the information process. In the presence of modulation phase noise, the quadrature components Q_k and I_k become

$$Q_k = \int_{kT}^{(k+1)T} V(t)\, V\left(t - T + \frac{\pi}{2\omega_o}\right) dt$$

$$\simeq \frac{1}{2} \int_{kT}^{(k+1)T} \cos\left(2\omega_o t + \gamma_k + \gamma_{k-1} + \sqrt{\epsilon}(w_t + w_{t-T}) + \frac{\pi}{2}\right) dt$$

$$+ \frac{1}{2} \int_{kT}^{(k+1)T} \cos\left(\gamma_k - \gamma_{k-1} + \sqrt{\epsilon}(w_t - w_{t-T}) - \frac{\pi}{2}\right) dt \quad (5.4.3)$$

and

$$
\begin{aligned}
I_k &= \int_{kT}^{(k+1)T} V(t)\, V(t-T)\, dt \\
&= \frac{1}{2} \int_{kT}^{(k+1)T} \cos\left(2\omega_o t + \gamma_k + \gamma_{k-1} + \sqrt{\epsilon}(w_t + w_{t-T})\right) dt \\
&\quad + \frac{1}{2} \int_{kT}^{(k+1)T} \cos\left(\gamma_k - \gamma_{k-1} + \sqrt{\epsilon}(w_t - w_{t-T})\right) dt \,.
\end{aligned}
$$

Since $\omega_o T \gg 1$, the first term in (5.4.3) is negligible, and therefore,

$$
Q_k \simeq \frac{1}{2} \int_{kT}^{(k+1)T} \sin\left(\gamma_k - \gamma_{k-1} + \sqrt{\epsilon}(w_t - w_{t-T})\right) dt \,. \tag{5.4.4}
$$

Similarly,

$$
I_k \simeq \frac{1}{2} \int_{kT}^{(k+1)T} \cos\left(\gamma_k - \gamma_{k-1} + \sqrt{\epsilon}(w_t - w_{t-T})\right) dt \,. \tag{5.4.5}
$$

The analysis continues under the assumption of equality in (5.4.4) and (5.4.5). A discussion of this approximation is left to Exercise 5.4.19.

The decision rule is still based on the phase $\Delta\hat{\gamma}_k$ of $I_k + iQ_k$ and is typically achieved by thresholding the ratio Q_k/I_k. Since the probability of error should not depend on the value of $(\gamma_k - \gamma_{k-1})$, it follows by symmetry that the decision rule partitions the unit circle $e^{i\theta}$, $\theta \in [-\pi, \pi]$ into the decision regions $\left(-\frac{\pi}{M}, \frac{\pi}{M}\right]$, $\left(\frac{\pi}{M}, \frac{3\pi}{M}\right]$, ..., $\left(-\frac{3\pi}{M}, -\frac{\pi}{M}\right]$.

Without loss of generality, assume now that $(\gamma_k - \gamma_{k-1}) = 0$ and $k = 1$. Then the error event is the event $A \cup B$, where

$$
\begin{aligned}
A &= \left\{ \omega : \Delta\hat{\gamma}_1 \in \left[-\frac{\pi}{M} - \pi, -\frac{\pi}{M}\right] \right\} \\
&= \left\{ \omega : I_k \sin\frac{\pi}{M} + Q_k \cos\frac{\pi}{M} \le 0 \right\} \\
&= \left\{ \omega : \int_{T}^{2T} \sin\left(\frac{\pi}{M} + \sqrt{\epsilon}(w_t - w_{t-T})\right) dt \le 0 \right\}
\end{aligned}
$$

and

$$
\begin{aligned}
B &= \left\{ \omega : \Delta\hat{\gamma}_1 \in \left(\frac{\pi}{M}, \frac{\pi}{M} + \pi\right) \right\} \\
&= \left\{ \omega : I_k \sin\frac{\pi}{M} - Q_k \cos\frac{\pi}{M} < 0 \right\} \\
&= \left\{ \omega : \int_{T}^{2T} \sin\left(\frac{\pi}{M} - \sqrt{\epsilon}(w_t - w_{t-T})\right) dt < 0 \right\} \,.
\end{aligned}
$$

Therefore, $A = \{\omega : F(\sqrt{\epsilon}w.) \leq 0\}$ where $F : C_0([0, 2T]) \to \mathbb{R}$ is given by

$$F(\phi) = \int_T^{2T} \sin\left(\phi_t - \phi_{t-T} + \frac{\pi}{M}\right) dt .$$

Similarly, $B = \{\omega : F(-\sqrt{\epsilon}w.) < 0\}$. The asymptotics as $\epsilon \to 0$ of $P_e^\epsilon \triangleq P(A \cup B)$, the probability of error per symbol, are stated in the following theorem.

Theorem 5.4.6

$$-\lim_{\epsilon \to 0} \epsilon \log P_e^\epsilon = \inf_{\phi \in \Phi} I_w(\phi) \triangleq I_0 < \infty ,$$

where

$$I_w(\phi) = \left\{ \begin{array}{ll} \frac{1}{2} \int_0^{2T} \dot{\phi}_t^2 \, dt, & \phi \in H_1([0, 2T]) \\ \infty & \text{otherwise} \end{array} \right.$$

and

$$\Phi \triangleq \{\phi \in C_0([0, 2T]) : F(\phi) \leq 0\} .$$

Proof: Let $\hat{\phi}_t = -\pi t/MT$. Since $F(\hat{\phi}) = 0$, it follows that

$$I_0 \leq I_w(\hat{\phi}) = \frac{\pi^2}{M^2 T} < \infty . \tag{5.4.7}$$

Now, note that

$$\begin{aligned} P(F(\sqrt{\epsilon}w.) \leq 0) \leq P_e^\epsilon &\leq P(F(\sqrt{\epsilon}w.) \leq 0) + P(F(-\sqrt{\epsilon}w.) < 0) \\ &\leq 2\,P(F(\sqrt{\epsilon}w.) \leq 0) , \end{aligned}$$

where the last inequality is due to the symmetry around zero of the Brownian motion. Therefore,

$$\lim_{\epsilon \to 0} |\epsilon \log P_e^\epsilon - \epsilon \log \nu_\epsilon(\Phi)| = 0 ,$$

where ν_ϵ is the law of $\sqrt{\epsilon}w.$ on $[0, 2T]$. Since $F(\cdot)$ is continuous with respect to convergence in the supremum norm, the set Φ is a closed set containing the open subset $\{\phi : F(\phi) < 0\}$. Hence, by the version of Schilder's theorem described in Exercise 5.2.8,

$$\begin{aligned} -\inf_{\{\phi : F(\phi) < 0\}} I_w(\phi) &\leq \liminf_{\epsilon \to 0} \epsilon \log P_e^\epsilon \tag{5.4.8} \\ &\leq \limsup_{\epsilon \to 0} \epsilon \log P_e^\epsilon \leq -\inf_{\phi \in \Phi} I_w(\phi) = -I_0 . \end{aligned}$$

Now fix $\phi \in H_1([0, 2T])$ and let

$$\psi_t = -(t - T) \cos\left(\phi_t - \phi_{t-T} + \frac{\pi}{M}\right) 1_{[T, 2T]}(t) .$$

Note that $\psi \in H_1([0, 2T])$ and

$$\frac{d}{dx} F(\phi + x\psi)_{|x=0} = -\int_T^{2T} (t - T) \cos^2 \left(\phi_t - \phi_{t-T} + \frac{\pi}{M} \right) dt \leq 0 \,,$$

with equality only if for some integer m, $\phi_t = \phi_{t-T} - \pi/M + \pi/2 + \pi m$, for all $t \in [T, 2T]$. In particular, equality implies that $F(\phi) \neq 0$. Thus, if $F(\phi) = 0$, it follows that $F(\phi + \eta\psi) < 0$ for all $\eta > 0$ small enough. Since $I_w(\phi + \eta\psi) \to I_w(\phi)$ as $\eta \to 0$, the lower and upper bounds in (5.4.8) coincide and the proof is complete. ☐

Unfortunately, an exact analytic evaluation of I_0 seems impossible to obtain. On the other hand, there exists a path that minimizes the good rate function $I_w(\cdot)$ over the closed set Φ. Whereas a formal study of this minimization problem is deferred to Exercise 5.4.18, bounds on I_0 and information on its limit as $M \to \infty$ are presented in the following theorem.

Theorem 5.4.9

$$\frac{\pi^2}{2M^2T} \leq I_0 \leq \frac{\pi^2}{M^2T} \,. \tag{5.4.10}$$

Moreover,

$$\lim_{M \to \infty} TM^2 I_0 = \frac{3}{4}\pi^2 \,. \tag{5.4.11}$$

Proof: The upper bound on I_0 is merely (5.4.7), and it implies that

$$I_0 = \inf_{\phi \in \hat{\Phi}} I_w(\phi) \,,$$

where

$$\hat{\Phi} = \Phi \cap \left\{ \phi : I_w(\phi) \leq \frac{\pi^2}{M^2T} \right\} \,.$$

Fix $\phi \in \hat{\Phi} \subset H_1([0, 2T])$. Then for $t \in [T, 2T]$, it follows from the Cauchy–Schwartz inequality that

$$\begin{aligned} I_w(\phi) &= \frac{1}{2}\int_0^{2T} \dot\phi_s^2 \, ds \geq \frac{1}{2}\int_{t-T}^t \dot\phi_s^2 \, ds \geq \frac{1}{2T} \left(\int_{t-T}^t \dot\phi_s \, ds \right)^2 \\ &= \frac{1}{2T}|\phi_t - \phi_{t-T}|^2 \,. \end{aligned} \tag{5.4.12}$$

To complete the proof of (5.4.10), note that $F(\phi) \leq 0$ implies that

$$\tau \overset{\triangle}{=} \inf\{t \geq T : |\phi_t - \phi_{t-T}| \geq \frac{\pi}{M}\} \leq 2T \,,$$

and apply the inequality (5.4.12) for $t = \tau$.

Turning now to the proof of (5.4.11), observe that for all $\phi \in \hat{\Phi}$,

$$\frac{\sqrt{2}\pi}{M} \geq \sqrt{2TI_w(\phi)} \geq \sup_{t \in [T,2T]} |\phi_t - \phi_{t-T}| .$$

Since $|\sin x - x| \leq |x|^3/6$, it follows that

$$\sup_{t \in [T,2T]} \left| \sin\left(\frac{\pi}{M} + \phi_t - \phi_{t-T}\right) - \left(\frac{\pi}{M} + \phi_t - \phi_{t-T}\right) \right| \leq \frac{(1+\sqrt{2})^3 \pi^3}{6M^3} .$$

Hence,

$$\int_T^{2T} \left(\phi_t - \phi_{t-T} + \frac{\pi}{M}\right) dt - \frac{(1+\sqrt{2})^3 \pi^3 T}{6M^3} \leq F(\phi)$$

$$\leq \int_T^{2T} \cdot \left(\phi_t - \phi_{t-T} + \frac{\pi}{M}\right) dt + \frac{(1+\sqrt{2})^3 \pi^3 T}{6M^3} . \qquad (5.4.13)$$

Let

$$\tilde{\Phi}(\alpha) = \left\{ \phi \in C_0([0,2T]) : \int_T^{2T} (\phi_t - \phi_{t-T} + \alpha)\, dt \leq 0, I_w(\phi) \leq \frac{\pi^2}{M^2 T} \right\} ;$$

then (5.4.13) implies that

$$\tilde{\Phi}\left(\frac{\pi}{M} + \frac{(1+\sqrt{2})^3 \pi^3}{6M^3}\right) \subset \hat{\Phi} \subset \tilde{\Phi}\left(\frac{\pi}{M} - \frac{(1+\sqrt{2})^3 \pi^3}{6M^3}\right) .$$

Consequently,

$$\tilde{I}\left(\frac{\pi}{M} - \frac{(1+\sqrt{2})^3 \pi^3}{6M^3}\right) \leq I_0 \leq \tilde{I}\left(\frac{\pi}{M} + \frac{(1+\sqrt{2})^3 \pi^3}{6M^3}\right) , \qquad (5.4.14)$$

where

$$\tilde{I}(\alpha) \stackrel{\triangle}{=} \inf_{\phi \in \tilde{\Phi}(\alpha)} I_w(\phi) .$$

Therefore, (5.4.11) follows from (5.4.14) and Lemma 5.4.15. $\square$

Lemma 5.4.15 *If* $|\alpha| \leq 2\pi/\sqrt{3}M$, *then*

$$\tilde{I}(\alpha) = \frac{3\,\alpha^2}{4\,T} .$$

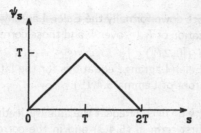

Figure 5.4.2: The function ψ_s.

Proof: Let $\psi_s = \min\{2T, T + s\} - \max\{T, s\}$ (see Fig. 5.4.2), and note that for any $\phi \in H_1([0, 2T])$,

$$\int_T^{2T} (\phi_t - \phi_{t-T})\, dt = \int_T^{2T} \int_{t-T}^{t} \dot{\phi}_s\, ds\, dt = \int_0^{2T} \dot{\phi}_s \psi_s\, ds.$$

Therefore,

$$\tilde{\Phi}(\alpha) = \{\phi \in H_1([0, 2T]) : \int_0^{2T} \dot{\phi}_s \psi_s ds \leq -\alpha T, \ \frac{1}{2}\int_0^{2T} \dot{\phi}_s^2 ds \leq \frac{\pi^2}{M^2 T}\}.$$

Hence, for $\phi \in \tilde{\Phi}(\alpha)$,

$$\left(\int_0^{2T} \dot{\phi}_s^2\, ds\right)\left(\int_0^{2T} \psi_s^2\, ds\right) \geq \left(\int_0^{2T} \dot{\phi}_s \psi_s\, ds\right)^2 \geq \alpha^2 T^2,$$

implying that

$$I_w(\phi) \geq \frac{\alpha^2 T^2}{2\int_0^{2T} \psi_s^2\, ds} = \frac{\alpha^2 T^2}{4\int_0^{T} s^2 ds} = \frac{3\alpha^2}{4T}. \tag{5.4.16}$$

Thus,

$$\tilde{I}(\alpha) = \inf_{\phi \in \tilde{\Phi}(\alpha)} I_w(\phi) \geq \frac{3\alpha^2}{4T}. \tag{5.4.17}$$

Observe now that the inequality (5.4.16) holds with equality if $\dot{\phi}_t = \lambda \psi_t$, where $\lambda = -\alpha T/(\int_0^{2T} \psi_s^2\, ds)$. This results in $\phi \in \tilde{\Phi}(\alpha)$, provided that $3\alpha^2/4T \leq \pi^2/M^2 T$. Hence, for these values of α, (5.4.17) also holds with equality and the proof of the lemma is complete. $\qquad\square$

Exercise 5.4.18 Write down formally the Euler–Lagrange equations that correspond to the minimization of $I_w(\cdot)$ over Φ and those corresponding to the minimization over $\{\phi \in H_1([0, 2T]) : \int_0^{2T} \dot\phi_s \psi_s ds \leq -\alpha T\}$. Show that $\lambda \int_0^t \psi_s ds$ is the solution to the Euler–Lagrange equations for the latter (where ψ_s and λ are as defined in the proof of Lemma 5.4.15).

Exercise 5.4.19 Let k be fixed. Repeat the analysis leading to Theorem 5.4.6 without omitting the first term in (5.4.3) and in the corresponding expression for I_k. Show that Theorem 5.4.6 continues to hold true when $\omega_o \to \infty$.
Hint: Take $\omega_o = 2\pi m/T, m = 1, 2, \ldots$ and show that Q_k of (5.4.3) are exponentially good approximations (in m) of the Q_k of (5.4.4).

5.5 Large Exceedances in $\mathbb{R}^d$

In some applications related to DNA sequence matching, queuing networks analysis, and abrupt change detection in dynamical systems, the following problem is relevant: Let X_i, $i = 1, 2, \ldots$, be i.i.d. $\mathbb{R}^d$-valued random variables, with zero mean and logarithmic moment generating function $\Lambda(\lambda)$, which is finite. Let $b \neq 0$ be a given drift vector in $\mathbb{R}^d$ and define

$$Z_n = \sum_{i=1}^n (X_i - b).$$

The increments $Z_n - Z_m$ of the random walk are of the form $Z_n - Z_m \sim -(n - m)b + \xi_{n-m}$, where ξ_{n-m} has zero mean and covariance of order $(n - m)$. Thus, with $n - m$ large, it becomes unlikely that $Z_n - Z_m$ is far away from $-(n-m)b$. Interest then lies in the probability of the rare events $\{Z_n - Z_m \in A/\epsilon\}$, for small ϵ, where A is a closed set that does not intersect the typical ray $\{-\alpha b\}_{\alpha \geq 0}$.

Some specific applications where the preceding problem occurs are as follows. (For references, see the historical notes at the end of this chapter.) First, the false alarm rate in the sequential detection of change points by the commonly used generalized likelihood ratio test (also known as the CUSUM method) corresponds to the location of the first such unlikely event, where $X_i - b$ is the log-likelihood score for the ith variable, $A = [1, \infty)$ and $1/\epsilon$ is the log-likelihood decision threshold. The same question also appears in the analysis of long DNA sequences where $X_i - b$ are letter scores measuring properties of biological importance. A third application involves general one-server queues, where the events $\{Z_n - Z_m \in A/\epsilon\}$ are related to unusually long waiting times for completion of service. Here, too, $A = [1, \infty)$, while $X_i - b$ is the net difference between the service time and inter-arrival time for the ith customer, and the condition $b > 0$ ensures that the queue is stable.

To be able to handle this question, it is helpful to rescale variables. Thus, define the rescaled time $t = \epsilon n$, and the rescaled variables

$$Y_t^\epsilon = \epsilon Z_{[\frac{t}{\epsilon}]} = \epsilon \sum_{i=1}^{[\frac{t}{\epsilon}]} (X_i - b), \quad t \in [0, \infty).$$

The rare events of interest correspond to events of the form $Y_t^\epsilon - Y_\tau^\epsilon \in A$. To formalize the results, define the following random times:

$$T_\epsilon \stackrel{\triangle}{=} \inf\{t : \ \exists s \in [0, t) \text{ such that } \ Y_t^\epsilon - Y_s^\epsilon \in A\},$$

$$\tau_\epsilon \stackrel{\triangle}{=} \sup\{s \in [0, T_\epsilon) : \ Y_{T_\epsilon}^\epsilon - Y_s^\epsilon \in A\} - \epsilon, \quad L_\epsilon \stackrel{\triangle}{=} T_\epsilon - \tau_\epsilon. \quad (5.5.1)$$

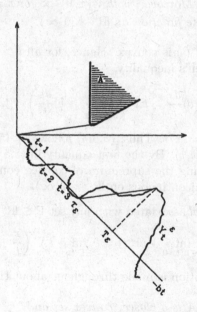

Figure 5.5.1: Typical A, Y_t^ϵ, T_ϵ, τ_ϵ.

The main results of this section are that, under appropriate conditions, as $\epsilon \to 0$, both $\epsilon \log T_\epsilon$ and L_ϵ converge in probability. More notations are now introduced in order to formalize the conditions under which such convergence takes place. First, let $I_t : L_\infty([0, t]) \to [0, \infty]$ be the rate functions associated with the trajectories Y^ϵ, namely,

$$I_t(\phi) \triangleq \begin{cases} \int_0^t \Lambda^*(\dot\phi_s + b)\, ds, & \text{if } \phi \in \mathcal{AC}^t, \phi_0 = 0, \\ \infty & \text{otherwise}, \end{cases}$$

where $\mathcal{AC}^t$ is the space of absolutely continuous functions $\phi : [0,t] \to \mathbb{R}^d$. (See Exercise 5.1.22.) The *cost* associated with a termination point $x \in \mathbb{R}^d$ at time $t \in (0,\infty)$ is given by

$$V(x,t) \triangleq \inf_{\{\phi \in \mathcal{AC}^t:\ \phi_0 = 0, \phi_t = x\}} I_t(\phi). \tag{5.5.2}$$

The following lemma motivates the preceding definition of $V(x,t)$.

Lemma 5.5.3 *For all $x \in \mathbb{R}^d, t > 0$,*

$$V(x,t) = \sup_{\lambda \in \mathbb{R}^d} \{\langle \lambda, x \rangle + t[\langle \lambda, b \rangle - \Lambda(\lambda)]\} = t\Lambda^*\left(\frac{x}{t} + b\right), \tag{5.5.4}$$

and $V(x,t) = I_t(\frac{s}{t}x)$. Moreover, with $V(x,0) \triangleq \infty$ for $x \neq 0$ and $V(0,0) = 0$, $V(x,t)$ is a convex rate function on $\mathbb{R}^d \times [0,\infty)$.

Proof: Recall that $\Lambda^*(\cdot)$ is convex. Hence, for all $t > 0$ and any $\phi \in \mathcal{AC}^t$ with $\phi_0 = 0$, by Jensen's inequality,

$$I_t(\phi) = t\int_0^t \Lambda^*(\dot\phi_s + b)\frac{ds}{t} \geq t\Lambda^*\left(\int_0^t (\dot\phi_s + b)\frac{ds}{t}\right) = t\Lambda^*\left(\frac{\phi_t - \phi_0}{t} + b\right),$$

with equality for $\phi_s = sx/t$. Thus, (5.5.4) follows by (5.5.2). Since $\Lambda^*(\cdot)$ is nonnegative, so is $V(x,t)$. By the first equality in (5.5.4), which also holds for $t = 0$, $V(x,t)$, being the supremum of convex, continuous functions is convex and lower semicontinuous on $\mathbb{R}^d \times [0,\infty)$. $\qquad\square$

The *quasi-potential* associated with any set $\Gamma \subset \mathbb{R}^d$ is defined as

$$V_\Gamma \triangleq \inf_{x \in \Gamma, t > 0} V(x,t) = \inf_{x \in \Gamma, t > 0} t\Lambda^*\left(\frac{x}{t} + b\right).$$

The following assumption is made throughout about the set A.

Assumption 5.5.5 *A is a closed, convex set and*
(A-1) *$V_A = V_{A^\circ} \in (0,\infty)$ (where A° denotes the interior of A).*
(A-2) *There is a unique pair $\overline{x} \in A$, $\overline{t} \in (0,\infty)$ such that $V_A = V(\overline{x},\overline{t})$. Moreover, the straight line $\phi_s^* \triangleq (s/\overline{t})\overline{x}$ is the unique path for which the value of $V(\overline{x},\overline{t})$ is achieved.*
(A-3)

$$\lim_{r \to \infty} V_{A \cap \{x:|x|>r\}} > 2V_A.$$

The need for this assumption is evident in the statement and proof of the following theorem, which is the main result of this section.

Theorem 5.5.6 *Let Assumption 5.5.5 hold. Then*

$$V_A = \lim_{\epsilon \to 0} \epsilon \log T_\epsilon \ \ in \ probability \tag{5.5.7}$$

and

$$\bar{t} = \lim_{\epsilon \to 0} L_\epsilon \ \ in \ probability. \tag{5.5.8}$$

Let $\hat{Y}_s^\epsilon \triangleq Y_{\tau_\epsilon + s}^\epsilon - Y_{\tau_\epsilon}^\epsilon$. Then

$$\lim_{\epsilon \to 0} \sup_{0 \le s \le \bar{t}} |\hat{Y}_s^\epsilon - \phi_s^*| = 0 \ \ in \ probability. \tag{5.5.9}$$

Finally, with $p_\epsilon = P(T_\epsilon \le \theta_\epsilon)$ and any $\theta_\epsilon \to \infty$ such that $\epsilon \log \theta_\epsilon \to 0$, the random variables $\theta_\epsilon^{-1} p_\epsilon T_\epsilon$ converge in distribution to an Exponential(1) random variable.

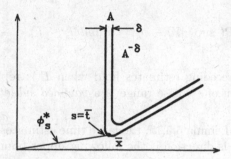

Figure 5.5.2: Minimizing path for A.

Proof: The following properties of the cost function, whose proofs are deferred to the end of the section, are needed.

Lemma 5.5.10

$$\lim_{r \to \infty} \inf_{x \in A, t \ge r} V(x, t) > 2V_A . \tag{5.5.11}$$

For all $T > \bar{t}$,

$$\lim_{\delta \to 0} \inf_{x \in A^{-\delta}, t \in [0, T]} V(x, t) = \inf_{x \in A^\circ, t \in [0, T]} V(x, t) = V_A , \tag{5.5.12}$$

where

$$A^{-\delta} \triangleq \{x : \inf_{y \notin A} |x - y| > \delta\}. \tag{5.5.13}$$

For all $\delta > 0$,

$$\inf_{x \in A, |t - \bar{t}| \ge \delta} V(x, t) > V_A. \tag{5.5.14}$$

The main difficulty in proving Theorem 5.5.6 is that it involves events on an infinite time scale; this excludes directly using the large deviations bounds of Section 5.1. The proof is therefore based on decoupling the infinite time horizon to finite time intervals that are weakly coupled. In these intervals, we use the following short time estimates, which are a specialization of Mogulskii's theorem (Theorem 5.1.2), and whose proof is deferred to the end of this section.

Lemma 5.5.15 *For every fixed $T \in (\bar{t}, \infty)$,*

$$\lim_{\epsilon \to 0} \epsilon \log P(T_\epsilon \leq T) = -V_A \ . \tag{5.5.16}$$

Moreover, for all $\delta > 0$,

$$\limsup_{\epsilon \to 0} \epsilon \log P(|L_\epsilon - \bar{t}| \geq \delta \text{ and } T_\epsilon \leq T) < -V_A \tag{5.5.17}$$

and

$$\limsup_{\epsilon \to 0} \epsilon \log P(\sup_{0 \leq s \leq \bar{t}} |\hat{Y}_s^\epsilon - \phi_s^*| \geq \delta \text{ and } T_\epsilon \leq T) < -V_A. \tag{5.5.18}$$

Remark: The preceding estimates hold when T is replaced by any deterministic function of ϵ whose range is a *bounded subset* of $(\bar{t}, \infty)$ that is bounded away from $\bar{t}$.

Equipped with Lemma 5.5.15, the large time estimates may be handled. The first step in this direction is the following upper bound on T_ϵ.

Lemma 5.5.19 *For any $\delta > 0$,*

$$\lim_{\epsilon \to 0} P(T_\epsilon > e^{(V_A + \delta)/\epsilon}) = 0 \ .$$

Proof: Let $\Delta = [(\bar{t} + 2)/\epsilon]\epsilon$ and split the time interval $[0, e^{(V_A + \delta)/\epsilon}]$ into disjoint intervals of length Δ each. Let N_ϵ be the (integer part of the) number of such intervals. Observe that

$$P(T_\epsilon > e^{(V_A + \delta)/\epsilon}) \ \leq \ P(Y_{k\Delta + t}^\epsilon - Y_{k\Delta + s}^\epsilon \notin A \,,$$
$$0 \leq s \leq t \leq \Delta \,, \ k = 0, \ldots, N_\epsilon - 1) \,.$$

The preceding events are independent for different values of k as they correspond to disjoint segments of the original random walk (Z_n). Moreover, they are of equal probability because the joint law of the increments of Y^ϵ is invariant under any shift of the form $n\epsilon$ with n integer. Hence,

$$P(T_\epsilon > e^{(V_A + \delta)/\epsilon}) \leq [1 - P(T_\epsilon \leq \Delta)]^{N_\epsilon} \ .$$

Note that $\Delta \in [\,\bar{t} + 1, \bar{t} + 2\,]$ for all $\epsilon < 1$. Hence, for some $0 < c < \infty$ (independent of ϵ) and all $\epsilon < 1$,

$$N_\epsilon \geq c e^{(V_A + \delta)/\epsilon} ,$$

while for all $\epsilon > 0$ small enough,

$$P(T_\epsilon \leq \Delta) \geq P(T_\epsilon \leq \bar{t} + 1) \geq e^{-(V_A + \delta/2)/\epsilon} ,$$

where the second inequality follows from (5.5.16). Combining all these bounds yields

$$\limsup_{\epsilon \to 0} P(T_\epsilon > e^{(V_A + \delta)/\epsilon}) \leq \limsup_{\epsilon \to 0} (1 - e^{-(V_A + \delta/2)/\epsilon})^{c e^{(V_A + \delta)/\epsilon}}$$

$$= \limsup_{\epsilon \to 0} e^{-c e^{\delta/2\epsilon}} = 0 . \qquad \square$$

Lemma 5.5.19 is not enough yet, for the upper bounds on T_ϵ are unbounded (as $\epsilon \to 0$). The following lemma allows for restricting attention to increments within finite time lags.

Lemma 5.5.20 *There exists a constant $C < \infty$ such that*

$$\lim_{\epsilon \to 0} P(L_\epsilon \geq C) = 0 .$$

Proof: Observe that the process Y^ϵ is constant over intervals of size ϵ, and T_ϵ and τ_ϵ are always the starting points of such intervals. Let $K_\epsilon \triangleq [\epsilon^{-1} e^{(V_A + \delta)/\epsilon}]$ denote the number of such points in $(0, e^{(V_A + \delta)/\epsilon}]$, and define

$$Q_{\epsilon,\delta} \triangleq \sum_{k,\ell=0, k-\ell \geq C/\epsilon}^{K_\epsilon} P(Y_{k\epsilon}^\epsilon - Y_{\ell\epsilon}^\epsilon \in A) .$$

It suffices to show that $\lim_{\delta \to 0} \lim_{\epsilon \to 0} Q_{\epsilon,\delta} = 0$ for large enough constant C, for then the proof of the lemma is completed by applying the union of events bound together with Lemma 5.5.19. To this end, the invariance of the law of increments of Y^ϵ to shifts of $n\epsilon$ and the identity $Y_{n\epsilon}^\epsilon = \epsilon Z_n$ for all integer n yield the following upper bound on $Q_{\epsilon,\delta}$.

$$Q_{\epsilon,\delta} \leq K_\epsilon \sum_{n \geq C/\epsilon}^{K_\epsilon} P(Y_{n\epsilon}^\epsilon \in A) \leq \epsilon^{-2} e^{2(V_A + \delta)/\epsilon} \max_{n \geq C/\epsilon} P(\epsilon Z_n \in A) . \quad (5.5.21)$$

Let $\hat{S}_n$ denote the empirical mean of $X_1, \ldots, X_n$. Observe that

$$\{\epsilon Z_n \in A\} \equiv \{\hat{S}_n \in \tilde{A}\} ,$$

where $\tilde{A} \triangleq \{z : \epsilon n(z - b) \in A\}$ is a convex, closed set (because A is). Hence, by the min–max theorem (see Exercise 2.2.38 for details),

$$
\begin{aligned}
\epsilon \log \mathrm{P}(\epsilon Z_n \in A) &= \epsilon \log \mathrm{P}(\hat{S}_n \in \tilde{A}) \leq -n\epsilon \inf_{z \in \tilde{A}} \Lambda^*(z) \\
&= -n\epsilon \inf_{x \in A} \Lambda^*\left(\frac{x}{n\epsilon} + b\right) = -\inf_{x \in A} V(x, n\epsilon) .
\end{aligned}
$$

Combining this inequality and (5.5.21) yields, for all $C, \delta, \epsilon > 0$,

$$
Q_{\epsilon,\delta} \leq \epsilon^{-2} e^{-\left[\inf_{x \in A, t \geq C} V(x,t) - 2(V_A + \delta)\right]/\epsilon} .
$$

By (5.5.11), C can be chosen large enough so that for all δ small enough and all ϵ small enough,

$$
\inf_{x \in A, t \geq C} V(x, t) > 2V_A + 3\delta - 2\epsilon \log \epsilon ,
$$

yielding $Q_{\epsilon,\delta} \leq e^{-\delta/\epsilon}$.   □

Returning now to the proof of the theorem, let C be the constant from Lemma 5.5.20 and define $C_\epsilon \triangleq \epsilon \lceil 1 + C/\epsilon \rceil \geq C$. For any integer n, define the decoupled random times

$$
T_{\epsilon,n} \triangleq \inf\{t : Y_t^\epsilon - Y_s^\epsilon \in A \text{ for some } t > s \geq 2nC_\epsilon \lfloor t/(2nC_\epsilon) \rfloor\} .
$$

Lemma 5.5.22

$$
\lim_{n \to \infty} \lim_{\epsilon \to 0} \mathrm{P}(T_{\epsilon,n} \neq T_\epsilon) = 0 .
$$

Proof: Divide $[C_\epsilon, \infty)$ into the disjoint intervals $I_\ell \triangleq [(2\ell - 1)C_\epsilon, (2\ell + 1)C_\epsilon)$, $\ell = 1, \ldots$. Define the events

$$
J_\ell \triangleq \{Y_t^\epsilon - Y_\tau^\epsilon \in A \text{ for some } \tau \leq t, \ t, \tau \in I_\ell\} ,
$$

and the stopping time

$$
N = \inf\{\ell \geq 1 : J_\ell \text{ occurs}\} .
$$

By the translation invariance, with respect to n, of $Z_{n+m} - Z_n$ and the fact that C_ϵ is an integer multiple of ϵ, the events J_ℓ are independent and equally likely. Let $p = \mathrm{P}(J_\ell)$. Then $\mathrm{P}(N = \ell) = p(1 - p)^{\ell-1}$ for $\ell \in \mathbb{Z}_+$. Hence,

$$
\begin{aligned}
\mathrm{P}(\{T_\epsilon < T_{\epsilon,n}\} \cap \{L_\epsilon < C_\epsilon\}) &\leq \mathrm{P}\left(\bigcup_{k=1}^{\infty} \{N = kn\}\right) \\
&= \sum_{k=1}^{\infty} p(1-p)^{kn-1} = \frac{p(1-p)^{n-1}}{1 - (1-p)^n} \leq \frac{1}{n} .
\end{aligned}
$$

Since by definition $T_\epsilon \leq T_{\epsilon,n}$, and $C_\epsilon \geq C$, the proof is completed by applying the union of events bound and using Lemma 5.5.20. $\qquad\square$

Returning to the proof of the theorem, it is enough to consider the rare events of interest with respect to the decoupled times for n large enough (and finite). This procedure results with a sequence of i.i.d. random variables corresponding to disjoint segments of Y^ϵ of length $2nC_\epsilon$ each. The short time estimates of Lemma 5.5.15 can then be applied. In particular, with $N_\epsilon \triangleq \left[(2nC_\epsilon)^{-1}e^{(V_A-\delta)/\epsilon}\right] + 1$ denoting the number of such segments in $[0, e^{(V_A-\delta)/\epsilon}]$, the following lower bound on $T_{\epsilon,n}$ is obtained

$$
\begin{aligned}
\mathrm{P}(T_{\epsilon,n} < e^{(V_A-\delta)/\epsilon}) \;&\leq\; \sum_{k=0}^{N_\epsilon-1} \mathrm{P}\Big(\Big[\frac{T_{\epsilon,n}}{2nC_\epsilon}\Big] = k\Big) \\
&\leq\; N_\epsilon \mathrm{P}(T_{\epsilon,n} < 2nC_\epsilon) = N_\epsilon \mathrm{P}(T_\epsilon < 2nC_\epsilon) \\
&\leq\; \Big(\frac{e^{(V_A-\delta)/\epsilon}}{2nC} + 1\Big)\,\mathrm{P}(T_\epsilon \leq 2nC_\epsilon)\,.
\end{aligned}
$$

Therefore, with n large enough for $2nC > \bar{t}$, because $\lim_{\epsilon\to 0} C_\epsilon = C$, the short time estimate (5.5.16) implies that

$$
\lim_{\epsilon\to 0} \mathrm{P}(T_{\epsilon,n} < e^{(V_A-\delta)/\epsilon}) \leq \lim_{\epsilon\to 0} \frac{e^{(V_A-\delta)/\epsilon}}{2nC} e^{-(V_A-\delta/2)/\epsilon} = 0\,.
$$

Hence, for all $\delta > 0$,

$$
\lim_{\epsilon\to 0} \mathrm{P}(T_\epsilon < e^{(V_A-\delta)/\epsilon}) = \lim_{n\to\infty}\lim_{\epsilon\to 0} \mathrm{P}(T_{\epsilon,n} < e^{(V_A-\delta)/\epsilon}) = 0\,,
$$

and (5.5.7) results in view of the upper bound of Lemma 5.5.19. The proofs of (5.5.8) and (5.5.9) as consequences of the short time estimates of Lemma 5.5.15 are similar and thus left to Exercise 5.5.29.

Suppose θ_ϵ/ϵ are integers and $\theta_\epsilon \to \infty$. Let

$$
T_{\epsilon,\theta_\epsilon} \triangleq \inf\{t: Y_t^\epsilon - Y_s^\epsilon \in A \text{ for some } t > s \geq \theta_\epsilon[t/\theta_\epsilon]\}\,.
$$

By the same argument as in Lemma 5.5.22, as $\epsilon \to 0$,

$$
P(\{T_{\epsilon,\theta_\epsilon} < T_\epsilon\} \cap \{L_\epsilon < C_\epsilon\}) \leq \lfloor \theta_\epsilon/(2C_\epsilon)\rfloor^{-1} \to 0\,.
$$

Hence, by Lemma 5.5.20, also $P(T_{\epsilon,\theta_\epsilon} \neq T_\epsilon) \to 0$. Fix $y > 0$ and let $m_\epsilon = \lfloor y/p_\epsilon \rfloor$ and $y_\epsilon = p_\epsilon m_\epsilon$. The event $\{\theta_\epsilon^{-1}p_\epsilon T_{\epsilon,\theta_\epsilon} > y_\epsilon\}$ is merely the intersection of m_ϵ independent events, each of which occurs with probability $(1 - p_\epsilon)$. Consequently,

$$
P(\theta_\epsilon^{-1}p_\epsilon T_{\epsilon,\theta_\epsilon} > y_\epsilon) = (1 - p_\epsilon)^{m_\epsilon}\,.
$$

Since $\epsilon \log \theta_\epsilon \to 0$, it follows from (5.5.7) that $p_\epsilon \to 0$ and $y_\epsilon \to y$. Therefore, $(1 - p_\epsilon)^{m_\epsilon} \to e^{-y}$ and the Exponential limit law of $\theta_\epsilon^{-1} p_\epsilon T_\epsilon$ follows.

Since p_ϵ is not affected by replacing θ_ϵ with $\epsilon \lfloor \theta_\epsilon / \epsilon \rfloor$, the convergence in law extends to possibly non-integer $\theta_\epsilon / \epsilon$ (by the continuity of the Exponential law).   □

Of particular interest is the case where X_i are Normal random vectors with uncorrelated components and unit variance. Then $V(x, t) = |x + bt|^2 / 2t$ and is obtained uniquely by a straight line. The infimum (in t) of $V(x, t)$ is obtained uniquely at $t = |x| / |b|$, yielding $V_{\{x\}} = |x| |b| + \langle x, b \rangle$. $V_{\{x\}}$ is an homogeneous, continuous function, that is nonconstant on any ray other than $\{-\alpha b\}_{\alpha \geq 0}$. Therefore, the Assumptions (A-1)–(A-3) are satisfied, for example, when A is a closed, convex set of non-empty interior which, for $\eta > 0$ small enough, excludes the cone

$$C_\eta \stackrel{\triangle}{=} \left\{ x : 1 + \frac{\langle x, b \rangle}{|x| \, |b|} < \eta \right\} .$$

Due to the exponential equivalence of the random walk and the (scaled) Brownian motion, the following is proved exactly as Theorem 5.5.6. (See Exercise 5.5.32 for an outline of the derivation.)

Theorem 5.5.23 *Suppose A is a closed convex set of non-empty interior with $C_\eta \subset A^c$ for $\eta > 0$ small enough. Define*

$$Y_t^\epsilon = -bt + \sqrt{\epsilon} w_t , \qquad (5.5.24)$$

where $w_.$ is a d-dimensional Brownian motion. Then all the results of Theorem 5.5.6 hold true.

Remark: In the definition (5.5.1) of τ_ϵ, the $-\epsilon$ term compensates for discretization effects. It is omitted when considering the continuous time process Y^ϵ of (5.5.24).

The proof of the auxiliary lemmas used before concludes this section.

Proof of Lemma 5.5.10: To establish (5.5.11), observe that $\nabla \Lambda(0) = E(X_1) = 0$, so it follows by part (b) of Lemma 2.3.9 that $\Lambda^*(z) > 0$ for $z \neq 0$. Moreover, $\Lambda^*(\cdot)$ is a good rate function, so also

$$a \stackrel{\triangle}{=} \inf_{|z - b| \leq |b| / 2} \Lambda^*(z) > 0 .$$

Hence, by (5.5.4), for all $r > 0$,

$$\inf_{|x| \leq r} \inf_{t \geq 2r/|b|} V(x, t) \geq \inf_{t \geq 2r/|b|} \inf_{|x| \leq \frac{|b|}{2} t} t \Lambda^* \left(\frac{x}{t} + b \right) \geq \frac{2ra}{|b|} ,$$

implying that

$$\lim_{r\to\infty} \inf_{x\in A, t\geq r} V(x,t) \geq \lim_{r\to\infty} \min\{\inf_{x\in A, |x|>r} V_{\{x\}}, \frac{2ra}{|b|}\} > 2V_A,$$

where the last inequality follows by Assumption (A-3).

Similarly, there exists an $r < \infty$ such that

$$V_A = V_{A^\circ} = \inf_{x\in A^\circ, |x|\leq r, t\leq r} V(x,t), \tag{5.5.25}$$

and

$$\inf_{x\in A, t\geq r} V(x,t) > V_A. \tag{5.5.26}$$

Consider an arbitrary sequence (x_n, t_n) such that $x_n \in A$, $|x_n| \leq r$, $t_n \in [0, r]$ and $V(x_n, t_n) \to V_A$. Such a sequence has at least one limit point, e.g., (x, t), and by the lower semicontinuity of $V(\cdot, \cdot)$,

$$V_A = \lim_{n\to\infty} V(x_n, t_n) \geq V(x,t).$$

However, $x \in A$ and $t < \infty$, implying by Assumption (A-2) that $x = \bar{x}$, $t = \bar{t}$ (and for all $T > \bar{t}$, eventually $t_n \in [0, T]$). When combined with (5.5.25) (or (5.5.26)), it yields (5.5.12) (or (5.5.14), respectively). Indeed, A° may be replaced by $A^{-\delta}$, $\delta \to 0$, since for any $x \in A^\circ$, also $x \in A^{-\delta}$ for all δ small enough. $\qquad\square$

Proof of Lemma 5.5.15: Let $\tilde{Y}^\epsilon$ denote the polygonal approximation of Y^ϵ, namely,

$$\tilde{Y}_t^\epsilon \triangleq Y_t^\epsilon + (t - [t/\epsilon]\epsilon)(X_{[t/\epsilon]+1} - b),$$

and define

$$\tilde{T}_\epsilon \triangleq \inf\{t: \ \exists s \in [0, t) \text{ such that } \ \tilde{Y}_t^\epsilon - \tilde{Y}_s^\epsilon \in A\}.$$

Recall that for any $T \in (0, \infty)$, Y^ϵ satisfy the LDP on $L_\infty([0, T])$ with rate function $I_T(\cdot)$. (See Exercise 5.1.22.) Moreover, the continuous processes $\tilde{Y}^\epsilon$, being exponentially equivalent to Y^ϵ (see Lemma 5.1.4 and the proof of Theorem 5.1.19), satisfy the same LDP on $C_0([0, T])$. The proof is based on relating the events of interest with events determined by $\tilde{Y}^\epsilon$ and then applying the large deviations bounds for $\tilde{Y}^\epsilon$.

Starting with upper bounding T_ϵ, note that $\tilde{T}_\epsilon \leq T_\epsilon$ with probability one, and hence,

$$P(T_\epsilon \leq T) \leq P(\tilde{T}_\epsilon \leq T) = P(\tilde{Y}^\epsilon \in \Psi),$$

where

$$\Psi \stackrel{\triangle}{=} \{\psi \in C_0([0,T]) : \psi_t - \psi_\tau \in A \text{ for some } \tau \leq t \in [0,T]\}. \qquad (5.5.27)$$

As for lower bounding T_ϵ define for all $\delta > 0$

$$\tilde{T}_\epsilon^{-\delta} \stackrel{\triangle}{=} \inf\{t : \exists s \in [0,t) \text{ such that } \tilde{Y}_t^\epsilon - \tilde{Y}_s^\epsilon \in A^{-\delta}\},$$

where $A^{-\delta}$ is defined in (5.5.13), and define the sets $\Psi^{-\delta}$ in analogy with (5.5.27). Hence,

$$P(\tilde{Y}^\epsilon \in \Psi^{-\delta}) = P(\tilde{T}_\epsilon^{-\delta} \leq T) \leq P(T_\epsilon \leq T) + P(\| \tilde{Y}^\epsilon - Y^\epsilon \| \geq \delta),$$

where $\| \cdot \|$ denotes throughout the supremum norm over $[0,T]$.

As for the proof of (5.5.17) and (5.5.18), note that

$$P(|L_\epsilon - \bar{t}| \geq \delta \text{ and } T_\epsilon \leq T) \leq P(\tilde{Y}^\epsilon \in \Phi_\delta),$$

where

$$\Phi_\delta \stackrel{\triangle}{=} \{\psi \in C_0([0,T]) : \psi_t - \psi_\tau \in A \text{ for some } \tau \leq t \in [0,T],$$
$$\text{such that } t - \tau \in [0, \bar{t} - \delta] \cup [\bar{t} + \delta, T]\},$$

and

$$P(\sup_{0 \leq s \leq \bar{t}} |\hat{Y}_s^\epsilon - \phi_s^*| \geq \delta \text{ and } T_\epsilon \leq T) \leq P(\tilde{Y}^\epsilon \in \Phi_\delta^*),$$

where

$$\Phi_\delta^* \stackrel{\triangle}{=} \{\psi \in C_0([0,T+\bar{t}]) : \psi_t - \psi_\tau \in A, \sup_{0 \leq s \leq \bar{t}} |\psi_{s+\tau} - \psi_\tau - \phi_s^*| \geq \delta$$
$$\text{for some } \tau \leq t \in [0,T]\}.$$

By these probability bounds, the fact that $\tilde{Y}^\epsilon$ satisfy the LDP in $C_0([0,T])$ with rate $I_T(\cdot)$, and the exponential equivalence of the processes Y^ϵ and $\tilde{Y}^\epsilon$, (5.5.16)–(5.5.18) are obtained as a consequence of the following lemma, whose proof, being elementary, is omitted.

Lemma 5.5.28 *The set Ψ is closed, and so are Φ_δ and Φ_δ^* for all $\delta > 0$, while $\Psi^{-\delta}$ are open. Moreover,*

$$V_A = \inf_{\psi \in \Psi} I_T(\psi) = \lim_{\delta \to 0} \inf_{\psi \in \Psi^{-\delta}} I_T(\psi),$$

while for all $\delta > 0$,

$$\inf_{\psi \in \Phi_\delta} I_T(\psi) > V_A$$

and

$$\inf_{\psi \in \Phi_\delta^*} I_{T+\bar{t}}(\psi) > V_A.$$

Exercise 5.5.29 Define

$$\tau_{\epsilon,n} \overset{\triangle}{=} \sup\{s : s \in [0, T_{\epsilon,n}) \quad Y^\epsilon_{T_{\epsilon,n}} - Y^\epsilon_s \in A\} - \epsilon.$$

Observe that $T_{\epsilon,n} \geq T_\epsilon$, and if $T_{\epsilon,n} = T_\epsilon$, then also $\tau_{\epsilon,n} = \tau_\epsilon$. Complete the derivation of (5.5.8) and (5.5.9) according to the following outline.
(a) Check that for all n and all ϵ, the law governing the random segment $[\tau_{\epsilon,n}, T_{\epsilon,n}]$ is the same as the law of $[\tau_\epsilon, T_\epsilon]$ conditioned on $T_\epsilon \leq 2nC_\epsilon$.
(b) Check that the estimates of Lemma 5.5.15 imply that for all $\delta > 0$ and any n large enough,

$$\lim_{\epsilon \to 0} P(|L_\epsilon - \bar{t}| \geq \delta | T_\epsilon \leq 2nC_\epsilon) = 0$$

and

$$\lim_{\epsilon \to 0} P(\sup_{0 \leq s \leq \bar{t}} |\hat{Y}^\epsilon_s - \phi^*_s| \geq \delta | T_\epsilon \leq 2nC_\epsilon) = 0.$$

(c) Combine the preceding with Lemma 5.5.22 to deduce (5.5.8) and (5.5.9).

Exercise 5.5.30 Let X_i and $I(\cdot) = \Lambda^*(\cdot)$ be as in Exercise 5.1.27. Prove the analog of Theorem 5.5.6 for this setup.

Exercise 5.5.31 Prove that Theorem 5.5.6 still holds true if A is not convex, provided that Assumption (A-3) is replaced with

$$\lim_{r \to \infty} V_{\overline{\text{co}}(A) \cap \{x : |x| > r\}} > 2V_A,$$

where $\overline{\text{co}}(A)$ denotes the closed convex hull of A.
Hint: The convexity of A was only used in the proof of Lemma 5.5.20 when applying the min–max theorem. Check that under the preceding condition, you can still establish this lemma with A replaced by $\overline{\text{co}}(A)$ throughout the proof.

Exercise 5.5.32 The following outline leads to Theorem 5.5.23.
(a) Verify that if A is a convex, closed set of non-empty interior with $C_\eta \subset A^c$ for some $\eta > 0$, then Assumptions (A-1)–(A-3) hold for $\Lambda^*(x) = |x|^2/2$.
(b) Observe that the analytic Lemmas 5.5.3, 5.5.10, and 5.5.28 are unchanged. Combine them with the LDP derived in Theorem 5.2.3 (and Exercise 5.2.8) for $\sqrt{\epsilon} w_t$ to prove the short time estimates of Lemma 5.5.15.
(c) Observe that T_ϵ for the Brownian motion with drift is almost surely bounded above by the stopping time $\hat{T}_\epsilon$, associated with the random walk Z_n generated by the standard Normal variables $X_i \overset{\triangle}{=} \epsilon^{-1/2}(w_{i\epsilon} - w_{(i-1)\epsilon})$. Deduce that the statement of Lemma 5.5.19 is valid for the Brownian motion with drift.
(d) Combine a union of events bound with the short time estimate of Lemma 5.2.1 to prove that for any $\eta > 0$ and any $M < \infty$,

$$\lim_{\epsilon \to 0} P(\sup_{0 \leq t \leq e^{M/\epsilon}} |Y^\epsilon_t - Y^\epsilon_{[\frac{t}{\epsilon}]\epsilon}| > \eta) = 0.$$

(e) Use part (d) of this exercise to prove Lemma 5.5.20 for the Brownian motion with drift.

Hint: Observe that $Y^\epsilon_{[t/\epsilon]\epsilon}$ correspond to the (discrete time) random walk introduced in part (c) and apply Lemma 5.5.20 for this random walk and for the η closed blowups of A (with $\eta > 0$ sufficiently small so that indeed this lemma holds).

(f) Check that Lemma 5.5.22 holds true with a minor modification of $\tau_{\epsilon,n}$ paralleling the remark below Theorem 5.5.23. (You may let $C_\epsilon = C$, since there are now no discretization effects.) Conclude by proving that (5.5.7)–(5.5.9) hold.

5.6   The Freidlin–Wentzell Theory

The results of Section 5.2 are extended here to the case of strong solutions of stochastic differential equations. Note that these, in general, do not possess independent increments. However, some underlying independence exists in the process via the Brownian motion, which generates the diffusion. This is exploited in this section, where large deviations principles are derived by applying various contraction principles.

First consider the following relatively simple situation. Let $\{x^\epsilon_t\}$ be the diffusion process that is the unique solution of the stochastic differential equation

$$dx^\epsilon_t = b(x^\epsilon_t)dt + \sqrt{\epsilon}dw_t \qquad 0 \le t \le 1, \quad x^\epsilon_0 = 0, \qquad (5.6.1)$$

where $b : \mathbb{R} \to \mathbb{R}$ is a uniformly Lipschitz continuous function (namely, $|b(x) - b(y)| \le B|x - y|$). The existence and uniqueness of the strong solution $\{x^\epsilon_t\}$ of (5.6.1) is standard. (See Theorem E.7.) Let $\tilde{\mu}_\epsilon$ denote the probability measure induced by $\{x^\epsilon_t\}$ on $C_0([0,1])$. Then $\tilde{\mu}_\epsilon = \mu_\epsilon \circ F^{-1}$, where μ_ϵ is the measure induced by $\{\sqrt{\epsilon}w_t\}$, and the deterministic map $F : C_0([0,1]) \to C_0([0,1])$ is defined by $f = F(g)$, where f is the unique continuous solution of

$$f(t) = \int_0^t b(f(s))ds + g(t), \qquad t \in [0,1]. \qquad (5.6.2)$$

The LDP associated with x^ϵ_t is therefore a direct application of the contraction principle with respect to the map F.

Theorem 5.6.3 $\{x^\epsilon_t\}$ *satisfies the LDP in* $C_0([0,1])$ *with the good rate function*

$$I(f) \triangleq \begin{cases} \frac{1}{2}\int_0^1 |\dot{f}(t) - b(f(t))|^2 dt & , \quad f \in H_1 \\ \infty & , \quad f \notin H_1 . \end{cases} \qquad (5.6.4)$$

Proof: Theorem 4.2.1 is applicable here, as F is continuous on $C_0([0,1])$. Indeed, if $f_1 = F(g_1)$, $f_2 = F(g_2)$, then by (5.6.2),

$$f_1(t) - f_2(t) = \int_0^t [b(f_1(s)) - b(f_2(s))]ds + g_1(t) - g_2(t).$$

Define $e(t) \triangleq |f_1(t) - f_2(t)|$ and consider any two functions g_1, $g_2 \in C_0([0,1])$ such that $\| g_1 - g_2 \| \leq \delta$. Then

$$e(t) \leq \int_0^t |b(f_1(s)) - b(f_2(s))|ds + |g_1(t) - g_2(t)| \leq B \int_0^t e(s)ds + \delta,$$

and by Gronwall's lemma (Lemma E.6), it follows that $e(t) \leq \delta e^{Bt}$. Thus, $\| f_1 - f_2 \| \leq \delta e^B$ and the continuity of F is established. Combining Schilder's theorem (Theorem 5.2.3) and the contraction principle (Theorem 4.2.1), it follows that $\tilde{\mu}_\epsilon$ satisfies, in $C_0([0,1])$, an LDP with the good rate function

$$I(f) = \inf_{\{g \in H_1 : f = F(g)\}} \frac{1}{2} \int_0^1 \dot{g}^2(t)\, dt.$$

In order to identify $I(\cdot)$ with (5.6.4), observe that F is an injection, and further that $g \in H_1$ implies that $f = F(g)$ is differentiable a.e. with

$$\dot{f}(t) = b(f(t)) + \dot{g}(t), \; f(0) = 0.$$

Thus,

$$|\dot{f}(t)| \leq B \int_0^t |\dot{f}(s)|ds + |b(0)| + |\dot{g}(t)|,$$

and consequently, by Gronwall's lemma (Lemma E.6), $g \in H_1$ implies that $f = F(g) \in H_1$ as well. □

Now, let $\{x_t^\epsilon\}$ be the diffusion process that is the unique solution of the stochastic differential equation

$$dx_t^\epsilon = b(x_t^\epsilon)dt + \sqrt{\epsilon}\sigma(x_t^\epsilon)dw_t, \quad 0 \leq t \leq 1, \quad x_0^\epsilon = x, \qquad (5.6.5)$$

where $x \in \mathbb{R}^d$ is deterministic, $b : \mathbb{R}^d \to \mathbb{R}^d$ is a uniformly Lipschitz continuous function, all the elements of the diffusion matrix σ are bounded, uniformly Lipschitz continuous functions, and w is a standard Brownian motion in $\mathbb{R}^d$. The existence and uniqueness of the strong solution $\{x_t^\epsilon\}$ of (5.6.5) is standard. (See Theorem E.7.)

The map defined by the process x^ϵ on $C([0,1])$ is measurable but need not be continuous, and thus the proof of Theorem 5.6.3 does not apply directly. Indeed, this noncontinuity is strikingly demonstrated by the fact that the solution to (5.6.5), when w_t is replaced by its polygonal approximation, differs in the limit from x^ϵ by a nonzero (Wong–Zakai) correction term. On the other hand, this correction term is of the order of ϵ, so it is not expected to influence the large deviations results. Such an argument leads to the guess that the appropriate rate function for (5.6.5) is

$$I_x(f) = \inf_{\{g \in H_1 : f(t) = x + \int_0^t b(f(s))ds + \int_0^t \sigma(f(s))\dot{g}(s)ds\}} \frac{1}{2} \int_0^1 |\dot{g}(t)|^2 dt, \quad (5.6.6)$$

where the infimum over an empty set is taken as $+\infty$, and $|\cdot|$ denotes both the usual Euclidean norm on $\mathbb{R}^d$ and the corresponding operator norm of matrices. The spaces H_1, and $L_2([0,1])$ for $\mathbb{R}^d$-valued functions are defined using this norm.

Theorem 5.6.7 *If all the entries of b and σ are bounded, uniformly Lipschitz continuous functions, then $\{x_t^\epsilon\}$, the solution of (5.6.5), satisfies the LDP in $C([0,1])$ with the good rate function $I_x(\cdot)$ of (5.6.6).*

Remark: For $\sigma(\cdot)$, a square matrix, and nonsingular diffusions, namely, solutions of (5.6.5) with $a(\cdot) \triangleq \sigma(\cdot)\sigma'(\cdot)$ which is uniformly positive definite, the preceding formula for the rate function simplifies considerably to

$$I_x(f) = \begin{cases} \frac{1}{2} \int_0^1 (\dot{f}(t) - b(f(t)))' a^{-1}(f(t))(\dot{f}(t) - b(f(t)))\, dt & , \quad f \in H_1^x \\ \infty & , \quad f \notin H_1^x, \end{cases}$$

where $H_1^x \triangleq \{f : f(t) = x + \int_0^t \phi(s)ds,\ \phi \in L_2([0,1])\}$.

Proof: It suffices to prove the theorem for $x = 0$ (as x may always be moved to the origin by a translation of the coordinates). Then the measure $\tilde{\mu}_\epsilon$ of x^ϵ is supported on $C_0([0,1])$. The proof here is based on approximating the process x^ϵ in the sense of Theorem 4.2.23. To this end, let $x^{\epsilon,m}$, $m = 1, 2, \ldots$ be the solution of the stochastic differential equation

$$dx_t^{\epsilon,m} = b(x_{\frac{[mt]}{m}}^{\epsilon,m})dt + \sqrt{\epsilon}\sigma(x_{\frac{[mt]}{m}}^{\epsilon,m})dw_t, \quad 0 \le t \le 1, \quad x_0^{\epsilon,m} = 0, \quad (5.6.8)$$

in which the coefficients of (5.6.5) are frozen over the time intervals $[\frac{k}{m}, \frac{k+1}{m})$. Since x^ϵ and $x^{\epsilon,m}$ are strong solutions of (5.6.5) and (5.6.8), respectively, they are defined on the same probability space. The following lemma, whose proof is deferred to the end of the section, shows that $x^{\epsilon,m}$ are exponentially good approximations of x^ϵ.

Lemma 5.6.9 *For any $\delta > 0$,*

$$\lim_{m \to \infty} \limsup_{\epsilon \to 0} \epsilon \log P(\| x^{\epsilon,m} - x^\epsilon \| > \delta) = -\infty.$$

Let the map F^m be defined via $h = F^m(g)$, where

$$h(t) = h\Big(\frac{k}{m}\Big) + b\Big(h(\frac{k}{m})\Big)\Big(t - \frac{k}{m}\Big) + \sigma\Big(h(\frac{k}{m})\Big)\Big(g(t) - g(\frac{k}{m})\Big),$$

$$t \in \Big[\frac{k}{m}, \frac{k+1}{m}\Big], k = 0, \dots, m-1, \ h(0) = 0.$$

Now, observe that F^m is a map of $C_0([0,1])$ onto itself and that $x^{\epsilon,m} = F^m(\sqrt{\epsilon}w.)$. Let $g_1, g_2 \in C_0([0,1])$ and $e \triangleq |F^m(g_1) - F^m(g_2)|$. Then by the assumptions on $b(\cdot)$ and $\sigma(\cdot)$,

$$\sup_{t \in [\frac{k}{m}, \frac{k+1}{m}]} e(t) \le C\Big(e\Big(\frac{k}{m}\Big) + \|g_1 - g_2\| \Big),$$

where $C < \infty$ is some constant that depends only on $\|g_1\|$. Since $e(0) = 0$, the continuity of F^m with respect to the supremum norm follows by iterating this bound over $k = 0, \dots, m-1$.

Let F be defined on the space H_1 such that $f = F(g)$ is the unique solution of the integral equation

$$f(t) = \int_0^t b(f(s))ds + \int_0^t \sigma(f(s))\dot{g}(s)ds, \ 0 \le t \le 1.$$

The existence and uniqueness of the solution is a consequence of the Lipschitz continuity of b and σ and is standard. In view of Lemma 5.6.9, the proof of the theorem is completed by combining Schilder's theorem (Theorem 5.2.3) and Theorem 4.2.23, as soon as we show that for every $\alpha < \infty$,

$$\lim_{m \to \infty} \sup_{\{g: \|g\|_{H_1} \le \alpha\}} \| F^m(g) - F(g) \| = 0. \tag{5.6.10}$$

To this end, fix $\alpha < \infty$ and $g \in H_1$ such that $\|g\|_{H_1} \le \alpha$. Let $h = F^m(g)$, $f = F(g)$, and $e(t) = |f(t) - h(t)|^2$. Then for all $t \in [0,1]$,

$$h(t) = \int_0^t b\Big(h(\frac{[ms]}{m})\Big)ds + \int_0^t \sigma\Big(h(\frac{[ms]}{m})\Big)\dot{g}(s)ds.$$

By the Cauchy–Schwartz inequality and the boundedness of $b(\cdot)$ and $\sigma(\cdot)$,

$$\sup_{0 \le t \le 1} \Big|h(t) - h\big(\frac{[mt]}{m}\big)\Big| \le (\alpha + 1)\delta_m, \tag{5.6.11}$$

where δ_m are independent of g, and converge to zero as $m \to \infty$. Applying the Cauchy–Schwartz inequality again, it follows by the uniform Lipschitz continuity of $b(\cdot)$ and $\sigma(\cdot)$ that

$$|f(t) - h(t)| \le (\alpha + 1)C \Big[\int_0^t |f(s) - h(\frac{[ms]}{m})|^2 ds \Big]^{1/2}$$

for some constant $C < \infty$ independent of m and g. Thus, due to (5.6.11),

$$e(t) \leq K \int_0^t e(s)ds + K\delta_m^2, \quad e(0) = 0,$$

where K is some constant that depends only on C and α. Hence, by Gronwall's lemma (Lemma E.6), $e(t) \leq K\delta_m^2 e^{Kt}$. Consequently,

$$\| F(g) - F^m(g) \| \leq \sqrt{K}\delta_m e^{K/2},$$

which establishes (5.6.10) and completes the proof of the theorem.   □

The following theorem (whose proof is deferred to the end of the section) strengthens Theorem 5.6.7 by allowing for ϵ dependent initial conditions.

Theorem 5.6.12 *Assume the conditions of Theorem 5.6.7. Let* $\{X_t^{\epsilon,y}\}$ *denote the solution of (5.6.5) for the initial condition* $X_0 = y$. *Then:*
(a) For any closed $F \subset C([0,1])$,

$$\limsup_{\substack{\epsilon \to 0 \\ y \to x}} \epsilon \log P(X^{\epsilon,y} \in F) \leq - \inf_{\phi \in F} I_x(\phi). \tag{5.6.13}$$

(b) For any open $G \subset C([0,1])$,

$$\liminf_{\substack{\epsilon \to 0 \\ y \to x}} \epsilon \log P(X^{\epsilon,y} \in G) \geq - \inf_{\phi \in G} I_x(\phi). \tag{5.6.14}$$

The following corollary is needed in Section 5.7.

Corollary 5.6.15 *Assume the conditions of Theorem 5.6.7. Then for any compact* $K \subset \mathbb{R}^d$ *and any closed* $F \subset C([0,1])$,

$$\limsup_{\epsilon \to 0} \epsilon \log \sup_{y \in K} P(X^{\epsilon,y} \in F) \leq - \inf_{\substack{\phi \in F \\ y \in K}} I_y(\phi). \tag{5.6.16}$$

Similarly, for any open $G \subset C([0,1])$,

$$\liminf_{\epsilon \to 0} \epsilon \log \inf_{y \in K} P(X^{\epsilon,y} \in G) \geq - \sup_{y \in K} \inf_{\phi \in G} I_y(\phi). \tag{5.6.17}$$

Proof: Let $-I_K$ denote the right side of (5.6.16). Fix $\delta > 0$ and let $I_K^\delta \triangleq \min\{I_K - \delta, 1/\delta\}$. Then from (5.6.13), it follows that for any $x \in K$, there exists an $\epsilon_x > 0$ such that for all $\epsilon \leq \epsilon_x$,

$$\epsilon \log \sup_{y \in B_{x,\epsilon_x}} P(X^{\epsilon,y} \in F) \leq -I_K^\delta.$$

Let $x_1, \ldots, x_m \in K$ be such that the compact set K is covered by the set $\cup_{i=1}^{m} B_{x_i, \epsilon_{x_i}}$. Then for $\epsilon \le \min_{i=1}^{m} \epsilon_{x_i}$,

$$\epsilon \log \sup_{y \in K} P(X^{\epsilon, y} \in F) \le -I_K^{\delta} .$$

By first considering $\epsilon \to 0$ and then $\delta \to 0$, (5.6.16) follows. The inequality (5.6.17) is obtained from (5.6.14) by a similar argument. $\qquad\square$

The proofs of Lemma 5.6.9 and Theorem 5.6.12 are based on the following lemma.

Lemma 5.6.18 *Let b_t, σ_t be progressively measurable processes, and let*

$$dz_t = b_t dt + \sqrt{\epsilon} \sigma_t dw_t , \qquad (5.6.19)$$

where z_0 is deterministic. Let $\tau_1 \in [0, 1]$ be a stopping time with respect to the filtration of $\{w_t, t \in [0, 1]\}$. Suppose that the coefficients of the diffusion matrix σ are uniformly bounded, and for some constants M, B, ρ and any $t \in [0, \tau_1]$,

$$|\sigma_t| \le M \left(\rho^2 + |z_t|^2\right)^{1/2}$$
$$|b_t| \le B \left(\rho^2 + |z_t|^2\right)^{1/2} . \qquad (5.6.20)$$

Then for any $\delta > 0$ and any $\epsilon \le 1$,

$$\epsilon \log P(\sup_{t \in [0, \tau_1]} |z_t| \ge \delta) \le K + \log \left(\frac{\rho^2 + |z_0|^2}{\rho^2 + \delta^2}\right) ,$$

where $K = 2B + M^2 (2 + d)$.

Proof: Let $u_t = \phi(z_t)$, where $\phi(y) = (\rho^2 + |y|^2)^{1/\epsilon}$. By Itô's formula, u_t is the strong solution of the stochastic differential equation

$$du_t = \nabla \phi(z_t)' dz_t + \frac{\epsilon}{2} \text{Trace} \left[\sigma_t \sigma_t' D^2 \phi(z_t)\right] dt \stackrel{\triangle}{=} g_t dt + \tilde{\sigma}_t dw_t , \qquad (5.6.21)$$

where $D^2 \phi(y)$ is the matrix of second derivatives of $\phi(y)$, and for a matrix (vector) v, v' denotes its transpose. Note that

$$\nabla \phi(y) = \frac{2 \phi(y)}{\epsilon(\rho^2 + |y|^2)} y ,$$

which in view of (5.6.20) implies

$$|\nabla \phi(z_t)' b_t| \le \frac{2B |z_t| \phi(z_t)}{\epsilon(\rho^2 + |z_t|^2)^{1/2}} \le \frac{2B}{\epsilon} u_t .$$

Similarly, for $\epsilon \leq 1$,

$$\frac{\epsilon}{2}\mathrm{Trace}[\sigma_t \sigma'_t D^2 \phi(z_t)] \leq \frac{M^2(2+d)}{\epsilon} u_t .$$

These bounds imply, in view of (5.6.21), that for any $t \in [0, \tau_1]$,

$$g_t \leq \frac{K u_t}{\epsilon} , \qquad (5.6.22)$$

where $K = 2B + M^2(2 + d) < \infty$.

Fix $\delta > 0$ and define the stopping time $\tau_2 \overset{\triangle}{=} \inf\{t : |z_t| \geq \delta\} \wedge \tau_1$. Since $|\tilde{\sigma}_t| \leq 2M u_t / \sqrt{\epsilon}$ is uniformly bounded on $[0, \tau_2]$, it follows that the stochastic Itô integral $u_t - \int_0^t g_s ds$ is a continuous martingale up to τ_2. Therefore, Doob's theorem (Theorem E.1) is applicable here, yielding

$$E[u_{t \wedge \tau_2}] = u_0 + E\left[\int_0^{t \wedge \tau_2} g_s ds\right] .$$

Hence, by (5.6.22) and the nonnegativity of $u_\cdot$,

$$
\begin{aligned}
E[u_{t \wedge \tau_2}] &\leq u_0 + \frac{K}{\epsilon} E\left[\int_0^{t \wedge \tau_2} u_s ds\right] = u_0 + \frac{K}{\epsilon} E\left[\int_0^{t \wedge \tau_2} u_{s \wedge \tau_2} ds\right] \\
&\leq u_0 + \frac{K}{\epsilon} \int_0^t E[u_{s \wedge \tau_2}] ds .
\end{aligned}
$$

Consequently, by Gronwall's lemma (Lemma E.6),

$$E[u_{\tau_2}] = E[u_{1 \wedge \tau_2}] \leq u_0 e^{K/\epsilon} = \phi(z_0) e^{K/\epsilon} .$$

Note that $\phi(y)$ is positive and monotone increasing in $|y|$. Therefore, by Chebycheff's inequality,

$$P(|z_{\tau_2}| \geq \delta) = P[\phi(z_{\tau_2}) \geq \phi(\delta)] \leq \frac{E[\phi(z_{\tau_2})]}{\phi(\delta)} = \frac{E[u_{\tau_2}]}{\phi(\delta)} .$$

Combining the preceding two inequalities yields

$$\epsilon \log P(|z_{\tau_2}| \geq \delta) \leq K + \epsilon \log \phi(z_0) - \epsilon \log \phi(\delta) = K + \log\left(\frac{\rho^2 + |z_0|^2}{\rho^2 + \delta^2}\right) .$$

The proof of the lemma is completed as $\sup_{t \in [0, \tau_1]} |z_t| \geq \delta$ iff $|z_{\tau_2}| \geq \delta$. $\square$

Proof of Lemma 5.6.9: Fix $\delta > 0$. Let $z_t \overset{\triangle}{=} x_t^{\epsilon, m} - x_t^\epsilon$, and for any $\rho > 0$, define the stopping time

$$\tau_1 = \inf\{t : |x_t^{\epsilon, m} - x_{\frac{[mt]}{m}}^{\epsilon, m}| \geq \rho\} \wedge 1 .$$

The process z_t is of the form (5.6.19), with $z_0 = 0$, $b_t \triangleq b(x^{\epsilon,m}_{[mt]/m}) - b(x^\epsilon_t)$
and $\sigma_t \triangleq \sigma(x^{\epsilon,m}_{[mt]/m}) - \sigma(x^\epsilon_t)$. Thus, by the uniform Lipschitz continuity of $b(\cdot)$
and $\sigma(\cdot)$ and the definition of τ_1, it follows that Lemma 5.6.18 is applicable
here, yielding for any $\delta > 0$ and any $\epsilon \leq 1$,

$$\epsilon \log P(\sup_{t \in [0, \tau_1]} |x^{\epsilon,m}_t - x^\epsilon_t| \geq \delta) \leq K + \log\left(\frac{\rho^2}{\rho^2 + \delta^2}\right),$$

with $K < \infty$ independent of ϵ, δ, ρ, and m. Hence, by considering first
$\epsilon \to 0$ and then $\rho \to 0$,

$$\lim_{\rho \to 0} \sup_{m \geq 1} \limsup_{\epsilon \to 0} \epsilon \log P(\sup_{t \in [0, \tau_1]} |x^{\epsilon,m}_t - x^\epsilon_t| \geq \delta) = -\infty.$$

Now, since

$$\{\| x^{\epsilon,m} - x^\epsilon \| > \delta\} \subset \{\tau_1 < 1\} \cup \{\sup_{t \in [0, \tau_1]} |x^{\epsilon,m}_t - x^\epsilon_t| \geq \delta\},$$

the lemma is proved as soon as we show that for all $\rho > 0$,

$$\lim_{m \to \infty} \limsup_{\epsilon \to 0} \epsilon \log P(\sup_{0 \leq t \leq 1} |x^{\epsilon,m}_t - x^{\epsilon,m}_{\frac{[mt]}{m}}| \geq \rho) = -\infty. \tag{5.6.23}$$

To this end, observe first that

$$\left|x^{\epsilon,m}_t - x^{\epsilon,m}_{\frac{[mt]}{m}}\right| \leq C\left[\frac{1}{m} + \sqrt{\epsilon} \max_{k=0}^{m-1} \sup_{0 \leq s \leq \frac{1}{m}} |w_{s + \frac{k}{m}} - w_{\frac{k}{m}}|\right],$$

where C is the bound on $|\sigma(\cdot)|$ and $|b(\cdot)|$. Therefore, for all $m > C/\rho$,

$$P(\sup_{0 \leq t \leq 1} |x^{\epsilon,m}_t - x^{\epsilon,m}_{\frac{[mt]}{m}}| \geq \rho) \leq mP\left(\sup_{0 \leq s \leq \frac{1}{m}} |w_s| \geq \frac{\rho - C/m}{\sqrt{\epsilon}C}\right)$$

$$\leq 4dme^{-m(\rho - C/m)^2/2d\epsilon C^2},$$

where the second inequality is the bound of Lemma 5.2.1. Hence, (5.6.23)
follows, and the lemma is established.   □

Proof of Theorem 5.6.12: By Theorem 4.2.13, it suffices to prove that
$X^{\epsilon,x}$ are exponentially equivalent in $C([0,1])$ to X^{ϵ,x_ϵ} whenever $x_\epsilon \to x$.
Fix $x_\epsilon \to x$ and let $z_t \triangleq X^{\epsilon,x_\epsilon}_t - X^{\epsilon,x}_t$. Then z_t is of the form (5.6.19), with
$z_0 = x_\epsilon - x$, $\sigma_t \triangleq \sigma(X^{\epsilon,x_\epsilon}_t) - \sigma(X^{\epsilon,x}_t)$ and $b_t \triangleq b(X^{\epsilon,x_\epsilon}_t) - b(X^{\epsilon,x}_t)$. Hence, the
uniform Lipschitz continuity of $b(\cdot)$ and $\sigma(\cdot)$ implies that (5.6.20) holds for
any $\rho > 0$ and with $\tau_1 = 1$. Therefore, Lemma 5.6.18 yields

$$\epsilon \log P(\| X^{\epsilon,x_\epsilon} - X^{\epsilon,x} \| \geq \delta) \leq K + \log\left(\frac{\rho^2 + |x_\epsilon - x|^2}{\rho^2 + \delta^2}\right)$$

for any $\delta > 0$ and any $\rho > 0$ (where $K < \infty$ is independent of ϵ, δ, and ρ). Considering first $\rho \to 0$ and then $\epsilon \to 0$ yields

$$\limsup_{\epsilon \to 0} \epsilon \log \mathrm{P}(\| X^{\epsilon,x_\epsilon} - X^{\epsilon,x} \| \geq \delta) \leq K + \limsup_{\epsilon \to 0} \log \left(\frac{|x_\epsilon - x|^2}{\delta^2} \right) = -\infty \,.$$

With X^{ϵ,x_ϵ} and $X^{\epsilon,x}$ exponentially equivalent, the theorem is established. $\square$

Exercise 5.6.24 Prove that Theorem 5.6.7 holds when $b(\cdot)$ is Lipschitz continuous but possibly unbounded.

Exercise 5.6.25 Extend Theorem 5.6.7 to the time interval $[0,T]$, $T < \infty$. **Hint:** The relevant rate function is now

$$I_x(f) = \inf_{\{g \in H_1([0,T]) \, : \, f(t) = x + \int_0^t b(f(s))ds + \int_0^t \sigma(f(s))\dot{g}(s)ds\}} \frac{1}{2} \int_0^T |\dot{g}(t)|^2 dt \,.$$

$$(5.6.26)$$

Exercise 5.6.27 Find $I(\cdot)$ for x_t^ϵ, where

$$dx_t^\epsilon = y_t^\epsilon dt \,, \quad x_0^\epsilon = 0 \,,$$

$$dy_t^\epsilon = b(x_t^\epsilon, y_t^\epsilon)dt + \sqrt{\epsilon}\sigma(x_t^\epsilon, y_t^\epsilon)dw_t \,, \quad y_0^\epsilon = 0 \,,$$

and $b, \sigma : \mathbb{R}^2 \to \mathbb{R}$ are uniformly Lipschitz continuous, bounded functions, such that $\inf_{x,y} \sigma(x,y) > 0$.

Exercise 5.6.28 Suppose that x_0^ϵ, the initial conditions of (5.6.5), are random variables that are independent of the Brownian motion $\{w_t, \ t \geq 0\}$. Prove that Theorem 5.6.7 holds if there exists $x \in \mathbb{R}^d$ such that for any $\delta > 0$,

$$\limsup_{\epsilon \to 0} \epsilon \log \mathrm{P}(|x_0^\epsilon - x| \geq \delta) = -\infty \,.$$

Hint: Modify the arguments in the proof of Theorem 5.6.12.

5.7 The Problem of Diffusion Exit from a Domain

Consider the system

$$dx_t^\epsilon = b(x_t^\epsilon)dt + \sqrt{\epsilon}\sigma(x_t^\epsilon)dw_t, \quad x_t^\epsilon \in \mathbb{R}^d, \quad x_0^\epsilon = x \,, \qquad (5.7.1)$$

in the open, bounded domain G, where $b(\cdot)$ and $\sigma(\cdot)$ are uniformly Lipschitz continuous functions of appropriate dimensions and w is a standard Brownian motion. The following assumption prevails throughout this section.

Assumption (A-1) *The unique stable equilibrium point in G of the d-dimensional ordinary differential equation*

$$\dot{\phi}_t = b(\phi_t) \qquad (5.7.2)$$

is at $0 \in G$, and

$$\phi_0 \in G \;\Rightarrow\; \forall t > 0, \; \phi_t \in G \; and \; \lim_{t \to \infty} \phi_t = 0 \,.$$

When ϵ is small, it is reasonable to guess that the system (5.7.1) tends to stay inside G. Indeed, suppose that the boundary of G is smooth enough for

$$\tau^\epsilon \overset{\triangle}{=} \inf\{t > 0 : \; x_t^\epsilon \in \partial G\}$$

to be a well-defined stopping time. It may easily be shown that under mild conditions, $P(\tau^\epsilon < T) \xrightarrow[\epsilon \to 0]{} 0$ for any $T < \infty$. (A proof of this fact follows from the results of this section; *c.f.* Theorem 5.7.3.) From an engineering point of view, (5.7.1) models a tracking loop in which some parasitic noise exists. The parasitic noise may exist because of atmospheric noise (e.g., in radar and astronomy), or because of a stochastic element in the signal model (e.g., in a phase lock loop). From that point of view, exiting the domain at ∂G is an undesirable event, for it means the loss of lock. An important question (both in the analysis of a given system and in the design of new systems) would be how probable is the loss of lock. For a detailed analysis of such a tracking loop, the reader is referred to Section 5.8.

In many interesting systems, the time to lose lock is measured in terms of a large multiple of the natural time constant of the system. For example, in modern communication systems, where the natural time constant is a bit duration, the error probabilities are in the order of 10^{-7} or 10^{-9}. In such situations, asymptotic computations of the exit time become meaningful.

Another important consideration in designing such systems is the question of where the exit occurs on ∂G, for it may allow design of modified loops, error detectors, etc.

It should be noted that here, as in Section 5.6, the choice of the model (5.7.1) is highly arbitrary. In particular, the same type of theory can be developed for Poisson processes, or more generally for Lévy processes. However, beware of using "natural" approximations, as those need not necessarily have similar large deviations behavior. (Such a situation is described in Section 5.8.)

Throughout, E_x denotes expectations with respect to the diffusion process (5.7.1), where $x_0^\epsilon = x$. The following classical theorem (see [KS88, page 365]) characterizes such expectations, for *any* ϵ, in terms of the solutions of appropriate partial differential equations.

Theorem 5.7.3 *Assume that for any $y \in \partial G$, there exists a ball $B(y)$ such that $\overline{G} \cap \overline{B(y)} = \{y\}$, and for some $\eta > 0$ and all $x \in G$, the matrices $\sigma(x)\sigma'(x) - \eta I$ are positive definite. Then for any Hölder continuous function g (on G) and any continuous function f (on ∂G), the function*

$$u(x) \stackrel{\triangle}{=} E_x \left[f(x^\epsilon_{\tau^\epsilon}) + \int_0^{\tau^\epsilon} g(x^\epsilon_t)dt \right]$$

has continuous second derivatives on G, is continuous on $\overline{G}$, and is the unique solution of the partial differential equation

$$L^\epsilon u = -g \quad in \quad G,$$
$$u = f \quad on \quad \partial G,$$

where the differential operator L^ϵ is defined via

$$L^\epsilon v \stackrel{\triangle}{=} \frac{\epsilon}{2} \sum_{i,j} (\sigma\sigma')_{ij}(x) \frac{\partial^2 v}{\partial x_i \partial x_j} + b(x)' \nabla v.$$

The following corollary is of particular interest.

Corollary 5.7.4 *Assume the conditions of Theorem 5.7.3. Let $u_1(x) = E_x(\tau^\epsilon)$. Then u_1 is the unique solution of*

$$L^\epsilon u_1 = -1, \quad in \quad G \quad ; \quad u_1 = 0, \quad on \quad \partial G. \qquad (5.7.5)$$

Further, let $u_2(x) = E_x(f(x^\epsilon_{\tau^\epsilon}))$. Then for any f continuous, u_2 is the unique solution of

$$L^\epsilon u_2 = 0, \quad in \quad G \quad ; \quad u_2 = f, \quad on \quad \partial G. \qquad (5.7.6)$$

Proof: To establish (5.7.5), specialize Theorem 5.7.3 to $f \equiv 0$ and $g \equiv 1$. Similarly, let $g \equiv 0$ in Theorem 5.7.3 to establish (5.7.6). $\qquad \square$

Remark: Formally, for $f(x) = \delta_{x_a}(x)$ and $x_a \in \partial G$, the fundamental solution (Green function) u_2 of (5.7.6), as a function of x_a, describes the exit density on ∂G.

In principle, Corollary 5.7.4 enables the computation of the quantities of interest for any ϵ. (Specific examples for $d = 1$ are presented in Exercises 5.7.32 and 5.7.35.) However, in general for $d \geq 2$, neither (5.7.5) nor (5.7.6) can be solved explicitly. Moreover, the numerical effort required in solving these equations is considerable, in particular when the solution over a range of values of ϵ is of interest. In view of that, the exit behavior analysis from an asymptotic standpoint is crucial.

Since large deviations estimates are for neighborhoods rather than for points, it is convenient to extend the definition of (5.7.1) to $\mathbb{R}^d$. From here on, it is assumed that the original domain G is smooth enough to allow for such an extension preserving the uniform Lipschitz continuity of $b(\cdot)$, $\sigma(\cdot)$. Throughout this section, $B < \infty$ is large enough to bound $\sup_{x \in \overline{G}} |b(x)|$ and $\sup_{x \in \overline{G}} |\sigma(x)|$ as well as the Lipschitz constants associated with $b(\cdot)$ and $\sigma(\cdot)$. Other notations used throughout are $B_\rho \triangleq \{x : |x| \leq \rho\}$ and $S_\rho \triangleq \{x : |x| = \rho\}$, where, without explicitly mentioning it, $\rho > 0$ is always small enough so that $B_\rho \subset G$.

Motivated by Theorem 5.6.7, define the cost function

$$V(y,z,t) \quad \triangleq \quad \inf_{\{\phi \in C([0,t]): \phi_t = z\}} I_{y,t}(\phi) \tag{5.7.7}$$

$$= \quad \inf_{\{u. \in L_2([0,t]): \phi_t = z \text{ where } \phi_s = y + \int_0^s b(\phi_\theta) d\theta + \int_0^s \sigma(\phi_\theta) u_\theta d\theta\}} \frac{1}{2} \int_0^t |u_s|^2 ds ,$$

where $I_{y,t}(\cdot)$ is the good rate function of (5.6.26), which controls the LDP associated with (5.7.1). This function is also denoted in as $I_y(\cdot)$, $I_t(\cdot)$ or $I(\cdot)$ if no confusion may arise. Heuristically, $V(y,z,t)$ is the cost of forcing the system (5.7.1) to be at the point z at time t when starting at y. Define

$$V(y,z) \triangleq \inf_{t>0} V(y,z,t) .$$

The function $V(0,z)$ is called the *quasi-potential*. The treatment to follow is guided by the heuristics that as $\epsilon \to 0$, the system (5.7.1) wanders around the stable point $x = 0$ for an exponentially long time, during which its chances of hitting any closed set $N \subset \partial G$ are determined by $\inf_{z \in N} V(0,z)$. The rationale here is that any excursion off the stable point $x = 0$ has an overwhelmingly high probability of being pulled back there, and it is not the time spent near any part of ∂G that matters but the *a priori* chance for a direct, fast exit due to a rare segment in the Brownian motion's path. Caution, however, should be exercised, as there are examples where this rationale fails.

Subsets of the following additional assumptions are used in various parts of this section.

Assumption (A-2) *All the trajectories of the deterministic system (5.7.2) starting at $\phi_0 \in \partial G$ converge to 0 as $t \to \infty$.*

Assumption (A-3) $\overline{V} \triangleq \inf_{z \in \partial G} V(0,z) < \infty$.

Assumption (A-4) *There exists an $M < \infty$ such that, for all $\rho > 0$ small enough and all x, y with $|x - z| + |y - z| \leq \rho$ for some $z \in \partial G \cup \{0\}$,* *there is a function u satisfying that* $\|u\| < M$ *and* $\phi_{T(\rho)} = y$, *where*

$$\phi_t = x + \int_0^t b(\phi_s)ds + \int_0^t \sigma(\phi_s)u_s ds$$

and $T(\rho) \to 0$ *as* $\rho \to 0$.

Assumption (A-2) prevents consideration of situations in which ∂G is the *characteristic boundary* of the domain of attraction of 0. Such boundaries arise as the separating curves of several isolated minima, and are of meaningful engineering and physical relevance. Some of the results that follow extend to characteristic boundaries, as shown in Corollary 5.7.16. However, caution is needed in that case. Assumption (A-3) is natural, for otherwise all points on ∂G are equally unlikely on the large deviations scale. Assumption (A-4) is related to the controllability of the system (5.7.1) (where a smooth control replaces the Brownian motion). Note, however, that this is a relatively mild assumption. In particular, in Exercise 5.7.29, one shows that if the matrices $\sigma(x)\sigma'(x)$ are positive definite for $x = 0$, and uniformly positive definite on ∂G, then Assumption (A-4) is satisfied.

Assumption (A-4) implies the following useful continuity property.

Lemma 5.7.8 *Assume (A-4). For any $\delta > 0$, there exists $\rho > 0$ small enough such that*

$$\sup_{x,y\in B_\rho} \inf_{t\in[0,1]} V(x,y,t) < \delta \qquad (5.7.9)$$

and

$$\sup_{\{x,y:\inf_{z\in\partial G}(|y-z|+|x-z|)\leq\rho\}} \inf_{t\in[0,1]} V(x,y,t) < \delta. \qquad (5.7.10)$$

Proof: Observe that the function ϕ_s described in Assumption (A-4) results in the upper bound $V(x,y,t) \leq M^2 t/2$, where $t = T(|x - y|) \to 0$ as $|x - y| \to 0$. Equations (5.7.9) and (5.7.10) follow from this bound. $\square$

The main result of this section is the following theorem, which yields the precise exponential growth rate of τ^ϵ, as well as valuable estimates on the exit measure.

Theorem 5.7.11 *Assume (A-1)–(A-4).*
(a) For all $x \in G$ and all $\delta > 0$,

$$\lim_{\epsilon\to0} P_x(e^{(\overline{V}+\delta)/\epsilon} > \tau^\epsilon > e^{(\overline{V}-\delta)/\epsilon}) = 1. \qquad (5.7.12)$$

Moreover, for all $x \in G$,

$$\lim_{\epsilon\to0} \epsilon \log E_x(\tau^\epsilon) = \overline{V}. \qquad (5.7.13)$$

(b) If $N \subset \partial G$ is a closed set and $\inf_{z \in N} V(0, z) > \overline{V}$, then for any $x \in G$,

$$\lim_{\epsilon \to 0} P_x(x_{\tau^\epsilon}^\epsilon \in N) = 0 . \tag{5.7.14}$$

In particular, if there exists $z^* \in \partial G$ such that $V(0, z^*) < V(0, z)$ for all $z \neq z^*$, $z \in \partial G$, then

$$\forall \delta > 0, \forall x \in G, \quad \lim_{\epsilon \to 0} P_x(|x_{\tau^\epsilon}^\epsilon - z^*| < \delta) = 1 . \tag{5.7.15}$$

Often, there is interest in the characteristic boundaries for which Assumption (A-2) is violated. This is the case when there are multiple stable points of the dynamical system (5.7.2), and G is just the attraction region of one of them. The exit measure analysis used for proving part (b) of the preceding theorem could in principle be incorrect. That is because the sample path that spends increasingly large times inside G, while avoiding the neighborhood of the stable point $x = 0$, could contribute a non-negligible probability. As stated in the following corollary, the exit time estimates of Theorem 5.7.11 hold true even without assuming (A-2).

Corollary 5.7.16 *Part (a) of Theorem 5.7.11 holds true under Assumptions (A-1), (A-3), and (A-4). Moreover, it remains true for the exit time of any processes $\{\tilde{x}^\epsilon\}$ (not necessarily Markov) that satisfy for any T', δ fixed, and any stopping times $\{T_\epsilon\}$ (with respect to the natural filtration $\{\mathcal{F}_t\}$), the condition*

$$\limsup_{\epsilon \to 0} \epsilon \log P \left(\sup_{t \in [0, T']} |x_t^\epsilon - \tilde{x}_{t+T_\epsilon}^\epsilon| > \delta \, \Big| \, \mathcal{F}_{T_\epsilon}, |x_0^\epsilon - \tilde{x}_{T_\epsilon}^\epsilon| < \frac{\delta}{2} \right) = -\infty .$$

$$\tag{5.7.17}$$

Remarks:
(a) Actually, part (b) of Theorem 5.7.11 also holds true without assuming (A-2), although the proof is more involved. (See [Day90a] for details.)

(b) When the quasi-potential $V(0, \cdot)$ has multiple minima on ∂G, then the question arises as to where the exit occurs. In symmetrical cases (like the angular tracker treated in Section 5.8), it is easy to see that each minimum point of $V(0, \cdot)$ is equally likely. In general, by part (b) of Theorem 5.7.11, the exit occurs from a neighborhood of the set of minima of the quasi-potential. However, refinements of the underlying large deviations estimates are needed for determining the exact weight among the minima.
(c) The results of this section can be, and were indeed, extended in various ways to cover general Lévy processes, dynamical systems perturbed by wide-band noise, queueing systems, partial differential equations, etc. The

reader is referred to the historical notes at the end of this chapter for a
partial guide to references.

The following lemmas are instrumental for the proofs of Theorem 5.7.11
and Corollary 5.7.16. Their proof follows the proof of the theorem.

The first lemma gives a uniform lower bound on the probability of an
exit from G.

Lemma 5.7.18 *For any $\eta > 0$ and any $\rho > 0$ small enough, there exists a
$T_0 < \infty$ such that*

$$\liminf_{\epsilon \to 0} \epsilon \log \inf_{x \in B_\rho} P_x(\tau^\epsilon \leq T_0) > -(\overline{V} + \eta).$$

Next, it is shown that the probability that the diffusion (5.7.1) wanders in
G for an arbitrarily long time, without hitting a small neighborhood of 0,
is exponentially negligible. Note that the proof of this lemma is not valid
when ∂G is a characteristic boundary (or in general when Assumption (A-2)
is violated).

Lemma 5.7.19 *Let*

$$\sigma_\rho \stackrel{\triangle}{=} \inf\{t \ : \ t \geq 0, x_t^\epsilon \in B_\rho \cup \partial G\}, \tag{5.7.20}$$

where $B_\rho \subset G$. Then

$$\lim_{t \to \infty} \limsup_{\epsilon \to 0} \epsilon \log \sup_{x \in G} P_x(\sigma_\rho > t) = -\infty.$$

The following upper bound relates the quasi-potential with the probability
that an excursion starting from a small sphere around 0 hits a given subset
of ∂G before hitting an even smaller sphere.

Lemma 5.7.21 *For any closed set $N \subset \partial G$,*

$$\lim_{\rho \to 0} \limsup_{\epsilon \to 0} \epsilon \log \sup_{y \in S_{2\rho}} P_y(x_{\sigma_\rho}^\epsilon \in N) \leq - \inf_{z \in N} V(0, z),$$

where σ_ρ is the stopping time defined in (5.7.20).

In order to extend the upper bound to hold for every $x_0^\epsilon \in G$, observe that
with high probability the process x^ϵ is attracted to an arbitrarily small
neighborhood of 0 without hitting ∂G on its way, i.e.,

Lemma 5.7.22 *For every $\rho > 0$ such that $B_\rho \subset G$ and all $x \in G$,*

$$\lim_{\epsilon \to 0} P_x(x_{\sigma_\rho}^\epsilon \in B_\rho) = 1.$$

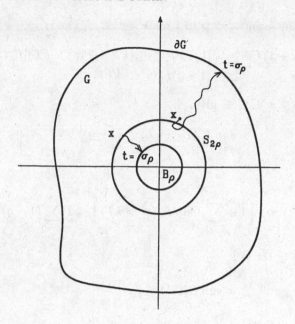

Figure 5.7.1: σ_ρ hitting times for different trajectories.

Finally, a useful uniform estimate, which states that over short time intervals the process x^ϵ has an exponentially negligible probability of getting too far from its starting point, is required.

Lemma 5.7.23 *For every $\rho > 0$ and every $c > 0$, there exists a constant $T(c,\rho) < \infty$ such that*

$$\limsup_{\epsilon \to 0} \epsilon \log \sup_{x \in G} P_x(\sup_{t \in [0,T(c,\rho)]} |x_t^\epsilon - x| \geq \rho) < -c \ .$$

All the preliminary steps required for the proof of Theorem 5.7.11 have now been completed.

Proof of Theorem 5.7.11: (a) To upper bound τ^ϵ, fix $\delta > 0$ arbitrarily small. Let $\eta = \delta/2$ and ρ and T_0 be as in Lemma 5.7.18. By Lemma 5.7.19, there exists a $T_1 < \infty$ large enough that

$$\limsup_{\epsilon \to 0} \epsilon \log \sup_{x \in G} P_x(\sigma_\rho > T_1) < 0 \ .$$

Let $T = (T_0 + T_1)$. Then there exists some $\epsilon_0 > 0$ such that for all $\epsilon \leq \epsilon_0$,

$$q \stackrel{\triangle}{=} \inf_{x \in G} P_x(\tau^\epsilon \leq T) \ \geq \ \inf_{x \in G} P_x(\sigma_\rho \leq T_1) \inf_{x \in B_\rho} P_x(\tau^\epsilon \leq T_0)$$

$$\geq \ e^{-(\overline{V} + \eta)/\epsilon} \ . \tag{5.7.24}$$

Considering the events $\{\tau^\epsilon > kT\}$ for $k = 1, 2, \ldots$ yields

$$
\begin{aligned}
P_x(\tau^\epsilon > (k+1)T) &= [1 - P_x(\tau^\epsilon \le (k+1)T | \tau^\epsilon > kT)] P_x(\tau^\epsilon > kT) \\
&\le (1-q) P_x(\tau^\epsilon > kT).
\end{aligned}
$$

Iterating over $k = 1, 2, \ldots$ gives

$$
\sup_{x \in G} P_x(\tau^\epsilon > kT) \le (1-q)^k.
$$

Therefore,

$$
\sup_{x \in G} E_x(\tau^\epsilon) \le T \Big[1 + \sum_{k=1}^\infty \sup_{x \in G} P_x(\tau^\epsilon > kT) \Big] \le T \sum_{k=0}^\infty (1-q)^k = \frac{T}{q},
$$

and since $q \ge e^{-(\overline{V}+\eta)/\epsilon}$,

$$
\sup_{x \in G} E_x(\tau^\epsilon) \le T e^{(\overline{V}+\eta)/\epsilon}. \tag{5.7.25}
$$

Now, by Chebycheff's bound,

$$
P_x(\tau^\epsilon \ge e^{(\overline{V}+\delta)/\epsilon}) \le e^{-(\overline{V}+\delta)/\epsilon} E_x(\tau^\epsilon) \le T e^{-\delta/2\epsilon},
$$

and the announced upper bound on τ^ϵ is established by considering $\epsilon \to 0$.

To prove the lower bound on τ^ϵ, let $\rho > 0$ be small enough that $S_{2\rho} \subset G$ (ρ is to be specified later). Let $\theta_0 = 0$ and for $m = 0, 1, \ldots$ define the stopping times

$$
\begin{aligned}
\tau_m &= \inf\{t : t \ge \theta_m, \, x_t^\epsilon \in B_\rho \cup \partial G\}, \\
\theta_{m+1} &= \inf\{t : t > \tau_m, \, x_t^\epsilon \in S_{2\rho}\},
\end{aligned}
$$

with the convention that $\theta_{m+1} = \infty$ if $\tau_m \in \partial G$. Each time interval $[\tau_m, \tau_{m+1}]$ represents one significant excursion off B_ρ. Note that necessarily $\tau^\epsilon = \tau_m$ for some integer m. Moreover, since τ_m are stopping times and x_t^ϵ is a strong Markov process, the process $z_m \triangleq x_{\tau_m}^\epsilon$ is a Markov chain.

For $\overline{V} = 0$, the lower bound on τ^ϵ in (5.7.12) is an easy consequence of Lemmas 5.7.22 and 5.7.23. Hence, assume hereafter that $\overline{V} > 0$, and fix $\delta > 0$ arbitrarily small. Note that ∂G is a closed set and choose $\rho > 0$ small enough as needed by Lemma 5.7.21 for

$$
\limsup_{\epsilon \to 0} \epsilon \log \sup_{y \in S_{2\rho}} P_y(x_{\sigma_\rho}^\epsilon \in \partial G) < -\overline{V} + \frac{\delta}{2}.
$$

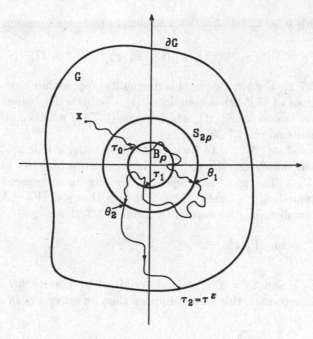

Figure 5.7.2: Stopping times θ_m, τ_m.

Now, let $c = \overline{V}$ and let $T_0 = T(c, \rho)$ be as determined by Lemma 5.7.23. Then there exists $\epsilon_0 > 0$ such that for all $\epsilon \le \epsilon_0$ and all $m \ge 1$,

$$\sup_{x \in G} P_x(\tau^\epsilon = \tau_m) \le \sup_{y \in S_{2\rho}} P_y(x^\epsilon_{\sigma_\rho} \in \partial G) \le e^{-(\overline{V} - \delta/2)/\epsilon} \qquad (5.7.26)$$

and

$$\sup_{x \in G} P_x(\theta_m - \tau_{m-1} \le T_0) \le \sup_{x \in G} P_x\big(\sup_{t \in [0, T_0]} |x^\epsilon_t - x| \ge \rho\big) \le e^{-(\overline{V} - \delta/2)/\epsilon} .$$
$$(5.7.27)$$

The event $\{\tau^\epsilon \le kT_0\}$ implies that either one of the first $k + 1$ among the mutually exclusive events $\{\tau^\epsilon = \tau_m\}$ occurs, or else that at least one of the first k excursions $[\tau_m, \tau_{m+1})$ off B_ρ is of length at most T_0. Thus, by the union of events bound, utilizing the preceding worst-case estimates, for all $x \in G$ and any integer k,

$$\begin{aligned} P_x(\tau^\epsilon \le kT_0) &\le \sum_{m=0}^k P_x(\tau^\epsilon = \tau_m) + P_x\big(\min_{1 \le m \le k}\{\theta_m - \tau_{m-1}\} \le T_0\big) \\ &\le P_x(\tau^\epsilon = \tau_0) + 2ke^{-(\overline{V} - \delta/2)/\epsilon} . \end{aligned}$$

Recall the identity $\{\tau^\epsilon = \tau_0\} \equiv \{x^\epsilon_{\sigma_\rho} \notin B_\rho\}$ and apply the preceding inequality with $k = [T_0^{-1} e^{(\overline{V} - \delta)/\epsilon}] + 1$ to obtain (for small enough ϵ)

$$P_x(\tau^\epsilon \leq e^{(\overline{V} - \delta)/\epsilon}) \leq P_x(\tau^\epsilon \leq kT_0) \leq P_x(x_{\sigma_\rho}^\epsilon \notin B_\rho) + 4T_0^{-1} e^{-\delta/2\epsilon}.$$

By Lemma 5.7.22, the left side of this inequality approaches zero as $\epsilon \to 0$; hence, the proof of (5.7.12) is complete. By Chebycheff's bound, (5.7.12) implies a lower bound on $E_x(\tau^\epsilon)$, which yields (5.7.13) when complemented by the upper bound of (5.7.25).

(b) Fix a closed set $N \subset \partial G$ such that $V_N = \inf_{z \in N} V(0, z) > \overline{V}$. (If $V_N = \infty$, then simply use throughout the proof an arbitrarily large finite constant as V_N.) The proof is a repeat of the argument presented before when lower bounding τ^ϵ. Indeed, fix $\eta > 0$ such that $\eta < (V_N - \overline{V})/3$, and set $\rho, \epsilon_0 > 0$ small enough as needed by Lemma 5.7.21 for

$$\sup_{y \in S_{2\rho}} P_y(x_{\sigma_\rho}^\epsilon \in N) \leq e^{-(V_N - \eta)/\epsilon}, \quad \forall \epsilon \leq \epsilon_0.$$

Let $c = V_N - \eta$ and $T_0 = T(c, \rho)$ as determined by Lemma 5.7.23. Then reducing ϵ_0 if necessary, this lemma implies that for every $\epsilon \leq \epsilon_0$ and every integer ℓ,

$$\sup_{x \in G} P_x(\tau_\ell \leq \ell T_0) \leq \ell \sup_{x \in G} P_x(\sup_{t \in [0, T_0]} |x_t^\epsilon - x| \geq \rho) \leq \ell e^{-(V_N - \eta)/\epsilon}.$$

Decomposing the event $\{x_{\tau^\epsilon}^\epsilon \in N\}$, for all $y \in B_\rho$ (for which $\tau^\epsilon > \tau_0 = 0$), and all integer ℓ, yields

$$
\begin{aligned}
P_y(x_{\tau^\epsilon}^\epsilon \in N) \;\leq\;& P_y(\tau^\epsilon > \tau_\ell) + \sum_{m=1}^{\ell} P_y(\tau^\epsilon > \tau_{m-1}) P_y(z_m \in N | \tau^\epsilon > \tau_{m-1}) \\
\leq\;& P_y(\tau^\epsilon > \ell T_0) + P_y(\tau_\ell \leq \ell T_0) \\
&+ \sum_{m=1}^{\ell} P_y(\tau^\epsilon > \tau_{m-1}) E_y[P_{x_{\theta_m}^\epsilon}(x_{\sigma_\rho}^\epsilon \in N) | \tau^\epsilon > \tau_{m-1}] \\
\leq\;& P_y(\tau^\epsilon > \ell T_0) + P_y(\tau_\ell \leq \ell T_0) + \ell \sup_{x \in S_{2\rho}} P_x(x_{\sigma_\rho}^\epsilon \in N) \\
\leq\;& P_y(\tau^\epsilon > \ell T_0) + 2\ell e^{-(V_N - \eta)/\epsilon}.
\end{aligned}
$$

Further reducing ϵ_0 as needed, one can guarantee that the inequality (5.7.25) holds for some $T < \infty$ and all $\epsilon \leq \epsilon_0$. Choosing $\ell = [e^{(\overline{V} + 2\eta)/\epsilon}]$, it then follows that

$$\limsup_{\epsilon \to 0} \sup_{y \in B_\rho} P_y(x_{\tau^\epsilon}^\epsilon \in N) \leq \limsup_{\epsilon \to 0} \left(\frac{T}{\ell T_0} e^{(\overline{V} + \eta)/\epsilon} + 2\ell e^{-(V_N - \eta)/\epsilon} \right) = 0.$$

(Recall that $0 < \eta < (V_N - \overline{V})/3$.) The proof of (5.7.14) is now completed by combining Lemma 5.7.22 and the inequality

$$P_x(x_{\tau^\epsilon}^\epsilon \in N) \leq P_x(x_{\sigma_\rho}^\epsilon \notin B_\rho) + \sup_{y \in B_\rho} P_y(x_{\tau^\epsilon}^\epsilon \in N).$$

Specializing (5.7.14) to $N = \{z \in \partial G : |z - z^*| \geq \delta\}$ yields (5.7.15). $\square$

Proof of Corollary 5.7.16: Let $G^{-\rho} \triangleq \{x \in G : |x - z| > \rho \ \forall z \in \partial G\}$. Observe that $G^{-\rho}$ are open sets with $\overline{G^{-\rho}} \subset G$. Hence, Assumption (A-2), holds for these sets, and for $\rho > 0$ small enough, so does Assumption (A-4). Therefore, Theorem 5.7.11 is applicable for the sets $G^{-\rho}$. The stopping times τ^ϵ related to $G^{-\rho}$ are monotonically decreasing in ρ. This procedure results with the announced lower bound on τ^ϵ because the continuity of the quasi-potential as $\rho \to 0$ is implied by (5.7.10). The upper bound on τ^ϵ is derived directly for G exactly as in Theorem 5.7.11. It only needs to be checked that the uniform bound of (5.7.24) holds. With G replaced by $G^{-\rho}$, this bound can be derived exactly as before (since Assumption (A-2) is not required for Lemma 5.7.18, and Lemma 5.7.19 is now valid for $x \in G^{-\rho}$). Moreover, when x_0^ϵ is near ∂G, the probability of a (fast) direct exit of G is large enough for reestablishing the uniform lower bound of (5.7.24). Specifically, by a construction similar to the one used in the proof of Lemma 5.7.18, utilizing (5.7.10) and the compactness of $\overline{G} \backslash G^{-\rho}$, for any $\eta > 0$ there exists $\rho > 0$ such that

$$\liminf_{\epsilon \to 0} \epsilon \log \inf_{x \in G \backslash G^{-\rho}} P_x(\tau^\epsilon \leq 1) > -\eta .$$

Finally, by Theorem 4.2.13, x_t^ϵ and $\tilde{x}_t^\epsilon$ share the same large deviations properties on any fixed time interval. In the proof of (5.7.12), the Markov structure of x_t^ϵ is used over *exponentially growing time horizon* twice. First, it is used in (5.7.24) and in the equation following it when deriving the upper bound on τ^ϵ. Note, however, that by relating the exit from G for $\tilde{x}_t^\epsilon$ to the exit from G^δ for x_t^ϵ, it still holds that for any k (even ϵ-dependent),

$$\inf_{x \in G} P_x(\tau^\epsilon \leq (k+1)T | \tau^\epsilon > kT) \geq e^{-(\overline{V} + \eta/2)/\epsilon},$$

and the rest of the proof of the upper bound is unchanged. The Markov property of x_t^ϵ is used again when deriving the worst-case estimates of (5.7.26) and (5.7.27), where the strong Markov property of the chain z_m is of importance. Under condition (5.7.17), these estimates for $\tilde{x}_t^\epsilon$ can be related to those of x_t^ϵ. The details are left to Exercise 5.7.31. $\square$

The proofs of the five preceding lemmas are now completed.

Proof of Lemma 5.7.18: Fix $\eta > 0$, and let $\rho > 0$ be small enough for $B_\rho \subset G$ and for Lemma 5.7.8 to hold with $\delta = \eta/3$. Then for all

$x \in B_\rho$, there exists a continuous path ψ^x of length $t_x \leq 1$ such that $I(\psi^x) \leq \eta/3$, $\psi_0^x = x$, and $\psi_{t_x}^x = 0$. By (5.7.10) and Assumption (A-3), there exist $z \notin \overline{G}$, $T_1 < \infty$ and $\phi \in C([0,T_1])$ such that $I_{0,T_1}(\phi) \leq \overline{V} + \eta/3$, $\phi_0 = 0$, and $\phi_{T_1} = z$. The distance Δ between $z \notin \overline{G}$ and the compact set $\overline{G}$ is positive. Let ϕ^x denote the path obtained by concatenating ψ^x and ϕ (in that order) and extending the resulting function to be of length $T_0 = T_1 + 1$ by following the trajectory of (5.7.2) after reaching z. Since the latter part does not contribute to the rate function, it follows that $I_{x,T_0}(\phi^x) \leq \overline{V} + 2\eta/3$.

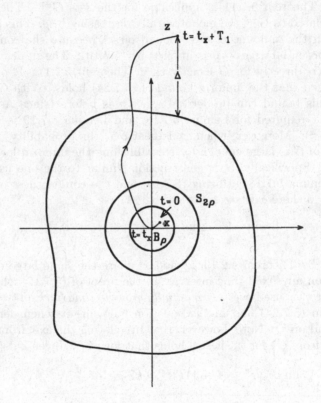

Figure 5.7.3: The exit path ϕ^x.

Consider the set

$$\Psi \stackrel{\triangle}{=} \bigcup_{x \in B_\rho} \{\psi \in C([0,T_0]) : \|\psi - \phi^x\| < \Delta/2\} .$$

Observe that Ψ is an open subset of $C([0,T_0])$ that contains the functions

$\{\phi^x\}_{x \in B_\rho}$. Therefore, by Corollary 5.6.15,

$$\liminf_{\epsilon \to 0} \epsilon \log \inf_{x \in B_\rho} P_x(x^\epsilon \in \Psi) \geq - \sup_{x \in B_\rho} \inf_{\psi \in \Psi} I_{x,T_0}(\psi)$$

$$\geq - \sup_{x \in B_\rho} I_{x,T_0}(\phi^x) > -(\overline{V} + \eta) \,.$$

If $\psi \in \Psi$, then $\psi_t \notin \overline{G}$ for some $t \in [0, T_0]$. Hence, for $x_0^\epsilon = x \in B_\rho$, the event $\{x^\epsilon \in \Psi\}$ is contained in $\{\tau^\epsilon \leq T_0\}$, and the proof is complete. $\quad\square$

Proof of Lemma 5.7.19: If $x_0^\epsilon = x \in B_\rho$, then $\sigma_\rho = 0$ and the lemma trivially holds. Otherwise, consider the closed sets

$$\Psi_t \overset{\triangle}{=} \{\phi \in C([0,t]) : \phi_s \in \overline{G \backslash B_\rho} \;\; \forall s \in [0,t]\} \,,$$

and observe that, for $x_0^\epsilon = x \in G$, the event $\{\sigma_\rho > t\}$ is contained in $\{x^\epsilon \in \Psi_t\}$. Recall that Corollary 5.6.15 yields, for all $t < \infty$,

$$\limsup_{\epsilon \to 0} \epsilon \log \sup_{x \in \overline{G \backslash B_\rho}} P_x(x^\epsilon \in \Psi_t) \leq - \inf_{\psi \in \Psi_t} I_t(\psi) \,,$$

where throughout this proof $I_t(\psi)$ stands for $I_{\psi_0,t}(\psi)$, with $\psi_0 \in \overline{G}$. Hence, in order to complete the proof of the lemma, it suffices to show that

$$\lim_{t \to \infty} \inf_{\psi \in \Psi_t} I_t(\psi) = \infty \,. \tag{5.7.28}$$

Let ϕ^x denote the trajectory of (5.7.2) starting at $\phi_0 = x \in \overline{G \backslash B_\rho}$. By Assumption (A-2), ϕ^x hits $S_{\rho/3}$ in a finite time, denoted T_x. Moreover, by the uniform Lipschitz continuity of $b(\cdot)$ and Gronwall's lemma (Lemma E.6), there exists an open neighborhood W_x of x such that, for all $y \in W_x$, the path ϕ^y hits $S_{2\rho/3}$ before T_x. Extracting a finite cover of $\overline{G \backslash B_\rho}$ by such sets (using the compactness of $\overline{G \backslash B_\rho}$), it follows that there exists a $T < \infty$ so that for all $y \in \overline{G \backslash B_\rho}$, the trajectory ϕ^y hits $S_{2\rho/3}$ before time T. Assume now that (5.7.28) does not hold true. Then for some $M < \infty$ and every integer n, there exists $\psi^n \in \Psi_{nT}$ such that $I_{nT}(\psi^n) \leq M$. Consequently, for some $\psi^{n,k} \in \Psi_T$,

$$M \geq I_{nT}(\psi^n) = \sum_{k=1}^{n} I_T(\psi^{n,k}) \geq n \min_{k=1}^{n} I_T(\psi^{n,k}) \,.$$

Hence, there exists a sequence $\phi^n \in \Psi_T$ with $\lim_{n \to \infty} I_T(\phi^n) = 0$. Since $\{\phi : I_{\phi_0,T}(\phi) \leq 1 \,, \phi_0 \in \overline{G}\}$ is a compact subset of $C([0,T])$, the sequence ϕ^n has a limit point ψ^* in Ψ_T. Consequently, $I_T(\psi^*) = 0$ by the lower semicontinuity of $I_T(\cdot)$, implying that ψ^* is a trajectory of (5.7.2). This trajectory, being in Ψ_T, remains outside of $S_{2\rho/3}$ throughout $[0, T]$, contradicting the earlier definition of T. In conclusion, (5.7.28) holds and the proof of the lemma is complete. □

Proof of Lemma 5.7.21: Fix a closed set $N \subset \partial G$. Fix $\delta > 0$ and define $V_N \triangleq (\inf_{z \in N} V(0, z) - \delta) \wedge 1/\delta$. By (5.7.9), for $\rho > 0$ small enough,

$$\inf_{y \in S_{2\rho}, z \in N} V(y, z) \geq \inf_{z \in N} V(0, z) - \sup_{y \in S_{2\rho}} V(0, y) \geq V_N \,.$$

Moreover, by Lemma 5.7.19, there exists a $T < \infty$ large enough for

$$\limsup_{\epsilon \to 0} \epsilon \log \sup_{y \in S_{2\rho}} P_y(\sigma_\rho > T) < -V_N \,.$$

Consider the following closed subset of $C([0, T])$:

$$\Phi \triangleq \{\phi \in C([0, T]) : \exists t \in [0, T] \text{ such that } \phi_t \in N \} \,.$$

Note that

$$\inf_{y \in S_{2\rho}, \phi \in \Phi} I_{y, T}(\phi) \geq \inf_{y \in S_{2\rho}, z \in N} V(y, z) \geq V_N \,,$$

and thus by Corollary 5.6.15,

$$\limsup_{\epsilon \to 0} \epsilon \log \sup_{y \in S_{2\rho}} P_y(x^\epsilon \in \Phi) \leq - \inf_{y \in S_{2\rho}, \phi \in \Phi} I_{y, T}(\phi) \leq -V_N \,.$$

Since

$$P_y(x^\epsilon_{\sigma_\rho} \in N) \leq P_y(\sigma_\rho > T) + P_y(x^\epsilon \in \Phi) \,,$$

it follows that

$$\limsup_{\epsilon \to 0} \epsilon \log \sup_{y \in S_{2\rho}} P_y(x^\epsilon_{\sigma_\rho} \in N) \leq -V_N \,.$$

Taking $\delta \to 0$ completes the proof of the lemma. □

Proof of Lemma 5.7.22: Let $\rho > 0$ be small enough so that $B_\rho \subset G$. For $x \in B_\rho$, there is nothing to prove. Thus, fix $x \in G \backslash B_\rho$, let ϕ denote the trajectory of (5.7.2) with initial condition $\phi_0 = x$, and let $T \triangleq \inf\{t : \phi_t \in S_{\rho/2}\} < \infty$. Since ϕ is a continuous path that does not hit the compact set ∂G, there exists a positive distance Δ between $\{\phi_t\}_{t \leq T}$ and ∂G. Let Δ be smaller, if necessary for $\Delta \leq \rho$, and let x^ϵ be the solution of (5.7.1), with $x^\epsilon_0 = x$. Then

$$\sup_{t \in [0, T]} |x^\epsilon_t - \phi_t| \leq \frac{\Delta}{2} \quad \Rightarrow \quad x^\epsilon_{\sigma_\rho} \in B_\rho \,.$$

By the uniform Lipschitz continuity of $b(\cdot)$,

$$|x_t^\epsilon - \phi_t| \le B \int_0^t |x_s^\epsilon - \phi_s| ds + \sqrt{\epsilon}\left| \int_0^t \sigma(x_s^\epsilon) dw_s \right|.$$

Hence, by Gronwall's lemma (Lemma E.6),

$$\sup_{t\in[0,T]} |x_t^\epsilon - \phi_t| \le \sqrt{\epsilon}e^{BT} \sup_{t\in[0,T]} \left| \int_0^t \sigma(x_s^\epsilon) dw_s \right|$$

and

$$
\begin{aligned}
P_x(x_{\sigma_\rho}^\epsilon \in \partial G) &\le P_x\left(\sup_{t\in[0,T]} |x_t^\epsilon - \phi_t| > \frac{\Delta}{2} \right) \\
&\le P_x\left(\sup_{t\in[0,T]} \left| \int_0^t \sigma(x_s^\epsilon) dw_s \right| > \frac{\Delta}{2\sqrt{\epsilon}}e^{-BT} \right) \\
&\le \epsilon c E_x\left(\int_0^T \mathrm{Trace}\,[\sigma(x_s^\epsilon)\sigma(x_s^\epsilon)'] ds \right) \xrightarrow[\epsilon\to 0]{} 0,
\end{aligned}
$$

where the last inequality is an application of the Burkholder–Davis–Gundy maximal inequality (Theorem E.3), and $c < \infty$ is independent of ϵ. $\qquad\square$

Proof of Lemma 5.7.23: For all $\rho > 0$ and all $T \le \rho/(2B)$,

$$P_x\left(\sup_{t\in[0,T]} |x_t^\epsilon - x| \ge \rho \right) \le P_x\left(\sqrt{\epsilon} \sup_{t\in[0,T]} |J_t| \ge \frac{\rho}{2} \right),$$

where

$$J_t \stackrel{\triangle}{=} \int_0^t \sigma(x_s^\epsilon) dw_s$$

is a continuous, square integrable martingale. By the union of events bound, it suffices to consider the one-dimensional case (possibly changing the constant $\rho/2$ into $\beta\rho$ with some dimensional constant β). Thus, from here on, $\sigma(x_s^\epsilon) = \sigma_s$ is a scalar function bounded by B, and w_t is a standard one-dimensional Brownian motion. By the time change in Theorem E.2, a standard one-dimensional Brownian motion, denoted ν_t, may be defined on the same probability space, such that almost surely $J_t = \nu_{\tau(t)}$, where

$$\tau(t) \stackrel{\triangle}{=} \int_0^t \sigma_\theta^2 d\theta.$$

Moreover, almost surely, $\tau(t) \leq B^2 t$ and is a continuous, increasing function. Therefore,

$$
P_x \left(\sqrt{\epsilon} \sup_{t \in [0,T]} |J_t| \geq \beta\rho \right) = P_x \left(\sqrt{\epsilon} \sup_{t \in [0,T]} |\nu_{\tau(t)}| \geq \beta\rho \right)
$$

$$
\leq P_x \left(\sqrt{\epsilon} \sup_{\tau \in [0,B^2 T]} |\nu_\tau| \geq \beta\rho \right) \leq 4 e^{-\beta^2 \rho^2 / (2\epsilon B^2 T)},
$$

where the last inequality follows by the estimate of Lemma 5.2.1. The proof is completed by choosing $T(c, \rho) < \min\{\rho/(2B), \beta^2 \rho^2/(2B^2 c)\}$. □

Exercise 5.7.29 (a) Suppose that $\sigma(x)\sigma'(x) - \eta I$ are positive definite for some $\eta > 0$ and all x on the line segment connecting y to z. Show that

$$
V(y, z, |z - y|) \leq K\eta^{-2}|z - y|,
$$

where $K < \infty$ depends only on the bounds on $|b(\cdot)|$ and $|\sigma(\cdot)|$.
Hint: Consider the path $\phi_s = y + s(z - y)/|z - y|$, and

$$
u_s = \sigma(\phi_s)'[\sigma(\phi_s)\sigma'(\phi_s)]^{-1}(\dot{\phi}_s - b(\phi_s)).
$$

Check that $|u_s| \leq K_1 \eta^{-1}$ for all $s \in [0, |z - y|]$ and some $K_1 < \infty$.
(b) Show that for all x, y, z,

$$
V(x, z) \leq V(x, y) + V(y, z).
$$

(c) Prove that if there exists $\eta > 0$ such that the matrices $\sigma(x)\sigma'(x) - \eta I$ are positive definite for all $x \in \partial G \cup \{0\}$, then Assumption (A-4) holds.
Hint: Observe that by the uniform Lipschitz continuity of $\sigma(\cdot)$, the positive definiteness of $\sigma(\cdot)\sigma(\cdot)'$ at a point x_0 implies that the matrices $\sigma(x)\sigma'(x) - \eta I$ are positive definite for all $|x - x_0| \leq \rho$ when $\eta, \rho > 0$ are small enough.

Exercise 5.7.30 Assume (A-1)–(A-4). Prove that for any closed set $N \subset \partial G$,

$$
\lim_{\rho \to 0} \limsup_{\epsilon \to 0} \epsilon \log \sup_{y \in B_\rho} P_y(x^\epsilon_{\tau^\epsilon} \in N) \leq -(\inf_{z \in N} V(0, z) - \overline{V}).
$$

Exercise 5.7.31 Complete the proof of Corollary 5.7.16.

Exercise 5.7.32 Consider the stochastic differential equations

$$
dx^\epsilon_t = \sqrt{\epsilon}\, dw_t, \quad x^\epsilon_t \in \mathbb{R}, \tag{5.7.33}
$$

$$
dx^\epsilon_t = -\alpha x^\epsilon_t dt + \sqrt{\epsilon}\, dw_t, \quad x^\epsilon_t \in \mathbb{R}, \quad \alpha > 0. \tag{5.7.34}
$$

Let $G = (a, b)$. Apply Theorem 5.7.3 for both cases to evaluate the following expressions and limits.

(a) Compute $E_x(\tau^\epsilon)$ and $\lim_{\epsilon \to 0} \epsilon \log E_x(\tau^\epsilon)$; compare the x dependence in both cases.

(b) Let $b \to \infty$. What happens to the asymptotic behavior of $E_x(\tau^\epsilon)$ in both cases? Compare to the case in which $\alpha < 0$ in (5.7.34).

(c) Compute $P_a(x) = P_x$ (exit at $x = a$). Compare (5.7.33) and (5.7.34) to see the influence of x when $\epsilon \to 0$.

Exercise 5.7.35 Viterbi [Vit66, page 86] proposes the following model for a first-order phased locked loop that tracks a constant phase:

$$d\phi_t = (\Delta - AK \sin \phi_t)dt + K dw_t,$$

where Δ is a constant (which reflects the demodulated frequency), $AK > \Delta$, and A, K are given constants.

A cycle slip is defined as an exit from the basin of attraction of $\phi_0 \triangleq \sin^{-1}(\Delta/AK)$. Compute the mean exit time. What happens when $K \to 0$?

Exercise 5.7.36 In this exercise, the Hamilton–Jacobi equations associated with the quasi-potential are derived.

(a) Consider the following optimization problem. Let $L(x, y, t)$ be a function that is convex in the variable y, C^∞ in all its variables and bounded in the variable x. Define

$$W(x, t) = \inf_{x_0 = 0, x_t = x} \int_0^t L(x_s, \dot{x}_s, s)ds$$

and

$$H(x, p, t) = \inf_y [L(x, y, t) - \langle p, y \rangle].$$

Note that $H(x, \cdot, t)$ is (up to sign) the Fenchel–Legendre transform of $L(x, \cdot, t)$ with respect to the variable y. Assume that W is C^1 with respect to (x, t). Prove that

$$\frac{\partial W}{\partial t} = H(x, W_x, t), \tag{5.7.37}$$

where $W_x \triangleq \partial W / \partial x$.

Hint: Use the optimality of W to write

$$W(x, t + \Delta t) - W(x, t)$$

$$\simeq \inf_{x'} \left[W(x', t) - W(x, t) + \int_t^{t+\Delta t} L\left(x, \frac{x - x'}{\Delta t}, s\right)ds \right],$$

and use the regularity assumption on W to get (5.7.37).

(b) Suppose that $L(x, y, t) = L(x, y)$, and define

$$W(x) = \inf_{t > 0, x_0 = 0, x_t = x} \int_0^t L(x_s, \dot{x}_s)ds.$$

Prove that

$$H(x, W_x) = 0\,. \tag{5.7.38}$$

Remark: Consider (5.7.1), where for some $\eta > 0$ the matrices $\sigma(x)\sigma(x)' - \eta I$ are positive definite for all x (uniform ellipticity), and $b(x), \sigma(x)$ are C^∞ functions. Then the quasi-potential $V(0, x) = W(x)$ satisfies (5.7.38), with

$$L(x, y) = \frac{1}{2}\langle y - b(x), (\sigma(x)\sigma(x)')^{-1}(y - b(x))\rangle\,,$$

and hence,

$$H(x, p) = -\langle p, b(x)\rangle - \frac{1}{2}|\sigma(x)'p|^2\,.$$

Exercise 5.7.39 Consider the system that is known as Langevin's equation:

$$
\begin{aligned}
dx_1 &= x_2 dt\,, &\tag{5.7.40}\\
dx_2 &= -(x_2 + U'(x_1))dt + \sqrt{2}\sqrt{\epsilon}dw_t\,,
\end{aligned}
$$

where w is a standard Brownian motion (and $U'(x) = dU/dx$). By adding a term $\beta\sqrt{\epsilon}dv_t$ to the right side of (5.7.40), where v_t is a standard Brownian motion independent of w_t, compute the Hamilton–Jacobi equation for $W(x) = V(0, x)$ and show that as $\beta \to 0$, the formula (5.7.38) takes the form

$$x_2 W_{x_1} - (x_2 + U'(x_1))W_{x_2} + W_{x_2}^2 = 0\,. \tag{5.7.41}$$

Show that $W(x_1, x_2) = U(x_1) + x_2^2/2$ is a solution of (5.7.41). Note that, although W can be solved explicitly, Green's fundamental solution is not explicitly computable.

Remark: Here, W is differentiable, allowing for the justification of asymptotic expansions (*c.f.* [Kam78]).

5.8 The Performance of Tracking Loops

In this section, two applications of the problem of exit from a domain to the analysis of tracking problems are considered. In both cases, the bounds of Section 5.7 are used to yield exponential bounds on the time to lose lock and thus on the performance of the tracking loops.

5.8.1 An Angular Tracking Loop Analysis

Consider a radar tracking the angular location of a target. The angular location of the target is modeled by the solution of the following stochastic differential equation

$$d\theta_t = m(\theta_t)\, dt + dw_t,\ \theta_0 = 0\,, \tag{5.8.1}$$

where w_t is a two-dimensional Brownian motion, $\theta \in \mathbb{R}^2$ consists of the elevation and azimuthal angles, and $m(\cdot)$ is assumed to be bounded and uniformly Lipschitz continuous. The random element in (5.8.1) models uncertainties in the dynamics, such as random evasive maneuvers of the target. The information available to the tracker comes from an observation device (antenna or lens). Denoting by $u_t \in \mathbb{R}^2$ its angle, this information may be modeled as the solution of the equation

$$dy_t = (\theta_t - u_t)\, dt + \epsilon\, dv_t\,, \qquad (5.8.2)$$

where v_t is a (two-dimensional) Brownian motion, independent of $w.$ and $\theta.$, which models the observation noise. Typical tracking loops are designed to achieve good tracking. In such loops, the observation noise is small compared to $\theta_t - u_t$, which is why the factor $\epsilon \ll 1$ appears in (5.8.2). On the other hand, the dynamics noise in (5.8.1) is not necessarily small.

Since the observation device has a limited field of view (which is measured in terms of its beam-width θ_{cr}), the diffusion equation (5.8.2) ceases to model the observation process as soon as the tracking error $\theta_t - u_t$ exits from the field of view of the tracking device. In general, this field of view can be seen as a domain D in $\mathbb{R}^2$. Two particular cases are of interest here:
(a) $D_1 = B_{0,\theta_{cr}}$. This is the case of a symmetric observation device.
(b) $D_2 = (-\theta_{cr}^x, \theta_{cr}^x) \times (-\theta_{cr}^y, \theta_{cr}^y)$. This corresponds normally to decoupled motions in the elevation and azimuthal axis, and often is related to a situation where $\theta_{cr}^y \gg \theta_{cr}^x$, or to situations where the elevation motion is not present, both in the dynamics model and in the observation, making the observation process one-dimensional. Therefore, consideration is given to the case $D_2 = (-\theta_{cr}, \theta_{cr}) \times \mathbb{R}$. The exit event is referred to as *track loss*. Based on the available observations $\{y_s,\ 0 \leq s \leq t\}$, the tracker may modify u_t in order to minimize the probability to lose track in a certain interval $[0, T]$, or to maximize the mean time to lose track. Such a notion of optimality may in general be ill-posed, as an optimal solution might not exist in the class of strong solutions of diffusions driven by the observation process $y.$ of (5.8.2). Therefore, the problem of designing an optimal control u_t is not dealt with here. Rather, the performance of a simple, sub-optimal design is analyzed. However, when the motions in the elevation and azimuthal angles are decoupled, this sub-optimal design actually maximizes the logarithmic rate associated both with the mean time to lose track and the probability to lose track in $[0, T]$. (See [ZZ92] for details.)

The tracking loop to be analyzed corresponds to u_t such that

$$du_t = m(u_t)dt + \frac{1}{\epsilon}\, dy_t\,, \quad u_0 = 0\,. \qquad (5.8.3)$$

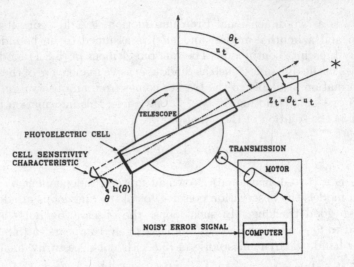

Figure 5.8.1: Tracking system.[1]

Since $m(\cdot)$ is uniformly Lipschitz continuous, a strong solution exists to the system of equations (5.8.1)–(5.8.3). Let $\tau = \inf\{s : |\theta_s - u_s| \geq \theta_{cr}\}$ be the time until the first loss of track starting with a perfect lock, where $|\cdot|$ is understood in case D_1 as the Euclidean norm and in case D_2 as the one-dimensional norm. The limiting behavior of τ, as $\epsilon \to 0$, is stated in the following theorem.

Theorem 5.8.4

$$\lim_{\epsilon \to 0} \epsilon \log E(\tau) = \frac{\theta_{cr}^2}{2}.$$

Proof: Let $z_t = \theta_t - u_t$ denote the angular tracking error. Then z_t satisfies the equation

$$dz_t = (m(\theta_t) - m(u_t))dt - \frac{1}{\epsilon} z_t dt - dv_t + dw_t, \;\; z_0 = 0,$$

which under the time change $t' = t/\epsilon$ becomes

$$dz_{t'} = \epsilon(m(\theta_{t'}) - m(u_{t'}))\, dt' - z_{t'} dt' + \sqrt{2}\epsilon d\tilde{w}_{t'},$$

where $\tilde{w}_{t'} \triangleq (w_{\epsilon t'} - v_{\epsilon t'})/\sqrt{2\epsilon}$ is a standard Brownian motion. Note that

$$\frac{\tau}{\epsilon} = \tau' \triangleq \inf\{s' : |z_{s'}| \geq \theta_{cr}\}.$$

Let $\hat{z}_{t'}$ be the solution of

$$d\hat{z}_{t'} = -\hat{z}_{t'} dt' + \sqrt{2\epsilon} d\tilde{w}_{t'}, \quad \hat{z}_0 = 0. \tag{5.8.5}$$

By Theorem 5.7.11 (applied to $\hat{\tau}' \triangleq \inf\{s' : |\hat{z}_{s'}| \geq \theta_{cr}\}$), it follows that

$$\lim_{\epsilon \to 0} 2\epsilon \log E(\hat{\tau}') = \inf_{T > 0} \inf_{\substack{\phi : \phi_0 = 0 \\ |\phi_T| = \theta_{cr}}} \frac{1}{2} \int_0^T |\dot{\phi}_t + \phi_t|^2 \, dt,$$

where again the norm is the two-dimensional Euclidean norm in case D_1 and the one-dimensional norm in case D_2. It is worthwhile to note that in both cases, the behavior of the model (5.8.5) is either a one-dimensional question (in case D_2) or may be reduced to such (by noting that $|\hat{z}_{t'}|$ is a one-dimensional Bessel process), and therefore can in principle be analyzed explicitly. The large deviations approach allows both a reduction to the model (5.8.5) and a simplification of the computations to be made.

The proof of the following lemma is left to Exercise 5.8.10.

Lemma 5.8.6 *In both cases D_1 and D_2,*

$$\inf_{T > 0} \inf_{\substack{\phi : \phi_0 = 0 \\ |\phi_T| = \theta_{cr}}} \frac{1}{2} \int_0^T |\dot{\phi}_t + \phi_t|^2 \, dt = \theta_{cr}^2. \tag{5.8.7}$$

Hence, it suffices to show that $\lim_{\epsilon \to 0} \epsilon |\log E(\hat{\tau}') - \log E(\tau')| = 0$ in order to complete the proof of the theorem. This is a consequence of the following lemma, which allows, for any $\eta > 0$, to bound τ' between the values of $\hat{\tau}'$ corresponding to $\theta_{cr} - \eta$ and $\theta_{cr} + \eta$, provided that ϵ is small enough.

Lemma 5.8.8

$$\sup_{t' \geq 0} |z_{t'} - \hat{z}_{t'}| \leq 2\epsilon \sup_{x \in \mathbb{R}^2} |m(x)| \quad \text{almost surely}.$$

Proof: Let $e_{t'} = z_{t'} - \hat{z}_{t'}$. Then

$$\dot{e}_{t'} = \epsilon(m(\theta_{t'}) - m(u_{t'})) - e_{t'}, \quad e_0 = 0.$$

Hence,

$$e_{t'} = \epsilon \int_0^{t'} (m(\theta_s) - m(u_s)) e^{-(t'-s)} ds.$$

Therefore,

$$|e_{t'}| \le 2\epsilon \sup_{x \in \mathbb{R}^2} |m(x)| \int_0^{t'} e^{-(t'-s)} ds \,. \qquad \Box$$

Exercise 5.8.9 Prove that Theorem 5.8.4 holds for any tracking system of the form

$$du_t = f(u_t)dt + \frac{1}{\epsilon} dy_t \,, \quad u_0 = 0$$

as long as $f(\cdot)$ is bounded and uniformly Lipschitz continuous.

Remark: In particular, the simple *universal* tracker $u_t = y_t/\epsilon$ can be used with no knowledge about the target drift $m(\cdot)$, yielding the same limit of $\epsilon \log E(\tau)$.

Exercise 5.8.10 Complete the proof of Lemma 5.8.6.
Hint: Observe that by symmetry it suffices to consider the one-dimensional version of (5.8.7) with the constraint $\phi_T = \theta_{cr}$. Then substitute $\dot{\psi}_t = \dot{\phi}_t + \phi_t$ and paraphrase the proof of Lemma 5.4.15.

5.8.2 The Analysis of Range Tracking Loops

An important component of radar trackers is their range tracking abilities. By transmitting a pulse $s(t)$ and analyzing its return from a target $s(t-\tau)$, one may estimate τ, the time it took the pulse to travel to the target and return. Dividing by twice the speed of light, an estimate of the distance to the target is obtained.

A range tracker keeps track of changes in τ. Suppose τ is itself a random process, modeled by

$$\tau_{k+1} = \tau_k - \epsilon T \beta \tau_k + \epsilon T^{1/2} v_k \,, \qquad (5.8.11)$$

where τ_k denotes the value of τ at the kth pulse transmission instant, $\{v_k\}$ denotes a sequence of zero mean i.i.d. random variables, T is a deterministic constant that denotes the time interval between successive pulses (so $1/T$ is the pulse repetition frequency), and β is a deterministic constant related to the speed in which the target is approaching the radar and to the target's motion bandwidth. The changes in the dynamics of the target are slow in comparison with the pulse repetition frequency, i.e., low bandwidth of its motion. This is indicated by the $\epsilon \ll 1$ factor in both terms in the right side of (5.8.11).

If $\{v_k\}$ are standard Normal, then (5.8.11) may be obtained by discretizing at T-intervals the solution of the stochastic differential equation $d\tau = -\epsilon \tilde{\beta} \tau dt + \epsilon \tilde{\alpha} dv_t$, with v_t a standard Brownian motion, $\beta = (1 - e^{-\epsilon \tilde{\beta} T})/\epsilon T \simeq \tilde{\beta}$ and $\tilde{\alpha}^2 = 2\epsilon \tilde{\beta} T/(1 - e^{-2\epsilon \tilde{\beta} T}) \simeq 1$.

The radar transmits pulses with shape $s(\cdot)$ at times kT, $k = 0, 1, \ldots$, where $s(t) = 0$ for $|t| \geq \frac{\delta}{2}$ and $\delta \ll T$. The kth pulse appears at the receiver as $s(t - kT - \tau_k)$, and to this noise is added. Hence, the receiver input is

$$dy_t = \sum_{k=0}^{\infty} s(t - kT - \tau_k)\, dt + N_0\, dw_t,$$

where w_t is a standard Brownian motion independent of $\{v_k\}$ and τ_0, while N_0 is a deterministic fixed constant reflecting the noise power level. Usually, T is chosen such that no ambiguity occurs between adjacent pulses, i.e., T is much larger then the dynamic range of the increments $(\tau_{k+1} - \tau_k)$. A

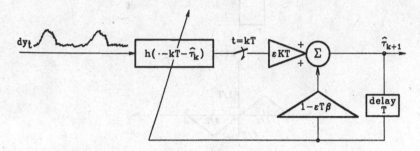

Figure 5.8.2: Block diagram of a range tracker.

typical radar receiver is depicted in Fig. 5.8.2. It contains a filter $h(\cdot)$ (the *range gate*) that is normalized such that $\int_{-\delta/2}^{\delta/2} |h(t)|^2\, dt = 1/\delta$ and is designed so that the function

$$g(x) = \int_{-\delta/2}^{\delta/2} s(t + x) h(t)\, dt$$

is bounded, uniformly Lipschitz continuous, with $g(0) = 0$, $g'(0) < 0$ and $xg(x) < 0$ for $0 < |x| < \delta$.

A typical example is

$$s(t) = 1_{[-\frac{\delta}{2}, \frac{\delta}{2}]}(t) \quad \text{and} \quad h(t) = \frac{1}{\delta}\,\text{sign}(t) 1_{[-\frac{\delta}{2}, \frac{\delta}{2}]}(t)$$

for which

$$g(x) = -\text{sign}(x)\left(\frac{|x|}{\delta} 1_{[0, \frac{\delta}{2}]}(|x|) + \left(1 - \frac{|x|}{\delta}\right) 1_{(\frac{\delta}{2}, \delta]}(|x|)\right).$$

The receiver forms the estimates

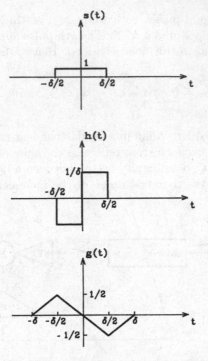

Figure 5.8.3: Typical range gate characteristics.

$$\hat{\tau}_{k+1} = \hat{\tau}_k - \epsilon T \beta \hat{\tau}_k + \epsilon KT \int_{\hat{\tau}_k + kT - \delta/2}^{\hat{\tau}_k + kT + \delta/2} h(t - \hat{\tau}_k - kT) dy_t, \qquad (5.8.12)$$

where $K > 0$ is the receiver gain. The correction term in the preceding estimates is taken of order ϵ to reduce the effective measurement noise to the same level as the random maneuvers of the target.

Since $T \gg \delta$, adjacent pulses do not overlap in (5.8.12), and the update formula for $\hat{\tau}_k$ may be rewritten as

$$\hat{\tau}_{k+1} = \hat{\tau}_k - \epsilon T \beta \hat{\tau}_k + \epsilon KT g(\hat{\tau}_k - \tau_k) + \epsilon \frac{N_0 KT}{\delta^{1/2}} w_k,$$

where $w_k \sim N(0, 1)$ is a sequence of i.i.d. Normal random variables. Assume that $\hat{\tau}_0 = \tau_0$, i.e., the tracker starts with perfect lock. Let $Z_k = \hat{\tau}_k - \tau_k$ denote the range error process. A loss of track is the event $\{|Z_k| \geq \delta\}$, and the asymptotic probability of such an event determines the performance of the range tracker. Note that Z_k satisfies the equation

$$Z_{k+1} = Z_k + \epsilon b(Z_k) + \epsilon \nu_k, \ Z_0 = 0, \qquad (5.8.13)$$

where $b(z) \triangleq - T\beta z + KT g(z)$ and $\nu_k \triangleq \frac{N_0 KT}{\delta^{1/2}} w_k - T^{1/2} v_k$.

In the absence of the noise sequence $\{\nu_k\}$ and when $\beta \geq 0$, the dynamical system (5.8.13) has 0 as its unique stable point in the interval $[-\delta, \delta]$ due to the design condition $zg(z) < 0$. (This stability extends to $\beta < 0$ if K is large enough.) Therefore, as $\epsilon \to 0$, the probability of loss of lock in any finite interval is small and it is reasonable to rescale time. Define the continuous time process $Z_\epsilon(t)$, $t \in [0, 1]$, via

$$Z_\epsilon(t) = Z_{[t/\epsilon]} + \epsilon^{-1} \left(t - \epsilon[t/\epsilon]\right) \left(Z_{[t/\epsilon]+1} - Z_{[t/\epsilon]}\right).$$

Let $\Lambda(\cdot)$ be the logarithmic moment generating function of the random variable ν_1 and assume that $\Lambda(\lambda) < \infty$ for all $\lambda \in \mathbb{R}$.

The large deviations behavior of $Z_\epsilon(\cdot)$ is stated in the following theorem.

Theorem 5.8.14 $\{Z_\epsilon(\cdot)\}$ *satisfies the LDP in* $C_0([0,1])$ *with the good rate function*

$$I(\phi) = \begin{cases} \int_0^1 \Lambda^*(\dot\phi(t) - b(\phi(t)))\, dt & \phi \in \mathcal{AC}, \phi(0) = 0 \\ \infty & \text{otherwise}. \end{cases} \qquad (5.8.15)$$

Proof: Note first that, by Theorem 5.1.19, the process

$$Y_\epsilon(t) = \epsilon \sum_{k=0}^{[t/\epsilon]-1} \nu_k$$

satisfies an LDP in $L_\infty([0,1])$ with the good rate function

$$I_\nu(\phi) = \begin{cases} \int_0^1 \Lambda^*(\dot\phi(t))dt, & \phi \in \mathcal{AC}, \phi(0) = 0 \\ \infty & \text{otherwise}. \end{cases} \qquad (5.8.16)$$

Let $\tilde{Y}_\epsilon(t)$ be the polygonal approximation of $Y_\epsilon(t)$, namely,

$$\tilde{Y}_\epsilon(t) \triangleq Y_\epsilon(t) + (t - \epsilon[t/\epsilon])\, \nu_{[t/\epsilon]}.$$

Then the same arguments as in the proof of Lemma 5.1.4 result in the exponential equivalence of $Y_\epsilon(\cdot)$ and $\tilde{Y}_\epsilon(\cdot)$ in $L_\infty([0,1])$. Hence, as $\tilde{Y}_\epsilon(t)$ takes values in $C_0([0,1])$, it satisfies the LDP there with the good rate function $I_\nu(\cdot)$ of (5.8.16).

Since $b(\cdot)$ is Lipschitz continuous, the map $\psi = F(\phi)$ defined by

$$\psi(t) = \int_0^t b(\psi(s))ds + \phi(t)$$

is a continuous map of $C_0([0,1])$ onto itself. Therefore, by the contraction principle (Theorem 4.2.1), $\tilde{Z}_\epsilon \triangleq F(\tilde{Y}_\epsilon)$ satisfies an LDP in $C_0([0,1])$ with the good rate function $I(\cdot)$ of (5.8.15). (For more details, compare with the proof of Theorem 5.6.3.) The proof is completed by applying Theorem 4.2.13, provided that $Z_\epsilon(\cdot)$ and $\hat{Z}_\epsilon(\cdot)$ are exponentially equivalent, namely, that for any $\eta' > 0$,

$$\lim_{\epsilon \to 0} \epsilon \log P(\| \hat{Z}_\epsilon - Z_\epsilon \| \geq \eta') = -\infty .$$

To this end, recall that

$$Z_\epsilon(t) - Z_\epsilon(\epsilon[t/\epsilon]) = (t - \epsilon[t/\epsilon]) \left(b(Z_{[t/\epsilon]}) + \nu_{[t/\epsilon]} \right) ,$$

and therefore,

$$\int_0^1 |Z_\epsilon(t) - Z_\epsilon(\epsilon[t/\epsilon])| dt \leq \epsilon \left(\max_{k=0}^{[1/\epsilon]} |\nu_k| + \sup_{z \in \mathbb{R}} |b(z)| \right) .$$

Since $b(\cdot)$ is bounded, it now follows that for any $\eta > 0$,

$$\lim_{\epsilon \to 0} \epsilon \log P \left(\int_0^1 |Z_\epsilon(t) - Z_\epsilon(\epsilon[t/\epsilon])| dt \geq \eta \right) = -\infty . \qquad (5.8.17)$$

(Recall Exercise 5.1.24.) Observe that by iterating (5.8.13),

$$Z_\epsilon(t) = \int_0^t b\left(Z_\epsilon(\epsilon[s/\epsilon]) \right) ds + \tilde{Y}_\epsilon(t) ,$$

while

$$\hat{Z}_\epsilon(t) = \int_0^t b(\hat{Z}_\epsilon(s)) ds + \tilde{Y}_\epsilon(t) .$$

Let $e(t) = |\hat{Z}_\epsilon(t) - Z_\epsilon(t)|$. Then

$$e(t) \leq \int_0^t \left| b(\hat{Z}_\epsilon(s)) - b\left(Z_\epsilon(\epsilon[s/\epsilon]) \right) \right| ds$$

$$\leq B \int_0^t e(s) ds + B \int_0^t |Z_\epsilon(s) - Z_\epsilon(\epsilon[s/\epsilon])| ds ,$$

where B is the Lipschitz constant of $b(\cdot)$. Hence, Gronwall's lemma (Lemma E.6) yields

$$\sup_{0 \leq t \leq 1} e(t) \leq Be^B \int_0^1 |Z_\epsilon(s) - Z_\epsilon(\epsilon[s/\epsilon])| ds ,$$

and the exponential equivalence of $Z_\epsilon(\cdot)$ and $\hat{Z}_\epsilon(\cdot)$ follows by (5.8.17). $\square$

Remarks:
(a) Observe that Theorem 5.8.14 applies for any random process that

evolves according to (5.8.13), provided that $b(\cdot)$ is bounded and uniformly Lipschitz continuous and ν_k are i.i.d. random variables whose logarithmic moment generating function $\Lambda(\cdot)$ is finite everywhere. The preceding proof does not use other special properties of the range tracking problem. Likewise, the theorem extends to finite intervals $[0, T]$ and to $\mathbb{R}^d$-valued processes.

(b) Focusing back on the range tracking application, since v_1 has zero mean and $b(0) = 0$, $I(\phi) = 0$ only when $\phi_t \equiv 0$. Moreover, by choosing K large enough such that $Kg'(0) < \beta$, the point 0 becomes a stable point of the deterministic ordinary differential equation

$$\dot{x}_s = b(x_s) = -T\beta x_s + TKg(x_s),$$

and it is the unique such point in some neighborhood of 0 that may be strictly included in the domain $G = (-\delta, \delta)$. However, for $\beta \geq 0$, indeed, $x = 0$ is the unique stable point in G and any trajectory with initial condition $x_0 \in G$ converges to 0 as $s \to \infty$. The analysis of the mean time until loss of track follows the Freidlin–Wentzell approach presented in Section 5.7 for diffusion processes.

(c) It is interesting to note that a common model in the literature is to approximate first τ_k by a continuous model (obtained after a time rescaling) and then compute the asymptotics of rare events. Since the time-rescaled continuous time model involves Brownian motion, this computation results in the rate function

$$I_N(\phi) = \begin{cases} \frac{c}{2} \int_0^1 (\dot{\phi}_t - b(\phi_t))^2 \, dt & , \quad \phi \in H_1 \\ \infty & , \quad \text{otherwise} \end{cases} \tag{5.8.18}$$

for some $c > 0$. As is obvious from the analysis, this approach is justified only if either the random maneuvers v_k are modeled by i.i.d. Normal random variables or if v_k are multiplied by a power of ϵ higher than 1, as in Exercise 5.8.19.

Exercise 5.8.19 Assume that τ_k satisfies the equation

$$\tau_{k+1} = \tau_k - \epsilon T \beta \tau_k + \epsilon^\alpha T^{1/2} v_k$$

for some $\alpha > 1$. Prove that then effectively $\nu_k = (N_0 K T / \sqrt{\delta}) w_k$ and the rate function controlling the LDP of $Z_\epsilon(\cdot)$ is $I_N(\cdot)$ of (5.8.18) with $c = (N_0 K T)^2 / \delta$.

Exercise 5.8.20 (a) Prove that Theorem 5.8.14 holds for any finite time interval $[0, T]$ with the rate function

$$I_T(\phi) = \begin{cases} \int_0^T \Lambda^*(\dot{\phi}(t) - b(\phi(t))) \, dt & , \quad \phi \in \mathcal{AC}^T, \phi(0) = 0 \\ \infty & , \quad \text{otherwise}. \end{cases}$$

(b) Let $\tau^* = \inf\{k : |Z_k| \geq \delta\}$ and suppose that $\beta \geq 0$. Repeating the arguments in Section 5.7, prove that

$$\lim_{\epsilon \to 0} \epsilon \log E(\tau^*) = \inf_{T>0} \quad \inf_{\{\phi: \phi(0)=0, |\phi(T)|=\delta\}} I_T(\phi) .$$

5.9   Historical Notes and References

Many of the large deviations results for sample path of stochastic processes, with the notable exception of Schilder's theorem, were developed by the Russian school.

The random walk results of Section 5.1 appear implicitly in [Var66], using a direct approach that did not make use of the projective limits discussed here. Building upon [Bor65, Bor67, Mog74], Mogulskii [Mog76] extends this result to accommodate different scalings and, using the Skorohod topology, accommodates also moment generating functions that are finite only in a neighborhood of the origin. See also [Pie81] for random variables with sub-exponential tails and [LyS87] for random variables possessing a one-sided heavy tail. See also [Mog93] for a general framework. Extensions to Banach space valued random walks are derived in [BoM80, DeZa95]. Using the approach of Section 5.1, [DeZa95] also relaxes the independence assumption.

Schilder [Sc66] derives Theorem 5.2.3 by a direct approach based on Girsanov's formula. A derivation related to ours may be found in [Puk94a].

As can be seen in Exercise 5.2.14, inequalities from the theory of Gaussian processes may be used in order to prove exponential tightness and large deviations for the sample path of such processes. A derivation of Schilder's theorem based on Fernique's inequality may be found in [DeuS89b]. A non-topological form of the LDP for centered Gaussian processes, which is based on the isoperimetric inequality, may be found in [BeLd93]. This version includes Schilder's theorem as a particular case.

The results of Section 5.3 first appeared in the first edition of this book. Large deviations for the Brownian sheet follow also from the Banach space version of Cramér's theorem described in Exercise 6.2.21, see [DoD86].

Freidlin and Wentzell derived both the sample path diffusion LDP and the asymptotics of the problem of exit from a domain in a series of papers beginning with [VF70, VF72] and culminating in the 1979 Russian edition of their book [FW84], soon to be updated. (See also [Wen90].) Their approach is based on a direct change of measure argument and explicit computations. Our treatment in Section 5.6, based on suitable "almost continuous" contractions, is a combination of the results of [Aze80, FW84,

Var84]. For other sample path results and extensions, see [Mik88, Bal91, DoS91, LP92, deA94a]. See also [MNS92, Cas93, BeC96] for analogous LDPs for stochastic flows.

An intimate relation exists between the sample path LDP for diffusions and Strassen's law of the iterated logarithm [Str64]. A large deviations proof of the latter in the case of Brownian motion is provided in [DeuS89b]. For the case of general diffusions and for references to the literature, see [Bal86, Gan93, Car98].

Sample path large deviations have been used as a starting point to the analysis of the heat kernel for diffusions, and to asymptotic expansions of exponential integrals of the form appearing in Varadhan's lemma. An early introduction may be found in Azencott's article in [As81]. The interested reader should consult [Aze80, As81, Bis84, Ben88, BeLa91, KuS91, KuS94] for details and references to the literature.

Sample path large deviations in the form of the results in Sections 5.1, 5.2, and 5.6 have been obtained for situations other than random walks or diffusion processes. For such results, see [AzR77, Fre85a, LyS87, Bez87a, Bez87b, DeuS89a, SW95, DuE97] and the comments relating to queueing systems at the end of this section.

The origin of the problem of exit from a domain lies in the reaction rate theory of chemical physics. Beginning with early results of Arrhenius [Arr89], the theory evolved until the seminal paper of Kramers [Kra40], who computed explicitly not only the exponential decay rate but also the precise pre-exponents. Extensive reviews of the physical literature and recent developments may be found in [Land89, HTB90]. A direct outgrowth of this line of thought has been the asymptotic expansions approach to the exit problem [MS77, BS82, MS82, MST83]. This approach, which is based on formal expansions, has been made rigorous in certain cases [Kam78, Day87, Day89]. A related point of view uses optimal stochastic control as the starting point for handling the singular perturbation problem [Fle78]. Theorem 5.7.3 was first proved in the one-dimensional case as early as 1933 [PAV89].

As mentioned before, the results presented in Section 5.7 are due to Freidlin and Wentzell [FW84], who also discuss the case of general Lévy processes, multiple minima, and quasi-stationary distributions. These results have been extended in numerous ways, and applied in diverse fields. In what follows, a partial list of such extensions and applications is presented. This list is by no means exhaustive.

The discrete time version of the Freidlin-Wentzell theory was developed by several authors, see [MoS89, Kif90b, KZ94, CatC97].

Kushner and Dupuis ([DuK86, DuK87] and references therein) discuss

problems where the noise is not necessarily a Brownian motion, and apply
the theory to the analysis of a Phased Lock Loop (PLL) example. Some of
their results may be derived by using the second part of Corollary 5.7.16. A
detailed analysis of the exit measure, including some interesting examples
of nonconvergence for characteristic boundaries, has been presented by Day
[Day90a, Day90b, Day92] and by Eizenberg and Kifer [Kif81, Eiz84, EK87].
Generalizations of diffusion models are presented in [Bez87b]. Large deviations and the exit problem for stochastic partial differential equations are
treated in [Fre85b, CM90, Io91a, DaPZ92].

For details about the DMPSK analysis in Section 5.4, including digital
communications background and simulation results, see [DGZ95] and the
references therein. For other applications to communication systems see
[Buc90].

The large exceedances statements in Section 5.5 are motivated by DNA
matching problems ([ArMW88, KaDK90]), the CUSUM method for the
detection of change points ([Sie85]), and light traffic queue analysis ([Igl72,
Ana88]). The case $d = 1$ has been studied extensively in the literature,
using its special renewal properties to obtain finer estimates than those
available via the LDP. For example, the exponential limit distribution for T_ϵ
is derived in [Igl72], strong limit laws for L_ϵ and $\hat{Y}^\epsilon$ are derived in [DeK91a,
DeK91b], and the CLT for rescaled versions of the latter are derived in
[Sie75, Asm82]. For an extension of the results of Section 5.5 to a class
of uniformly recurrent Markov-additive processes and stationary strong-
mixing processes, see [Zaj95]. See also [DKZ94c] for a similar analysis and
almost sure results for T_ϵ, with a continuous time multidimensional Lévy
process instead of a random walk.

The analysis of angular tracking was undertaken in [ZZ92], and in the
form presented there is based on the large deviations estimates for the nonlinear filtering problem contained in [Zei88]. For details about the analysis
of range tracking, motivated in part by [BS88], see [DeZ94] and the references therein.

A notable omission in our discussion of applications of the exit problem
is the case of queueing systems. As mentioned before, the exceedances
analysis of Section 5.5 is related to such systems, but much more can be
said. We refer the reader to the book [SW95] for a detailed description
of such applications. See also the papers [DuE92, IMS94, BlD94, DuE95,
Na95, AH98] and the book [DuE97].

Chapter 6

The LDP for Abstract Empirical Measures

One of the striking successes of the large deviations theory in the setting of finite dimensional spaces explored in Chapter 2 was the ability to obtain refinements, in the form of Cramér's theorem and the Gärtner–Ellis theorem, of the weak law of large numbers. As demonstrated in Chapter 3, this particular example of an LDP leads to many important applications; and in this chapter, the problem is tackled again in a more general setting, moving away from the finite dimensional world.

The general paradigm developed in this book, namely, the attempt to obtain an LDP by lifting to an infinite dimensional setting finite dimensional results, is applicable here, too. An additional ingredient, however, makes its appearance; namely, sub-additivity is exploited. Not only does this enable abstract versions of the LDP to be obtained, but it also allows for explicit mixing conditions that are sufficient for the existence of an LDP to be given, even in the $\mathbb{R}^d$ case.

6.1 Cramér's Theorem in Polish Spaces

A general version of Cramér's theorem for i.i.d. random variables is presented in this section. This theorem is then specialized to strengthen Cramér's theorem in $\mathbb{R}^d$. Sanov's theorem is derived in Section 6.2 as a consequence of the general formulation of this section. The core new idea in the derivation presented here, namely, the use of sub-additivity as a tool for proving the LDP, is applicable beyond the i.i.d. case.

A. Dembo, O. Zeitouni, *Large Deviations Techniques and Applications*, 251
Stochastic Modelling and Applied Probability 38,
DOI 10.1007/978-3-642-03311-7_6,
© Springer-Verlag Berlin Heidelberg 1998, corrected printing 2010

Let μ be a Borel probability measure on a locally convex, Hausdorff, topological real vector space $\mathcal{X}$. On the space $\mathcal{X}^*$ of continuous linear functionals on $\mathcal{X}$, define the logarithmic moment generating function

$$\Lambda(\lambda) \triangleq \log \int_{\mathcal{X}} e^{\langle \lambda, x \rangle} d\mu \,, \tag{6.1.1}$$

and let $\Lambda^*(\cdot)$ denote the Fenchel–Legendre transform of Λ.

For every integer n, suppose that $X_1, \ldots, X_n$ are i.i.d. random variables on $\mathcal{X}$, each distributed according to the law μ; namely, their joint distribution μ^n is the product measure on the space $(\mathcal{X}^n, (\mathcal{B}_{\mathcal{X}})^n)$. We would like to consider the partial averages

$$\hat{S}_n^m \triangleq \frac{1}{n-m} \sum_{\ell=m+1}^{n} X_\ell \,,$$

with $\hat{S}_n \triangleq \hat{S}_n^0$ being the empirical mean. Note that $\hat{S}_n^m$ are always measurable with respect to the σ-field $\mathcal{B}_{\mathcal{X}^n}$, because the addition and scalar multiplication are continuous operations on $\mathcal{X}^n$. In general, however, $(\mathcal{B}_{\mathcal{X}})^n \subset \mathcal{B}_{\mathcal{X}^n}$ and $\hat{S}_n^m$ may be non-measurable with respect to the product σ-field $(\mathcal{B}_{\mathcal{X}})^n$. When $\mathcal{X}$ is separable, $\mathcal{B}_{\mathcal{X}^n} = (\mathcal{B}_{\mathcal{X}})^n$, by Theorem D.4, and there is no need to further address this measurability issue. In most of the applications we have in mind, the measure μ is supported on a convex subset of $\mathcal{X}$ that is made into a Polish (and hence, separable) space in the topology induced by $\mathcal{X}$. Consequently, in this setup, for every $m, n \in \mathbb{Z}_+$, $\hat{S}_n^m$ is measurable with respect to $(\mathcal{B}_{\mathcal{X}})^n$.

Let μ_n denote the law induced by $\hat{S}_n$ on $\mathcal{X}$. In view of the preceding discussion, μ_n is a Borel measure as soon as the convex hull of the support of μ is separable. The following (technical) assumption formalizes the conditions required for our approach to Cramér's theorem.

Assumption 6.1.2 *(a)* $\mathcal{X}$ *is a locally convex, Hausdorff, topological real vector space.* $\mathcal{E}$ *is a closed, convex subset of* $\mathcal{X}$ *such that* $\mu(\mathcal{E}) = 1$ *and* $\mathcal{E}$ *can be made into a Polish space with respect to the topology induced by* $\mathcal{X}$.
(b) The closed convex hull of each compact $K \subset \mathcal{E}$ *is compact.*

The following is the extension of Cramér's theorem (Theorem 2.2.3).

Theorem 6.1.3 *Let Assumption 6.1.2 hold. Then* $\{\mu_n\}$ *satisfies in* $\mathcal{X}$ *(and* $\mathcal{E}$*) a weak LDP with rate function* Λ^*. *Moreover, for every open, convex subset* $A \subset \mathcal{X}$,

$$\lim_{n \to \infty} \frac{1}{n} \log \mu_n(A) = - \inf_{x \in A} \Lambda^*(x) \,. \tag{6.1.4}$$

Remarks:

(a) If, instead of part (b) of Assumption 6.1.2, both the exponential tightness of $\{\mu_n\}$ and the finiteness of $\Lambda(\cdot)$ are assumed, then the LDP for $\{\mu_n\}$ is a direct consequence of Corollary 4.6.14.

(b) By Mazur's theorem (Theorem B.13), part (b) of Assumption 6.1.2 follows from part (a) as soon as the metric $d(\cdot, \cdot)$ of $\mathcal{E}$ satisfies, for all $\alpha \in [0,1]$, $x_1, x_2, y_1, y_2 \in \mathcal{E}$, the convexity condition

$$d(\alpha x_1 + (1-\alpha)x_2, \alpha y_1 + (1-\alpha)y_2) \leq \max\{d(x_1, y_1), d(x_2, y_2)\}. \quad (6.1.5)$$

This condition is motivated by the two applications we have in mind, namely, either $\mathcal{X} = \mathcal{E}$ is a separable Banach space, or $\mathcal{X} = M(\Sigma), \mathcal{E} = M_1(\Sigma)$ (see Section 6.2 for details). It is straight forward to verify that (6.1.5) holds true in both cases.

(c) Observe that $\hat{S}_n^m$ are convex combinations of $\{X_\ell\}_{\ell=m}^n$, and hence with probability one belong to $\mathcal{E}$. Consider the sample space $\Omega = \mathcal{E}^{\mathbb{Z}_+}$ of semi-infinite sequences of points in $\mathcal{E}$ with the product topology inherited from the topology of $\mathcal{E}$. Since $\mathcal{E}$ is separable, the Borel σ-field on Ω is $\mathcal{B}_\Omega = (\mathcal{B}_\mathcal{E})^{\mathbb{Z}_+}$, allowing the *semi-infinite* sequence $X_1, \ldots, X_\ell, \ldots$ to be viewed as a random point in Ω, where the latter is equipped with the Borel product measure $\mu^{\mathbb{Z}_+}$, and with $\hat{S}_n$ being measurable maps from $(\Omega, \mathcal{B}_\Omega)$ to $(\mathcal{E}, \mathcal{B}_\mathcal{E})$. This viewpoint turns out to be particularly useful when dealing with Markov extensions of the results of this section.

The following direct corollary of Theorem 6.1.3 for $\mathcal{X} = \mathcal{E} = \mathbb{R}^d$ is a considerable strengthening of Cramér's theorem (Theorem 2.2.30), since it dispenses with the requirement that either $\mathcal{D}_\Lambda = \mathbb{R}^d$ or Λ be steep.

Corollary 6.1.6 *The sequence $\{\mu_n\}$ of the laws of empirical means of $\mathbb{R}^d$-valued i.i.d. random variables satisfies a weak LDP with the convex rate function Λ^*. Moreover, if $0 \in \mathcal{D}_\Lambda^o$, then $\{\mu_n\}$ satisfies the full LDP with the good, convex rate function Λ^*.*

Proof: The weak LDP is merely a specialization of Theorem 6.1.3. If $0 \in \mathcal{D}_\Lambda^o$, the full LDP follows, since $\{\mu_n\} \subset M_1(\mathbb{R}^d)$ is then exponentially tight. (See the inequality (2.2.33) and the discussion there.) $\qquad \square$

The proof of Theorem 6.1.3 is based on the following key lemmas.

Lemma 6.1.7 *Let part (a) of Assumption 6.1.2 hold true. Then, the sequence $\{\mu_n\}$ satisfies the weak LDP in $\mathcal{X}$ with a convex rate function $I(\cdot)$.*

Lemma 6.1.8 *Let Assumption 6.1.2 hold true. Then, for every open, convex subset $A \subset \mathcal{X}$,*

$$\lim_{n \to \infty} \frac{1}{n} \log \mu_n(A) = - \inf_{x \in A} I(x) \,,$$

where $I(\cdot)$ is the convex rate function of Lemma 6.1.7.

We first complete the proof of the theorem assuming the preceding lemmas established, and then devote most of the section to prove them, relying heavily on sub-additivity methods.

Proof of Theorem 6.1.3: As stated in Lemma 6.1.7, $\{\mu_n\}$ satisfies the weak LDP with a convex rate function $I(\cdot)$. Therefore, given Lemma 6.1.8, the proof of the theorem amounts to checking that all the conditions of Theorem 4.5.14 hold here, and hence $I(\cdot) = \Lambda^*(\cdot)$. By the independence of $X_1, X_2, \ldots, X_n$, it follows that for each $\lambda \in \mathcal{X}^*$ and every $t \in \mathbb{R}$, $n \in \mathbb{Z}_+$,

$$\frac{1}{n} \log E \left[e^{n \langle t\lambda, \hat{S}_n \rangle} \right] = \log E \left[e^{t \langle \lambda, X_1 \rangle} \right] = \Lambda_\lambda(t) \,.$$

Consequently, the limiting logarithmic moment generating function is just the function $\Lambda(\cdot)$ given in (6.1.1). Moreover, for every $\lambda \in \mathcal{X}^*$, the function $\Lambda_\lambda(\cdot)$ is the logarithmic moment generating function of the real valued random variable $\langle \lambda, X_1 \rangle$. Hence, by Fatou's lemma, it is lower semicontinuous.

It remains only to establish the inequality

$$\inf_{\{x : (\langle \lambda, x \rangle - a) > 0\}} I(x) \le \inf_{z > a} \Lambda_\lambda^*(z) \,, \quad \forall a \in \mathbb{R}, \ \forall \lambda \in \mathcal{X}^* \,. \tag{6.1.9}$$

To this end, first consider $\lambda = 0$. Then $\Lambda_\lambda^*(z) = 0$ for $z = 0$, and $\Lambda_\lambda^*(z) = \infty$ otherwise. Since $\mathcal{X}$ is an open, convex set with $\mu_n(\mathcal{X}) = 1$ for all n, it follows from Lemma 6.1.8 that $\inf_{x \in \mathcal{X}} I(x) = 0$. Hence, the preceding inequality holds for $\lambda = 0$ and every $a \in \mathbb{R}$. Now fix $\lambda \in \mathcal{X}^*$, $\lambda \ne 0$, and $a \in \mathbb{R}$. Observe that the open half-space $H_a \triangleq \{x : (\langle \lambda, x \rangle - a) > 0\}$ is convex, and therefore, by Lemma 6.1.8,

$$- \inf_{\{x : (\langle \lambda, x \rangle - a) > 0\}} I(x) \ = \ \lim_{n \to \infty} \frac{1}{n} \log \mu_n(H_a)$$

$$\ge \ \sup_{\delta > 0} \limsup_{n \to \infty} \frac{1}{n} \log \mu_n(\overline{H}_{a+\delta}) \,,$$

where the inequality follows from the inclusions $\overline{H}_{a+\delta} \subset H_a$ for all $\delta > 0$. Let $Y_\ell \triangleq \langle \lambda, X_\ell \rangle$, and

$$\hat{Z}_n \triangleq \frac{1}{n} \sum_{\ell=1}^n Y_\ell = \langle \lambda, \hat{S}_n \rangle \,.$$

Note that $\hat{S}_n \in \overline{H}_y$ iff $\hat{Z}_n \in [y, \infty)$. By Cramér's theorem (Theorem 2.2.3), the empirical means $\hat{Z}_n$ of the i.i.d. real valued random variables Y_ℓ satisfy the LDP with rate function $\Lambda_\lambda^*(\cdot)$, and by Corollary 2.2.19,

$$\sup_{\delta > 0} \limsup_{n \to \infty} \frac{1}{n} \log \mu_n(\overline{H}_{a+\delta}) = \sup_{\delta > 0} \left\{ - \inf_{z \geq a+\delta} \Lambda_\lambda^*(z) \right\} = - \inf_{z > a} \Lambda_\lambda^*(z) .$$

Consequently, the inequality (6.1.9) holds, and with all the assumptions of Theorem 4.5.14 verified, the proof is complete. $\square$

The proof of Lemma 6.1.7 is based on sub-additivity, which is defined as follows:

Definition 6.1.10 *A function* $f : \mathbb{Z}_+ \to [0, \infty]$ *is called sub-additive if* $f(n + m) \leq f(n) + f(m)$ *for all* $n, m \in \mathbb{Z}_+$.

Lemma 6.1.11 (Sub-additivity) *If* $f : \mathbb{Z}_+ \to [0, \infty]$ *is a sub-additive function such that* $f(n) < \infty$ *for all* $n \geq N$ *and some* $N < \infty$, *then*

$$\lim_{n \to \infty} \frac{f(n)}{n} = \inf_{n \geq N} \frac{f(n)}{n} < \infty .$$

Proof: Fix $m \geq N$ and let $M_m \triangleq \max\{f(r) : m \leq r \leq 2m\}$. By assumption, $M_m < \infty$. For each $n \geq m \geq N$, let $s = \lfloor n/m \rfloor \geq 1$ and $r = n - m(s - 1) \in \{m, \ldots, 2m\}$. Since f is sub-additive,

$$\frac{f(n)}{n} \leq \frac{(s-1)f(m)}{n} + \frac{f(r)}{n} \leq \frac{(s-1)f(m)}{n} + \frac{M_m}{n} .$$

Clearly, $(s - 1)/n \to 1/m$ as $n \to \infty$, and therefore,

$$\limsup_{n \to \infty} \frac{f(n)}{n} \leq \frac{f(m)}{m} .$$

With this inequality holding for every $m \geq N$, the proof is completed by considering the infimum over $m \geq N$. $\square$

The following observation is key to the application of sub-additivity.

Lemma 6.1.12 *Let part (a) of Assumption 6.1.2 hold true. Then, for every convex* $A \in \mathcal{B}_\mathcal{X}$, *the function* $f(n) \triangleq - \log \mu_n(A)$ *is sub-additive.*

Proof: Without loss of generality, it may be assumed that $A \subset \mathcal{E}$. Now,

$$\hat{S}_{m+n} = \frac{m}{m+n} \hat{S}_m + \frac{n}{m+n} \hat{S}_{m+n}^m .$$

Therefore, $\hat{S}_{m+n}$ is a convex combination (with deterministic coefficients) of the independent random variables $\hat{S}_m$ and $\hat{S}_{m+n}^m$. Thus, by the convexity of A,

$$\{\omega : \hat{S}_{m+n}^m(\omega) \in A\} \cap \{\omega : \hat{S}_m(\omega) \in A\} \subset \{\omega : \hat{S}_{m+n}(\omega) \in A\}.$$

Since, evidently,

$$\mu^{n+m}(\{\omega : \hat{S}_{m+n}^m(\omega) \in A\}) = \mu^n(\{\omega : \hat{S}_n(\omega) \in A\}),$$

it follows that

$$\mu_n(A)\mu_m(A) \leq \mu_{n+m}(A), \tag{6.1.13}$$

or alternatively, $f(n) = -\log\mu_n(A)$ is sub-additive. $\square$

The last tool needed for the proof of Lemma 6.1.7 is the following lemma.

Lemma 6.1.14 *Let part (a) of Assumption 6.1.2 hold true. If $A \subset \mathcal{E}$ is (relatively) open and $\mu_m(A) > 0$ for some m, then there exists an $N < \infty$ such that $\mu_n(A) > 0$ for all $n \geq N$.*

Proof: Let $A \subset \mathcal{E}$ be a fixed open set and suppose that $\mu_m(A) > 0$ for some $m < \infty$. Since $\mathcal{E}$ is a Polish space, by Prohorov's theorem (Theorem D.9), any finite family of probability measures on $\mathcal{E}$ is tight. In particular, there exists a compact $K \subset \mathcal{E}$, such that $\mu_r(K) > 0$, $r = 1, \ldots, m$.

Suppose that every $p \in A$ possesses a neighborhood B_p such that $\mu_m(B_p) = 0$. Because $\mathcal{E}$ has a countable base, there exists a countable union of the sets B_p that covers A. This countable union of sets of μ_m measure zero is also of μ_m measure zero, contradicting the assumption that $\mu_m(A) > 0$. Consequently, there exists a point $p_0 \in A$ such that every neighborhood of p_0 is of positive μ_m measure.

Consider the function $f(a, p, q) = (1-a)p + aq : [0,1] \times \mathcal{E} \times \mathcal{E} \to \mathcal{E}$. (The range of f is in $\mathcal{E}$, since $\mathcal{E}$ is convex.) Because $\mathcal{E}$ is equipped with the topology induced by the real topological vector space $\mathcal{X}$, f is a continuous map with respect to the product topology on $[0,1] \times \mathcal{E} \times \mathcal{E}$. Moreover, for any $q \in \mathcal{E}$, $f(0, p_0, q) = p_0 \in A$, and therefore, there exist $\epsilon_q > 0$, and two neighborhoods (in $\mathcal{E}$), W_q of q and U_q of p_0, such that

$$(1-\epsilon)U_q + \epsilon W_q \subset A, \quad \forall\, 0 \leq \epsilon \leq \epsilon_q.$$

The compact set K may be covered by a finite number of these W_q. Let $\epsilon^* > 0$ be the minimum of the corresponding ϵ_q values, and let U denote the finite intersection of the corresponding U_q. Clearly, U is a neighborhood of p_0, and by the preceding inequality,

$$(1-\epsilon)U + \epsilon K \subset A, \quad \forall\, 0 \leq \epsilon \leq \epsilon^*.$$

Since $\mathcal{E}$ is convex and $\mathcal{X}$ is locally convex, U contains a convex neighborhood, denoted V, of p_0, and by the preceding set inclusion,

$$\mu_n(A) \geq \mu_n((1-\epsilon)V + \epsilon K) \tag{6.1.15}$$

for $n = 1, 2, \ldots$ and for every $0 \leq \epsilon \leq \epsilon^*$. Let $N = m\lceil 1/\epsilon^* \rceil + 1 < \infty$ and represent each $n \geq N$ as $n = ms + r$ with $1 \leq r \leq m$. Since

$$\hat{S}_n = (1 - \frac{r}{n})\hat{S}_{ms} + \frac{r}{n}\hat{S}_n^{ms},$$

and by the choice of N, (6.1.15) holds for $\epsilon = r/n \leq \epsilon^*$, it follows that

$$\begin{aligned}
\mu_n(A) &\geq \mu^n(\{\hat{S}_{ms} \in V, \hat{S}_n^{ms} \in K\}) \\
&= \mu^n(\{\hat{S}_{ms} \in V\})\mu^n(\{\hat{S}_n^{ms} \in K\}) \\
&\geq \mu_{ms}(V)\mu_r(K),
\end{aligned}$$

where the independence of $\hat{S}_{ms}$ and $\hat{S}_n^{ms}$ has been used. Recall that K is such that for $r = 1, 2, \ldots, m$, $\mu_r(K) > 0$, while V is a convex neighborhood of p_0. Hence, $\mu_m(V) > 0$, and by (6.1.13), $\mu_{ms}(V) \geq \mu_m(V)^s > 0$. Thus, $\mu_n(A) > 0$ for all $n \geq N$. □

Proof of Lemma 6.1.7: Fix an open, convex subset $A \subset \mathcal{X}$. Since $\mu_n(A) = \mu_n(A \cap \mathcal{E})$ for all n, either $\mu_n(A) = 0$ for all n, in which case $\mathcal{L}_A = -\lim_{n \to \infty} \frac{1}{n} \log \mu_n(A) = \infty$, or else the limit

$$\mathcal{L}_A = -\lim_{n \to \infty} \frac{1}{n} \log \mu_n(A)$$

exists by Lemmas 6.1.11, 6.1.12, and 6.1.14.

Let $\mathcal{C}^o$ denote the collection of all open, convex subsets of $\mathcal{X}$. Define

$$I(x) \stackrel{\triangle}{=} \sup\{\mathcal{L}_A : x \in A, \; A \in \mathcal{C}^o\}.$$

Applying Theorem 4.1.11 for the base $\mathcal{C}^o$ of the topology of $\mathcal{X}$, it follows that μ_n satisfies the weak LDP with this rate function. To prove that $I(\cdot)$ is convex, we shall apply Lemma 4.1.21. To this end, fix $A_1, A_2 \in \mathcal{C}^o$ and let $A \stackrel{\triangle}{=} (A_1 + A_2)/2$. Then since $(\hat{S}_n + \hat{S}_{2n}^n)/2 = \hat{S}_{2n}$, it follows that

$$\mu_n(A_1)\mu_n(A_2) = \mu^{2n}(\{\omega : \hat{S}_n \in A_1\} \cap \{\omega : \hat{S}_{2n}^n \in A_2\}) \leq \mu_{2n}(A).$$

Thus, by taking n-limits, the convexity condition (4.1.22) is verified, namely,

$$\limsup_{n \to \infty} \frac{1}{n} \log \mu_n(A) \geq \limsup_{n \to \infty} \frac{1}{2n} \log \mu_{2n}(A) \geq -\frac{1}{2}(\mathcal{L}_{A_1} + \mathcal{L}_{A_2}).$$

With (4.1.22) established, Lemma 4.1.21 yields the convexity of I and the proof is complete. □

Proof of Lemma 6.1.8: In the proof of Lemma 6.1.7, it is shown that for any open, convex subset $A \in \mathcal{C}^o$, the limit $\mathcal{L}_A = -\lim_{n \to \infty} \frac{1}{n} \log \mu_n(A)$ exists, and by the large deviations lower bound $\mathcal{L}_A \leq \inf_{x \in A} I(x)$. Consequently, it suffices to show that for every $A \in \mathcal{C}^o$ and all $\delta > 0$,

$$\inf_{x \in A} I(x) \leq \mathcal{L}_A + 2\delta .$$

Fix $A \in \mathcal{C}^o$ and $\delta > 0$. Without loss of generality it may be assumed $\mathcal{L}_A < \infty$. Then there exists an $N \in \mathbb{Z}_+$ such that

$$-\frac{1}{N} \log \mu_N(A \cap \mathcal{E}) = -\frac{1}{N} \log \mu_N(A) \leq \mathcal{L}_A + \delta < \infty .$$

The relatively open set $A \cap \mathcal{E}$ can be made into a Polish space in the topology induced by $\mathcal{E}$. Hence, by Theorem D.7, there exists a compact set $C \subset A \cap \mathcal{E}$ such that

$$-\frac{1}{N} \log \mu_N(C) \leq -\frac{1}{N} \log \mu_N(A) + \delta \leq \mathcal{L}_A + 2\delta .$$

Since A is an open subset of the locally convex (regular) space $\mathcal{X}$, for every $x \in A$ there exists a convex neighborhood B_x such that $\overline{B_x} \subset A$. The compact set C may thus be covered by a finite number of such neighborhoods $\{B_i\}, i = 1, \ldots, k$. By Assumption 6.1.2, $\overline{\text{co}}(C)$ is compact. Since $\overline{B_i}$ are convex, the set

$$\tilde{C} \triangleq \bigcup_{i=1}^{k} (\overline{B_i} \cap \overline{\text{co}}(C)) = \overline{\text{co}}(C) \cap \left(\bigcup_{i=1}^{k} \overline{B_i} \right) ,$$

is a finite union of compact convex sets that contains C. Hence, $K \triangleq \text{co}(\tilde{C})$ contains C and is a closed subset of both A and $\overline{\text{co}}(C)$. Consequently, K is a convex and compact set which satisfies

$$-\frac{1}{N} \log \mu_N(K) \leq -\frac{1}{N} \log \mu_N(C) \leq \mathcal{L}_A + 2\delta .$$

Since K is convex, the function $g(n) = -\log \mu_n(K)$ is sub-additive. Thus,

$$
\begin{aligned}
-\limsup_{n \to \infty} \frac{1}{n} \log \mu_n(K) &\leq \liminf_{n \to \infty} \left[-\frac{1}{nN} \log \mu_{nN}(K) \right] \\
&\leq -\frac{1}{N} \log \mu_N(K) \leq \mathcal{L}_A + 2\delta .
\end{aligned}
$$

The weak LDP upper bound applied to the compact set $K \subset A$ yields

$$\inf_{x \in A} I(x) \leq \inf_{x \in K} I(x) \leq -\limsup_{n \to \infty} \frac{1}{n} \log \mu_n(K) ,$$

and the proof is completed by combining the preceding two inequalities. $\square$

Exercise 6.1.16 (a) Prove that (6.1.4) holds for any finite union of open convex sets.

(b) Construct a probability measure $\mu \in M_1(\mathbb{R}^d)$ and $K \subset \mathbb{R}^d$ compact and convex such that

$$\limsup_{n\to\infty} \frac{1}{n}\log \mu_n(K) > \liminf_{n\to\infty} \frac{1}{n}\log\mu_n(K).$$

Exercise 6.1.17 A function $f : \mathbb{Z}_+ \to [0,\infty]$ is Δ sub-additive if, for all n large enough, $f(n) < \infty$, and moreover, there exists a function $k : \mathbb{Z}_+ \to [0,\infty]$ such that $k(n)/n \to 0$, and for all $n, m \in \mathbb{Z}_+$,

$$f(n + m) \leq f(n) + f(m) + k(n \wedge m).$$

Prove that $\lim_{n\to\infty} f(n)/n$ exists for Δ sub-additive functions.

Exercise 6.1.18 Prove that Lemma 6.1.7 holds when X_i are dependent random variables in $\mathcal{E}$ that satisfy the following (very strong) mixing property: for any $n, m \in \mathbb{Z}_+$ and any $A, B \in \mathcal{B}_{\mathcal{E}}$,

$$\mathrm{P}(\hat{S}_m \in A, \; \hat{S}^m_{m+n} \in B) \geq k(m,n)\mathrm{P}(\hat{S}_m \in A)\mathrm{P}(\hat{S}_n \in B),$$

where $|\log k(m,n)|/(m \wedge n) \xrightarrow[(m\wedge n)\to\infty]{} 0$ and is independent of A and of B.

Hint: See Exercise 6.1.17.

Exercise 6.1.19 Let $(w_t^1, w_t^2, \ldots, w_t^n, \ldots)$ be i.i.d. random variables taking values in $C_0([0,1])$, each distributed according to the Wiener measure (i.e., w^i are independent standard Brownian motions). Note that the measure μ_n on $C_0([0,1])$ defined by $\hat{S}_n = (w_t^1 + w_t^2 + \cdots + w_t^n)/n$ is the same as the measure of $\sqrt{\epsilon}w_t^1$ for $\epsilon = 1/n$. This observation is used here to derive Schilder's theorem as a consequence of Theorem 6.1.3.

(a) Prove that μ_n are exponentially tight.

Hint: Use the compact sets

$$K_\alpha = \{\phi \in C_0([0,1]) : \sup_{0\leq s<t\leq 1} \frac{|\phi(t)-\phi(s)|}{|t-s|^{1/4}} \leq \alpha\},$$

and apply Borell's inequality (5.2.15).

(b) Check that

$$I_w(\phi) = \begin{cases} \frac{1}{2}\int_0^1 \dot{\phi}^2(t)\,dt & , \quad \phi \in H_1 \\ \infty & , \quad \text{otherwise} \end{cases}$$

is a convex, good rate function on $C([0,1])$, and moreover, for every bounded Borel measure ν on $[0,1]$,

$$\sup_{\phi\in C([0,1])} \left\{ \int_0^1 \phi(t)\nu(dt) - I_w(\phi) \right\} = \frac{1}{2}\int_0^1 |\nu((s,1])|^2 ds \stackrel{\triangle}{=} \Lambda(\nu).$$

(c) Verify that

$$\Lambda(\nu) = \frac{1}{2} \int_0^1 \int_0^1 (t \wedge s) \, \nu(dt)\nu(ds)$$

is the limiting logarithmic moment generating function associated with $\{\mu_n\}$, and apply the duality lemma (Lemma 4.5.8) to conclude that $I_w(\cdot) = \Lambda^*(\cdot)$. (Recall that, by Riesz's representation theorem, the topological dual of $C([0,1])$ is the space of all bounded Borel measures on $[0,1]$).
(d) Complete the proof of Schilder's theorem (Theorem 5.2.3).

6.2 Sanov's Theorem

This section is about the large deviations of the empirical law of a sequence of i.i.d. random variables; namely, let Σ be a Polish space and let $Y_1, \ldots, Y_n$ be a sequence of independent, Σ-valued random variables, identically distributed according to $\mu \in M_1(\Sigma)$, where $M_1(\Sigma)$ denotes the space of (Borel) probability measures on Σ. With δ_y denoting the probability measure degenerate at $y \in \Sigma$, the *empirical law* of $Y_1, \ldots, Y_n$ is

$$L_n^{\mathbf{Y}} \stackrel{\triangle}{=} \frac{1}{n} \sum_{i=1}^n \delta_{Y_i} \in M_1(\Sigma). \tag{6.2.1}$$

Sanov's theorem about the large deviations of $L_n^{\mathbf{Y}}$ is proved in Theorem 2.1.10 for a finite set Σ. Here, the general case is considered. The derivation follows two different approaches: first, the LDP with respect to the weak topology is deduced, based on Cramér's theorem (Theorem 6.1.3), and then, independently, the LDP is derived by using the projective limit approach of Section 4.6. The latter yields the LDP with respect to a somewhat stronger topology (the τ-topology). A third derivation based on the direct change of measure argument of Section 4.5.3 is outlined in Exercise 6.2.20.

To set up the framework for applying the results of Section 6.1, let $X_i = \delta_{Y_i}$ and observe that $X_1, \ldots, X_n$ are i.i.d. random variables taking values in the real vector space $M(\Sigma)$ of finite (signed) measures on Σ. Moreover, the empirical mean of $X_1, \ldots, X_n$ is $L_n^{\mathbf{Y}}$ and belongs to $M_1(\Sigma)$, which is a convex subset of $M(\Sigma)$. Hence, our program is to equip $\mathcal{X} = M(\Sigma)$ with an appropriate topology and $M_1(\Sigma) = \mathcal{E}$ with the relative topology induced by $\mathcal{X}$, so that all the assumptions of Cramér's theorem (Theorem 6.1.3) hold and a weak LDP for $L_n^{\mathbf{Y}}$ (in $\mathcal{E}$) follows. A full LDP is then deduced by proving that the laws of $L_n^{\mathbf{Y}}$ are exponentially tight in $\mathcal{E}$, and an explicit formula for the rate function in terms of relative entropy is derived by an auxiliary argument.

To this end, let $C_b(\Sigma)$ denote the collection of bounded continuous functions $\phi : \Sigma \to \mathbb{R}$, equipped with the supremum norm, i.e., $\|\phi\| = \sup_{x \in \Sigma} |\phi(x)|$. Equip $M(\Sigma)$ with the *weak topology* generated by the sets $\{U_{\phi,x,\delta}, \phi \in C_b(\Sigma), x \in \mathbb{R}, \delta > 0\}$, where

$$U_{\phi,x,\delta} \triangleq \{\nu \in M(\Sigma) : |\langle \phi, \nu \rangle - x| < \delta\} , \qquad (6.2.2)$$

and throughout, $\langle \phi, \nu \rangle \triangleq \int_\Sigma \phi d\nu$ for any $\phi \in C_b(\Sigma)$ and any $\nu \in M(\Sigma)$. The Borel σ-field generated by the weak topology is denoted $\mathcal{B}^w$.

Since the collection of linear functionals $\{\nu \mapsto \langle \phi, \nu \rangle : \phi \in C_b(\Sigma)\}$ is separating in $M(\Sigma)$, by Theorem B.8 this topology makes $M(\Sigma)$ into a locally convex, Hausdorff topological vector space, whose topological dual is the preceding collection, hereafter identified with $C_b(\Sigma)$. Moreover, $M_1(\Sigma)$ is a closed subset of $M(\Sigma)$, and $M_1(\Sigma)$ is a Polish space when endowed with the relative topology and the Lévy metric. (See Appendix D for some properties of $M_1(\Sigma)$ and for the definition of the Lévy metric.) Note that the topology thus induced on $M_1(\Sigma)$ corresponds to the weak convergence of probability measures, and that the Lévy metric satisfies the convexity condition (6.1.5).

The preceding discussion leads to the following immediate corollary of Theorem 6.1.3.

Corollary 6.2.3 *The empirical measures L_n^Y satisfy a weak LDP in $M_1(\Sigma)$ (equipped with the weak topology and $\mathcal{B} = \mathcal{B}^w$) with the convex rate function*

$$\Lambda^*(\nu) = \sup_{\phi \in C_b(\Sigma)} \{\langle \phi, \nu \rangle - \Lambda(\phi)\}, \quad \nu \in M_1(\Sigma) , \qquad (6.2.4)$$

where for $\phi \in C_b(\Sigma)$,

$$\Lambda(\phi) \triangleq \log E[e^{\langle \phi, \delta_{Y_1} \rangle}] = \log E[e^{\phi(Y_1)}] = \log \int_\Sigma e^\phi d\mu . \qquad (6.2.5)$$

The strengthening of the preceding corollary to a full LDP with a good rate function $\Lambda^*(\cdot)$ is accomplished by

Lemma 6.2.6 *The laws of L_n^Y of (6.2.1) are exponentially tight.*

Proof: Note that, by Theorem D.7, $\mu \in M_1(\Sigma)$ is tight, and in particular, there exist compact sets $\Gamma_\ell \subset \Sigma$, $\ell = 1, 2, \ldots$ such that

$$\mu(\Gamma_\ell^c) \leq e^{-2\ell^2}(e^\ell - 1) . \qquad (6.2.7)$$

The set of measures

$$K^\ell = \left\{\nu : \nu(\Gamma_\ell) \geq 1 - \frac{1}{\ell}\right\}$$

is closed, since, for any sequence $\{\nu_n\}$ that converges weakly in $M_1(\Sigma)$, $\liminf_{n\to\infty}\nu_n(\Gamma_\ell) \leq \nu(\Gamma_\ell)$ by the Portmanteau theorem (Theorem D.10). For $L = 1, 2, \ldots$ define

$$K_L \overset{\triangle}{=} \bigcap_{\ell=L}^{\infty} K^\ell .$$

Since $\nu(\Gamma_\ell) \geq 1 - \frac{1}{\ell}$ for every $\ell \geq L$, and every $\nu \in K_L$, by Prohorov's theorem (Theorem D.9), each K_L is a compact subset of $M_1(\Sigma)$. Now, by Chebycheff's bound,

$$
\begin{aligned}
P(L_n^{\mathbf{Y}} \notin K^\ell) &= P\left(L_n^{\mathbf{Y}}(\Gamma_\ell^c) > \frac{1}{\ell}\right) \leq E\left[e^{2n\ell^2\left(L_n^{\mathbf{Y}}(\Gamma_\ell^c)-\frac{1}{\ell}\right)}\right] \\
&= e^{-2n\ell}E\left[\exp\left(2\ell^2\sum_{i=1}^{n} 1_{Y_i\in\Gamma_\ell^c}\right)\right] \\
&= e^{-2n\ell}E\left[\exp(2\ell^2 1_{Y_1\in\Gamma_\ell^c})\right]^n \\
&= e^{-2n\ell}\left(\mu(\Gamma_\ell) + e^{2\ell^2}\mu(\Gamma_\ell^c)\right)^n \leq e^{-n\ell},
\end{aligned}
$$

where the last inequality is based on (6.2.7). Hence, using the union of events bound,

$$P(L_n^{\mathbf{Y}} \notin K_L) \leq \sum_{\ell=L}^{\infty}P(L_n^{\mathbf{Y}} \notin K^\ell) \leq \sum_{\ell=L}^{\infty}e^{-n\ell} \leq 2e^{-nL},$$

implying that

$$\limsup_{n\to\infty}\frac{1}{n}\log P(L_n^{\mathbf{Y}} \in K_L^c) \leq -L.$$

Thus, the laws of $L_n^{\mathbf{Y}}$ are exponentially tight. $\square$

Define the *relative entropy* of the probability measure ν with respect to $\mu \in M_1(\Sigma)$ as

$$H(\nu|\mu) \overset{\triangle}{=} \begin{cases} \int_\Sigma f \log f d\mu & \text{if } f \overset{\triangle}{=} \frac{d\nu}{d\mu} \text{ exists} \\ \infty & \text{otherwise}, \end{cases} \tag{6.2.8}$$

where $d\nu/d\mu$ stands for the Radon-Nikodym derivative of ν with respect to μ when it exists.

Remark: The function $H(\nu|\mu)$ is also referred to as *Kullback-Leibler distance* or *divergence* in the literature. It is worth noting that although $H(\nu|\mu)$ is called a distance, it is not a metric, for $H(\nu|\mu) \neq H(\mu|\nu)$. Moreover, even the symmetric sum $(H(\nu|\mu) + H(\mu|\nu))/2$ does not satisfy the triangle inequality.

In view of Sanov's theorem (Theorem 2.1.10) for finite Σ, $H(\cdot|\mu)$ is expected to be the rate function for the LDP associated with $L_n^{\mathbf{Y}}$. This amounts to proving that $\Lambda^*(\cdot) = H(\cdot|\mu)$. Since this identification of Λ^* is also instrumental for establishing the LDP for $L_n^{\mathbf{Y}}$ under stronger topologies on $M_1(\Sigma)$, it will be postponed until Lemma 6.2.13, where a more general setup is considered.

In the rest of this section, the LDP is established in a stronger topology on $M_1(\Sigma)$ than the preceding one, namely, in the τ-topology. The latter is the topology generated by the collection

$$W_{\phi,x,\delta} \stackrel{\triangle}{=} \{\nu \in M_1(\Sigma) : |\langle \phi, \nu \rangle - x| < \delta\}, \qquad (6.2.9)$$

where $x \in \mathbb{R}$, $\delta > 0$ and $\phi \in B(\Sigma)$ – the vector space of all bounded, Borel measurable functions on Σ. Indeed, the collection $\{U_{\phi,x,\delta}\}$ is a subset of the collection $\{W_{\phi,x,\delta}\}$, and hence the τ-topology is finer (stronger) than the weak topology. Unfortunately, $M_1(\Sigma)$ equipped with the τ-topology is neither metrizable nor separable, and thus the results of Section 6.1 do not apply directly. Moreover, the map $\delta_y : \Sigma \to M_1(\Sigma)$ need not be a measurable map from $\mathcal{B}_\Sigma$ to the Borel σ-field induced by the τ-topology. Similarly, $n^{-1}\sum_{i=1}^n \delta_{y_i}$ need not be a measurable map from $(\mathcal{B}_\Sigma)^n$ to the latter σ-field. Thus, a somewhat smaller σ-field will be used.

Definition: *For any $\phi \in B(\Sigma)$, let $p_\phi : M_1(\Sigma) \to \mathbb{R}$ be defined by $p_\phi(\nu) = \langle \phi, \nu \rangle = \int_\Sigma \phi d\nu$. The cylinder σ-field on $M_1(\Sigma)$, denoted $\mathcal{B}^{cy}$, is the smallest σ-field that makes all $\{p_\phi\}$ measurable.*

It is obvious that $L_n^{\mathbf{Y}}$ is measurable with respect to $\mathcal{B}^{cy}$. Moreover, since the weak topology makes $M_1(\Sigma)$ into a Polish space, $\mathcal{B}^w$ is the cylinder σ-field generated by $C_b(\Sigma)$, and as such it equals $\mathcal{B}^{cy}$. (For details, see Exercise 6.2.18.)

We next present a derivation of the LDP based on a projective limit approach, which may also be viewed as an alternative to the proof of Sanov's theorem for the weak topology via Cramér's theorem.

Theorem 6.2.10 (Sanov) *Let $\mathcal{B} = \mathcal{B}^{cy}$. The empirical measures $L_n^{\mathbf{Y}}$ satisfy the LDP (1.2.4) in $M_1(\Sigma)$ equipped with the τ-topology (and hence, also in the weak topology) with the good, convex rate function $H(\cdot|\mu)$.*

A few definitions and preliminary lemmas are presented first, culminating with the proof of this theorem. Let $\mathcal{X}$ denote the algebraic dual of $B(\Sigma)$ equipped with the $B(\Sigma)$-topology. Observe that $\mathcal{X}$ satisfies Assumption 4.6.7. Moreover, identifying probability measures with points in $\mathcal{X}$ via the map $\langle \phi, \nu \rangle = \int_\Sigma \phi d\nu$, it follows that the τ-topology is the relative topology induced on $M_1(\Sigma)$ by $\mathcal{X}$. However, $M_1(\Sigma)$ is neither an open nor a closed subset of $\mathcal{X}$. The definition (6.2.5) of $\Lambda(\phi)$ extends to $\phi \in B(\Sigma)$ in the obvious way, namely, $\Lambda(\phi) = \log \int_\Sigma e^\phi d\mu$ for all $\phi \in B(\Sigma)$. Define now $\Lambda^* : \mathcal{X} \to [0, \infty]$ via

$$\Lambda^*(\omega) = \sup_{\phi \in B(\Sigma)} \{\langle \phi, \omega \rangle - \Lambda(\phi)\}, \quad \omega \in \mathcal{X}, \qquad (6.2.11)$$

where $\langle \phi, \omega \rangle$ is the value of the linear functional $\omega : B(\Sigma) \to \mathbb{R}$ at the point ϕ. Note that the preceding definition, motivated by the definition used in Section 4.6, may in principle yield a function that is different from the one defined in (6.2.4). This is not the case (c.f. Lemma 6.2.13), and thus, with a slight abuse of notation, Λ^* denotes both functions.

Lemma 6.2.12 *Extend $H(\cdot|\mu)$ to $\mathcal{X}$ by setting $H(\cdot|\mu) = \infty$ outside $M_1(\Sigma)$. Then $H(\cdot|\mu)$ is a convex, good rate function on $\mathcal{X}$, and for all $\alpha < \infty$, $H(\cdot|\mu)$ is strictly convex on the compact, convex sets $\{\nu : H(\nu|\mu) \leq \alpha\}$.*

The proof of this lemma is deferred to the end of the section. The following identification of the rate function $\Lambda^*(\cdot)$ is a consequence of Lemma 6.2.12 and the duality lemma (Lemma 4.5.8).

Lemma 6.2.13 *The identity $H(\cdot|\mu) = \Lambda^*(\cdot)$ holds over $\mathcal{X}$. Moreover, the definitions (6.2.11) and (6.2.4) yield the same function over $M_1(\Sigma)$.*

Proof: Observe that $\mathcal{X}$ is, by Theorem B.8, a locally convex Hausdorff topological vector space whose (topological) dual $\mathcal{X}^*$ is $B(\Sigma)$. By combining Lemma 6.2.12 with the duality lemma (Lemma 4.5.8) (for $f = H(\cdot|\mu)$), the identity $H(\cdot|\mu) = \Lambda^*(\cdot)$ is obtained if for all $\phi \in B(\Sigma)$,

$$\Lambda(\phi) = \sup_{\omega \in \mathcal{X}} \{\langle \phi, \omega \rangle - H(\omega|\mu)\} . \qquad (6.2.14)$$

Note that $H(\omega|\mu) < \infty$ only for $\omega \in M_1(\Sigma)$ with density f with respect to μ, in which case $H(\omega|\mu) = \int_\Sigma f \log f d\mu$. Therefore, (6.2.14) is just

$$\log \int_\Sigma e^\phi d\mu = \sup_{f \in L_1(\mu),\, f \geq 0\ \mu-\text{a.e.},\, \int_\Sigma f d\mu = 1} \left\{ \int_\Sigma \phi f d\mu - \int_\Sigma f \log f d\mu \right\} . \qquad (6.2.15)$$

Choosing $f = e^\phi / \int_\Sigma e^\phi d\mu$, it can easily be checked that the left side in (6.2.15) can not exceed the right side. On the other hand, for any $f \geq 0$ such that $\int_\Sigma f d\mu = 1$, by Jensen's inequality,

$$\log \int_\Sigma e^\phi d\mu \geq \log \int_\Sigma 1_{\{f>0\}} \frac{e^\phi}{f} f d\mu \geq \int_\Sigma 1_{\{f>0\}} \log(e^\phi/f) f d\mu ,$$

and the proof of (6.2.15) is complete.

Fix $\nu \in M_1(\Sigma)$, $\phi \in B(\Sigma)$ and recall that there exist $\phi_n \in C_b(\Sigma)$ such that $\phi_n \to \phi$ both in $L_1(\mu)$ and in $L_1(\nu)$. A uniformly bounded sequence of such approximations results by truncating each ϕ_n to the bounded range of ϕ. Consequently, there exists a sequence $\{\phi_n\} \subset C_b(\Sigma)$ such that

$$\lim_{n \to \infty} \left[\int_\Sigma \phi_n d\nu - \log \int_\Sigma e^{\phi_n} d\mu \right] = \int_\Sigma \phi d\nu - \log \int_\Sigma e^\phi d\mu .$$

Since $\nu \in M_1(\Sigma)$ and $\phi \in B(\Sigma)$ are arbitrary, the definitions (6.2.4) and (6.2.11) coincide over $M_1(\Sigma)$. □

An important consequence of Lemma 6.2.13 is the "obvious fact" that $\mathcal{D}_{\Lambda^*} \subset M_1(\Sigma)$ (i.e., $\Lambda^*(\omega) = \infty$ for $\omega \in M_1(\Sigma)^c$), allowing for the application of the projective limit approach.

Proof of Theorem 6.2.10: The proof is based on applying part (a) of Corollary 4.6.11 for the sequence of random variables $\{L_n^\mathbf{Y}\}$, taking values in the topological vector space $\mathcal{X}$ (with ϵ replaced here by $1/n$). Indeed, $\mathcal{X}$ satisfies Assumption 4.6.8 for $\mathcal{W} = B(\Sigma)$ and $\mathcal{B} = \mathcal{B}^{cy}$, and $\Lambda(\cdot)$ of (4.6.12) is given here by $\Lambda(\phi) = \log \int_\Sigma e^\phi d\mu$. Note that this function is finite everywhere (in $B(\Sigma)$). Moreover, for every $\phi, \psi \in B(\Sigma)$, the function $\Lambda(\phi + t\psi) : \mathbb{R} \to \mathbb{R}$ is differentiable at $t = 0$ with

$$\frac{d}{dt} \Lambda(\phi + t\psi) \Big|_{t=0} = \frac{\int_\Sigma \psi e^\phi d\mu}{\int_\Sigma e^\phi d\mu}$$

(c.f. the statement and proof of part (c) of Lemma 2.2.5). Thus, $\Lambda(\cdot)$ is Gateaux differentiable. Consequently, all the conditions of part (a) of Corollary 4.6.11 are satisfied, and hence the sequence $\{L_n^\mathbf{Y}\}$ satisfies the LDP in $\mathcal{X}$ with the convex, good rate function $\Lambda^*(\cdot)$ of (6.2.11). In view of Lemma 6.2.13, this rate function is actually $H(\cdot|\mu)$, and moreover, $\mathcal{D}_{\Lambda^*} \subset M_1(\Sigma)$. Therefore, by Lemma 4.1.5, the same LDP holds in $M_1(\Sigma)$ when equipped with the topology induced by $\mathcal{X}$. As mentioned before, this is the τ-topology, and the proof of the theorem is complete. □

The next lemma implies Lemma 6.2.12 as a special case.

Lemma 6.2.16 *Suppose γ is a convex, good rate function on $\mathbb{R}$ such that $\gamma(x)/|x| \to \infty$ for $|x| \to \infty$. Then,*

$$I_\gamma(\nu) \triangleq \begin{cases} \int_\Sigma \gamma(f) d\mu & \text{if } f \triangleq \frac{d\nu}{d\mu} \text{ exists} \\ \infty & \text{otherwise}, \end{cases}$$

is a convex, good rate function on $M(\Sigma)$ equipped with the $B(\Sigma)$-topology.

Proof: Since $\gamma(\cdot) \geq 0$, also $I_\gamma(\cdot) \geq 0$. Fix an $\alpha < \infty$ and consider the set

$$\Psi_I(\alpha) \stackrel{\triangle}{=} \{f \in L_1(\mu) : \int_\Sigma \gamma(f)d\mu \leq \alpha\}.$$

Assume $f_n \in \Psi_I(\alpha)$ and $f_n \to f$ in $L_1(\mu)$. Passing to a subsequence along which $f_n \to f$ μ-a.e, by Fatou's lemma and lower semicontinuity of γ, also

$$\alpha \geq \liminf_{n \to \infty} \int_\Sigma \gamma(f_n)d\mu \geq \int_\Sigma (\liminf_{n \to \infty} \gamma(f_n))d\mu \geq \int_\Sigma \gamma(f)d\mu .$$

Consequently, the convex set $\Psi_I(\alpha)$ is closed in $L_1(\mu)$, and hence also weakly closed in $L_1(\mu)$ (see Theorem B.9). Since $\Psi_I(\alpha)$ is a uniformly integrable, bounded subset of $L_1(\mu)$, by Theorem C.7, it is weakly sequentially compact in $L_1(\mu)$, and by the Eberlein–Šmulian theorem (Theorem B.12), also weakly compact in $L_1(\mu)$. The mapping $\nu \mapsto d\nu/d\mu$ is a homeomorphism between the level set $\{\nu : I_\gamma(\nu) \leq \alpha\}$, equipped with the $B(\Sigma)$-topology, and $\Psi_I(\alpha)$, equipped with the weak topology of $L_1(\mu)$. Therefore, all level sets of $I_\gamma(\cdot)$, being the image of $\Psi_I(\alpha)$ under this homeomorphism, are compact. Fix $\nu_1 \neq \nu_2$ such that $I_\gamma(\nu_j) \leq \alpha$. Then, there exist $f_j \in L_1(\mu)$, distinct μ-a.e., such that for all $t \in [0,1]$,

$$I_\gamma(t\nu_1 + (1-t)\nu_2) = \int_\Sigma \gamma(tf_1 + (1-t)f_2)d\mu ,$$

and by the convexity of $\gamma(\cdot)$

$$I_\gamma(t\nu_1 + (1-t)\nu_2) \leq tI_\gamma(\nu_1) + (1-t)I_\gamma(\nu_2) .$$

Hence, $I_\gamma(\cdot)$ is a convex function, since all its level sets are convex. $\qquad\square$

Proof of Lemma 6.2.12: Consider the good rate function $\gamma(x) = x\log x - x + 1$ when $x \geq 0$, and $\gamma(x) = \infty$ otherwise. Since

$$\{\omega : H(\omega|\mu) \leq \alpha\} = \{\nu \in M(\Sigma) : I_\gamma(\nu) \leq \alpha\} \bigcap M_1(\Sigma) ,$$

all level sets of $H(\cdot|\mu)$ are convex, compact subsets of $M(\Sigma) \subset \mathcal{X}$. To complete the proof of Lemma 6.2.12 note that the strict convexity of $\gamma(x)$ on $[0,\infty)$ implies the strict convexity of $H(\cdot|\mu)$ on its level sets. $\qquad\square$

Exercise 6.2.17 Prove that $H(\nu|\mu) \geq \frac{1}{2} \parallel \nu - \mu \parallel_{var}^2$, where

$$\parallel m \parallel_{var} \stackrel{\triangle}{=} \sup_{u \in B(\Sigma), \parallel u \parallel \leq 1} \langle u, m \rangle$$

denotes the variational norm of $m \in M(\Sigma)$.

Hint: (a) Applying Jensen's inequality to the convex function $x \log x$, show that for every $\nu, \mu \in M_1(\Sigma)$ and every measurable $A \subset \Sigma$,

$$H(\nu|\mu) \geq \nu(A) \log \frac{\nu(A)}{\mu(A)} + (1 - \nu(A)) \log \frac{1 - \nu(A)}{1 - \mu(A)} \geq 2(\nu(A) - \mu(A))^2.$$

(b) Show that whenever $d\nu/d\mu = f$ exists, $\| \nu - \mu \|_{var} = 2(\nu(A) - \mu(A))$ for $A = \{x : f(x) \geq 1\}$.

Exercise 6.2.18 [From [BolS89], Lemma 2.1.]
Prove that on Polish spaces, $\mathcal{B}^w = \mathcal{B}^{cy}$.

Exercise 6.2.19 In this exercise, the non-asymptotic upper bound of part (b) of Exercise 4.5.5 is specialized to the setup of Sanov's theorem in the weak topology. This is done based on the computation of the *metric entropy* of $M_1(\Sigma)$ with respect to the Lévy metric. (See Theorem D.8.) Throughout this exercise, Σ is a compact metric space, whose metric entropy is denoted by $m(\Sigma, \delta)$, i.e., $m(\Sigma, \delta)$ is the minimal number of balls of radius δ that cover Σ.
(a) Use a net in Σ of size $m(\Sigma, \delta)$ corresponding to the centers of the balls in such a cover. Show that any probability measure in $M_1(\Sigma)$ may be approximated to within δ in the Lévy metric by a weighted sum of atoms located on this net, with weights that are integer multiples of $1/K(\delta)$, where $K(\delta) = \lceil m(\Sigma, \delta)/\delta \rceil$. Check that the number of such combinations with weights in the simplex (i.e., are nonnegative and sum to one) is

$$\binom{K(\delta) + m(\Sigma, \delta) - 1}{K(\delta)},$$

and show that

$$m(M_1(\Sigma), \delta) \leq \left(\frac{4}{\delta}\right)^{m(\Sigma, \delta)},$$

where $m(M_1(\Sigma), \delta)$ denotes the minimal number of balls of radius δ (in the Lévy metric) that cover $M_1(\Sigma)$.
Remark: This bound is quite tight. (For details, see [ZK95].)
(b) Check that for every $n \in \mathbb{Z}_+$ and every measurable $A \subset M_1(\Sigma)$, part (b) of Exercise 4.5.5 specializes to the upper bound

$$P(L_n^{\mathbf{Y}} \in A) \leq \inf_{\delta > 0} \left\{ m(M_1(\Sigma), \delta) \exp\left(-n \inf_{\nu \in A^\delta} H(\nu|\mu)\right) \right\},$$

where A^δ is the closed δ-blowup of A with respect to the Lévy metric.

Exercise 6.2.20 In this exercise, an alternative derivation of Sanov's theorem in the weak topology is provided based on Baldi's theorem (Theorem 4.5.20).
(a) Suppose that ν is such that $f = d\nu/d\mu$ exists, $f \in C_b(\Sigma)$, and f is

bounded away from zero. Use Lemma 6.2.13 to prove that ν is an exposed point of the function $\Lambda^*(\cdot)$ of (6.2.4), with $\lambda = \log f \in C_b(\Sigma)$ being the exposing hyperplane.

Hint: Observe that $H(\tilde{\nu}|\mu) - \langle \lambda, \tilde{\nu} \rangle = \int_\Sigma g \log(g/f) d\mu \geq 0$ if $H(\tilde{\nu}|\mu) < \infty$, where $g = d\tilde{\nu}/d\mu$.

(b) Show that the continuous, bounded away from zero and infinity μ-densities are dense in $\{f \in L_1(\mu) : f \geq 0 \ \mu - \text{a.e.}, \int_\Sigma f d\mu = 1, \int_\Sigma f \log f d\mu \leq \alpha\}$ for any $\alpha < \infty$. Conclude that (4.5.22) holds for any open set $G \subset M_1(\Sigma)$. Complete the proof of Sanov's theorem by applying Lemma 6.2.6 and Theorem 4.5.20.

Exercise 6.2.21 In this exercise, the compact sets K_L constructed in the proof of Lemma 6.2.6 are used to prove a version of Cramér's theorem in separable Banach spaces. Let $X_1, X_2, \ldots, X_n$ be a sequence of i.i.d. random variables taking values in a separable Banach space $\mathcal{X}$, with each X_i distributed according to $\mu \in M_1(\mathcal{X})$. Let μ_n denote the law of $\hat{S}_n = \frac{1}{n} \sum_{i=1}^n X_i$. Since $\mathcal{X}$ is a Polish space, by Theorem 6.1.3, $\{\mu_n\}$, as soon as it is exponentially tight, satisfies in $\mathcal{X}$ the LDP with the good rate function $\Lambda^*(\cdot)$. Assume that for all $\alpha < \infty$,

$$g(\alpha) \triangleq \log \int_{\mathcal{X}} e^{\alpha \|x\|} \mu(dx) < \infty,$$

where $\| \cdot \|$ denotes the norm of $\mathcal{X}$. Let Q_n denote the law of $L_n^{\mathbf{X}} = n^{-1} \sum_{i=1}^n \delta_{X_i}$, and define

$$\Gamma_L = \{\nu : \int_{\mathcal{X}} g^*(\| x \|) \nu(dx) \leq L\}, \quad L \in [0, \infty),$$

where $g^*(\cdot)$ is the Fenchel–Legendre transform of $g(\cdot)$.

(a) Recall that by (5.1.16), $\int_{\mathcal{X}} e^{\delta g^*(\|x\|)} \mu(dx) \leq 2/(1-\delta)$ for all $\delta < 1$. Using this and Chebycheff's inequality, conclude that, for all $\delta < 1$,

$$Q_n(\Gamma_L^c) \leq \left(\frac{2}{1-\delta}\right)^n e^{-n\delta L}. \qquad (6.2.22)$$

(b) Define the *mean* of ν, denoted $m(\nu)$, as the unique element $m(\nu) \in \mathcal{X}$ such that, for all $\lambda \in \mathcal{X}^*$,

$$\langle \lambda, m(\nu) \rangle = \int_{\mathcal{X}} \langle \lambda, x \rangle \nu(dx).$$

Show that $m(\nu)$ exists for every $\nu \in M_1(\mathcal{X})$ such that $\int_{\mathcal{X}} \| x \| d\nu < \infty$.

Hint: Let $\{A_i^{(n)}\}_{i=1}^\infty$ be partitions of $\mathcal{X}$ by sets of diameter at most $1/n$, such that the nth partition is a refinement of the $(n-1)$st partition. Show that $m_n = \sum_{i=1}^\infty x_i^{(n)} \nu(A_i^{(n)})$ is a well-defined Cauchy sequence as long as

$x_i^{(n)} \in A_i^{(n)}$, and let $m(\nu)$ be the limit of one such sequence.

(c) Prove that $m(\nu)$ is continuous with respect to the weak topology on the closed set Γ_L.

Hint: That Γ_L is closed follows from Theorem D.12. Let $\nu_n \in \Gamma_L$ converge to $\nu \in \Gamma_L$. Let $h_r : \mathbb{R} \to [0,1]$ be a continuous function such that $h_r(z) = 1$ for all $z \leq r$ and $h_r(z) = 0$ for all $z \geq r+1$. Then

$$\| m(\nu_n) - m(\nu) \| = \sup_{\lambda \in \mathcal{X}^*, \|\lambda\|_{\mathcal{X}^*} \leq 1} \{ \langle \lambda, m(\nu_n) \rangle - \langle \lambda, m(\nu) \rangle \}$$

$$\leq \left[\int_{\|x\| \geq r} \| x \| \, \nu_n(dx) + \int_{\|x\| \geq r} \| x \| \, \nu(dx) \right]$$

$$+ \sup_{\|\lambda\|_{\mathcal{X}^*} \leq 1} \left\{ \int_{\mathcal{X}} \langle \lambda, x \rangle h_r(\| x \|) \nu_n(dx) - \int_{\mathcal{X}} \langle \lambda, x \rangle h_r(\| x \|) \nu(dx) \right\}.$$

$$(6.2.23)$$

Use Theorem D.11 to conclude that, for each r fixed, the second term in (6.2.23) converges to 0 as $n \to \infty$. Show that the first term in (6.2.23) may be made arbitrarily small by an appropriate choice of r, since by Lemma 2.2.20, $\lim_{r \to \infty} (g^*(r)/r) = \infty$.

(d) Let K_L be the compact subsets of $M_1(\mathcal{X})$ constructed in Lemma 6.2.6. Define the following subsets of $\mathcal{X}$,

$$C_L \triangleq \{ m(\nu) : \nu \in K_L \bigcap \Gamma_{2L} \}.$$

Using (c), show that C_L is a compact subset of $\mathcal{X}$. Complete the proof of the exponential tightness using (6.2.22).

Exercise 6.2.24 [From [EicG98].]
This exercise demonstrates that exponential tightness (or in fact, tightness) might not hold in the context of Sanov's theorem for the τ-topology. Throughout, Σ denotes a Polish space.

(a) Suppose $K \subset M_1(\Sigma)$ is compact in the τ-topology (hence, compact also in the weak topology). Show that K is τ-sequentially compact.

(b) Prove that for all $\epsilon > 0$, there exist $\delta > 0$ and a probability measure $\nu_\epsilon \in M_1(\Sigma)$ such that for any $\Gamma \in \mathcal{B}_\Sigma$,

$$\nu_\epsilon(\Gamma) < \delta \quad \Longrightarrow \quad \mu(\Gamma) < \epsilon, \quad \forall \mu \in K.$$

Hint: Assume otherwise, and for some $\epsilon > 0$ construct a sequence of sets $\Gamma_\ell \in \mathcal{B}_\Sigma$ and probability measures $\mu_\ell \in K$ such that $\mu_j(\Gamma_\ell) < 2^{-\ell}$, $j = 1, \ldots, \ell$, and $\mu_{\ell+1}(\Gamma_\ell) > \epsilon$. Let $\nu \triangleq \sum_{\ell=1}^{\infty} 2^{-\ell} \mu_\ell$ so that each μ_ℓ is absolutely continuous with respect to ν, and check that $\nu(\Gamma_n) \to 0$ as $n \to \infty$. Pass to a subsequence μ_{ℓ_i}, $i = 1, 2, \ldots$, that converges in the τ-topology, and recall the Vitali-Hahn-Saks theorem [DunS58, Theorem III.7.2]: $\nu(\Gamma_n) \to_{n \to \infty} 0$

implies that $\sup_i \mu_{\ell_i}(\Gamma_n) \to 0$ as $n \to \infty$, a contradiction to the fact that $\mu_{n+1}(\Gamma_n) > \epsilon$ for all n.

(c) Let $\mathrm{Atm}(\nu)$ denote the set of atoms of a measure ν. Prove that the set $A_K \triangleq \cup_{\nu \in K} \mathrm{Atm}(\nu)$ is at most countable.

Hint: Show that any $\nu \in K$ is absolutely continuous with respect to the probability measure $\sum_{k=1}^{\infty} 2^{-k} \nu_{1/k}$.

(d) Let $Y_1, \ldots, Y_n$ be a sequence of independent random variables identically distributed according to a probability measure μ that possesses no atoms. Show that for any $K \subset M_1(\Sigma)$ compact in the τ-topolgy, and $n = 1, 2, \ldots$,

$$P(L_n^{\mathbf{Y}} \in K) \leq P(\{Y_1, \ldots, Y_n\} \subseteq A_K) = 0.$$

Exercise 6.2.25 [From [Scd98].]

The purpose of this exercise is to show that for any measurable $\phi : \Sigma \to \mathbb{R}_+$ and any $\alpha \in (0, \infty)$, the mapping

$$\nu \mapsto p_\phi(\nu) = \int_\Sigma \phi \, d\nu$$

is continuous on $\{\nu : H(\nu|\mu) \leq \alpha\}$ equipped with the topology induced by the τ-topology (in short, is τ-continuous), if and only if $\int_\Sigma \exp(\lambda\phi)d\mu < \infty$ for all $\lambda \in \mathbb{R}_+$.

(a) Show that $p_\phi(\cdot)$ is τ-continuous on compact $K \subset M_1(\Sigma)$ if and only if $\sup_{\nu \in K} p_{\phi - \phi_n}(\nu) \to 0$ as $n \to \infty$, where $\phi_n = \phi \wedge n$.

Hint: A uniform limit of τ-continuous functions on K compact is continuous.

(b) Show that for $\psi : \Sigma \to \mathbb{R}_+$ measurable, $\nu, \mu \in M_1(\Sigma)$ and $\lambda > 0$

$$\lambda^{-1} \log \int_\Sigma e^{\lambda\psi} d\mu + \lambda^{-1} H(\nu|\mu) \geq \int_\Sigma \psi \, d\nu . \qquad (6.2.26)$$

Fix $\alpha < \infty$ and assume that $\int_\Sigma \exp(\lambda\phi)d\mu < \infty$ for all $\lambda \in \mathbb{R}_+$. Show that p_ϕ is then τ-continuous on $\{\nu : H(\nu|\mu) \leq \alpha\}$.

Hint: Apply (6.2.26) for $\psi = \phi - \phi_n$. Take $n \to \infty$ followed by $\lambda \to \infty$.

(c) Hereafter, let $\psi = \phi_m - \phi_n$ for $m \geq n$. Note that $g_m(\lambda) \triangleq \Lambda(\lambda\psi)$: $\mathbb{R}_+ \to \mathbb{R}_+$ is a convex, non-decreasing, C^∞ function, such that $g_m(\lambda) \to g_\infty(\lambda) \triangleq \Lambda(\lambda(\phi - \phi_n))$ for $m \to \infty$. For $\nu_\lambda^m \in M_1(\Sigma)$ such that $d\nu_\lambda^m/d\mu = \exp(\lambda\psi - \Lambda(\lambda\psi))$, check that $H(\nu_\lambda^m|\mu) = \lambda g_m'(\lambda) - g_m(\lambda)$ is a continuous, non-decreasing function on $\mathbb{R}_+$, with $H(\nu_0^m|\mu) = 0$.

(d) Suppose $\int_\Sigma \exp(\lambda\phi)d\mu$ is finite for $\lambda < \lambda^*$ and infinite for $\lambda > \lambda^*$, some $\lambda^* = \lambda^*(\phi) \in (0, \infty)$. Show that $H(\nu_{\lambda_m}^m|\mu) = \alpha$ for some $m = m(n) \geq n$ and $\lambda_m = \lambda_m(n) \leq 2\lambda^*$.

Hint: $\lambda g_m'(\lambda) \geq 2(g_m(\lambda) - g_m(\lambda/2))$ implies that $H(\nu_\lambda^m|\mu) \to \infty$ for $m \to \infty$ and $\lambda \in (\lambda^*, 2\lambda^*)$.

(e) Since $\int_\Sigma \psi \, d\nu_\lambda^m = \Lambda'(\lambda\psi) \geq \lambda^{-1} H(\nu_\lambda^m|\mu)$, for $m \geq n$ and $\lambda_m \leq 2\lambda^*$ as in part (d),

$$\int_\Sigma (\phi - \phi_n) d\nu_{\lambda_m}^m \geq \int_\Sigma \psi \, d\nu_{\lambda_m}^m \geq \lambda_m^{-1}\alpha \geq C^{-1}\alpha > 0.$$

Conclude by part (a) that, then, $p_\phi(\nu)$ is not τ-continuous on the level set $\{\nu : H(\nu|\mu) \le \alpha\}$.

(f) In case $\lambda^*(\phi) = 0$, construct $\hat\phi \le \phi$ for which $\lambda^*(\hat\phi) > 0$. By part (e), $p_{\hat\phi}(\cdot)$ is not τ-continuous on $\{\nu : H(\nu|\mu) \le \alpha\}$. Use part (a) to conclude that $p_\phi(\cdot)$ cannot be τ-continuous on this set.

Exercise 6.2.27 In this exercise, Sanov's theorem provides the LDP for U-statistics and V-statistics.

(a) For Σ Polish and $k \ge 2$ integer, show that the function $\mu \mapsto \mu^k : M_1(\Sigma) \to M_1(\Sigma^k)$ is continuous when both spaces are equipped with the weak topology. **Hint:** See Lemma 7.3.12.

(b) Let $Y_1, \ldots, Y_n$ be a sequence of i.i.d. Σ-valued random variables of law $\mu \in M_1(\Sigma)$. Fix $k \ge 2$ and let $n_{(k)} = n!/(n-k)!$. Show that $V_n = (L_n^{\mathbf{Y}})^k$ and

$$U_n = n_{(k)}^{-1} \sum_{1 \le i_1 \ne \cdots \ne i_k \le n} \delta_{Y_{i_1}, \ldots, Y_{i_k}}$$

satisfy the LDP in $M_1(\Sigma^k)$ equipped with the weak topology, with the good rate function $I_k(\nu) = H(\nu_1|\mu)$ for $\nu = \nu_1^k$ and $I_k(\nu) = \infty$ otherwise. **Hint:** Show that $\{V_n\}$ and $\{U_n\}$ are exponentially equivalent.

(c) Show that the U-statistics $n_{(k)}^{-1} \sum_{1 \le i_1 \ne \cdots \ne i_k \le n} h(Y_{i_1}, \ldots, Y_{i_k})$, and the V-statistics $n^{-k} \sum_{1 \le i_1, \ldots, i_k \le n} h(Y_{i_1}, \ldots, Y_{i_k})$, satisfy the LDP in $\mathbb{R}$ when $h \in C_b(\Sigma^k)$, with the good rate function $I(x) = \inf\{H(\nu|\mu) : \int h d\nu^k = x\}$.

Remark: Note that $\nu \mapsto \nu^k$ is not continuous with respect to the τ-topology (see Exercise 7.3.18). Using either projective limits or generalized exponential approximations, [SeW97] and [EicS97] extend the results of parts (b) and (c) to the τ-topology and even to unbounded h satisfying certain exponential moment conditions.

Exercise 6.2.28 Let $I_\gamma(\cdot)$ be as in Lemma 6.2.16 with $\inf_x \gamma(x) < \infty$. For $\Sigma = \Sigma_1 \times \Sigma_2$ the product of two Polish spaces, this exercise provides a dual characterization of the existence of $\nu \in M(\Sigma)$ of specified marginals ν_i such that $I_\gamma(\nu) < \infty$.

(a) Check that for any $\phi \in C_b(\Sigma)$,

$$\int \gamma^*(\phi) d\mu = \sup_{\nu \in M(\Sigma)} \left(\int_\Sigma \phi d\nu - I_\gamma(\nu) \right).$$

Hint: $\gamma^*(\phi) \in C_b(\Sigma)$.

(b) Let $\mathcal{X} = M(\Sigma_1) \times M(\Sigma_2)$ equipped with the product of the corresponding C_b-topologies, and $p : M(\Sigma) \to \mathcal{X}$ denote the projection operator. Show that $I'_\gamma(\nu_1, \nu_2) = \inf\{I_\gamma(\nu) : \nu \in p^{-1}(\nu_1, \nu_2)\}$ is a convex, good rate function on the space $\mathcal{X}$. **Hint:** Use Lemma 6.2.16 and part (a) of Theorem 4.2.1.

(c) Let $\mathcal{Y} = \{\phi(y_1, y_2) = \phi_1(y_1) + \phi_2(y_2) : \phi_i \in C_b(\Sigma_i), i = 1, 2\}$. Note that the $\mathcal{Y}$ topology of $\mathcal{X}$ coincides with the original one, hence $\mathcal{Y} = \mathcal{X}^*$ by Theorem B.8, and that $\mathcal{Y}$ is a subset of $C_b(\Sigma)$. Show that for every $\phi \in \mathcal{Y}$,

$$\int_\Sigma \gamma^*(\phi) d\mu = \sup_{(\nu_1, \nu_2) \in \mathcal{X}} \left(\int_{\Sigma_1} \phi_1 d\nu_1 + \int_{\Sigma_2} \phi_2 d\nu_2 - I'_\gamma(\nu_1, \nu_2) \right),$$

and conclude by the duality lemma (Lemma 4.5.8) that for any $\nu_i \in M(\Sigma_i)$, $i = 1, 2$

$$I'_\gamma(\nu_1, \nu_2) = \sup_{\phi \in \mathcal{X}^*} \left(\int_{\Sigma_1} \phi_1 d\nu_1 + \int_{\Sigma_2} \phi_2 d\nu_2 - \int_\Sigma \gamma^*(\phi) d\mu \right).$$

6.3 LDP for the Empirical Measure—The Uniform Markov Case

This section presents a quick derivation, based on sub-additivity, of the LDP for a class of Markov chains that satisfy some strong uniformity conditions. The identification of the rate function is postponed to Section 6.5. The reader interested in the LDP for the class of Markov chains discussed here may skip directly to Theorem 6.5.4 after having read this section.

The results derived here are in the setup of discrete time Markov chains. The extension to continuous time is technically more involved but follows the same ideas, and the reader is referred to [DeuS89b] for the details. Alternatively the continuous time case can be handled by adapting the approach of Exercise 4.2.28.

Let Σ be a Polish space, and let $M_1(\Sigma)$ denote the space of Borel probability measures on Σ equipped with the Lévy metric, making it into a Polish space with convergence compatible with the weak convergence. $M(\Sigma)$ is the space of (signed) finite measures equipped with the weak convergence, which makes it into a locally convex Hausdorff topological vector space. Let $\pi(\sigma, \cdot)$ be a transition probability measure, i.e., for all $\sigma \in \Sigma$, $\pi(\sigma, \cdot) \in M_1(\Sigma)$. For any $n \geq 1$, define the measure $P_{n,\sigma} \in M_1(\Sigma^n)$ as the measure which assigns to any Borel measurable set $\Gamma \subset \Sigma^n$ the value

$$P_{n,\sigma}(\Gamma) = \int_\Gamma \prod_{i=1}^{n-1} \pi(x_i, dx_{i+1}) \pi(\sigma, dx_1).$$

Remark: As in Section 6.1, let $\Omega = \Sigma^{\mathbb{Z}_+}$ be the space of semi-infinite sequences with values in Σ, equipped with the product topology, and denote by Y_n the coordinates in the sequence, i.e., $Y_n(\omega_0, \omega_1, \ldots, \omega_n, \ldots) = \omega_n$. Ω

is a Polish space and its Borel σ-field is precisely $(\mathcal{B}_\Sigma)^{\mathbb{Z}_+}$. Let $\mathcal{F}_n$ denote the σ-field generated by $\{Y_m,\ 0 \leq m \leq n\}$. A measure P_σ on Ω can be uniquely constructed by the relations $P_\sigma(Y_0 = \sigma) = 1$, $P_\sigma(Y_{n+1} \in \Gamma | \mathcal{F}_n) = \pi(Y_n, \Gamma)$, a.s. P_σ for every $\Gamma \in \mathcal{B}_\Sigma$ and every $n \in \mathbb{Z}_+$. It follows that the restriction of P_σ to the first $n+1$ coordinates of Ω is precisely $P_{n,\sigma}$.

Define the (random) probability measure

$$L_n^{\mathbf{Y}} \stackrel{\triangle}{=} \frac{1}{n} \sum_{i=1}^{n} \delta_{Y_i} \in M_1(\Sigma) \,,$$

and denote by $\mu_{n,\sigma}$ the probability distribution of the $M_1(\Sigma)$-valued random variable $L_n^{\mathbf{Y}}$. We derive the LDP for $\mu_{n,\sigma}$, which, obviously, may also lead by contraction to the LDP for the empirical mean.

In analogy with Section 6.1, define

$$L_n^{\mathbf{Y},m} = \frac{1}{n-m} \sum_{i=m+1}^{n} \delta_{Y_i} \,.$$

The following is the first step towards the application of sub-additivity.

Lemma 6.3.1 *Let $A \in \mathcal{B}_{M(\Sigma)}$ be convex. Define*

$$\tilde{\mu}_n(A) \stackrel{\triangle}{=} \inf_{\sigma \in \Sigma} \mu_{n,\sigma}(A) \,.$$

Then $\tilde{\mu}_n(A)$ is super multiplicative, i.e.,

$$\tilde{\mu}_{n+m}(A) \geq \tilde{\mu}_n(A) \tilde{\mu}_m(A) \,.$$

Proof: Note that

$$
\begin{aligned}
\mu_{n+m,\sigma}(A) &= P_\sigma(L_{n+m}^{\mathbf{Y}} \in A) \geq P_\sigma(L_n^{\mathbf{Y}} \in A,\ L_{n+m}^{\mathbf{Y},n} \in A) \\
&= \int_{\Sigma^n} 1_{\{L_n^{\mathbf{Y}} \in A\}} \mu_{m,Y_n}(A)\, dP_{n,\sigma} \geq \mu_{n,\sigma}(A) \inf_{\tilde{\sigma} \in \Sigma} \mu_{m,\tilde{\sigma}}(A) \\
&\geq \tilde{\mu}_n(A) \tilde{\mu}_m(A) \,, &(6.3.2)
\end{aligned}
$$

where the first inequality follows from convexity. $\qquad\square$

Another ingredient needed for the applicability of the sub-additivity lemma (Lemma 6.1.11) is the requirement that if $\tilde{\mu}_m(A) > 0$ for some finite m, then $\tilde{\mu}_n(A) > 0$ for all n large enough. Unlike the i.i.d. case, this can not be proven directly for all open sets (or even for all open, convex sets), and some smoothing procedure is called for. In order to carry through this program, consider the sets

$$U_{f,x,\delta} = \{\nu \in M(\Sigma) :\ |\langle f, \nu \rangle - x| < \delta\} \,,$$

where $f \in C_b(\Sigma)$, $x \in \mathbb{R}$, and $\delta > 0$. Finite intersections of these sets are convex and by definition they form a base for the topology of weak convergence on $M(\Sigma)$. Let Θ denote this base. With a slight abuse of notations, denote by A_δ an arbitrary set from Θ, i.e., $A_\delta = \cap_{i=1}^N U_{f_i,x_i,\delta_i}$, and by $A_{\delta/2}$ the set $\cap_{i=1}^N U_{f_i,x_i,\delta_i/2}$. The reason for the introduction of these notations lies in the following lemma.

Lemma 6.3.3 *For any $A_\delta \in \Theta$, either*

$$\tilde{\mu}_n(A_{\delta/2}) = 0, \quad \forall n \in \mathbb{Z}_+,$$

or

$$\tilde{\mu}_n(A_\delta) > 0, \quad \forall n \geq n_0(A_\delta).$$

Proof: Let $A_\delta = \cap_{i=1}^N U_{f_i,x_i,\delta_i}$. Assume that $\tilde{\mu}_m(A_{\delta/2}) > 0$ for some m. For any $n \geq m$, let $q_n = [n/m]$, $r_n = n - q_n m$. Then

$$\mu_{n,\sigma}(A_\delta) \geq P_\sigma(L_n^{\mathbf{Y},r_n} \in A_{\delta/2}) \quad \text{and}$$
$$|\langle f_i, L_n^{\mathbf{Y}} \rangle - \langle f_i, L_n^{\mathbf{Y},r_n} \rangle| < \frac{\delta_i}{2}, i = 1,\ldots,N) .$$

However,

$$
\begin{aligned}
\langle f_i, L_n^{\mathbf{Y}} \rangle - \langle f_i, L_n^{\mathbf{Y},r_n} \rangle &= \frac{1}{n}\sum_{k=1}^n f_i(Y_k) - \frac{1}{q_n m}\sum_{k=r_n+1}^n f_i(Y_k) \\
&= \frac{1}{n}\sum_{k=1}^{r_n} f_i(Y_k) + \frac{q_n m - n}{n q_n m}\sum_{k=r_n+1}^n f_i(Y_k) .
\end{aligned}
$$

Therefore,

$$|\langle f_i, L_n^{\mathbf{Y}} \rangle - \langle f_i, L_n^{\mathbf{Y},r_n} \rangle| \leq 2\frac{r_n}{n}\|f_i\| \leq 2\frac{m}{n}\|f_i\|,$$

and thus, for $n > 4m \max_{i=1}^N \{\|f_i\|/\delta_i\}$,

$$\mu_{n,\sigma}(A_\delta) \geq P_\sigma(L_n^{\mathbf{Y},r_n} \in A_{\delta/2}) \geq \tilde{\mu}_{q_n m}(A_{\delta/2}) \geq \left[\tilde{\mu}_m(A_{\delta/2})\right]^{q_n} > 0 ,$$

where the last inequality is due to Lemma 6.3.1. The arbitrariness of σ completes the proof. $\qquad\square$

Define

$$\Theta' \overset{\triangle}{=} \{A_\delta \in \Theta : \tilde{\mu}_m(A_{\delta/2}) > 0 \text{ for some } m, \text{ or } \tilde{\mu}_n(A_\delta) = 0, \ \forall n \in \mathbb{Z}_+\} .$$

Note that Θ' is a base for the topology of weak convergence because $A_\delta \notin \Theta'$ implies that $A_{\delta/2} \in \Theta'$. By Lemmas 6.3.1 and 6.3.3, and the sub-additivity lemma (Lemma 6.1.11), it follows that for all $A = A_\delta \in \Theta'$, the limit

$$\mathcal{L}_A \stackrel{\triangle}{=} - \lim_{n \to \infty} \frac{1}{n} \log \tilde{\mu}_n(A) \tag{6.3.4}$$

exists (with ∞ as a possible value).

Unfortunately, the results obtained so far are only partial because they involve taking infimum over the initial conditions. The following uniformity hypothesis is introduced in order to progress to large deviations statements.
Assumption (U)   *There exist integers* $0 < \ell \le N$ *and a constant* $M \ge 1$ *such that* $\forall \sigma, \tau \in \Sigma$,

$$\pi^\ell(\sigma, \cdot) \le \frac{M}{N} \sum_{m=1}^N \pi^m(\tau, \cdot),$$

where $\pi^m(\tau, \cdot)$ *is the m-step transition probability for initial condition* τ,

$$\pi^{m+1}(\tau, \cdot) = \int_\Sigma \pi^m(\xi, \cdot) \pi(\tau, d\xi).$$

Remark: This assumption clearly holds true for every finite state irreducible Markov chain.

The reason for the introduction of Assumption (U) lies in the following lemma.

Lemma 6.3.5 *Assume (U). For any* $A_\delta \in \Theta$, *there exists an* $n_0(A_\delta)$ *such that for all* $n > n_0(A_\delta)$,

$$\sup_{\sigma \in \Sigma} \mu_{n,\sigma}(A_{\delta/2}) \le M \tilde{\mu}_n(A_\delta). \tag{6.3.6}$$

Proof: Let ℓ be as in (U), and $n = q_n \ell + r_n$ where $0 \le r_n \le \ell - 1$. Then with $A_\delta = \cap_{i=1}^{N'} U_{f_i, x_i, \delta_i}$,

$$\mu_{n,\sigma}(A_{\delta/2}) = P_\sigma \left(\frac{1}{n} \sum_{k=1}^n \delta_{Y_k} \in A_{\delta/2} \right) \tag{6.3.7}$$

$$\le P_\sigma \left(\frac{1}{(q_n - 1)\ell} \sum_{k=\ell+1}^{q_n \ell} \delta_{Y_k} \in A_{3\delta/4} \right)$$

$$+ \sum_{i=1}^{N'} P_\sigma \left(|\langle f_i, \frac{1}{n} \sum_{k=1}^\ell \delta_{Y_k} + \frac{1}{n} \sum_{k=q_n\ell+1}^n \delta_{Y_k} + \left(\frac{1}{n} - \frac{1}{(q_n - 1)\ell} \right) \sum_{k=\ell+1}^{q_n \ell} \delta_{Y_k} \rangle| \ge \frac{\delta_i}{4} \right).$$

For n large enough, the last term in the preceding inequality equals zero. Consequently, for such n,

$$\mu_{n,\sigma}(A_{\delta/2}) \ \leq \ P_\sigma\Big(\frac{1}{(q_n-1)\ell}\sum_{k=\ell+1}^{q_n\ell}\delta_{Y_k}\in A_{3\delta/4}\Big)$$

$$= \ \int_\Sigma \pi^\ell(\sigma,d\xi)P_\xi\Big(\frac{1}{(q_n-1)\ell}\sum_{k=1}^{(q_n-1)\ell}\delta_{Y_k}\in A_{3\delta/4}\Big).$$

Therefore, by Assumption (U), for all $\tau\in\Sigma$,

$$\mu_{n,\sigma}(A_{\delta/2})$$

$$\leq \frac{M}{N}\sum_{m=1}^N\int_\Sigma\pi^m(\tau,d\xi)P_\xi\Big(\frac{1}{(q_n-1)\ell}\sum_{k=1}^{(q_n-1)\ell}\delta_{Y_k}\in A_{3\delta/4}\Big)$$

$$= \frac{M}{N}\sum_{m=1}^N P_\tau\Big(\frac{1}{(q_n-1)\ell}\sum_{k=m+1}^{m+(q_n-1)\ell}\delta_{Y_k}\in A_{3\delta/4}\Big)$$

$$\leq \frac{M}{N}\sum_{m=1}^N P_\tau\Big(\frac{1}{n}\sum_{k=1}^n\delta_{Y_k}\in A_\delta\Big) = MP_\tau\Big(\frac{1}{n}\sum_{k=1}^n\delta_{Y_k}\in A_\delta\Big),$$

where the last inequality holds for all n large enough by the same argument as in (6.3.7). $\square$

In order to pursue a program similar to the one laid out in Section 6.1, let

$$I(\nu)\overset{\triangle}{=}\sup_{\{A\in\Theta':\nu\in A\}}\mathcal{L}_A.$$

Since Θ' form a base of the topology, (6.3.4) and (6.3.6) may be used to apply Lemma 4.1.15 and conclude that, for all $\sigma\in\Sigma$, the measures $\mu_{n,\sigma}$ satisfy the weak LDP in $M(\Sigma)$ with the (same) rate function $I(\cdot)$ defined before. Moreover, $I(\cdot)$ is convex as follows by extending (6.3.2) to

$$\mu_{2n,\sigma}\Big(\frac{A_1+A_2}{2}\Big)\geq\tilde\mu_n(A_1)\tilde\mu_n(A_2)$$

and then following an argument similar to the proof of Lemma 4.1.21.

Theorem 6.3.8 *Assume (U). Then the following limits exist:*

$$\Lambda(f)=\lim_{n\to\infty}\frac{1}{n}\log E_\sigma\Big(\exp(\sum_{i=1}^n f(Y_i))\Big), \qquad (6.3.9)$$

where $f \in C_b(\Sigma)$. Moreover, $\{\mu_{n,\sigma}\}$ satisfies a full LDP in $M_1(\Sigma)$ with the good, convex rate function

$$I(\nu) = \Lambda^*(\nu) = \sup_{f \in C_b(\Sigma)} \{\langle f, \nu \rangle - \Lambda(f)\}, \qquad (6.3.10)$$

where $\Lambda^*(\cdot)$ does not depend on the initial point $\sigma \in \Sigma$.

Remark: Actually, $\{\mu_{n,\sigma}\}$ satisfies the LDP uniformly in σ. For details, see Exercise 6.3.13.

Proof: Note that, since f is bounded,

$$\limsup_{n \to \infty} \frac{1}{n} \log E_\sigma\Big(\exp(\sum_{i=1}^n f(Y_i))\Big) \leq \|f\| < \infty.$$

The reader will prove the exponential tightness of $\{\mu_{n,\sigma}\}$ for each $\sigma \in \Sigma$ in Exercise 6.3.12. Due to the preceding considerations and the exponential tightness of $\mu_{n,\sigma}$, an application of Theorem 4.5.10 establishes both the existence of the limits (6.3.9) and the validity of (6.3.10). Moreover, since I does not depend on σ, the limit (6.3.9) also does not depend of σ. The transformation of the LDP from $M(\Sigma)$ to its closed subset $M_1(\Sigma)$ is standard by now. $\qquad \square$

Exercise 6.3.11 Let $g \in B(\mathbb{R})$ with $|g| \leq 1$. Let $\nu \in M_1(\mathbb{R})$, and assume that ν possesses a density f with respect to Lebesgue measure, such that

$$\sup_{|y| \leq 2, \, x \in \mathbb{R}} \left\{ \frac{f(x+y)}{f(x)} \right\} < \infty.$$

Let $\{v_n\}_{n=1}^\infty$ be i.i.d. real-valued random variables, each distributed according to ν. Prove that the Markov chain

$$Y_{n+1} = g(Y_n) + v_n$$

satisfies Assumption (U), with $N = \ell = 1$.

Exercise 6.3.12 Assume (U). Let $\mu(\cdot) = \frac{1}{N} \sum_{m=1}^N \pi^m(\tau, \cdot)$, where N is the same as in Assumption (U), and $\tau \in \Sigma$ is arbitrary.
(a) Show that for all $f \in B(\Sigma)$, all $n \in \mathbb{Z}_+$, and every $\sigma \in \Sigma$,

$$E_\sigma\Big[\exp(\sum_{i=1}^n f(Y_i))\Big] \leq \prod_{k=1}^\ell E_\sigma\Big[e^{\ell f(Y_k)}\Big]^{1/\ell} \Big[M \int_\Sigma e^{\ell f} d\mu\Big]^{\lceil n/\ell \rceil - 1},$$

where ℓ and M are as in Assumption (U).

Remark: In particular, note that for every $p \in M_1(\Sigma)$,

$$\limsup_{n \to \infty} \frac{1}{n} \log E_p\left(\exp(\sum_{i=1}^{n} f(Y_i))\right) \leq \frac{1}{\ell} \log \int_{\Sigma} e^{\ell f} d\mu + \frac{1}{\ell} \log M \,,$$

and that μ may also be replaced by any q such that $\int_{\Sigma} q(d\sigma)\pi(\sigma, \cdot) = q(\cdot)$.
(b) Prove the exponential tightness of $\{\mu_{n,\sigma}\}$ by a construction similar to that of Lemma 6.2.6.

Exercise 6.3.13 Assume (U).
(a) Show that for every closed set $F \subset M_1(\Sigma)$,

$$\limsup_{n \to \infty} \frac{1}{n} \log \sup_{\sigma \in \Sigma} \mu_{n,\sigma}(F) \leq - \inf_{\nu \in F} \Lambda^*(\nu) \,,$$

and for every open set $G \subset M_1(\Sigma)$,

$$\liminf_{n \to \infty} \frac{1}{n} \log \inf_{\sigma \in \Sigma} \mu_{n,\sigma}(G) \geq - \inf_{\nu \in G} \Lambda^*(\nu) \,.$$

Hint: Paraphrasing the proof of Lemma 6.3.5, show that for every closed set F, every $\delta > 0$, and every n large enough,

$$\sup_{\sigma \in \Sigma} \mu_{n,\sigma}(F) \leq M\tilde{\mu}_n(F^\delta) \,,$$

where F^δ are the closed blowups of F with respect to the Lévy metric. Then deduce the upper bound by applying Theorem 6.3.8 and Lemma 4.1.6.
(b) Let $p \in M_1(\Sigma)$, and let $\mu_{n,p}$ denote the measure induced by L_n^Y when the initial state of the Markov chain is distributed according to p. Prove that the full LDP holds for $\mu_{n,p}$. Prove also that

$$\Lambda(f) = \lim_{n \to \infty} \frac{1}{n} \log E_p\left(\exp(\sum_{i=1}^{n} f(Y_i))\right) \,.$$

6.4   Mixing Conditions and LDP

The goal of this section is to establish the LDP for stationary processes satisfying a certain mixing condition. Bryc's theorem (Theorem 4.4.10) is applied in Section 6.4.1 to establish the LDP of the empirical mean for a class of stationary processes taking values in a convex compact subset of $\mathbb{R}^d$. This result is combined in Section 6.4.2 with the projective limit approach to yield the LDP for the empirical measures of a class of stationary processes taking values in Polish spaces.

6.4.1 LDP for the Empirical Mean in $\mathbb{R}^d$

Let $X_1, \ldots, X_n, \ldots$ be a stationary process taking values in a convex, compact set $K \subset \mathbb{R}^d$. (Note that such a K may be found as soon as the support of the law of X_1 is bounded.) Let

$$\hat{S}_n^m = \frac{1}{n-m} \sum_{i=m+1}^{n} X_i \, ,$$

with $\hat{S}_n = \hat{S}_n^0$ and μ_n denoting the law of $\hat{S}_n$. The following mixing assumption prevails throughout this section.

Assumption 6.4.1 *For any continuous $f : K \to [0,1]$, there exist $\beta(\ell) \geq 1$, $\gamma(\ell) \geq 0$ and $\delta > 0$ such that*

$$\lim_{\ell \to \infty} \gamma(\ell) = 0 \ , \quad \limsup_{\ell \to \infty} \left(\beta(\ell) - 1 \right) \ell (\log \ell)^{1+\delta} < \infty \, , \tag{6.4.2}$$

and when ℓ and $n + m$ are large enough,

$$E[f(\hat{S}_n)^n f(\hat{S}_{n+m+\ell}^{n+\ell})^m] \geq E[f(\hat{S}_n)^n] E[f(\hat{S}_m)^m]$$
$$-\gamma(\ell) \left\{ E[f(\hat{S}_n)^n] E[f(\hat{S}_m)^m] \right\}^{1/\beta(\ell)} \tag{6.4.3}$$

The main result of this section is the following.

Theorem 6.4.4 *Let Assumption 6.4.1 hold. Then $\{\mu_n\}$ satisfies the LDP in $\mathbb{R}^d$ with the good convex rate function $\Lambda^*(\cdot)$, which is the Fenchel–Legendre transform of*

$$\Lambda(\lambda) = \lim_{n \to \infty} \frac{1}{n} \log E[e^{n\langle \lambda, \hat{S}_n \rangle}] \, . \tag{6.4.5}$$

In particular, the limit (6.4.5) exists.

Remark: Assumption 6.4.1, and hence Theorem 6.4.4, hold when $X_1, \ldots, X_n, \ldots$ is a bounded, ψ-mixing process. (See [Bra86] for the definition.) Other strong mixing conditions that suffice for Theorem 6.4.4 to hold are provided in [BryD96].

Proof: Note that $\hat{S}_n \in K$ for all $n \in \mathbb{Z}_+$, and hence the sequence $\{\mu_n\}$ is exponentially tight. Consequently, when combined with Lemma 4.4.8 and Theorem 4.4.10, the following lemma, whose proof is deferred, implies that μ_n satisfies the LDP in $\mathbb{R}^d$ with a good rate function $I(\cdot)$.

Lemma 6.4.6 *Let Assumption 6.4.1 hold. For any concave, continuous, bounded above function $g : \mathbb{R}^d \to \mathbb{R}$, the following limit exists*

$$\Lambda_g \overset{\triangle}{=} \lim_{n\to\infty} \frac{1}{n} \log E[e^{ng(\hat{S}_n)}] .$$

The following lemma, whose proof is deferred, is needed in order to prove that the rate function $I(\cdot)$ is convex.

Lemma 6.4.7 *Let assumption 6.4.1 hold. Suppose $\delta > 0$ and x_1, x_2 are such that for $i = 1, 2$,*

$$\liminf_{n\to\infty} \frac{1}{n} \log \mu_n(B_{x_i, \delta/2}) > -\infty .$$

Then for all n large enough,

$$\mu_{2n}(B_{(x_1+x_2)/2, \delta}) \geq \frac{1}{2} \mu_n(B_{x_1, \delta/2}) \mu_n(B_{x_2, \delta/2}) . \qquad (6.4.8)$$

Since the collection $\{B_{y,\delta}\}$ of all balls is a base for the topology of $\mathbb{R}^d$, it follows by Theorem 4.1.18 that, for all $x \in \mathbb{R}^d$,

$$-I(x) = \inf_{\delta > 0, y \in B_{x,\delta}} \liminf_{n\to\infty} \frac{1}{n} \log \mu_n(B_{y,\delta}) \qquad (6.4.9)$$

$$= \inf_{\delta > 0, y \in B_{x,\delta}} \limsup_{n\to\infty} \frac{1}{n} \log \mu_n(B_{y,\delta}) .$$

Fix x_1, x_2 such that $I(x_1) < \infty$ and $I(x_2) < \infty$. Then due to (6.4.9), Lemma 6.4.7 holds for x_1, x_2 and all $\delta > 0$. Note that $y \in B_{(x_1+x_2)/2, \delta}$ implies the inclusion $B_{(x_1+x_2)/2, \delta'} \subset B_{y,\delta}$ for all $\delta' > 0$ small enough. Hence, by (6.4.8) and (6.4.9),

$$-I((x_1 + x_2)/2) = \inf_{\delta > 0} \limsup_{n\to\infty} \frac{1}{n} \log \mu_n(B_{(x_1+x_2)/2, \delta})$$

$$\geq \inf_{\delta > 0} \liminf_{n\to\infty} \frac{1}{2n} \log \mu_{2n}(B_{(x_1+x_2)/2, \delta})$$

$$\geq \frac{1}{2} \inf_{\delta > 0} \liminf_{n\to\infty} \frac{1}{n} \log \mu_n(B_{x_1, \delta/2})$$

$$+ \frac{1}{2} \inf_{\delta > 0} \liminf_{n\to\infty} \frac{1}{n} \log \mu_n(B_{x_2, \delta/2}) \geq -\frac{I(x_1) + I(x_2)}{2} .$$

Thus, for all $x_1, x_2 \in \mathbb{R}^d$,

$$I((x_1 + x_2)/2) \leq \frac{1}{2} I(x_1) + \frac{1}{2} I(x_2) .$$

By iterations, this inequality extends to all convex combinations $\alpha x_1 + (1 - \alpha)x_2$ with $\alpha = k/2^m$ for some integers k, m. The convexity of the function $I(\cdot)$ then follows by its lower semicontinuity.

Since $\hat{S}_n \in K$ for all $n \in \mathbb{Z}_+$, it follows that for all $\lambda \in \mathbb{R}^d$,

$$\limsup_{n \to \infty} \frac{1}{n} \log E(e^{n\langle \lambda, \hat{S}_n \rangle}) < \infty \,,$$

and the proof of the theorem is completed by applying Theorem 4.5.10. $\square$

We shall now prove Lemmas 6.4.6 and 6.4.7.

Proof of Lemma 6.4.6: Fix $g : \mathbb{R}^d \to \mathbb{R}$ bounded above, continuous, and concave. Then $g(\cdot)$ is Lipschitz continuous when restricted to the compact set K, i.e., $|g(x) - g(y)| \le G|x - y|$ for all $x, y \in K$. Without loss of generality, assume that $-\infty < -B \le g(x) \le 0$ for all $x \in K$. Denoting $C = \sup_{x \in K} |x|$, note that

$$\left| \hat{S}_{n+m} - \left(\frac{n}{n+m} \hat{S}_n + \frac{m}{n+m} \hat{S}_{n+m+\ell}^{n+\ell} \right) \right|$$

$$= \frac{1}{n+m} \left| \sum_{i=n+1}^{n+\ell} X_i - \sum_{i=n+m+1}^{n+m+\ell} X_i \right| \le \frac{2\ell C}{n+m} \,.$$

Hence,

$$\left| g(\hat{S}_{n+m}) - g\left(\frac{n}{n+m} \hat{S}_n + \frac{m}{n+m} \hat{S}_{n+m+\ell}^{n+\ell} \right) \right| \le \frac{2\ell C G}{n+m} \,,$$

and by the concavity of $g(\cdot)$,

$$(n+m)g(\hat{S}_{n+m}) \ge ng(\hat{S}_n) + mg(\hat{S}_{n+m+\ell}^{n+\ell}) - 2\ell C G \,.$$

Define $h(n) \triangleq -\log E[e^{ng(\hat{S}_n)}]$. Then

$$h(n+m) \le 2\ell C G - \log E\left[e^{ng(\hat{S}_n)} e^{mg(\hat{S}_{n+m+\ell}^{n+\ell})} \right] \,.$$

Applying Assumption 6.4.1 for the continuous function $f(\cdot) = e^{g(\cdot)}$, whose range is $[0, 1]$, it follows that for all ℓ large enough and all integers n, m such that $n + m$ is large enough,

$$\frac{E\left[e^{ng(\hat{S}_n)} e^{mg(\hat{S}_{n+m+\ell}^{n+\ell})} \right]}{E[e^{ng(\hat{S}_n)}] E[e^{mg(\hat{S}_m)}]} \ge 1 - \gamma(\ell) \left\{ E[e^{ng(\hat{S}_n)}] E[e^{mg(\hat{S}_m)}] \right\}^{(\frac{1}{\beta(\ell)} - 1)}$$

$$\ge 1 - \gamma(\ell) e^{B(n+m)(\beta(\ell)-1)/\beta(\ell)} \,.$$

Hence,

$$h(n+m) \le h(n) + h(m) + 2\ell CG - \log\{[1 - \gamma(\ell)e^{B(n+m)(\beta(\ell)-1)/\beta(\ell)}] \vee 0\}.$$

Choosing $\ell = [(n+m)\{\log(n+m)^{-(1+\delta)}\}]$ with $\delta > 0$ corresponding to Assumption 6.4.1 for $f(\cdot)$, it follows that $(n+m)(\beta(\ell)-1)/\beta(\ell)$ are uniformly, in $n+m$, bounded above, and hence,

$$\gamma(\ell)e^{B(n+m)(\beta(\ell)-1)/\beta(\ell)} \to 0.$$

From the preceding considerations, for all $n+m$ large enough,

$$h(n+m) \le h(n) + h(m) + (2CG+1)(n+m)\{\log(n+m)\}^{-(1+\delta)},$$

and the proof of the lemma is completed by the following approximate sub-additivity lemma.

Lemma 6.4.10 (Approximate sub-additivity) *Assume* $f : \mathbb{Z}_+ \to \mathbb{R}$ *is such that for all* $n, m \ge 1$,

$$f(n+m) \le f(n) + f(m) + \epsilon(n+m), \qquad (6.4.11)$$

where for some $\delta > 0$,

$$\limsup_{n\to\infty} \left[\frac{\epsilon(n)}{n}(\log n)^{1+\delta}\right] < \infty. \qquad (6.4.12)$$

Then $\bar{f} = \lim_{n\to\infty} [f(n)/n]$ *exists.*

Remark: With a somewhat more involved proof, Hammersley in [Ham62] relaxes (6.4.12) for $\epsilon(n)$ non-decreasing to

$$\sum_{r=1}^{\infty} \frac{\epsilon(r)}{r(r+1)} < \infty,$$

showing it then to be also necessary for the existence of $\bar{f} < \infty$. Explicit upper bounds on $\bar{f} - f(m)/m$ for every $m \ge 1$ are also provided there.

Proof of Lemma 6.4.10: Fix $s \in \mathbb{Z}_+$, $s \ge 2$. Observe that for all $m \ge 1$, $1 \le r \le (s-1)$,

$$f(ms+r) \le f(ms) + f(r) + \epsilon(ms+r).$$

Hence,

$$\frac{f(ms+r)}{ms+r} \le \frac{f(ms)}{ms}(1 - \frac{r}{ms+r}) + \frac{f(r)}{ms+r} + \frac{\epsilon(ms+r)}{ms+r},$$

i.e.,

$$\sup_{0 \le r \le (s-1)} \frac{f(ms+r)}{ms+r} \le \frac{1}{ms} \max_{r=0}^{s-1}[f(r) \vee 0]$$

$$+ \max\{\frac{f(ms)}{ms}, \frac{f(ms)}{ms}(1 - \frac{1}{m})\}$$

$$+ \max_{r=0}^{s-1} \frac{\epsilon(ms+r)}{ms+r}.$$

Consequently,

$$\limsup_{n\to\infty} \frac{f(n)}{n} = \limsup_{m\to\infty} \sup_{0 \le r \le (s-1)} \frac{f(ms+r)}{ms+r}$$

$$\le \limsup_{m\to\infty} \frac{f(ms)}{ms} + \limsup_{n\to\infty} \frac{\epsilon(n)}{n}.$$

Since the second term in the right-hand side is zero, it follows that for all $2 \le s < \infty$,

$$\limsup_{n\to\infty} \frac{f(n)}{n} \le \limsup_{m\to\infty} \frac{f(ms)}{ms} \stackrel{\triangle}{=} \gamma_s.$$

Recall that by (6.4.12) there exists $s_0, C < \infty$ such that for all $s \ge s_0$ and all $m \in \mathbb{Z}_+$,

$$\frac{\epsilon(ms)}{ms} \le C[\log_2(ms)]^{-(1+\delta)}.$$

Fix $s \ge s_0$, and define

$$p_s(k) \stackrel{\triangle}{=} \sup_{1 \le m \le 2^k} \left\{ \frac{f(ms)}{ms} \right\}, \quad k = 0, 1, \dots$$

Observe that $\gamma_s \le \lim_{k\to\infty} p_s(k)$, while for all $k \ge 1$,

$$p_s(k) \le p_s(k-1) + \sup_{2^{k-1} \le m \le 2^k} \left\{ \frac{\epsilon(ms)}{ms} \right\}.$$

Consequently,

$$p_s(k) - p_s(k-1) \le \sup_{2^{k-1} \le m \le 2^k} C[\log_2(ms)]^{-(1+\delta)}$$

$$\le C[\log_2 s + (k-1)]^{-(1+\delta)},$$

yielding the inequality

$$p_s(k) \le p_s(0) + \sum_{i=1}^{k} C[\log_2 s + (i-1)]^{-(1+\delta)} \le p_s(0) + C \sum_{j=[\log_2 s]}^{\infty} j^{-(1+\delta)}.$$

Since $p_s(0) = f(s)/s$, it follows that for all $s \geq s_0$,

$$\gamma_s \leq \lim_{k \to \infty} p_s(k) \leq \frac{f(s)}{s} + C \sum_{j=[\log_2 s]}^{\infty} j^{-(1+\delta)} .$$

Therefore,

$$\limsup_{n \to \infty} \frac{f(n)}{n} \quad \leq \quad \liminf_{s \to \infty} \gamma_s$$

$$\leq \quad \liminf_{s \to \infty} \frac{f(s)}{s} + C \limsup_{s \to \infty} \sum_{j=[\log_2 s]}^{\infty} j^{-(1+\delta)}$$

$$= \quad \liminf_{n \to \infty} \frac{f(n)}{n} . \qquad \square$$

Proof of Lemma 6.4.7: Observe that by our assumption there exists $M < \infty$ such that for all n large enough,

$$\mu_n(B_{x_1,\delta/2})\mu_n(B_{x_2,\delta/2}) \geq e^{-Mn} .$$

Thus, the proof is a repeat of the proof of Lemma 6.4.6. Specifically, for any integer ℓ,

$$\mu_{2n}(B_{(x_1+x_2)/2,\delta}) = \mathrm{P}\left(\left| \hat{S}_{2n} - \frac{x_1+x_2}{2} \right| < \delta \right)$$

$$\geq \mathrm{P}\left(\left| \frac{\hat{S}_n - x_1}{2} + \frac{\hat{S}_{2n+\ell}^{n+\ell} - x_2}{2} \right| < \delta - \frac{C\ell}{n} \right)$$

$$\geq \mathrm{P}\left(|\hat{S}_n - x_1| < \delta - \frac{C\ell}{n} , \ |\hat{S}_{2n+\ell}^{n+\ell} - x_2| < \delta - \frac{C\ell}{n} \right) .$$

In particular, for $\ell = \delta n/2C$, one has by Assumption 6.4.1 that for all n large enough,

$$\mu_{2n}(B_{(x_1+x_2)/2,\delta}) \geq \mathrm{P}\left(|\hat{S}_n - x_1| < \frac{\delta}{2} , \ |\hat{S}_{2n+\ell}^{n+\ell} - x_2| < \frac{\delta}{2} \right)$$

$$\geq \mu_n(B_{x_1,\frac{\delta}{2}})\mu_n(B_{x_2,\frac{\delta}{2}}) \left(1 - \gamma(\ell)[\mu_n(B_{x_1,\frac{\delta}{2}})\mu_n(B_{x_2,\frac{\delta}{2}})]^{(\frac{1}{\beta(\ell)}-1)} \right) .$$

Note that there exists $C < \infty$ such that for n large enough, $Mn(\beta(\ell) - 1)/\beta(\ell) \leq C$, implying

$$\gamma(\ell)\left(\mu_n(B_{x_1,\frac{\delta}{2}})\mu_n(B_{x_2,\frac{\delta}{2}})\right)^{(\frac{1-\beta(\ell)}{\beta(\ell)})} \leq \gamma(\ell)e^{Mn(\beta(\ell)-1)/\beta(\ell)} \leq \gamma(\ell)e^C \leq \frac{1}{2} ,$$

where the last inequality follows by (6.4.2). Hence, (6.4.8) follows and the proof is complete. $\qquad \square$

Exercise 6.4.13 [Suggested by Y. Peres] Let $f(n) = n \sin(\log(2 + \log n))$. Verify that there exists a $C < \infty$ such that $f(n)$ satisfies (6.4.11) for $\epsilon(n) = Cn/\log n$. Hence, the rate condition (6.4.12) is almost optimal.
Hint: Note that $|d \sin(\log(2 + x))/dx| \leq 1/(2 + x)$ and thus by the mean value theorem,

$$\frac{f(n + m) - f(n) - f(m)}{n + m} \leq \frac{2 + \log(n + m)}{(n + m)} g_{n+m}(n) - 1$$

where

$$g_x(n) = \frac{n}{2 + \log n} + \frac{x - n}{2 + \log(x - n)}, \quad n = 1, 2, \ldots, x - 1.$$

Check that for fixed x, the maximum of $g_x(n)$ is obtained at $n = x/2$.

6.4.2 Empirical Measure LDP for Mixing Processes

The previous results coupled with Corollary 4.6.11 allow one to deduce the LDP for quite a general class of processes. Let Σ be a Polish space, and $B(\Sigma)$ the space of all bounded, measurable real-valued functions on Σ, equipped with the supremum norm. As in Section 6.2, define $M_1(\Sigma) \subset M(\Sigma) \subset B(\Sigma)'$.

Let $\Omega = \Sigma^{\mathbb{Z}_+}$, let P be a stationary and ergodic measure on Ω, and let $Y_1, \ldots, Y_n, \ldots$ denote its realization. Throughout, P_n denotes the nth marginal of P, i.e., the measure on Σ^n whose realization is $Y_1, \ldots, Y_n$. As in Section 6.2, $L_n^{\mathbf{Y}} = \frac{1}{n} \sum_{i=1}^n \delta_{Y_i} \in M_1(\Sigma)$, and μ_n denotes the probability measure induced on $(B(\Sigma)', \mathcal{B}^{cy})$ by $L_n^{\mathbf{Y}}$.

Theorem 6.4.14 (a) Let $\{g_j\}_{j=1}^d \in B(\Sigma)$. Define the $\mathbb{R}^d$-valued station-ary process $X_1, \ldots, X_n, \ldots$ by $X_i = (g_1(Y_i), \ldots, g_d(Y_i))$. Suppose that, for any d and $\{g_j\}_{j=1}^d$, the process $\{X_i\}$ satisfies Assumption 6.4.1. Then $\{\mu_n\}$ satisfies the LDP in the space $B(\Sigma)'$ equipped with the $B(\Sigma)$-topology and the σ-field $\mathcal{B}^{cy}$. This LDP is governed by the good rate function

$$\Lambda^*(\omega) \stackrel{\triangle}{=} \sup_{f \in B(\Sigma)} \{\langle f, \omega \rangle - \Lambda(f)\}, \tag{6.4.15}$$

where for all $f \in B(\Sigma)$,

$$\Lambda(f) = \lim_{n \to \infty} \frac{1}{n} \log E_P \left[e^{\sum_{i=1}^n f(Y_i)} \right].$$

In particular, the preceding limit exists.
(b) Assume further that, for some constants $\gamma, M > 0$ and all $f \in B(\Sigma)$,

$$\Lambda(f) \leq \frac{1}{\gamma} \log \int_\Sigma e^{\gamma f(x)} P_1(dx) + \log M. \tag{6.4.16}$$

Then the LDP in part (a) can be restricted to $M_1(\Sigma)$ equipped with the τ-topology.

Proof: The proof relies on part (b) of Corollary 4.6.11. Indeed, the triplet $(B(\Sigma)', \mathcal{B}^{cy}, \mu_n)$ satisfies Assumption 4.6.8 with $\mathcal{W} = B(\Sigma)$. Moreover, by Theorem 6.4.4, for any $g_1, \ldots, g_d \in B(\Sigma)$, the vectors $(\langle g_1, L_n^{\mathbf{Y}} \rangle, \langle g_2, L_n^{\mathbf{Y}} \rangle, \ldots, \langle g_d, L_n^{\mathbf{Y}} \rangle)$ satisfy the LDP in a compact subset of $\mathbb{R}^d$ with a good convex rate function. Therefore, Corollary 4.6.11 applies and yields both the existence of $\Lambda(f)$ and part (a) of the theorem.

To see part (b) of the theorem, note that by (6.4.16),

$$
\begin{aligned}
\Lambda^*(\omega) &\geq \sup_{f \in B(\Sigma)} \left\{ \langle f, \omega \rangle - \frac{1}{\gamma} \log E_P \left[e^{\gamma f(Y_1)} \right] \right\} - \log M \\
&= \frac{1}{\gamma} \sup_{f \in B(\Sigma)} \{ \langle f, \omega \rangle - \log E_P[e^{f(Y_1)}] \} - \log M \\
&= \frac{1}{\gamma} H(\omega | P_1) - \log M,
\end{aligned}
$$

where the last equality is due to Lemma 6.2.13. Hence, $\mathcal{D}_{\Lambda^*} \subset M_1(\Sigma)$ (since $H(\omega | P_1) = \infty$ for $\omega \notin M_1(\Sigma)$), and the proof is completed by Lemma 4.1.5. □

In the rest of this section, we present the conditions required both for the LDP and for the restriction to $M_1(\Sigma)$ in a slightly more transparent way that is reminiscent of mixing conditions. To this end, for any given integers $r \geq k \geq 2$, $\ell \geq 1$, a family of functions $\{f_i\}_{i=1}^k \in B(\Sigma^r)$ is called ℓ-*separated* if there exist k disjoint intervals $\{a_i, a_i + 1, \ldots, b_i\}$ with $a_i \leq b_i \in \{1, \ldots, r\}$ such that $f_i(\sigma_1, \ldots, \sigma_r)$ is actually a bounded measurable function of $\{\sigma_{a_i}, \ldots, \sigma_{b_i}\}$ and for all $i \neq j$ either $a_i - b_j \geq \ell$ or $a_j - b_i \geq \ell$.

Assumption (H-1)   *There exist $\ell, \alpha < \infty$ such that, for all $k, r < \infty$, and any ℓ-separated functions $f_i \in B(\Sigma^r)$,*

$$
E_P(|f_1(Y_1, \ldots, Y_r) \cdots f_k(Y_1, \ldots, Y_r)|) \leq \prod_{i=1}^k E_P(|f_i(Y_1, \ldots, Y_r)|^\alpha)^{1/\alpha}.
$$

$$(6.4.17)$$

Assumption (H-2)   *There exist a constant ℓ_0 and functions $\beta(\ell) \geq 1$, $\gamma(\ell) \geq 0$ such that, for all $\ell > \ell_0$, all $r < \infty$, and any two ℓ-separated functions $f, g \in B(\Sigma^r)$,*

$$
|E_P(f(Y_1, \ldots, Y_r)) E_P(g(Y_1, \ldots, Y_r)) - E_P(f(Y_1, \ldots, Y_r) g(Y_1, \ldots, Y_r))|
$$

$$
\leq \gamma(\ell) E_P \left(|f(Y_1, \ldots, Y_r)|^{\beta(\ell)} \right)^{1/\beta(\ell)} E_P \left(|g(Y_1, \ldots, Y_r)|^{\beta(\ell)} \right)^{1/\beta(\ell)},
$$

and $\lim_{\ell \to \infty} \gamma(\ell) = 0$, $\limsup_{\ell \to \infty} (\beta(\ell) - 1)\ell(\log \ell)^{1+\delta} < \infty$ for some $\delta > 0$.

Remarks:

(a) Conditions of the type (H-1) and (H-2) are referred to as *hypermixing* conditions. Hypermixing is tied to analytical properties of the semigroup in Markov processes. For details, consult the excellent exposition in [DeuS89b]. Note, however, that in (H-2) of the latter, a less stringent condition is put on $\beta(\ell)$, whereas in (H-1) there, $\alpha(\ell)$ converges to one.

(b) The particular case of $\beta(\ell) = 1$ in Assumption (H-2) corresponds to ψ-mixing [Bra86], with $\gamma(\ell) = \psi(\ell)$.

Lemma 6.4.18 *Let* $Y_1, \ldots, Y_n, \ldots$ *be the stationary process defined before.*
(a) Assume that (H-1) holds and $\Lambda(f)$ *exists for all* $f \in B(\Sigma)$. *Then the inequality (6.4.16) holds true.*
(b) Assume that (H-2) holds. Then Assumption 6.4.1 holds true for the process $X_1, \ldots, X_n, \ldots$ *which appears in the statement of Theorem 6.4.14.*

Remark: Consequently, when both (H-1) and (H-2) hold, L_n^Y satisfies the LDP in $M_1(\Sigma)$ with the good rate function $\Lambda^*(\cdot)$ of (6.4.15).

Proof: (a) Since $\Lambda(f)$ exists, it is enough to consider limits along the sequence $n = m\ell$. By Jensen's inequality,

$$E_P\left[e^{\sum_{i=1}^{m\ell} f(Y_i)}\right] = E_P\left[e^{\ell^{-1} \sum_{k=1}^{\ell} \ell \sum_{j=0}^{m-1} f(Y_{k+j\ell})}\right]$$

$$\leq \frac{1}{\ell} \sum_{k=1}^{\ell} E_P\left[e^{\ell \sum_{j=0}^{m-1} f(Y_{k+j\ell})}\right].$$

Note that the various functions in the exponent of the last expression are ℓ-separated. Therefore,

$$E_P\left[e^{\sum_{i=1}^{m\ell} f(Y_i)}\right] \leq \frac{1}{\ell} \sum_{k=1}^{\ell} \prod_{j=0}^{m-1} E_P\left[e^{\alpha \ell f(Y_{k+j\ell})}\right]^{1/\alpha}$$

$$= E_P\left[e^{\alpha \ell f(Y_1)}\right]^{m/\alpha},$$

where Assumption (H-1) was used in the inequality and the stationarity of P in the equality. Hence,

$$\frac{1}{m\ell} \log E_P\left[e^{\sum_{i=1}^{m\ell} f(Y_i)}\right] \leq \frac{1}{\ell\alpha} \log E_P[e^{\alpha \ell f(Y_1)}],$$

and (6.4.16) follows with $\gamma = \ell\alpha$ and $M = 1$.

(b) Fix d and $\{g_j\}_{j=1}^{d} \in B(\Sigma)$, and let $X_i = (g_1(Y_i), \ldots, g_d(Y_i))$. Let

$K \subset \mathbb{R}^d$ be the convex, compact hypercube $\Pi_{i=1}^d [-\|g_i\|, \|g_i\|]$. Fix an arbitrary continuous function $f : K \to [0,1]$, any two integers n, m, and any $\ell > \ell_0$. Let $r = n + m + \ell$, and define

$$f_1(Y_1, \ldots, Y_r) = f(\frac{1}{n} \sum_{i=1}^n X_i)^n$$

$$f_2(Y_1, \ldots, Y_r) = f(\frac{1}{m} \sum_{i=n+1+\ell}^{n+m+\ell} X_i)^m .$$

Clearly, $f_1, f_2 : \Sigma^r \to [0,1]$ are ℓ-separated, and therefore, by the stationarity of P, Assumption (H-2) implies that (6.4.3) holds.   □

Exercise 6.4.19 Let $|\alpha| < 1$, $x_0 \sim \mathrm{Normal}(0, 1/(1 - \alpha^2))$, and ξ_n be a sequence of i.i.d. standard Normal random variables. Check that the process

$$x_{n+1} = \alpha x_n + \xi_n ,$$

satisfies Assumptions (H-1) and (H-2) but not Assumption (U).

Hint: Let $\| \cdot \|_p$ denote the $L_p(P_1)$ norm. For any $f \in B(\Sigma)$, let $\pi^\ell f(x) \triangleq \int_\Sigma \pi^\ell(x, dy) f(y)$. Using Hölder's inequality and the explicit form of the transition probability measure, show that there exist positive constants β, γ such that for all ℓ large enough,

$$\|\pi^\ell f\|_{q(\ell)} \le \|f\|_2, \quad \|\pi^\ell(f - \bar{f})\|_2 \le e^{-\gamma \ell} \|f\|_2, \quad \|\pi^\ell f\|_2 \le \|f\|_{p(\ell)},$$

where $\bar{f} = E_{P_1} f$, $p(\ell) = 1 + e^{-\beta \ell}$, and $q(\ell) = 1 + e^{\beta \ell}$. (These estimates are known as *hypercontractive estimates*.)

Now let f, g be nonnegative functions of one variable. Using the preceding estimates and the Markov property, check that for ℓ large,

$$
\begin{aligned}
E_P(f(x_1)g(x_{\ell+1})) &= E_{P_1}(f\pi^\ell g) = E_{P_1}(\pi^\ell(f\pi^\ell g)) \le \|\pi^\ell(f\pi^\ell g)\|_2 \\
&\le \|f\pi^\ell g\|_{p(\ell)} \le \|f\|_4 \|\pi^\ell g\|_2 \\
&\le \|f\|_4 \|g\|_2 \le \|f\|_4 \|g\|_4 .
\end{aligned}
$$

Prove Assumption (H-1) by repeating the preceding steps in the general situation. To see Assumption (H-2), let f, g be as before (but not necessarily nonnegative), and paraphrasing the preceding arguments, show that

$$
\begin{aligned}
|E_P[g(x_1)(f(x_{3\ell+1}) - \bar{f})]| &\le \|g\|_{p(\ell)} \|\pi^{3\ell}(f - \bar{f})\|_{q(\ell)} \\
&\le \|g\|_{p(\ell)} \|\pi^{2\ell}(f - \bar{f})\|_2 \\
&\le e^{-\gamma \ell} \|g\|_{p(\ell)} \|\pi^\ell f\|_2 \\
&\le e^{-\gamma \ell} \|g\|_{p(\ell)} \|f\|_{p(\ell)} .
\end{aligned}
$$

Extend these considerations to the general situation.

6.5 LDP for Empirical Measures of Markov Chains

Sections 6.3 and 6.4 enable LDPs to be obtained for the empirical measures associated with various Markov processes. The expression for the rate function involves, under both the uniform (U) and the hypermixing (H-1), (H-2) conditions, taking Fenchel–Legendre transforms, although over different spaces and with somewhat different logarithmic moment generating functions. Motivated by the results of Section 3.1, alternative, more tractable expressions for this rate function are derived here.

Note that, as in the finite state setup, life may become simpler if the pair empirical measure is dealt with. With this in mind, the rate functions for the LDP of the empirical measure L_n^Y, and for the LDP of the k-tuple empirical measure are treated separately. Since L_n^Y measures what fraction of the total "time" interval $(1, 2, \ldots, n)$ the process actually spends in a particular Borel set, and since the continuous time analog of L_n^Y clearly involves exactly the notion of time spent at a Borel set, the empirical measure L_n^Y is often referred to as *occupation time*. In order to distinguish L_n^Y from its k-tuple counterpart, this name is retained in this section.

6.5.1 LDP for Occupation Times

Throughout this section, let $\pi(x, dy)$ be a transition probability measure (also called Markov or transition kernel) on the Polish space Σ. Consider the measure $P \in M_1(\Sigma^{\mathbb{Z}_+})$ generated by $\pi(\cdot, \cdot)$ from the initial measure $P_1 \in M_1(\Sigma)$; that is, let the marginals $P_n \in M_1(\Sigma^n)$ be such that for any $n \geq 1$ and any $\Gamma \in \mathcal{B}_{\Sigma^n}$,

$$P(\{\sigma \in \Sigma^{\mathbb{Z}_+} : (\sigma_1, \ldots, \sigma_n) \in \Gamma\}) = \int_\Gamma P_1(dx_1) \prod_{i=1}^{n-1} \pi(x_i, dx_{i+1})$$

$$= \int_\Gamma P_n(dx_1, \ldots, dx_n).$$

To motivate the derivation, recall that if P satisfies the hypermixing conditions (H-1) and (H-2) of Section 6.4.2, then, by Theorem 6.4.14 and Lemma 6.4.18, the random variables L_n^Y satisfy the LDP in the τ-topology of $M_1(\Sigma)$, with the good rate function being the Fenchel–Legendre transform (with respect to $B(\Sigma)$) of

$$\Lambda(f) = \limsup_{n \to \infty} \frac{1}{n} \log E_P\left[e^{\sum_{i=1}^n f(Y_i)}\right].$$

In this section, a program for the explicit identification of this rate function and related ones is embarked on. It turns out that this identification procedure does not depend upon the existence of the LDP. A similar identification for the (different) rate function of the LDP derived in Section 6.3 under the uniformity assumption (U) is also presented, and although it is not directly dependent upon the existence of an LDP, it does depend on structural properties of the Markov chains involved. (See Theorem 6.5.4.)

Throughout this section, let $\pi_f : B(\Sigma) \to B(\Sigma)$ denote the linear, non-negative operator

$$(\pi_f u)(\sigma) = e^{f(\sigma)} \int_\Sigma u(\tau)\pi(\sigma, d\tau) \,,$$

where $f \in B(\Sigma)$, and let $\pi u \triangleq \pi_0 u = \int_\Sigma u(\tau)\pi(\cdot, d\tau)$. Note that

$$E_P\left[e^{\sum_{i=1}^n f(Y_i)}\right] = E_P\left[\left(e^{f(Y_n)}\int_\Sigma \pi(Y_n, d\tau)\right)e^{\sum_{i=1}^{n-1} f(Y_i)}\right]$$

$$= E_P\left[(\pi_f 1)(Y_n)e^{\sum_{i=1}^{n-1} f(Y_i)}\right] = E_P\left[\{(\pi_f)^2 1\}(Y_{n-1})e^{\sum_{i=1}^{n-2} f(Y_i)}\right]$$

$$= E_P[\{(\pi_f)^n 1\}(Y_1)] = \langle(\pi_f)^n 1, P_1\rangle \,,$$

and hence, the logarithmic moment generating function associated with $\pi(\cdot, \cdot)$ and P_1 is

$$\Lambda(f) \triangleq \limsup_{n\to\infty} \frac{1}{n} \log(\langle(\pi_f)^n 1, P_1\rangle) \,. \qquad (6.5.1)$$

Definition $\mu \in M_1(\Sigma)$ *is an invariant measure for* $\pi(\cdot, \cdot)$ *if*

$$\mu\pi(\cdot) \triangleq \int_\Sigma \mu(d\sigma)\pi(\sigma, \cdot) = \mu(\cdot) \,.$$

If P_1 is an invariant measure for $\pi(\cdot, \cdot)$, then P is a stationary measure.

The existence of an invariant measure makes the computation of the Fenchel–Legendre transform particularly transparent. Namely,

Theorem 6.5.2 *Let* P_1 *be an invariant measure for* $\pi(\cdot, \cdot)$. *Then for all* $\nu \in M_1(\Sigma)$,

$$\sup_{f\in B(\Sigma)} \{\langle\nu, f\rangle - \Lambda(f)\}$$

$$= \begin{cases} \sup_{u\in B(\Sigma), u\geq 1} \left\{-\int_\Sigma \log\left(\frac{\pi u}{u}\right) d\nu\right\} & , \quad \frac{d\nu}{dP_1} \text{ exists} \\ \infty & , \quad \text{otherwise.} \end{cases} \qquad (6.5.3)$$

Remark: That one has to consider the two separate cases in (6.5.3) above is demonstrated by the following example. Let $\pi(x, A) = 1_A(x)$ for every Borel set A. It is easy to check that any measure P_1 is invariant, and that L_n^Y satisfies the LDP with the rate function

$$I(\nu) = \begin{cases} 0 & , \quad d\nu/dP_1 \text{ exists} \\ \infty & , \quad \text{otherwise}. \end{cases}$$

Proof: (a) Suppose that $d\nu/dP_1$ does not exist, i.e., there exists a set $A \in \mathcal{B}_\Sigma$ such that $P_1(A) = 0$ while $\nu(A) > 0$. For any $\alpha > 0$, the function $\alpha 1_A$ belongs to $B(\Sigma)$, and by the union of events bound and the stationarity of P, it follows that for all n,

$$E_P\left[e^{\alpha \sum_{i=1}^n 1_A(Y_i)}\right] \le 1 + n P_1(A) e^{\alpha n} = 1.$$

Therefore, $\Lambda(\alpha 1_A) \le 0$, implying that

$$\begin{aligned}
\sup_{f \in B(\Sigma)} \{\langle \nu, f \rangle - \Lambda(f)\} &\ge \sup_{\alpha > 0} \{\alpha \langle 1_A, \nu \rangle - \Lambda(\alpha 1_A)\} \\
&\ge \sup_{\alpha > 0} \{\alpha \nu(A)\} = \infty.
\end{aligned}$$

(b) Suppose now that $d\nu/dP_1$ exists. For each $u \in B(\Sigma)$ with $u \ge 1$, let $f = \log(u/\pi u)$ and observe that $f \in B(\Sigma)$ as $\pi u \ge \pi 1 = 1$. Since $\pi_f u = u$ and $\pi_f 1 = u/\pi u \le u$, it follows that $(\pi_f)^n 1 \le u$. Hence,

$$\Lambda(f) \le \limsup_{n \to \infty} \frac{1}{n} \log \langle u, P_1 \rangle \le \limsup_{n \to \infty} \frac{1}{n} \log \|u\| = 0.$$

Therefore,

$$\langle f, \nu \rangle = -\int_\Sigma \log\left(\frac{\pi u}{u}\right) d\nu \le \langle f, \nu \rangle - \Lambda(f),$$

implying that

$$\sup_{u \in B(\Sigma), u \ge 1} \left\{-\int_\Sigma \log\left(\frac{\pi u}{u}\right) d\nu\right\} \le \sup_{f \in B(\Sigma)} \{\langle f, \nu \rangle - \Lambda(f)\}.$$

To establish the opposite inequality, fix $f \in B(\Sigma)$, $\Lambda(f) < \alpha < \infty$, and let $u_n \triangleq \sum_{m=0}^n e^{-m\alpha} (\pi_f)^m 1$. Note that $u_n \in B(\Sigma)$ and $u_n \ge 1$ for all n. Moreover,

$$e^{-\alpha} \pi_f u_n = \sum_{m=1}^{n+1} e^{-m\alpha} (\pi_f)^m 1 = u_{n+1} - 1 = u_n + v_n - 1,$$

where $v_n \stackrel{\triangle}{=} u_{n+1} - u_n = e^{-(n+1)\alpha}(\pi_f)^{n+1}1$. Therefore,

$$-\log\left(\frac{\pi u_n}{u_n}\right) = f - \alpha - \log\left(\frac{u_n + v_n - 1}{u_n}\right) \geq f - \alpha - \log\left(1 + \frac{v_n}{u_n}\right).$$

Observe that

$$v_n = e^{-\alpha}\pi_f v_{n-1} \leq e^{\|f\|-\alpha}v_{n-1} \leq e^{\|f\|-\alpha}u_n.$$

Thus, with $u_n \geq 1$,

$$\log\left(1 + \frac{v_n}{u_n}\right) \leq \frac{v_n}{u_n} \leq e^{(\|f\|-\alpha)} \wedge v_n,$$

implying that for all $\delta > 0$,

$$\int_\Sigma \log\left(1 + \frac{v_n}{u_n}\right) d\nu \leq \delta + e^{(\|f\|-\alpha)}\nu(\{v_n \geq \delta\}).$$

Since $\alpha > \Lambda(f)$, it follows that $\langle v_n, P_1 \rangle \to 0$ as $n \to \infty$. Consequently, for any $\delta > 0$ fixed, by Chebycheff's inequality, $P_1(\{v_n \geq \delta\}) \to 0$ as $n \to \infty$, and since $d\nu/dP_1$ exists, also $\nu(\{v_n \geq \delta\}) \to 0$ as $n \to \infty$. Hence,

$$\sup_{u \in B(\Sigma), u \geq 1}\left\{-\int_\Sigma \log\left(\frac{\pi u}{u}\right) d\nu\right\} \geq \limsup_{n \to \infty}\left\{-\int_\Sigma \log\left(\frac{\pi u_n}{u_n}\right) d\nu\right\}$$

$$\geq \left[\langle f, \nu \rangle - \alpha - \liminf_{n \to \infty}\int_\Sigma \log\left(1 + \frac{v_n}{u_n}\right) d\nu\right] \geq \langle f, \nu \rangle - \alpha - \delta.$$

Considering first $\delta \searrow 0$, then $\alpha \searrow \Lambda(f)$, and finally taking the supremum over $f \in B(\Sigma)$, the proof is complete. $\square$

Having treated the rate function corresponding to the hypermixing assumptions (H-1) and (H-2), attention is next turned to the setup of Section 6.3. It has already been observed that under the uniformity assumption (U), $L_n^{\mathbf{Y}}$ satisfies the LDP, in the weak topology of $M_1(\Sigma)$, regardless of the initial measure P_1, with the good rate function

$$I^U(\nu) \stackrel{\triangle}{=} \sup_{f \in C_b(\Sigma)}\left\{\langle f, \nu \rangle - \lim_{n \to \infty}\frac{1}{n}\log E_\sigma\left[e^{\sum_{i=1}^n f(Y_i)}\right]\right\},$$

which is independent of σ. It is our goal now to provide an alternative expression for $I^U(\cdot)$.

Theorem 6.5.4 *Assume that (U) of Section 6.3 holds.*
Then, for all $\nu \in M_1(\Sigma)$,

$$I^U(\nu) = \sup_{u \in C_b(\Sigma), u \geq 1}\left\{-\int_\Sigma \log\left(\frac{\pi u}{u}\right) d\nu\right\}. \tag{6.5.5}$$

Proof: Recall that Assumption (U) implies that for any $\mu \in M_1(\Sigma)$ and any $\sigma \in \Sigma$,

$$\lim_{n \to \infty} \frac{1}{n} \log E_\sigma \left[e^{\sum_{i=1}^n f(Y_i)} \right] = \lim_{n \to \infty} \frac{1}{n} \log \int_\Sigma E_\xi \left[e^{\sum_{i=1}^n f(Y_i)} \right] \mu(d\xi) .$$

(See Exercise 6.3.13.) Thus, repeating the computation leading to (6.5.1), it follows that for any $\nu \in M_1(\Sigma)$,

$$I^U(\nu) = \sup_{f \in C_b(\Sigma)} \left\{ \langle f, \nu \rangle - \limsup_{n \to \infty} \frac{1}{n} \log \langle (\pi_f)^n 1, \nu \rangle \right\} .$$

Assuming that π is Feller continuous, i.e., that $\pi f \in C_b(\Sigma)$ if $f \in C_b(\Sigma)$, the theorem is proved by a repeat of part (b) of the proof of Theorem 6.5.2, with $C_b(\Sigma)$ replacing $B(\Sigma)$ and ν replacing P_1 throughout the proof. Indeed, $\pi_f : C_b(\Sigma) \to C_b(\Sigma)$ by the Feller continuity of π, and moreover, with P_1 replaced by ν, obviously $\langle v_n, \nu \rangle \to 0$ for all $\nu \in M_1(\Sigma)$. Following [DeuS89b], the assumption that π is Feller continuous is removed in Exercise 6.5.7 by an approximation argument.   □

Remark: Note that (6.5.5) is not exactly the same as the expression (3.1.7) of Theorem 3.1.6. Indeed, in the finite state space discussed there, it was more convenient to consider the operator $u\pi$, where by Exercise 3.1.11, the resulting two expressions are identical. At least under condition (U), when Σ is not finite, the analog of (3.1.7) is the expression

$$I^U(\nu) = \sup_{\mu \in M_1(\Sigma), \log\left(\frac{d\mu\pi}{d\mu}\right) \in B(\Sigma)} \left\{ - \int_\Sigma \log \left(\frac{d\mu\pi}{d\mu} \right) d\nu \right\} , \qquad (6.5.6)$$

where $\mu\pi(\cdot) = \int_\Sigma \mu(dx)\pi(x, \cdot)$. For details, see Exercise 6.5.9.

Exercise 6.5.7 (a) Check that for every $\nu \in M_1(\Sigma)$ and every transition probability measure $\pi(\cdot, \cdot)$,

$$\sup_{u \in C_b(\Sigma), u \geq 1} \left\{ - \int_\Sigma \log \left(\frac{\pi u}{u} \right) d\nu \right\} = \sup_{u \in B(\Sigma), u \geq 1} \left\{ - \int_\Sigma \log \left(\frac{\pi u}{u} \right) d\nu \right\} .$$

Hint: Show that if $1 \leq u_n \to u$ boundedly and pointwise, then

$$- \int_\Sigma \log \left(\frac{\pi u_n}{u_n} \right) d\nu \to - \int_\Sigma \log \left(\frac{\pi u}{u} \right) d\nu .$$

Recall that by the monotone class theorem, if $\mathcal{H} \subset \{u : u \in B(\Sigma), u \geq 1\}$ contains $\{u : u \in C_b(\Sigma), u \geq 1\}$ and is closed under bounded pointwise convergence, then $\mathcal{H} = \{u : u \in B(\Sigma), u \geq 1\}$.

(b) Assume that Assumption (U) of Section 6.3 holds, and let $\{f_n\}$ be a uniformly bounded sequence in $B(\Sigma)$ such that $f_n \to f$ pointwise. Show that $\limsup_{n\to\infty} \Lambda(f_n) \le \Lambda(f)$.

Hint: Use part (a) of Exercise 6.3.12, and the convexity of $\Lambda(\cdot)$.

(c) Conclude that Theorem 6.5.4 holds under Assumption (U) even when π is not Feller continuous.

Exercise 6.5.8 (a) Using Jensen's inequality and Lemma 6.2.13, prove that for any $\nu \in M_1(\Sigma)$ and any transition probability measure $\pi(\cdot, \cdot)$,

$$J_\pi(\nu) \triangleq \sup_{u \in C_b(\Sigma), u \ge 1} \left\{ -\int_\Sigma \log\left(\frac{\pi u}{u}\right) d\nu \right\} \ge H(\nu | \nu\pi) \ .$$

(b) Recall that, under Assumption (U), $J_\pi(\nu) = I^U(\nu)$ is the good rate function controlling the LDP of $\{L_n^Y\}$. Prove that there then exists at least one invariant measure for $\pi(\cdot, \cdot)$.

(c) Conclude that when Assumption (U) holds, the invariant measure is unique.

Hint: Invariant measures for π are also invariant measures for $\Pi \triangleq 1/N \sum_{i=1}^N \pi^i$. Hence, it suffices to consider the latter Markov kernel, which by Assumption (U) satisfies $\inf_{x \in \Sigma} \Pi(x, \cdot) \ge (1/M)\nu(\cdot)$ for some $\nu \in M_1(\Sigma)$, and some $M \ge 1$. Therefore,

$$\tilde{\Pi}((b, \sigma), \{b'\} \times \Gamma) = \frac{b'}{M}\nu(\Gamma) + 1_{\{b'=-1\}}\Pi(\sigma, \Gamma)$$

is a Markov kernel on $\{-1, 1\} \times \Sigma$ (equipped with the product σ-field). Let $\tilde{P}_{(b,\sigma)}$ be the Markov chain with transition $\tilde{\Pi}$ and initial state (b, σ), with $\{(b_n, X_n)\}$ denoting its realization. Check that for every $\sigma \in \Sigma$ and $\Gamma \in \mathcal{B}_\Sigma$,

$$
\begin{aligned}
\Pi^n(\sigma, \Gamma) &= \tilde{\Pi}^n((b, \sigma), \{-1, 1\} \times \Gamma) \\
&= q_n(\Gamma) + \tilde{P}_{(b,\sigma)}(X_n \in \Gamma, b_1 = b_2 = \cdots = b_n = -1) \ ,
\end{aligned}
$$

where $q_n(\Gamma)$ is independent of σ. Conclude that for any $\sigma, \sigma' \in \Sigma$, and any $\Gamma \in \mathcal{B}_\Sigma$,

$$|\Pi^n(\sigma, \Gamma) - \Pi^n(\sigma', \Gamma)| \le 2(1 - 1/M)^n \ .$$

Exercise 6.5.9 In this exercise, you will prove that (6.5.6) holds whenever Assumption (U) holds true.

(a) Prove that for any $\mu \in M_1(\Sigma)$,

$$I^U(\nu) = \sup_{f \in B(\Sigma)} \left\{ \langle \nu, f \rangle - \limsup_{n\to\infty} \frac{1}{n} \log\langle \mu \tilde{\pi}_f^n, 1 \rangle \right\} \ ,$$

where

$$\mu \tilde{\pi}_f(d\tau) = e^{f(\tau)} \int_\Sigma \pi(\sigma, d\tau)\mu(d\sigma).$$

(b) Let μ be given with $\bar{f} \triangleq \log(d\mu/d\mu\pi) \in B(\Sigma)$. Show that $\mu\tilde{\pi}_{\bar{f}} = \mu$, and thus, for every $\nu \in M_1(\Sigma)$,

$$I^U(\nu) \geq \langle \nu, \bar{f} \rangle - 0 = -\int_{\Sigma} \log\left(\frac{d\mu\pi}{d\mu}\right) \nu(d\sigma).$$

(c) Fix $v \in B(\Sigma)$ such that $v(\cdot) \geq 1$, note that $\phi \triangleq \pi v \geq 1$, and define the transition kernel

$$\hat{\pi}(\sigma, d\tau) = \pi(\sigma, d\tau)\frac{v(\tau)}{\phi(\sigma)}.$$

Prove that $\hat{\pi}$ is a transition kernel that satisfies Assumption (U). Applying Exercise 6.5.8, conclude that it possesses an invariant measure, denoted $\rho(\cdot)$. Define $\mu(d\sigma) = \rho(d\sigma)/c\phi(\sigma)$, with $c = \int_{\Sigma} \rho(d\sigma)/\phi(\sigma)$. Prove that

$$\log\left(\frac{d\mu\pi}{d\mu}\right) = \log\left(\frac{\pi v}{v}\right) \in B(\Sigma),$$

and apply part (a) of Exercise 6.5.7 to conclude that (6.5.6) holds.

6.5.2 LDP for the k-Empirical Measures

The LDP obtained for the occupation time of Markov chains may easily be extended to the empirical measure of k-tuples, i.e.,

$$L_{n,k}^Y \triangleq \frac{1}{n} \sum_{i=1}^n \delta_{(Y_i, Y_{i+1}, \dots, Y_{i+k-1})} \in M_1(\Sigma^k),$$

where throughout this section $k \geq 2$. This extension is motivated by the fact that at least when $|\Sigma| < \infty$, the resulting LDP has an explicit, simple rate function as shown in Section 3.1.3. The corresponding results are derived here for any Polish state space Σ.

The starting point for the derivation of the LDP for $L_{n,k}^Y$ lies in the observation that if the sequence $\{Y_n\}$ is a Markov chain with state space Σ and transition kernel $\pi(x, dy)$, then the sequence $\{(Y_n, \dots, Y_{n+k-1})\}$ is a Markov chain with state space Σ^k and transition kernel

$$\pi_k(x, dy) = \pi(x_k, dy_k)\prod_{i=1}^{k-1} \delta_{x_{i+1}}(y_i),$$

where $y = (y_1, \dots, y_k)$, $x = (x_1, \dots, x_k) \in \Sigma^k$. Moreover, it is not hard to check that if π satisfies Assumption (U) of Section 6.3 (or π satisfies Assumptions (H-1) and (H-2) of Section 6.4 with some π-invariant measure P_1), so does π_k (now with the π_k-invariant measure $P_k(dx) \triangleq P_1(dx_1)\pi(x_1, dx_2) \cdots \pi(x_{k-1}, dx_k)$). Hence, the following corollary summarizes the results of Sections 6.3, 6.4.2, and 6.5.1 when applied to $L_{n,k}^Y$.

Corollary 6.5.10 *(a) Assume π satisfies Assumption (U) of Section 6.3. Then $L_{n,k}^{\mathbf{Y}}$ satisfies (in the weak topology of $M_1(\Sigma^k)$) the LDP with the good rate function*

$$I_k^U(\nu) \triangleq \sup_{u \in B(\Sigma^k), u \geq 1} \left\{ -\int_{\Sigma^k} \log\left(\frac{\pi_k u}{u}\right) d\nu \right\} .$$

(b) Assume that (H-1) and (H-2) hold for P that is generated by π and the π-invariant measure P_1. Then $L_{n,k}^{\mathbf{Y}}$ satisfies (in the τ-topology of $M_1(\Sigma^k)$) the LDP with the good rate function

$$I_k(\nu) \triangleq \begin{cases} I_k^U(\nu), & \frac{d\nu}{dP_k} \text{ exists} \\ \infty, & \text{otherwise}. \end{cases}$$

To further identify $I_k^U(\cdot)$ and $I_k(\cdot)$, the following definitions and notations are introduced.

Definition 6.5.11 *A measure $\nu \in M_1(\Sigma^k)$ is called shift invariant if, for any $\Gamma \in \mathcal{B}_{\Sigma^{k-1}}$,*

$$\nu(\{\sigma \in \Sigma^k : (\sigma_1, \ldots, \sigma_{k-1}) \in \Gamma\}) = \nu(\{\sigma \in \Sigma^k : (\sigma_2, \ldots, \sigma_k) \in \Gamma\}) .$$

Next, for any $\mu \in M_1(\Sigma^{k-1})$, define the probability measure $\mu \otimes_k \pi \in M_1(\Sigma^k)$ by

$$\mu \otimes_k \pi(\Gamma) = \int_{\Sigma^{k-1}} \mu(dx) \int_{\Sigma} \pi(x_{k-1}, dy) 1_{\{(x,y) \in \Gamma\}} , \quad \forall \Gamma \in \mathcal{B}_{\Sigma^k} .$$

Theorem 6.5.12 *For any transition kernel π, and any $k \geq 2$,*

$$I_k^U(\nu) = \begin{cases} H(\nu | \nu_{k-1} \otimes_k \pi) , & \nu \text{ shift invariant} \\ \infty & , \text{ otherwise}, \end{cases}$$

where ν_{k-1} denotes the marginal of ν on the first $(k-1)$ coordinates.

Remark: The preceding identification is independent of the existence of the LDP. Moreover, it obviously implies that

$$I_k(\nu) = \begin{cases} H(\nu | \nu_{k-1} \otimes_k \pi) , & \nu \text{ shift invariant, } \frac{d\nu}{dP_k} \text{ exists} \\ \infty & , \text{ otherwise} \end{cases}$$

Proof: Assume first that ν is not shift invariant. A $\psi \in B(\Sigma^{k-1})$ may be found such that $\psi \geq 0$ and

$$\int_{\Sigma^k} \psi(x_1, \ldots, x_{k-1}) \nu(dx) - \int_{\Sigma^k} \psi(x_2, \ldots, x_k) \nu(dx) \geq 1 .$$

For any $\alpha > 0$, define $u_\alpha \in B(\Sigma^k)$ by

$$u_\alpha(x) = e^{\alpha \psi(x_1, \ldots, x_{k-1})} \geq 1 .$$

Therefore, $(\pi_k u_\alpha)(x) = e^{\alpha \psi(x_2, \ldots, x_k)}$, and

$$- \int_{\Sigma^k} \log \left(\frac{\pi_k u_\alpha}{u_\alpha} \right) d\nu$$

$$= \alpha \left\{ \int_{\Sigma^k} \psi(x_1, \ldots, x_{k-1}) \nu(dx) - \int_{\Sigma^k} \psi(x_2, \ldots, x_k) \nu(dx) \right\} \geq \alpha .$$

Considering $\alpha \to \infty$, it follows that $I_k^U(\nu) = \infty$. Next, note that for ν shift invariant, by Lemma 6.2.13,

$$H(\nu | \nu_{k-1} \otimes_k \pi)$$

$$= \sup_{\phi \in B(\Sigma^k)} \left\{ \langle \phi, \nu \rangle - \log \int_{\Sigma^{k-1}} \nu(dx_1, \ldots, dx_{k-1}, \Sigma) \int_\Sigma e^{\phi(x_1, \ldots, x_{k-1}, y)} \pi(x_{k-1}, dy) \right\}$$

$$= \sup_{\phi \in B(\Sigma^k)} \left\{ \langle \phi, \nu \rangle - \log \int_{\Sigma^{k-1}} \nu(\Sigma, dx_2, \ldots, dx_k) \int_\Sigma e^{\phi(x_2, \ldots, x_k, y)} \pi(x_k, dy) \right\} .$$

$$(6.5.13)$$

Fix $\phi \in B(\Sigma^k)$ and let $u \in B(\Sigma^k)$ be defined by $u(x) = c e^{\phi(x)}$, where $c > 0$ is chosen such that $u \geq 1$. Then

$$- \log \left(\frac{\pi_k u}{u} \right) = \phi(x) - \log \int_\Sigma e^{\phi(x_2, \ldots, x_k, y)} \pi(x_k, dy).$$

Thus, by Jensen's inequality,

$$- \int_{\Sigma^k} \log \left(\frac{\pi_k u}{u} \right) d\nu$$

$$= \langle \phi, \nu \rangle - \int_{\Sigma^{k-1}} \nu(\Sigma, dx_2, \ldots, dx_k) \log \int_\Sigma e^{\phi(x_2, \ldots, x_k, y)} \pi(x_k, dy)$$

$$\geq \langle \phi, \nu \rangle - \log \int_{\Sigma^k} e^{\phi(x_2, \ldots, x_k, y)} \nu(\Sigma, dx_2, \ldots, dx_k) \pi(x_k, dy),$$

and it follows from (6.5.13) that $I_k^U(\nu) \geq H(\nu | \nu_{k-1} \otimes_k \pi)$.

To see the reverse inequality, let $\nu^x(\cdot)$ denote the regular conditional probability distribution of x_k given the σ-field generated by the restriction of Σ^k to the first $(k-1)$ coordinates. (Such a conditional probability exists because Σ^k is Polish; *c.f.* Appendix D.) Assume with no loss of generality that $H(\nu | \nu_{k-1} \otimes_k \pi) < \infty$. Then

$$\nu(\{x \in \Sigma^{k-1} : \frac{d\nu^x}{d\pi(x_{k-1}, \cdot)} \text{ exists }\}) = 1,$$

and

$$H(\nu|\nu_{k-1} \otimes_k \pi)$$

$$= \int_{\Sigma^{k-1}} \nu(dx_1, \ldots, dx_{k-1}, \Sigma) \int_{\Sigma} d\nu^x \log \frac{d\nu^x}{d\pi(x_{k-1}, \cdot)}$$

$$= \int_{\Sigma^{k-1}} \nu(dx_1, \ldots, dx_{k-1}, \Sigma) \sup_{\phi \in B(\Sigma^k)} \left\{ \int_{\Sigma} \phi(x_1, \ldots, x_k) \nu^x(dx_k) \right.$$

$$\left. - \log \int_{\Sigma} e^{\phi(x_1, \ldots, x_{k-1}, y)} \pi(x_{k-1}, dy) \right\}$$

$$\geq \sup_{\phi \in B(\Sigma^k)} \left\{ \int_{\Sigma^k} \phi(x) \nu(dx) - \int_{\Sigma^{k-1}} \nu(dx_1, \ldots, dx_{k-1}, \Sigma) \right.$$

$$\left. \log \int_{\Sigma} e^{\phi(x_1, \ldots, x_{k-1}, y)} \pi(x_{k-1}, dy) \right\}$$

$$= \sup_{\phi \in B(\Sigma^k)} \left\{ \int_{\Sigma^k} \phi d\nu - \int_{\Sigma^{k-1}} \nu(\Sigma, dx_2, \ldots, dx_k) \right.$$

$$\left. \log \int_{\Sigma} e^{\phi(x_2, \ldots, x_k, y)} \pi(x_k, dy) \right\}$$

$$= \sup_{\phi \in B(\Sigma^k)} \left\{ \int_{\Sigma^k} \phi d\nu - \int_{\Sigma^k} \log(\pi_k e^\phi) d\nu \right\}$$

$$= \sup_{\phi \in B(\Sigma^k)} \left\{ -\int_{\Sigma^k} \log\left(\frac{\pi_k e^\phi}{e^\phi}\right) d\nu \right\} = I_k^U(\nu) .$$

$\square$

6.5.3   Process Level LDP for Markov Chains

The LDP derived in Section 6.5.2 enables the deviant behavior of empirical means of fixed length sequences to be dealt with as the number of terms n in the empirical sum grows. Often, however, some information is needed on the behavior of sequences whose length is not bounded with n. It then becomes useful to consider sequences of infinite length. Formally, one could form the empirical measure

$$L_{n,\infty}^{\mathbf{Y}} \triangleq \frac{1}{n} \sum_{i=1}^{n} \delta_{T^i \mathbf{Y}} ,$$

where $\mathbf{Y} = (Y_1, Y_2, \ldots)$ and $T^i \mathbf{Y} = (Y_{i+1}, Y_{i+2}, \ldots)$, and inquire about the LDP of the random variable $L_{n,\infty}^{\mathbf{Y}}$ in the space of probability measures on $\Sigma^{\mathbb{Z}_+}$. Since such measures may be identified with probability measures on processes, this LDP is referred to as *process level* LDP.

A natural point of view is to consider the infinite sequences $\mathbf{Y}$ as limits of finite sequences, and to use a projective limit approach. Therefore, the discussion on the process level LDP begins with some topological preliminaries and an exact definition of the probability spaces involved. Since the projective limit approach necessarily involves weak topology, only the weak topologies of $M_1(\Sigma)$ and $M_1(\Sigma^{\mathbb{Z}_+})$ will be considered.

As in the beginning of this section, let Σ be a Polish space, equipped with the metric d and the Borel σ-field $\mathcal{B}_\Sigma$ associated with it, and let Σ^k denote its kth-fold product, whose topology is compatible with the metric $d_k(\sigma, \sigma') = \sum_{i=1}^{k} d(\sigma_i, \sigma'_i)$. The sequence of spaces Σ^k with the obvious projections $p_{m,k} : \Sigma^m \to \Sigma^k$, defined by $p_{m,k}(\sigma_1, \ldots, \sigma_m) = (\sigma_1, \ldots, \sigma_k)$ for $k \leq m$, form a projective system with projective limit that is denoted $\Sigma^{\mathbb{Z}_+}$, and canonical projections $p_k : \Sigma^{\mathbb{Z}_+} \to \Sigma^k$. Since Σ^k are separable spaces and $\Sigma^{\mathbb{Z}_+}$ is countably generated, it follows that $\Sigma^{\mathbb{Z}_+}$ is separable, and the Borel σ-field on $\Sigma^{\mathbb{Z}_+}$ is the product of the appropriate Borel σ-fields. Finally, the projective topology on $\Sigma^{\mathbb{Z}_+}$ is compatible with the metric

$$d_\infty(\sigma, \sigma') = \sum_{k=1}^{\infty} \frac{1}{2^k} \left[\frac{d_k(p_k\sigma, p_k\sigma')}{1 + d_k(p_k\sigma, p_k\sigma')} \right],$$

which makes $\Sigma^{\mathbb{Z}_+}$ into a Polish space. Consider now the spaces $M_1(\Sigma^k)$, equipped with the weak topology and the projections $p_{m,k} : M_1(\Sigma^m) \to M_1(\Sigma^k)$, $k \leq m$, such that $p_{m,k}\nu$ is the marginal of $\nu \in M_1(\Sigma^m)$ with respect to its first k coordinates. The projective limit of this projective system is merely $M_1(\Sigma^{\mathbb{Z}_+})$ as stated in the following lemma.

Lemma 6.5.14 *The projective limit of $(M_1(\Sigma^k), p_{m,k})$ is homeomorphic to the space $M_1(\Sigma^{\mathbb{Z}_+})$ when the latter is equipped with the weak topology.*

Proof: By Kolmogorov's extension theorem, for any sequence $\nu_n \in M_1(\Sigma^n)$ such that $p_{m,k}\nu_m = \nu_k$ for all $k \leq m$, there exists a $\nu \in M_1(\Sigma^{\mathbb{Z}_+})$ such that $\nu_n = p_n\nu$, i.e., $\{\nu_n\}$ are the marginals of ν. The converse being trivial, it follows that the space $M_1(\Sigma^{\mathbb{Z}_+})$ may be identified with the projective limit of $(M_1(\Sigma^k), p_{m,k})$, denoted $\mathcal{X}$. It remains only to prove that the bijection $\nu \mapsto \{p_n\nu\}$ between $M_1(\Sigma^{\mathbb{Z}_+})$ and $\mathcal{X}$ is a homeomorphism. To this end, recall that by the Portmanteau theorem, the topologies of $M_1(\Sigma^{\mathbb{Z}_+})$ and $\mathcal{X}$ are, respectively, generated by the sets

$$U_{f,x,\delta} = \left\{ \nu \in M_1(\Sigma^{\mathbb{Z}_+}) : |\int_{\Sigma^{\mathbb{Z}_+}} f d\nu - x| < \delta \right\},$$
$$f \in C_u(\Sigma^{\mathbb{Z}_+}), x \in \mathbb{R}, \delta > 0,$$

and

$$\hat{U}_{\hat{f},x,\delta,k} = \left\{ \nu \in \mathcal{X} : |\int_{\Sigma^k} \hat{f} dp_k \nu - x| < \delta \right\},$$

$$k \in \mathbb{Z}_+, \hat{f} \in C_u(\Sigma^k), x \in \mathbb{R}, \delta > 0,$$

where $C_u(\Sigma^{\mathbb{Z}+})$ $(C_u(\Sigma^n))$ denotes the space of bounded, real valued, uniformly continuous functions on $\Sigma^{\mathbb{Z}+}$ (Σ^n, respectively). It is easy to check that each $\hat{U}_{\hat{f},x,\delta,k} \subset \mathcal{X}$ is mapped to $U_{f,x,\delta}$, where $f \in C_u(\Sigma^{\mathbb{Z}+})$ is the natural extension of $\hat{f}$, i.e., $f(\sigma) = \hat{f}(\sigma_1, \ldots, \sigma_k)$. Consequently, the map $\nu \mapsto \{p_n\nu\}$ is continuous. To show that its inverse is continuous, fix $U_{f,x,\delta} \subset M_1(\Sigma^{\mathbb{Z}+})$ and $\nu \in U_{f,x,\delta}$. Let $\hat{\delta} = (\delta - |\int_{\Sigma^{\mathbb{Z}+}} f d\nu - x|)/3 > 0$. Observe that $d_\infty(\sigma, \sigma') \le 2^{-k}$ whenever $\sigma_1 = \sigma'_1, \ldots, \sigma_k = \sigma'_k$. Hence, by the uniform continuity of f,

$$\lim_{k \to \infty} \sup_{\sigma, \sigma' \in \Sigma^{\mathbb{Z}+}: (\sigma_1, \ldots, \sigma_k) = (\sigma'_1, \ldots, \sigma'_k)} |f(\sigma) - f(\sigma')| = 0,$$

and for $k < \infty$ large enough, $\sup_{\sigma \in \Sigma^{\mathbb{Z}+}} |f(\sigma) - \hat{f}(\sigma_1, \ldots, \sigma_k)| \le \hat{\delta}$, where

$$\hat{f}(\sigma_1, \ldots, \sigma_k) = f(\sigma_1, \ldots, \sigma_k, \sigma^*, \sigma^*, \ldots)$$

for some arbitrary $\sigma^* \in \Sigma$. Clearly, $\hat{f} \in C_u(\Sigma^k)$. Moreover, for $\hat{x} = \int_{\Sigma^k} \hat{f} dp_k \nu$, the image of $U_{f,x,\delta}$ contains the neighborhood $\hat{U}_{\hat{f},\hat{x},\hat{\delta},k}$ of $\{p_k\nu\}$ in $\mathcal{X}$. Hence, the image of each $U_{f,x,\delta}$ is open, and the bijection $\nu \mapsto \{p_k\nu\}$ possesses a continuous inverse.   □

Returning to the empirical process, observe that for each k, $p_k(L_{n,\infty}^{\mathbf{Y}}) = L_{n,k}^{\mathbf{Y}}$. The following is therefore an immediate consequence of Lemma 6.5.14, the Dawson–Gärtner theorem (Theorem 4.6.1), and the results of Section 6.5.2.

Corollary 6.5.15 *Assume that (U) of section 6.3 holds. Then the sequence* $\{L_{n,\infty}^{\mathbf{Y}}\}$ *satisfies the LDP in* $M_1(\Sigma^{\mathbb{Z}+})$ *(equipped with the weak topology) with the good rate function*

$$I_\infty(\nu) = \begin{cases} \sup_{k \ge 2} H(p_k\nu | p_{k-1}\nu \otimes_k \pi) & , \quad \nu \text{ shift invariant} \\ \infty & , \quad \text{otherwise} \end{cases}$$

where $\nu \in M_1(\Sigma^{\mathbb{Z}+})$ *is called shift invariant if, for all* $k \in \mathbb{Z}_+$, $p_k\nu$ *is shift invariant in* $M_1(\Sigma^k)$.

Our goal now is to derive an explicit expression for $I_\infty(\cdot)$. For $i = 0, 1$, let $\mathbb{Z}_i = \mathbb{Z} \cap (-\infty, i]$ and let $\Sigma^{\mathbb{Z}_i}$ be constructed similarly to $\Sigma^{\mathbb{Z}+}$ via projective limits. For any $\mu \in M_1(\Sigma^{\mathbb{Z}+})$ shift invariant, consider the measures

$\mu_i^* \in M_1(\Sigma^{\mathbb{Z}_i})$ such that for every $k \geq 1$ and every $\mathcal{B}_{\Sigma^k}$ measurable set Γ,

$$\mu_i^*(\{(\ldots, \sigma_{i+1-k}, \ldots, \sigma_i) : (\sigma_{i+1-k}, \ldots, \sigma_i) \in \Gamma\}) = p_k\mu(\Gamma).$$

Such a measure exists and is unique by the consistency condition satisfied by μ and Kolmogorov's extension theorem. Next, for any $\mu_0^* \in M_1(\Sigma^{\mathbb{Z}_0})$, define the Markov extension of it, denoted $\mu_0^* \otimes \pi \in M_1(\Sigma^{\mathbb{Z}_1})$, such that for any $\phi \in B(\Sigma^{k+1})$, $k \geq 1$,

$$\int_{\Sigma^{\mathbb{Z}_1}} \phi(\sigma_{-(k-1)}, \ldots, \sigma_0, \sigma_1) d\mu_0^* \otimes \pi$$

$$= \int_{\Sigma^{\mathbb{Z}_0}} \int_{\Sigma} \phi(\sigma_{-(k-1)}, \ldots, \sigma_0, \tau) \pi(\sigma_0, d\tau) d\mu_0^*.$$

In these notations, for any shift invariant $\nu \in M_1(\Sigma^{\mathbb{Z}_+})$,

$$H(p_k\nu|p_{k-1}\nu \otimes_k \pi) = H(\bar{p}_k\nu_1^*|\bar{p}_k(\nu_0^* \otimes \pi)),$$

where for any $\mu \in M_1(\Sigma^{\mathbb{Z}_1})$, and any $\Gamma \in \mathcal{B}_{\Sigma^k}$,

$$\bar{p}_k\mu(\Gamma) = \mu(\{(\sigma_{k-2}, \ldots, \sigma_0, \sigma_1) \in \Gamma\}).$$

The characterization of $I_\infty(\cdot)$ is a direct consequence of the following lemma.

Lemma 6.5.16 (Pinsker) *Let Σ be Polish and $\nu, \mu \in M_1(\Sigma^{\mathbb{Z}_1})$. Then*

$$H(\bar{p}_k\nu|\bar{p}_k\mu) \nearrow H(\nu|\mu) \text{ as } k \to \infty.$$

Proof of Lemma 6.5.16: Recall that

$$H(\bar{p}_k\nu|\bar{p}_k\mu) = \sup_{\phi \in C_b(\Sigma^k)} \left\{ \int_{\Sigma^k} \phi d\bar{p}_k\nu - \log\left(\int_{\Sigma^k} e^\phi d\bar{p}_k\mu\right) \right\},$$

whereas

$$H(\nu|\mu) = \sup_{\phi \in C_b(\Sigma^{\mathbb{Z}_1})} \left\{ \int_{\Sigma^{\mathbb{Z}_1}} \phi d\nu - \log\left(\int_{\Sigma^{\mathbb{Z}_1}} e^\phi d\mu\right) \right\}.$$

It clearly follows that $H(\bar{p}_k\nu|\bar{p}_k\mu) \leq H(\bar{p}_{k+1}\nu|\bar{p}_{k+1}\mu) \leq H(\nu|\mu)$. On the other hand, by the same construction used in the proof of Lemma 6.5.14, it is clear that any function $\phi \in C_u(\Sigma^{\mathbb{Z}_1})$ must satisfy

$$\lim_{k \to \infty} \sup_{\sigma, \sigma' \in \Sigma^{\mathbb{Z}_1} : (\sigma_{-(k-1)}, \ldots, \sigma_1) = (\sigma'_{-(k-1)}, \ldots, \sigma'_1)} |\phi(\sigma) - \phi(\sigma')| = 0.$$

The lemma follows by considering, for each $\phi \in C_b(\Sigma^{\mathbb{Z}_1})$, an approximating (almost everywhere ν, μ) sequence in $C_u(\Sigma^{\mathbb{Z}_1})$, and then approximating each $\phi \in C_u(\Sigma^{\mathbb{Z}_1})$ by the sequence $\phi(\ldots, \sigma^*, \sigma_{-(k-2)}, \ldots, \sigma_1)$, where σ^* is arbitrary in Σ. $\square$

Combining Corollary 6.5.15, Lemma 6.5.16 and the preceding discussion, the following identification of $I_\infty(\cdot)$ is obtained.

Corollary 6.5.17

$$I_\infty(\nu) = \begin{cases} H(\nu_1^*|\nu_0^* \otimes \pi) & , \quad \nu \text{ shift invariant} \\ \infty & , \quad \text{otherwise}. \end{cases}$$

Exercise 6.5.18 Check that the metric defined on $\Sigma^{\mathbb{Z}_+}$ is indeed compatible with the projective topology.

Exercise 6.5.19 Assume that (U) of Section 6.3 holds. Deduce the occupation times and the kth order LDP from the process level LDP of this section.

Exercise 6.5.20 Deduce the process level LDP (in the weak topology) for Markov processes satisfying Assumptions (H-1) and (H-2).

6.6 A Weak Convergence Approach to Large Deviations

The variational characterization of the relative entropy functional is key to an alternative approach for establishing the LDP in Polish spaces. This approach, developed by Dupuis and Ellis, is particularly useful in obtaining the LDP for the occupation measure of Markov chains with discontinuous statistics, and is summarized in the monograph [DuE97]. It is based on transforming large deviations questions to questions about weak convergence of associated *controlled* processes. Here, we present the essential idea in a particularly simple setup, namely that of Sanov's theorem (Corollary 6.2.3).

The key ingredient in the approach of Dupuis and Ellis is the following observation, based on the duality representation of the relative entropy.

Theorem 6.6.1 (Dupuis-Ellis) *Suppose $I(\cdot)$ is a good rate function on a Polish space $\mathcal{X}$ such that for a family of probability measures $\{\tilde{\mu}_\epsilon\}$ on $\mathcal{X}$ and any $f \in C_b(\mathcal{X})$,*

$$\Lambda_f \stackrel{\triangle}{=} \sup_{x \in \mathcal{X}} \{f(x) - I(x)\} = \lim_{\epsilon \to 0} \sup_{\tilde{\nu} \in M_1(\mathcal{X})} \{\langle f, \tilde{\nu} \rangle - \epsilon H(\tilde{\nu}|\tilde{\mu}_\epsilon)\}. \tag{6.6.2}$$

Then, $\{\tilde{\mu}_\epsilon\}$ satisfies the LDP in $\mathcal{X}$ with good rate function $I(\cdot)$.

Proof: The duality in Lemma 6.2.13 (see in particular (6.2.14)) implies that for all $f \in C_b(\mathcal{X})$,

$$\log \int_{\mathcal{X}} e^{f(x)/\epsilon} \tilde{\mu}_\epsilon(dx) = \sup_{\tilde{\nu} \in M_1(\mathcal{X})} \{\epsilon^{-1} \langle f, \tilde{\nu} \rangle - H(\tilde{\nu}|\tilde{\mu}_\epsilon)\}.$$

Coupled with (6.6.2) this implies that

$$\Lambda_f = \lim_{\epsilon \to 0} \epsilon \log \int_{\mathcal{X}} e^{f(x)/\epsilon} \tilde{\mu}_\epsilon(dx) = \sup_{x \in \mathcal{X}} \{f(x) - I(x)\}.$$

The conclusion of the theorem follows from Theorem 4.4.13. □

We turn to the following important application of Theorem 6.6.1. Let Σ be a Polish space and $\mathcal{X} = M_1(\Sigma)$ equipped with the topology of weak convergence and the corresponding Borel σ-field. To fix notations related to the decomposition of measures on Σ^n, for given $i \in \{1, \ldots, n\}$ and $\nu^{(n)} \in M_1(\Sigma^n)$, let $\nu^{(n)}_{[1,i]} \in M_1(\Sigma^i)$ denote the restriction of $\nu^{(n)}$ to the first i coordinates, such that $\nu^{(n)}_{[1,i]}(\Gamma) = \nu^{(n)}(\Gamma \times \Sigma^{n-i})$, for all $\Gamma \in \mathcal{B}_{\Sigma^i}$. For $i \geq 2$, writing $\Sigma^i = \Sigma^{i-1} \times \Sigma$, let $\nu^{(n),\mathbf{y}}_i \in M_1(\Sigma)$ denote the r.c.p.d. of $\nu^{(n)}_{[1,i]}$ given the projection $\pi_{i-1} : \Sigma^i \to \Sigma^{i-1}$ on the first $(i-1)$ coordinates of $\mathbf{y}$, so that for any $\Gamma_{i-1} \in \mathcal{B}_{\Sigma^{i-1}}$ and $\Gamma \in \mathcal{B}_\Sigma$,

$$\nu^{(n)}_{[1,i]}(\Gamma_{i-1} \times \Gamma) = \int_{\Gamma_{i-1}} \nu^{(n),\mathbf{y}}_i(\Gamma) \nu^{(n)}_{[1,i-1]}(d\mathbf{y})$$

(see Definition D.2 and Theorem D.3 for the existence of such r.c.p.d.). To simplify notations, we also use $\nu^{(n),\mathbf{y}}_i$ to denote $\nu^{(n)}_{[1,i]}$ in case $i = 1$. For any $n \in \mathbb{Z}_+$, let $\mu^{(n)} \in M_1(\Sigma^n)$ denote the law of the Σ^n-valued random variable $\mathbf{Y} = (Y_1, \ldots, Y_n)$, with $\tilde{\mu}_n \in M_1(\mathcal{X})$ denoting the law of $L_n^{\mathbf{Y}}$ induced on $\mathcal{X}$ via the measurable map $\mathbf{y} \to L_n^{\mathbf{y}}$.

The following lemma provides an alternative representation of the right-side of (6.6.2). It is essential for proving the convergence in (6.6.2), and for identifying the limit as Λ_f.

Lemma 6.6.3 *For any $f \in C_b(\mathcal{X})$ and $\mu^{(n)} \in M_1(\Sigma^n)$,*

$$\Lambda_{f,n} \stackrel{\triangle}{=} \sup_{\tilde{\nu} \in M_1(\mathcal{X})} \{\langle f, \tilde{\nu} \rangle - \frac{1}{n} H(\tilde{\nu}|\tilde{\mu}_n)\}$$

$$= \sup_{\nu^{(n)} \in M_1(\Sigma^n)} \{\int_{\Sigma^n} f(L_n^{\mathbf{y}}) \nu^{(n)}(d\mathbf{y}) - \frac{1}{n} H(\nu^{(n)}|\mu^{(n)})\} \qquad (6.6.4)$$

$$= \sup_{\nu^{(n)} \in M_1(\Sigma^n)} \int_{\Sigma^n} [f(L_n^{\mathbf{y}}) - \frac{1}{n} \sum_{i=1}^n H(\nu^{(n),\mathbf{y}}_i(\cdot)|\mu^{(n),\mathbf{y}}_i(\cdot))] \nu^{(n)}(d\mathbf{y}).$$

Proof: Fix $f \in C_b(\mathcal{X})$, $n \in \mathbb{Z}_+$ and $\mu^{(n)} \in M_1(\Sigma^n)$. For computing $\Lambda_{f,n}$ it suffices to consider $\tilde{\nu} = \tilde{\nu}_n$ such that $\rho_n = d\tilde{\nu}_n / d\tilde{\mu}_n$ exists. Fixing such $\tilde{\nu}_n \in M_1(\mathcal{X})$, since $\tilde{\mu}_n(\{L_n^{\mathbf{y}} : \mathbf{y} \in \Sigma^n\}) = 1$, also $\tilde{\nu}_n(\{L_n^{\mathbf{y}} : \mathbf{y} \in \Sigma^n\}) = 1$. Consequently, $\tilde{\nu}_n$ can be represented as a law on $\mathcal{X}$ induced by some $\nu^{(n)} \in M_1(\Sigma^n)$ via the continuous mapping $\mathbf{y} \mapsto L_n^{\mathbf{y}}$. Hence, by Lemma 6.2.13,

$$
\begin{aligned}
H(\tilde{\nu}_n | \tilde{\mu}_n) &= \sup_{g \in C_b(\mathcal{X})} \{\langle g, \tilde{\nu}_n \rangle - \log \int_{\mathcal{X}} e^{g(x)} \tilde{\mu}_n(dx)\} \\
&= \sup_{g \in C_b(\mathcal{X})} \{\int_{\Sigma^n} g(L_n^{\mathbf{y}}) \nu^{(n)}(d\mathbf{y}) - \log \int_{\Sigma^n} e^{g(L_n^{\mathbf{y}})} \mu^{(n)}(d\mathbf{y})\} \\
&\leq \sup_{h \in C_b(\Sigma^n)} \{\int_{\Sigma^n} h(\mathbf{y}) \nu^{(n)}(d\mathbf{y}) - \log \int_{\Sigma^n} e^{h(\mathbf{y})} \mu^{(n)}(d\mathbf{y})\} \\
&= H(\nu^{(n)} | \mu^{(n)}) ,
\end{aligned}
$$

with equality when $\nu^{(n)}$ is such that $d\nu^{(n)}/d\mu^{(n)} = \rho_n \circ (L_n^{\mathbf{y}})^{-1}$. Since $\langle f, \tilde{\nu}_n \rangle = \int_{\Sigma^n} f(L_n^{\mathbf{y}}) \nu^{(n)}(d\mathbf{y})$ and ρ_n is an arbitrary $\tilde{\mu}_n$-probability density,

$$
\Lambda_{f,n} = \sup_{\nu^{(n)} \in M_1(\Sigma^n)} \{\int_{\Sigma^n} f(L_n^{\mathbf{y}}) \nu^{(n)}(d\mathbf{y}) - \frac{1}{n} H(\nu^{(n)} | \mu^{(n)})\} . \tag{6.6.5}
$$

Applying Theorem D.13 sequentially for $\pi_{i-1} : \Sigma^{i-1} \times \Sigma \to \Sigma^{i-1}$, $i = n, n-1, \ldots, 2$, we have the identity

$$
H(\nu^{(n)} | \mu^{(n)}) = \sum_{i=1}^{n} \int_{\Sigma^n} H(\nu_i^{(n),\mathbf{y}}(\cdot) | \mu_i^{(n),\mathbf{y}}(\cdot)) \nu^{(n)}(d\mathbf{y}) . \tag{6.6.6}
$$

Combining (6.6.5) and (6.6.6) completes the proof of the lemma. □

We demonstrate this approach by providing yet another proof of Sanov's theorem (with respect to the weak topology, that is, Corollary 6.2.3). This proof is particularly simple because of the ease in which convexity can be used in demonstrating the upper bound. See the historical notes at the end of this chapter for references to more challenging applications.

Proof of Corollary 6.2.3 (Weak convergence method): The fact that $H(\cdot | \mu)$ is a good rate function is a consequence of Lemma 6.2.12. Fixing $f \in C_b(\mathcal{X})$, by Theorem 6.6.1, with $\epsilon = 1/n$, suffices for showing the LDP to prove the convergence of $\Lambda_{f,n}$ of (6.6.4) to Λ_f of (6.6.2). With $\mu^{(n)} = \mu^n$ a product measure, it follows that $\mu_i^{(n),\mathbf{y}} = \mu$ for $1 \leq i \leq n$. Choose in the third line of (6.6.4) a product measure $\nu^{(n)} = \nu^n$ such that $\nu_i^{(n),\mathbf{y}} = \nu$ for $1 \leq i \leq n$, to get the lower bound

$$
\Lambda_{f,n} \geq \int_{\Sigma^n} f(L_n^{\mathbf{y}}) \nu^n(d\mathbf{y}) - H(\nu | \mu) .
$$

Let $b = 1 + \sup_{x \in \mathcal{X}} |f(x)| < \infty$. Fix $\eta \in (0, b)$ and consider the open set $G_\eta = \{x : f(x) > f(\nu) - \eta\} \subset \mathcal{X}$, which contains a neighborhood of ν from the base of the topology of $\mathcal{X}$, that is, a set $\cap_{i=1}^m U_{\phi_i, \langle \phi_i, \nu \rangle, \delta}$ as in (6.2.2) for some $m \in \mathbb{Z}_+$, $\delta > 0$ and $\phi_i \in C_b(\Sigma)$. If $Y_1, \ldots, Y_n$ is a sequence of independent, Σ-valued random variables, identically distributed according to $\nu \in M_1(\Sigma)$, then for any $\phi \in C_b(\Sigma)$ fixed, $\langle \phi, L_n^{\mathbf{Y}} \rangle \to \langle \phi, \nu \rangle$ in probability by the law of large numbers. In particular,

$$
\int_{\Sigma^n} f(L_n^{\mathbf{y}}) \nu^n(d\mathbf{y}) \geq f(\nu) - \eta - 2b\tilde{\nu}_n(G_\eta^c)
$$

$$
\geq f(\nu) - \eta - 2b \sum_{i=1}^m \nu^n(\{\mathbf{y} : |\langle \phi_i, L_n^{\mathbf{y}} \rangle - \langle \phi_i, \nu \rangle| > \delta\})
$$

$$
\xrightarrow[n \to \infty]{} f(\nu) - \eta.
$$

Therefore, considering $\eta \to 0$ and $\nu \in M_1(\Sigma)$ arbitrary,

$$
\liminf_{n \to \infty} \Lambda_{f,n} \geq \sup_{\nu \in M_1(\Sigma)} \{f(\nu) - H(\nu|\mu)\} = \Lambda_f . \tag{6.6.7}
$$

We turn to the proof of the complementary upper bound. By the convexity of $H(\cdot|\mu)$, for any $\nu^{(n)} \in M_1(\Sigma^n)$,

$$
\frac{1}{n} \sum_{i=1}^n H(\nu_i^{(n),\mathbf{y}}|\mu) \geq H(\nu_n^{\mathbf{y}}|\mu) ,
$$

where $\nu_n^{\mathbf{y}} = n^{-1} \sum_{i=1}^n \nu_i^{(n),\mathbf{y}}$ is a *random* probability measure on Σ. Thus, by Lemma 6.6.3 there exist $\nu^{(n)} \in M_1(\Sigma^n)$ such that

$$
\bar{\Lambda}_f \triangleq \limsup_{n \to \infty} \Lambda_{f,n} \leq \limsup_{n \to \infty} \int_{\Sigma^n} [f(L_n^{\mathbf{y}}) - H(\nu_n^{\mathbf{y}}|\mu)] \nu^{(n)}(d\mathbf{y}) . \tag{6.6.8}
$$

For proving that $\bar{\Lambda}_f \leq \Lambda_f$, we may and will assume that $\bar{\Lambda}_f > -\infty$. Consider the random variables $(\nu_n^{\mathbf{Y}}, L_n^{\mathbf{Y}}) \in \mathcal{X}^2$ corresponding to sampling independently at each n, according to the law $\nu^{(n)}$ on Σ^n, a sample $\mathbf{Y} = (Y_1, \ldots, Y_n)$. This construction allows for embedding the whole sequence $\{(\nu_n^{\mathbf{Y}}, L_n^{\mathbf{Y}})\}$ in one probability space $(\Omega, \mathcal{F}, P)$, where $P \in M_1\left((\mathcal{X}^2)^{\mathbb{Z}_+}\right)$ is the law of the sequence $\{(\nu_n^{\mathbf{Y}}, L_n^{\mathbf{Y}})\}$. Applying Fatou's lemma for the nonnegative functions $H(\nu_n^{\mathbf{y}}|\mu) - f(L_n^{\mathbf{y}}) + b$, by (6.6.8), it follows that

$$
\bar{\Lambda}_f \leq \limsup_{n \to \infty} \int \left(f(L_n^{\mathbf{Y}}) - H(\nu_n^{\mathbf{Y}}|\mu) \right) dP
$$

$$
\leq \int \limsup_{n \to \infty} \left(f(L_n^{\mathbf{Y}}) - H(\nu_n^{\mathbf{Y}}|\mu) \right) dP . \tag{6.6.9}
$$

Let $\mathcal{F}_{i,n} = \{\Gamma \times \Sigma^{n-i} : \Gamma \in \mathcal{B}_{\Sigma^i}\}$ for $1 \leq i \leq n$ and $\mathcal{F}_{0,n} = \{\emptyset, \Sigma^n\}$. Fix $\phi \in C_b(\Sigma)$ and $n \in \mathbb{Z}_+$. Note that $\langle \phi, L_n^{\mathbf{Y}} \rangle - \langle \phi, \nu_n^{\mathbf{Y}} \rangle = n^{-1} S_n$, where for the filtration $\mathcal{F}_{k,n}$,

$$S_k \stackrel{\triangle}{=} \sum_{i=1}^{k} \Big(\phi(Y_i) - E(\phi(Y_i)|\mathcal{F}_{i-1,n}) \Big), \quad k = 1, \ldots, n, \quad S_0 = 0,$$

is a discrete time martingale of bounded differences, null at 0. By (2.4.10), for some $c > 0$ and any $\delta > 0$, $n \in \mathbb{Z}_+$,

$$P(|\langle \phi, L_n^{\mathbf{Y}} \rangle - \langle \phi, \nu_n^{\mathbf{Y}} \rangle| > \delta) \leq 2 e^{-cn\delta^2},$$

so applying the Borel-Cantelli lemma, it follows that

$$\langle \phi, L_n^{\mathbf{Y}} \rangle - \langle \phi, \nu_n^{\mathbf{Y}} \rangle \to 0, \quad P - \text{almost surely.}$$

With $M_1(\Sigma)$ possessing a countable convergence determining class $\{\phi_i\}$, it follows that P-almost surely, the (random) set of limit points of the sequence $\{L_n^{\mathbf{Y}}\}$ is the same as that of the sequence $\{\nu_n^{\mathbf{Y}}\}$. Since $\bar{\Lambda}_f > -\infty$ and $f(\cdot)$ is bounded, P-almost surely $H(\nu_n^{\mathbf{Y}}|\mu)$ is a bounded sequence and then with $H(\cdot|\mu)$ a good rate function, every subsequence of $\{\nu_n^{\mathbf{Y}}\}$ has at least one limit point in $M_1(\Sigma)$. Therefore, by the continuity of $f(\cdot)$ and lower semi-continuity of $H(\cdot|\mu)$, P-almost surely,

$$\limsup_{n \to \infty} \{f(L_n^{\mathbf{Y}}) - H(\nu_n^{\mathbf{Y}}|\mu)\} \leq \sup_{\nu \in M_1(\Sigma)} \{f(\nu) - H(\nu|\mu)\} = \Lambda_f.$$

Consequently, $\bar{\Lambda}_f \leq \Lambda_f$ by (6.6.9). The LDP then follows by (6.6.7) and Theorem 6.6.1.	□

6.7　Historical Notes and References

The approach taken in this chapter is a combination of two powerful ideas: sub-additivity methods and the projective limit approach. The sub-additivity approach for proving the existence of the limits appearing in the various definitions of logarithmic moment generating functions may be traced back to Ruelle and to Lanford [Rue65, Rue67, Lan73]. In the context of large deviations, they were applied by Bahadur and Zabell [BaZ79], who used it to derive both the sharp form of Cramér's theorem discussed in Section 6.1 and the weak form of Sanov's theorem in Corollary 6.2.3. In the same article, they also derive the LDP for the empirical mean of Banach valued i.i.d. random variables described in Exercise 6.2.21. The latter was also considered in [DV76, Bol84, deA85b] and in [Bol86, Bol87a], where Laplace's method is used and the case of degeneracy is resolved. Assumption 6.1.2 borrows from both [BaZ79] and [DeuS89b]. In view of the example in [Wei76, page 323], some restriction of the form of part (b) of Assumption 6.1.2 is necessary in order to prove Lemma 6.1.8. (Note that this contradicts the claim in the appendix of [BaZ79]. This correction of the latter does not, however, affect other results there and, moreover, it is hard to imagine an interesting situation in which this restriction is not satisfied.) Exercise 6.1.19 follows the exposition in [DeuS89b]. The moderate deviations version of Cramér's theorem, namely the infinite dimensional extension of Theorem 3.7.1, is considered in [BoM78, deA92, Led92].

Long before the "general" theory of large deviations was developed, Sanov proved a version of Sanov's theorem for real valued random variables [San57]. His work was extended in various ways by Sethuraman, Hoeffding, Hoadley, and finally by Donsker and Varadhan [Set64, Hoe65, Hoa67, DV76], all in the weak topology. A derivation based entirely on the projective limit approach may be found in [DaG87], where the Daniell–Stone theorem is applied for identifying the rate function in the weak topology. The formulation and proof of Sanov's theorem in the τ-topology is due to Groeneboom, Oosterhoff, and Ruymgaart [GOR79], who use a version of what would later become the projective limit approach. These authors also consider the application of the results to various statistical problems. Related refinements may be found in [Gro80, GrS81]. For a different approach to the strong form of Sanov's theorem, see [DeuS89b]. See also [DiZ92] for the exchangeable case. Combining either the approximate or inverse contraction principles of Section 4.2 with concentration inequalities that are based upon those of Section 2.4, it is possible to prove Sanov's theorem for topologies stronger than the τ-topology. For example, certain nonlinear and possibly unbounded functionals are part of the topology considered by [SeW97, EicS97], whereas in the topologies of [Wu94, DeZa97] convergence is uniform over certain classes of linear functionals.

Asymptotic expansions in the general setting of Cramér's theorem and Sanov's theorem are considered in [EinK96] and [Din92], respectively. These infinite dimensional extensions of Theorem 3.7.4 apply to open convex sets, relying upon the concept of dominating points first introduced in [Ney83].

As should be clear from an inspection of the proof of Theorem 6.2.10, which was based on projective limits, the assumption that Σ is Polish played no role whatsoever, and this proof could be made to work as soon as Σ is a Hausdorff space. In this case, however, the measurability of L_n^Y with respect to either $\mathcal{B}^w$ or $\mathcal{B}^{cy}$ has to be dealt with because Σ is not necessarily separable. Examples of such a discussion may be found in [GOR79], Proposition 3.1, and in [Cs84], Remark 2.1. With some adaptations one may even dispense of the Hausdorff assumption, thus proving Sanov's theorem for Σ

an arbitrary measurable space. See [deA94b] for such a result.

Although Sanov's theorem came to be fully appreciated only in the 1970s, relative entropy was introduced into statistics by Kullback and Leibler in the early 1950s [KL51], and came to play an important role in what became information theory. For an account of the use of relative entropy in this context, see [Ga68, Ber71].

The fundamental reference for the empirical measure LDP for Markov chains satisfying certain regularity conditions is Donsker and Varadhan [DV75a, DV75b, DV76], who used a direct change of measure argument to derive the LDP, both in discrete and in continuous time. As can be seen from the bibliography, the applications of their result have been far-reaching, explored first by them and then by a variety of other authors. Another early reference on the same subject is Gärtner [Gär77]. A completely different approach was proposed by Ney and Nummelin [NN87a, NN87b], who evaluate the LDP for additive functionals of Markov chains, and hence need to consider mainly real valued random variables, even for Markov chains taking values in abstract spaces. They also are able to obtain a local LDP for quite general irreducible Markov chains. Extensions and refinements of these results to (abstract valued) additive functionals are presented in [deA88], again under a quite general irreducibility assumption. See also [deA90, Jai90] for a refinement of this approach and [DiN95] for its application to the LDP for empirical measures. These latter extensions, while providing sharper results, make use of a technically more involved machinery. The LDP might hold for a Markov chain with a good rate function other than that of Donsker and Varadhan, possibly even non-convex. Such examples are provided in Section 3.1.1, borrowing from [Din93] (where the issue of identification of the rate function is studied in much generality). An example of such a behavior for an ergodic Markov chain on $[0,1]$ is provided in [DuZ96]. Note in this context that [BryS93] constructs a bounded additive functional of a Markov chain which satisfies mixing conditions, converges exponentially fast to its expectation, but does not satisfy the LDP. Such a construction is provided in [BryD96], with the additional twist of the LDP now holding for any atomic initial measure of the chain but not for its stationary measure. See also [BxJV91, BryD96] for Markov chains such that every bounded additive functional satisfies the LDP while the empirical measure does not satisfy the LDP.

Stroock [St84] was the first to consider the uniform case (U) with $\ell = N = 1$. The general uniform case was derived by Ellis [Ell88]. (See also [ElW89].) Another type of uniformity was considered by [INN85], based on the additive functional approach. While the uniformity assumptions are quite restrictive (in particular (U) implies Doëblin recurrence), they result in a stronger uniform LDP statement while allowing for a transparent derivation, avoiding the heavier machinery necessary for handling general irreducible chains. Uniformity assumptions also result with finer results for additive functionals, such as dominating points in [INN85] and asymptotic expansions in [Jen91].

The possibilities of working in the τ-topology and dispensing with the Feller continuity were observed by Bolthausen [Bol87b].

For the moderate deviations of empirical measures of Markov chains see [Wu95, deA97b].

Our derivation in Section 6.3 of the LDP in the weak topology for the uniform case and the identification of the rate function presented in Section 6.5, are based on Chapter 4.1 of [DeuS89b]. Our notion of convergence in the discussion of the nonuniform case is different than that of [DeuS89b] and the particular form of (6.5.3) seems to be new.

Building upon the approximate subadditive theorem of [Ham62], the τ-topology result of Section 6.4 is a combination of ideas borrowed from the projective limit approach of Dawson and Gärtner, Bryc's inverse Varadhan lemma (which may be thought of as a way to generate an approximate identity on open convex sets using convex functions) and the mixing hypotheses of [CK88]. See also [BryD96] for the same result under weaker assumptions and its relation with strong mixing conditions.

Relations between hypermixing assumptions and hypercontractivity have been studied by Stroock [St84]. (See also [DeuS89b] for an updated account.) In particular, for Markov processes, there are intimate links between the hypermixing assumptions (H-1) and (H-2) and Logarithmic Sobolev inequalities. Exercise 6.4.19 is based on the continuous time Orenstein–Uhlenbeck example discussed in [DeuS89b].

Many of the preceding references also treat the continuous time case. For a description of the changes required in transforming the discrete time results to continuous time ones, see [DeuS89b]. It is hardly surprising that in the continuous time situation, it is often easier to compute the resulting rate function. In this context, see the evaluation of the rate function in the symmetric case described in [DeuS89b], and the explicit results of Pinsky [Pin85a, Pin85b]. See also [KuT84, BrM91, BolDT95, BolDT96] for related asymptotic expansions in the context of Laplace's method.

The first to consider the LDP for the empirical process were Donsker and Varadhan [DV83]. As pointed out, in the uniform case, the kth-fold empirical measure computation and its projective extension to the process level LDP may be found in [ElW89] and in the book by Deuschel and Stroock [DeuS89b]. Lemma 6.5.16 is due to Pinsker [Pi64].

Section 6.6 is motivated by the weak convergence approach of [DuE97].

Large deviations for empirical measures and processes have been considered in a variety of situations. A partial list follows.

The case of stationary processes satisfying appropriate mixing conditions is handled by [Num90] (using the additive functional approach), by [Sch89, BryS93] for exponential convergence of tail probabilities of the empirical mean, and by [Ore85, Oll87, OP88] at the process level. For an excellent survey of mixing conditions, see [Bra86]. As pointed out by Orey [Ore85], the empirical process LDP is related to the Shannon–McMillan theorem of information theory. For some extensions of the latter and early large deviations statements, see [Moy61, Föl73, Barr85]. See also [EKW94, Kif95, Kif96] for other types of ergodic averages.

The LDP for Gaussian empirical processes was considered by Donsker and Varadhan [DV85], who provide an explicit evaluation of the resulting rate function in terms of the spectral density of the process. See also [BryD95, BxJ96] for similar results in continuous time and for Gaussian processes with bounded spectral density for which the LDP does not hold. For a critical LDP for the "Gaussian free field" see [BolD93].

Process level LDPs for random fields and applications to random fields occurring from Gibbs conditioning are discussed in [DeuSZ91, Bry92]. The Gibbs random field case has also been considered by [FO88, Oll88, Com89, Sep93], whereas the LDP for the $\mathbb{Z}^d$-indexed Gaussian field is discussed in [SZ92]. See also the historical notes of Chapter 7 for some applications to statistical mechanics.

An abstract framework for LDP statements in dynamical systems is described in [Tak82] and [Kif92]. LDP statements for branching processes are described in [Big77, Big79].

Several estimates on the time to relaxation to equilibrium of a Markov chain are closely related to large deviations estimates. See [LaS88, HoKS89, JS89, DiaS91, JS93, DiaSa93, Mic95] and the accounts [Sal97, Mart98] for further details.

The analysis of simulated annealing is related to both the Freidlin–Wentzell theory and the relaxation properties of Markov chains. For more on this subject, see [Haj88, HoS88, Cat92, Mic92] and the references therein.

Chapter 7

Applications of Empirical Measures LDP

In this chapter, we revisit three applications considered in Chapters 2 and 3 in the finite alphabet setup. Equipped with Sanov's theorem and the projective limit approach, the general case (Σ Polish) is treated here.

7.1 Universal Hypothesis Testing

7.1.1 A General Statement of Test Optimality

Suppose a random variable Z that assumes values in a topological space $\mathcal{Y}$ is observed. Based on this observation, a choice may be made between the null hypothesis H_0 (where Z was drawn according to a Borel probability measure μ_0) and the alternative hypothesis H_1 (where Z was drawn according to another Borel probability measure, denoted μ_1). Whenever the laws μ_0 and μ_1 are known *a priori*, and $d\mu_1/d\mu_0$ exists, it may be proved that the likelihood ratio test is optimal in the Neyman–Pearson sense. To state precisely what this statement means, define a test as a Borel measurable map $\mathcal{S} : \mathcal{Y} \to \{0,1\}$ with the interpretation that when $Z = z$ is observed, then H_0 is accepted (H_1 rejected) if $\mathcal{S}(z) = 0$, while H_1 is accepted (H_0 rejected) if $\mathcal{S}(z) = 1$. Associated with each test are the error probabilities of first and second kind,

$$\alpha(\mathcal{S}) \stackrel{\triangle}{=} \mu_0(\{\mathcal{S}(z) = 1\}),$$
$$\beta(\mathcal{S}) \stackrel{\triangle}{=} \mu_1(\{\mathcal{S}(z) = 0\}).$$

A. Dembo, O. Zeitouni, *Large Deviations Techniques and Applications*, Stochastic Modelling and Applied Probability 38, DOI 10.1007/978-3-642-03311-7_7, © Springer-Verlag Berlin Heidelberg 1998, corrected printing 2010

$\beta(\mathcal{S})$ may always be minimized by choosing $\mathcal{S}(\cdot) \equiv 1$ at the expense of $\alpha(\mathcal{S}) = 1$. The *Neyman-Pearson* criterion for optimality involves looking for a test $\mathcal{S}$ that minimizes $\beta(\mathcal{S})$ subject to the constraint $\alpha(\mathcal{S}) \leq \gamma$. Now define the likelihood ratio test with threshold η by

$$\tilde{\mathcal{S}}(z) = \begin{cases} 0 & \text{if } \frac{d\mu_1}{d\mu_0}(z) \leq \eta \\ 1 & \text{otherwise.} \end{cases}$$

Exact optimality for the likelihood ratio test is given in the following classical lemma. (For proofs, see [CT91] and [Leh59].)

Lemma 7.1.1 (Neyman–Pearson) *For any $0 \leq \gamma \leq 1$, there exists a $\eta(\gamma, \mu_0, \mu_1)$ such that $\alpha(\tilde{\mathcal{S}}) \leq \gamma$, and any other test $\mathcal{S}$ that satisfies $\alpha(\mathcal{S}) \leq \gamma$ must satisfy $\beta(\mathcal{S}) \geq \beta(\tilde{\mathcal{S}})$.*

The major deficiency of the likelihood ratio test lies in the fact that it requires perfect knowledge of the measures μ_0 and μ_1, both in forming the likelihood ratio and in computing the threshold η. Thus, it is not applicable in situations where the alternative hypotheses consist of a family of probability measures. To overcome this difficulty, other tests will be proposed. However, the notion of optimality needs to be modified to allow for asymptotic analysis. This is done in the same spirit of Definition 3.5.1.

In order to relate the optimality of tests to asymptotic computations, the following assumption is now made.

Assumption 7.1.2 *A sequence of random variables Z_n with values in a metric space $(\mathcal{Y}, d)$ is observed. Under the null hypothesis (H_0), the Borel law of Z_n is $\mu_{0,n}$, where $\{\mu_{0,n}\}$ satisfies the LDP in $\mathcal{Y}$ with the good rate function $I(\cdot)$, which is known a priori.*

A test is now defined as a *sequence* of Borel measurable maps $\mathcal{S}^n : \mathcal{Y} \to \{0, 1\}$. To any test $\{\mathcal{S}^n\}$, associate the error probabilities of first and second kind:

$$\alpha_n(\mathcal{S}^n) \triangleq \mu_{0,n}(\{\mathcal{S}^n(Z_n) = 1\}),$$
$$\beta_n(\mathcal{S}^n) \triangleq \mu_{1,n}(\{\mathcal{S}^n(Z_n) = 0\}),$$

where $\mu_{1,n}$ is the (unknown) Borel law of Z_n under the alternative hypothesis (H_1). Note that it is implicitly assumed here that Z_n is a sufficient statistics given $\{Z_k\}_{k=1}^n$, and hence it suffices to consider tests of the form $\mathcal{S}^n(Z_n)$. For stating the optimality criterion and presenting an optimal test, let $(\mathcal{S}_0^n, \mathcal{S}_1^n)$ denote the partitions induced on $\mathcal{Y}$ by the test $\{\mathcal{S}^n\}$ (i.e., $z \in \mathcal{S}_0^n$ iff $\mathcal{S}^n(z) = 0$). Since in the general framework discussed here, pointwise bounds on error probabilities are not available, smooth versions of the maps $\mathcal{S}^n$ are considered. Specifically, for each $\delta > 0$, let $\mathcal{S}^{n,\delta}$ denote the δ-smoothing of the map $\mathcal{S}^n$ defined via

$$\mathcal{S}_1^{n,\delta} \triangleq \{y \in \mathcal{Y} : d(y, \mathcal{S}_1^n) < \delta\} \quad , \quad \mathcal{S}_0^{n,\delta} \triangleq \mathcal{Y} \setminus \mathcal{S}_1^{n,\delta},$$

i.e., the original partition is smoothed by using for $\mathcal{S}_1^{n,\delta}$ the open δ-blowup of the set $\mathcal{S}_1^n$. Define the δ-smoothed rate function

$$J_\delta(z) \triangleq \inf_{x \in \tilde{B}_{z,2\delta}} I(x),$$

where the closed sets $\tilde{B}_{z,2\delta}$ are defined by

$$\tilde{B}_{z,2\delta} \triangleq \{y \in \mathcal{Y} : d(z,y) \leq 2\delta\}.$$

Note that the goodness of $I(\cdot)$ implies the lower semicontinuity of $J_\delta(\cdot)$.

The following theorem is the main result of this section, suggesting asymptotically optimal tests based on $J_\delta(\cdot)$.

Theorem 7.1.3 *Let Assumption 7.1.2 hold. For any $\delta > 0$ and any $\eta \geq 0$, let $\mathcal{S}_\delta^*$ denote the test*

$$\mathcal{S}_\delta^*(z) = \begin{cases} 0 & \text{if } J_\delta(z) < \eta \\ 1 & \text{otherwise,} \end{cases}$$

with $\mathcal{S}^{*,\delta}$ denoting the δ-smoothing of $\mathcal{S}_\delta^*$. Then

$$\limsup_{n \to \infty} \frac{1}{n} \log \alpha_n(\mathcal{S}^{*,\delta}) \leq -\eta, \tag{7.1.4}$$

and any other test $\mathcal{S}^n$ that satisfies

$$\limsup_{n \to \infty} \frac{1}{n} \log \alpha_n(\mathcal{S}^{n,4\delta}) \leq -\eta \tag{7.1.5}$$

must satisfy

$$\liminf_{n \to \infty} \left(\frac{\beta_n(\mathcal{S}^{n,\delta})}{\beta_n(\mathcal{S}^{*,\delta})} \right) \geq 1. \tag{7.1.6}$$

Remarks:

(a) Note the factor 4δ appearing in (7.1.5). Although the number 4 is quite arbitrary (and is tied to the exact definition of $J_\delta(\cdot)$), some margin is needed in the definition of optimality to allow for the rough nature of the large deviations bounds.

(b) The test $\mathcal{S}_\delta^*$ consists of maps that are independent of n and depend on $\mu_{0,n}$ solely via the rate function $I(\cdot)$. This test is universally asymptotically optimal, in the sense that (7.1.6) holds *with no assumptions* on the Borel law $\mu_{1,n}$ of Z_n under the alternative hypothesis (H_1).

Proof: The proof is based on the following lemmas:

Lemma 7.1.7 *For all $\delta > 0$,*

$$c^\delta \stackrel{\triangle}{=} \inf_{z \in \overline{\mathcal{S}_1^{*,\delta}}} I(z) \geq \eta.$$

Lemma 7.1.8 *For any test satisfying (7.1.5), and all $n > n_0(\delta)$,*

$$\mathcal{S}_0^{*,\delta} \subseteq \mathcal{S}_0^{n,\delta}. \tag{7.1.9}$$

To prove (7.1.4), deduce from the large deviations upper bound and Lemma 7.1.7 that

$$
\begin{aligned}
\limsup_{n \to \infty} \frac{1}{n} \log \alpha_n(\mathcal{S}^{*,\delta}) &= \limsup_{n \to \infty} \frac{1}{n} \log \mu_{0,n}(\mathcal{S}_1^{*,\delta}) \\
&\leq - \inf_{z \in \mathcal{S}_1^{*,\delta}} I(z) \leq -\eta.
\end{aligned}
$$

To conclude the proof of the theorem, note that, by Lemma 7.1.8, $\beta_n(\mathcal{S}^{n,\delta}) \geq \beta_n(\mathcal{S}^{*,\delta})$ for any test satisfying (7.1.5), any n large enough, and any sequence of probability measures $\mu_{1,n}$. $\qquad\square$

Proof of Lemma 7.1.7: Without loss of generality, assume that $c^\delta < \infty$, and hence (with $I(\cdot)$ a good rate function), there exists a $z_0 \in \overline{\mathcal{S}_1^{*,\delta}}$ such that $I(z_0) = c^\delta$. By the definition of $\mathcal{S}^{*,\delta}$, there exists a $y \in \mathcal{Y}$ such that $z_0 \in \tilde{B}_{y,2\delta}$ and $J_\delta(y) \geq \eta$. It follows that

$$c^\delta = I(z_0) \geq \inf_{x \in \tilde{B}_{y,2\delta}} I(x) = J_\delta(y) \geq \eta. \qquad\square$$

Proof of Lemma 7.1.8: Assume that $\{\mathcal{S}^n\}$ is a test for which (7.1.9) does not hold. Then for some infinite subsequence $\{n_m\}$, there exist $z_m \in \mathcal{S}_0^{*,\delta}$ such that $z_m \in \mathcal{S}_1^{n_m,\delta}$. Therefore, $J_\delta(z_m) < \eta$, and hence there exist $y_m \in \tilde{B}_{z_m,2\delta}$ with $I(y_m) < \eta$. Since $I(\cdot)$ is a good rate function, it follows that y_m has a limit point y^*. Hence, along some infinite subsequence $\{n_{m'}\}$,

$$B_{y^*,\delta/2} \subseteq B_{z_{m'},3\delta} \subseteq \mathcal{S}_1^{n_{m'},4\delta}.$$

Therefore, by the large deviations lower bound,

$$\limsup_{n\to\infty} \frac{1}{n} \log \alpha_n(\mathcal{S}^{n,4\delta}) \geq \liminf_{m'\to\infty} \frac{1}{n_{m'}} \log \mu_{0,n_{m'}}(\mathcal{S}_1^{n_{m'},4\delta})$$

$$\geq \liminf_{n\to\infty} \frac{1}{n} \log \mu_{0,n}(B_{y^*,\delta/2}) \geq -I(y_{m_0}) > -\eta,$$

where m_0 is large enough for $y_{m_0} \in B_{y^*,\delta/2}$. In conclusion, any test $\{\mathcal{S}^n\}$ for which (7.1.9) fails can not satisfy (7.1.5). $\qquad\square$

Another frequently occurring situation is when only partial information about the sequence of probability measures $\mu_{0,n}$ is given *a priori*. One possible way to model this situation is by a composite null hypothesis $H_0 \triangleq \cup_{\theta\in\Theta} H_\theta$, as described in the following assumption.

Assumption 7.1.10 *A sequence of random variables Z_n with values in a metric space $(\mathcal{Y}, d)$ is observed. Under the null hypothesis (H_0), the law of Z_n is $\mu_{\theta,n}$ for some $\theta \in \Theta$ that is independent of n. For each $\theta \in \Theta$, the sequence $\{\mu_{\theta,n}\}$ satisfies the LDP in $\mathcal{Y}$ with the good rate function $I_\theta(\cdot)$, and these functions are known a priori.*

The natural candidate for optimal test now is again $\mathcal{S}_\delta^*$, but with

$$J_\delta(z) \triangleq \inf_{\theta\in\Theta} \inf_{x\in\bar{B}_{z,2\delta}} I_\theta(x).$$

Indeed, the optimality of $\mathcal{S}_\delta^*$ is revealed in the following theorem.

Theorem 7.1.11 *Let Assumption 7.1.10 hold.*
(a) For all $\theta \in \Theta$,

$$\limsup_{n\to\infty} \frac{1}{n} \log \mu_{\theta,n}(\mathcal{S}_1^{*,\delta}) \leq -\eta. \tag{7.1.12}$$

(b) Let $\{\mathcal{S}^n\}$ be an alternative test such that for all $\theta \in \Theta$,

$$\limsup_{n\to\infty} \frac{1}{n} \log \mu_{\theta,n}(\mathcal{S}_1^{n,4\delta}) \leq -\eta. \tag{7.1.13}$$

Assume that either:

$$\text{The set } \{z : \inf_{\theta\in\Theta} I_\theta(z) < \eta\} \text{ is pre-compact} \tag{7.1.14}$$

or

$$\mathcal{S}^n \equiv \mathcal{S} \text{ is independent of } n. \tag{7.1.15}$$

Then for any sequence $\{\mu_{1,n}\}$,

$$\liminf_{n\to\infty} \left(\frac{\mu_{1,n}(\mathcal{S}_0^{n,\delta})}{\mu_{1,n}(\mathcal{S}_0^{*,\delta})} \right) \geq 1. \tag{7.1.16}$$

Remark: If neither (7.1.14) nor (7.1.15) holds, a somewhat weaker statement may still be proved for exponentially tight $\{\mu_{1,n}\}$. For details, *c.f.* Exercise 7.1.17.

Proof: (a) By a repeat of the proof of Lemma 7.1.7, it is concluded that for all $\delta > 0$ and all $\theta \in \Theta$,

$$c_\theta^\delta \stackrel{\triangle}{=} \inf_{z \in \mathcal{S}_1^{*,\delta}} I_\theta(z) \geq \eta.$$

Therefore, (7.1.12) follows by the large deviations upper bound.

(b) Suppose that the set inclusion (7.1.9) fails. Then by a repeat of the proof of Lemma 7.1.8, there exist $n_m \to \infty$, $z_m \in \mathcal{S}_1^{n_m,\delta}$ with $y_m \in \tilde{B}_{z_m,2\delta}$, and $\theta_m \in \Theta$ with $I_{\theta_m}(y_m) < \eta$. Hence, if (7.1.14) holds, then the sequence y_m possesses a limit point y^* and (7.1.13) fails for $\theta = \theta_{m_0}$ such that $y_{m_0} \in B_{y^*,\delta/2}$.

Alternatively, if $\mathcal{S}^n \equiv \mathcal{S}$ and (7.1.9) fails, then there exists a $z^* \in \mathcal{S}_1^\delta$ such that $J_\delta(z^*) < \eta$. Therefore, there exist a $y^* \in \tilde{B}_{z^*,2\delta}$ and a $\theta^* \in \Theta$ such that $I_{\theta^*}(y^*) < \eta$. Since

$$B_{y^*,\delta/2} \subseteq B_{z^*,3\delta} \subseteq \mathcal{S}_1^{4\delta},$$

it follows from this inclusion and the large deviations lower bound that

$$\limsup_{n \to \infty} \frac{1}{n} \log \mu_{\theta^*,n}(\mathcal{S}_1^{4\delta}) \;\geq\; \liminf_{n \to \infty} \frac{1}{n} \log \mu_{\theta^*,n}(B_{y^*,\delta/2})$$
$$\geq\; -I_{\theta^*}(y^*) > -\eta \;,$$

contradicting (7.1.13). Consequently, when either (7.1.14) or (7.1.15) holds, then (7.1.9) holds for any test that satisfies (7.1.13). Since (7.1.16) is an immediate consequence of (7.1.9), the proof is complete. □

Exercise 7.1.17 (a) Prove that for any test satisfying (7.1.13) and any compact set $K \subset \mathcal{Y}$,

$$\mathcal{S}_0^{*,\delta} \cap K \subseteq \mathcal{S}_0^{n,\delta}, \qquad \forall n > n_0(\delta) \;. \tag{7.1.18}$$

(b) Use part (a) to show that even if both (7.1.14) and (7.1.15) fail, still

$$\liminf_{n \to \infty} \frac{1}{n} \log \mu_{1,n}(\mathcal{S}_0^{n,\delta}) \geq \liminf_{n \to \infty} \frac{1}{n} \log \mu_{1,n}(\mathcal{S}_0^{*,\delta})$$

for any test $\{\mathcal{S}^n\}$ that satisfies (7.1.13), and any exponentially tight $\{\mu_{1,n}\}$.

7.1.2 Independent and Identically Distributed Observations

The results of Theorems 7.1.3 and 7.1.11 are easily specialized to the particular case of i.i.d. observations. Specifically, let $Y_1, Y_2, \ldots, Y_n$ be an observed sequence of Σ-valued random variables, where Σ is a Polish space. Under both hypotheses, these variables are i.i.d., with μ_j denoting the probability measure of Y_1 under H_j, $j = 0, 1$. By a proof similar to the one used in Lemma 3.5.3, for each integer n, the random variable $Z_n = L_n^{\mathbf{Y}} = n^{-1} \sum_{i=1}^n \delta_{Y_i}$, which takes values in $\mathcal{Y} = M_1(\Sigma)$, is a sufficient statistics in the sense of Lemma 3.5.3. Thus, it is enough to consider tests of the form $\mathcal{S}^n(Z_n)$. The space $M_1(\Sigma)$ is a metric space when equipped with the weak topology, and by Sanov's theorem (Theorem 6.2.10), $\{Z_n\}$ satisfies under H_0 the LDP with the good rate function $I(\cdot) = H(\cdot|\mu_0)$. Hence, Assumption 7.1.2 holds, and consequently so does Theorem 7.1.3. When μ_0 is known only to belong to the set $A \triangleq \{\mu_\theta\}_{\theta \in \Theta}$, then Theorem 7.1.11 holds. Moreover:

Theorem 7.1.19 *If $\overline{A}$ is a compact subset of $M_1(\Sigma)$, then (7.1.14) holds, and the test $\mathcal{S}^*$ is optimal.*

Remark: When Σ is a compact space, so is $M_1(\Sigma)$ (when equipped with the weak topology; see Theorem D.8), and (7.1.14) trivially holds.

Proof: Assume that the set of measures $\overline{A}$ is compact. Then by Prohorov's theorem (Theorem D.9), for each $\delta > 0$ there exists a compact set $K_\delta \subseteq \Sigma$ such that $\mu(K_\delta^c) < \delta$ for all $\mu \in A$. Fix $\eta < \infty$ and define $B \triangleq \{\nu : \inf_{\mu \in A} H(\nu|\mu) < \eta\}$. Fix $\nu \in B$ and let $\mu \in A$ be such that $H(\nu|\mu) < \eta$. Then by Jensen's inequality,

$$\eta > \int_{K_\delta} d\nu \log \frac{d\nu}{d\mu} + \int_{K_\delta^c} d\nu \log \frac{d\nu}{d\mu}$$

$$\geq \nu(K_\delta) \log \frac{\nu(K_\delta)}{\mu(K_\delta)} + \nu(K_\delta^c) \log \frac{\nu(K_\delta^c)}{\mu(K_\delta^c)} \geq -\log 2 + \nu(K_\delta^c) \log \frac{1}{\delta},$$

where the last inequality follows from the bound

$$\inf_{p \in [0,1]} \{p \log p + (1-p) \log(1-p)\} \geq -\log 2.$$

Therefore, $\nu(K_\delta^c) \leq (\eta + \log 2)/\log(1/\delta)$ for all $\nu \in B$ and all $\delta > 0$, implying that B is tight. Hence, again by Prohorov's theorem (Theorem D.9), B is pre-compact and this is precisely the condition (7.1.14). $\square$

Exercise 7.1.20 Prove that the empirical measure $L_n^{\mathbf{Y}}$ is a sufficient statistics in the sense of Section 3.5.

7.2　Sampling Without Replacement

Consider the following setup, which is an extension of the one discussed in Section 2.1.3. Let Σ be a Polish space. Out of an initial deterministic pool of m items $\mathbf{y} \triangleq (y_1, \ldots, y_m) \in \Sigma^m$, an n-tuple $\mathbf{Y} \triangleq (y_{i_1}, y_{i_2}, \ldots, y_{i_n})$ is sampled without replacement, namely, $i_1 \neq i_2 \neq \cdots \neq i_n$ and each choice of $i_1 \neq i_2 \neq \cdots \neq i_n \in \{1, \ldots, m\}$ is equally likely (and independent of the sequence $\mathbf{y}$).

Suppose now that, as $m \to \infty$, the deterministic empirical measures $L_m^{\mathbf{y}} = m^{-1} \sum_{i=1}^m \delta_{y_i}$ converge weakly to some $\mu \in M_1(\Sigma)$. Let $\mathbf{Y}$ be a random vector obtained by the sampling without replacement of n out of m elements as described before. Such a situation occurs when one surveys a small part of a large population and wishes to make statistical inference based on this sample. The next theorem provides the LDP for the (random) empirical measure $L_n^{\mathbf{Y}}$ associated with the vector $\mathbf{Y}$, where $n/m(n) \to \beta \in (0,1)$ as $n \to \infty$.

Theorem 7.2.1 *The sequence $L_n^{\mathbf{Y}}$ satisfies the LDP in $M_1(\Sigma)$ equipped with the weak topology, with the convex good rate function*

$$I(\nu | \beta, \mu) = \begin{cases} H(\nu | \mu) + \frac{1-\beta}{\beta} H\left(\frac{\mu - \beta\nu}{1-\beta} \Big| \mu\right) & \text{if } \frac{\mu - \beta\nu}{1-\beta} \in M_1(\Sigma) \\ \\ \infty & \text{otherwise}. \end{cases} \quad (7.2.2)$$

Remark: Consider the probability space $(\Omega_1 \times \Omega_2, \mathcal{B} \times \mathcal{B}_\Sigma^{\mathbb{Z}_+}, P_1 \times P_2)$, with $\Omega_2 = \Sigma^{\mathbb{Z}_+}$, P_2 stationary and ergodic with marginal μ on Σ, and $(\Omega_1, \mathcal{B}, P_1)$ representing the randomness involved in the sub-sampling. Let $\omega_2 = (y_1, y_2, \ldots, y_m, \ldots)$ be a realization of an infinite sequence under the measure P_2. Since Σ is Polish, by the ergodic theorem the empirical measures $L_m^{\mathbf{y}}$ converge to μ weakly for (P_2) almost every ω_2. Hence, Theorem 7.2.1 applies for almost every ω_2, yielding the same LDP for $L_n^{\mathbf{Y}}$ under the law P_1 for almost every ω_2. Note that for P_2 a product measure (corresponding to an i.i.d. sequence), the LDP for $L_n^{\mathbf{Y}}$ under the law $P_1 \times P_2$ is given by Sanov's theorem (Theorem 6.2.10) and admits a different rate function!

The first step in the proof of Theorem 7.2.1 is to derive the LDP for a sequence of empirical measures of deterministic positions and random weights which is much simpler to handle. As shown in the sequel, this LDP also provides an alternative proof for Theorem 5.1.2.

Theorem 7.2.3 *Let X_i be real-valued i.i.d. random variables with $\Lambda_X(\lambda) = \log E[e^{\lambda X_1}]$ finite everywhere and $y_i \in \Sigma$ non-random such that $L_m^{\mathbf{y}} \to \mu$*

weakly in $M_1(\Sigma)$. Then, for $n = n(m)$ such that $n/m \to \beta \in (0, \infty)$, the sequence $L'_n = n^{-1} \sum_{i=1}^m X_i \delta_{y_i}$ satisfies the LDP in $M(\Sigma)$ equipped with the $C_b(\Sigma)$-topology, with the convex good rate function

$$
I_X(\nu) = \begin{cases} \int_\Sigma \frac{1}{\beta} \Lambda_X^*(\beta f) d\mu & \text{if } f = \frac{d\nu}{d\mu} \text{ exists} \\ \infty & \text{otherwise.} \end{cases} \tag{7.2.4}
$$

Remark: In case X_i are non-negative, $L'_n \in M_+(\Sigma)$, the subset of $M(\Sigma)$ consisting of all positive finite Borel measures on Σ. Thus, the LDP of Theorem 7.2.3 holds in $M_+(\Sigma)$ equipped with the induced topology, by part (b) of Lemma 4.1.5.

Proof: Note that $\Lambda_X^*(\cdot)$ is a convex good rate function and by Lemma 2.2.20, $\Lambda_X^*(x)/|x| \to \infty$ as $|x| \to \infty$. Hence, by Lemma 6.2.16, $I_X(\cdot)$ is a convex, good rate function on $M(\Sigma)$ equipped with the $B(\Sigma)$-topology. The same applies for the weaker $C_b(\Sigma)$-topology, by considering the identity map on $M(\Sigma)$.

By part (c) of Lemma 2.2.5, $\Lambda_X(\cdot)$ is differentiable. Hence, for $\phi \in C_b(\Sigma)$ we have that $\Lambda_X(\phi(\cdot)) \in C_b(\Sigma)$ and

$$
\log E[\exp(n \int_\Sigma \phi dL'_n)] = \log E[\exp(\sum_{i=1}^m X_i \phi(y_i))] = \sum_{i=1}^m \Lambda_X(\phi(y_i)).
$$

Therefore,

$$
\Lambda(\phi) \stackrel{\triangle}{=} \lim_{n \to \infty} \frac{1}{n} \log E[\exp(n \int_\Sigma \phi dL'_n)] = \frac{1}{\beta} \int_\Sigma \Lambda_X(\phi) d\mu.
$$

Thus, for any fixed collection $\phi_1, \ldots, \phi_k \in C_b(\Sigma)$, the function $g(\lambda) = \Lambda(\sum_{i=1}^k \lambda_i \phi_i)$ is finite. Moreover, applying dominated convergence (Theorem C.10) to justify the change of order of differentiation and integration, $g(\cdot)$ is differentiable in $\lambda = (\lambda_1, \ldots, \lambda_k)$ throughout $\mathbb{R}^k$. By part (a) of Corollary 4.6.11, the sequence L'_n then satisfies the LDP in $\mathcal{X}$, the algebraic dual of $C_b(\Sigma)$ equipped with the $C_b(\Sigma)$-topology, with the good rate function

$$
\Lambda^*(\vartheta) = \sup_{\phi \in C_b(\Sigma)} \{\langle \phi, \vartheta \rangle - \Lambda(\phi)\}, \quad \vartheta \in \mathcal{X}
$$

Identifying $M(\Sigma)$ as a subset of $\mathcal{X}$, set $I_X(\cdot) = \infty$ outside $M(\Sigma)$. Observe that for every $\phi \in C_b(\Sigma)$ and $\nu \in M(\Sigma)$ such that $f = \frac{d\nu}{d\mu}$ exists,

$$
\int_\Sigma \phi d\nu - I_X(\nu) - \Lambda(\phi) = \frac{1}{\beta} \int [\phi \beta f - \Lambda_X^*(\beta f) - \Lambda_X(\phi)] d\mu \le 0. \tag{7.2.5}
$$

Since, by (2.2.10), the choice $\tilde{f} = \frac{1}{\beta}\Lambda'_X(\phi) \in B(\Sigma)$ results with equality in (7.2.5), it follows that

$$\Lambda(\phi) = \sup_{\nu \in M(\Sigma)} \left\{ \int_\Sigma \phi d\nu - I_X(\nu) \right\} = \sup_{\vartheta \in \mathcal{X}} \left\{ \langle \phi, \vartheta \rangle - I_X(\vartheta) \right\},$$

implying by the duality lemma (Lemma 4.5.8) that $I_X(\cdot) = \Lambda^*(\cdot)$. In particular, L'_n thus satisfies the LDP in $M(\Sigma)$ (see part (b) of Lemma 4.1.5) with the convex good rate function $I_X(\cdot)$.   □

Proof of Theorem 7.2.1: Fix $\beta \in (0,1)$. By part (b) of Exercise 2.2.23, for X_i i.i.d. Bernoulli(β) random variables

$$\frac{1}{\beta}\Lambda_X^*(\beta f) = \begin{cases} f \log f + \frac{1-\beta}{\beta} g_\beta(f) \log g_\beta(f) & 0 \leq f \leq \frac{1}{\beta} \\ \infty & \text{otherwise} \end{cases}$$

where $g_\beta(x) = (1 - \beta x)/(1 - \beta)$. For this form of $\Lambda_X^*(\cdot)$ and for every $\nu \in M_1(\Sigma)$, $I_X(\nu)$ of (7.2.4) equals to $I(\nu|\beta,\mu)$ of (7.2.2). Hence, $I(\cdot|\beta,\mu)$ is a convex good rate function. Use $(y_1, y_2, \ldots)$ to generate the sequence L'_n as in Theorem 7.2.3. Let V_n denote the number of i-s such that $X_i = 1$, i.e., $V_n = nL'_n(\Sigma)$. The key to the proof is the following coupling. If $V_n > n$ choose (by sampling without replacement) a random subset $\{i_1, \ldots, i_{V_n-n}\}$ among those indices with $X_i = 1$ and set X_i to zero on this subset. Similarly, if $V_n < n$ choose a random subset $\{i_1, \ldots, i_{n-V_n}\}$ among those indices with $X_i = 0$ and set X_i to one on this subset. Re-evaluate L'_n using the modified X_i values and denote the resulting (random) probability measure by Z_n. Note that Z_n has the same law as L_n^Y which is also the law of L'_n conditioned on the event $\{V_n = n\}$. Since V_n is a Binomial(m, β) random variable, and $n/m \to \beta \in (0,1)$ it follows that

$$\liminf_{n \to \infty} \frac{1}{n} \log P(V_n = n) = \liminf_{n \to \infty} \frac{1}{n} \log \left[\binom{m}{n} \beta^n (1-\beta)^{m-n} \right] = 0.$$

$$(7.2.6)$$

Fix a closed set $F \subset M_1(\Sigma)$. Since $P(L_n^Y \in F) = P(Z_n \in F) \leq P(L'_n \in F)/P(V_n = n)$, it follows that $\{L_n^Y\}$ satisfies the large deviations upper bound (1.2.12) in $M_1(\Sigma)$ with the good rate function $I(\cdot|\beta,\mu)$. Recall that the Lipschitz bounded metric $d_{LU}(\cdot, \cdot)$ is compatible with the weak topology on $M_1(\Sigma)$ (see Theorem D.8). Note that for any $\nu \in M_1(\Sigma)$,

$$d_{LU}(Z_n, L'_n) = n^{-1}|V_n - n| = |L'_n(\Sigma) - 1| \leq d_{LU}(L'_n, \nu).$$

Therefore, by the triangle inequality for $d_{LU}(\cdot, \cdot)$,

$$\begin{aligned} P(d_{LU}(L'_n, \nu) < 2\delta) &= P(d_{LU}(Z_n, L'_n) < 2\delta, d_{LU}(L'_n, \nu) < 2\delta) \\ &\leq P(d_{LU}(Z_n, \nu) < 4\delta) = P(d_{LU}(L_n^Y, \nu) < 4\delta). \end{aligned}$$

Fix a neighborhood G of ν in $M_1(\Sigma)$ and $\delta \in (0,1)$ small enough so that $\{\tilde{\nu} \in M_1(\Sigma) : d_{LU}(\tilde{\nu}, \nu) < 4\delta\} \subset G$. Fix $\phi_1 = 1$, $\phi_2, \ldots, \phi_k : \Sigma \to [-1,1]$ continuous and $0 < \epsilon < \delta$ such that for any $\tilde{\nu} \in M_1(\Sigma)$

$$\max_{i=2}^{k} |\int_\Sigma \phi_i d\tilde{\nu} - \int_\Sigma \phi_i d\nu| < 2\epsilon \quad \text{implies} \quad d_{LU}(\tilde{\nu}, \nu) < \delta \, .$$

For any $\hat{\nu} \in M_+(\Sigma)$,

$$d_{LU}(\hat{\nu}, \nu) \leq d_{LU}(\hat{\nu}/\hat{\nu}(\Sigma), \nu) + |\hat{\nu}(\Sigma) - 1| \, .$$

Hence, for any $\hat{\nu} \in M_+(\Sigma)$,

$$\max_{i=1}^{k} |\int_\Sigma \phi_i d\hat{\nu} - \int_\Sigma \phi_i d\nu| < \epsilon \quad \text{implies} \quad d_{LU}(\hat{\nu}, \nu) < 2\delta \, ,$$

and one concludes that $\{\hat{\nu} \in M_+(\Sigma) : d_{LU}(\hat{\nu}, \nu) < 2\delta\}$ contains a neighborhood of ν in $M_+(\Sigma)$. Therefore, by the LDP of Theorem 7.2.3,

$$\liminf_{n\to\infty} \frac{1}{n} \log P(L_n^{\mathbf{Y}} \in G) \geq \liminf_{n\to\infty} \frac{1}{n} \log P(d_{LU}(L_n', \nu) < 2\delta) \geq -I_X(\nu) \, .$$

Since $\nu \in M_1(\Sigma)$ and its neighborhood G are arbitrary, this completes the proof of the large deviations lower bound (1.2.8). □

As an application of Theorem 7.2.3, we provide an alternative proof of Theorem 5.1.2 in case X_i are real-valued i.i.d. random variables. A similar approach applies to $\mathbb{R}^d$-valued X_i and in the setting of Theorem 5.3.1.

Proof of Theorem 5.1.2 $(d = 1)$: Let $y_i = y_i^{(m)} = (i - 1)/n$, $i = 1, \ldots, m(n) = n + 1$. Clearly, $L_m^{\mathbf{Y}}$ converge weakly to Lebesgue measure on $\Sigma = [0,1]$. By Theorem 7.2.3, $L_n' = n^{-1} \sum_{i=0}^{n} X_i \delta_{i/n}$ then satisfies the LDP in $M([0,1])$ equipped with the $C([0,1])$-topology. Let $D([0,1])$ denote the space of functions continuous from the right and having left limits, equipped with the supremum norm topology, and let $f : M([0,1]) \to D([0,1])$ be such that $f(\nu)(t) = \nu([0,t])$. Then, the convex good rate function $I_X(\cdot)$ of (7.2.4) equals $I(f(\cdot))$ for $I(\cdot)$ of (5.1.3) and $f(L_n') = Z_n(\cdot)$ of (5.1.1). Since $f(\cdot)$ is not a continuous mapping, we apply the approximate contraction principle (Theorem 4.2.23) for the continuous maps $f_k : M([0,1]) \to D([0,1])$ such that $f_k(\nu)(t) = f(\nu)(t) + \int_{t+}^{t+k^{-1}} (1 - k(s - t))d\nu(s)$. To this end, note that

$$\sup_{t\in[0,1]} |f_k(L_n')(t) - f(L_n')(t)| \leq \sup_{t\in[0,1)} |L_n'|((t, t+k^{-1}]) \leq \frac{1}{n} \max_{j=0}^{n-1} \sum_{i=j+1}^{j+\lceil n/k\rceil} |X_i| \, .$$

By Cramér's theorem for the random variables

$$\hat{S}_n = n^{-1} \sum_{i=1}^{n} |X_i| \, ,$$

we thus have that for $\delta > 0$ and k large,

$$\limsup_{n \to \infty} \frac{1}{n} \log P(\sup_{t \in [0,1]} |f_k(L_n')(t) - f(L_n')(t)| > 2\delta)$$

$$\leq \frac{1}{k} \limsup_{r \to \infty} \frac{1}{r} \log P(\hat{S}_r \geq k\delta) \leq -k^{-1}\Lambda_{|X|}^*(k\delta)$$

(see (2.2.12)). Since $k^{-1}\Lambda_{|X|}^*(\delta k) \to \infty$ as $k \to \infty$, it follows that $f_k(L_n')$ are exponentially good approximations of $f(L_n')$ in $D([0,1])$. If $I_X(\nu) \leq \alpha$, then ν is absolutely continuous with respect to Lebesgue's measure, and $\phi = f(\nu) \in \mathcal{AC}$. With $M(c) \triangleq \inf_{|x| \geq c} \Lambda_X^*(x)/|x|$, for such ν and all $t \in [0,1]$,

$$\int_t^{t+k^{-1}} |\dot{\phi}| 1_{|\dot{\phi}| \geq c} ds \leq \frac{1}{M(c)} \int_t^{t+k^{-1}} \Lambda_X^*(\dot{\phi}(s)) ds \leq \frac{\alpha}{M(c)},$$

implying that

$$\sup_{t \in [0,1]} |f_k(\nu)(t) - f(\nu)(t)| \leq \sup_{t \in [0,1]} \int_t^{t+k^{-1}} |\dot{\phi}(s)| ds \leq \frac{c}{k} + \frac{\alpha}{M(c)}.$$

Considering $k \to \infty$ followed by $c \to \infty$ we verify condition (4.2.24) and conclude by applying Theorem 4.2.23. $\qquad\square$

Exercise 7.2.7 Deduce from Theorem 7.2.1 that for every $\phi \in C_b(\Sigma)$,

$$\Lambda_\beta(\phi) = \lim_{n \to \infty} \frac{1}{n} \log E[\exp(\sum_{i=1}^n \phi(Y_i^m))] = \sup_{\nu \in M_1(\Sigma)} \{\int_\Sigma \phi d\nu - I(\nu|\beta, \mu)\}$$

and show that

$$\Lambda_\beta(\phi) = \frac{1}{\beta} \int_\Sigma [\phi + \log(\beta + \lambda e^{-\phi})] d\mu + \frac{1-\beta}{\beta} \log((1-\beta)/\lambda),$$

where $\lambda > 0$ is the unique solution of $\int_\Sigma (\beta + \lambda e^{-\phi})^{-1} d\mu = 1$.
Hint: Try first $d\nu/d\mu = (\beta + \lambda e^{-\phi})^{-1}$. To see the other direction, let $g_u(z) \triangleq -z \log(uz) - \frac{1-\beta z}{\beta} \log \frac{1-\beta z}{1-\beta}$ for $u \triangleq \lambda e^{-\phi}/(1-\beta)$ and check that

$$\sup_{\nu \in M_1(\Sigma)} \{\int_\Sigma \phi d\nu - I(\nu|\beta, \mu)\} \leq \sup_{f \in B(\Sigma), \frac{1}{\beta} \geq f \geq 0} \int_\Sigma g_u(f) d\mu + \log(\lambda/(1-\beta)),$$

whereas $g_u(z)$ is maximal at $z = 1/(\beta + u(1-\beta))$.

Exercise 7.2.8 In this exercise you derive the LDP for sampling with replacement from the deterministic sequence $(y_1, \ldots, y_m)$.

(a) Suppose X_i are i.i.d. Poisson(β) for some $\beta \in (0,\infty)$. Check that (7.2.6) holds for $V_n = nL'_n(\Sigma)$ a Poisson($m\beta$) random variable, and that $I_X(\cdot) = H(\cdot|\mu)$ on $M_1(\Sigma)$.
Hint: Use part (a) of Exercise 2.2.23.
(b) View $\{X_i\}_{i=1}^m$ as the number of balls in m distinct urns. If $V_n > n$, remove $V_n - n$ balls, with each ball equally likely to be removed. If $V_n < n$, add $n - V_n$ new balls independently, with each urn equally likely to receive each added ball. Check that the re-evaluated value of L'_n, denoted Z_n, has the same law as L_n^Y in case of sampling with replacement of n values out of $(y_1, \ldots, y_m)$, and that $d_{LU}(Z_n, L'_n) = n^{-1}|V_n - n|$.
(c) Conclude that if $L_m^Y \to \mu$ weakly and $n/m \to \beta \in (0,\infty)$ then in case of sampling with replacement L_n^Y satisfies the LDP in $M_1(\Sigma)$ (equipped with the weak topology), with the good rate function $H(\cdot|\mu)$ of Sanov's theorem.

Remark: L_n^Y corresponds to the bootstrapped empirical measure.

7.3 The Gibbs Conditioning Principle

In this section, the problem dealt with in Section 3.3 is considered in much greater generality. The motivation for the analysis lies in the following situation: Let Σ be a Polish space and $Y_1, Y_2, \ldots, Y_n$ a sequence of Σ-valued i.i.d. random variables, each distributed according to the law $\mu \in M_1(\Sigma)$. Let $L_n^Y \in M_1(\Sigma)$ denote the empirical measure associated with these variables. Given a functional $\Phi : M_1(\Sigma) \to \mathbb{R}$ (the *energy* functional), we are interested in computing the law of Y_1 under the constraint $\Phi(L_n^Y) \in D$, where D is some measurable set in $\mathbb{R}$ and $\{\Phi(L_n^Y) \in D\}$ is of positive probability. This situation occurs naturally in statistical mechanics, where Y_i denote some attribute of independent particles (e.g., their velocity), Φ is some constraint on the ensemble of particles (e.g., an average energy per particle constraint), and one is interested in making predictions on individual particles based on the existence of the constraint. The distribution of Y_1 under the energy conditioning alluded to before is then called the *micro-canonical* distribution of the system.

As in Section 3.3, note that for every measurable set $A \subset M_1(\Sigma)$ such that $\{L_n^Y \in A\}$ is of positive probability, and every bounded measurable function $f : \Sigma \to \mathbb{R}$, due to the exchangeability of the Y_i-s,

$$
\begin{aligned}
E(f(Y_1)|L_n^Y \in A) &= E(f(Y_i)|L_n^Y \in A) \\
&= E(\frac{1}{n}\sum_{i=1}^n f(Y_i)|L_n^Y \in A) \\
&= E(\langle f, L_n^Y \rangle|L_n^Y \in A). \quad (7.3.1)
\end{aligned}
$$

Thus, for $A \triangleq \{\nu : \Phi(\nu) \in D\}$, computing the conditional law of Y_1 under the conditioning $\{\Phi(L_n^{\mathbf{Y}}) \in D\} = \{L_n^{\mathbf{Y}} \in A\}$ is equivalent to the computation of the conditional expectation of $L_n^{\mathbf{Y}}$ under the same constraint. It is this last problem that is treated in the rest of this section, in a slightly more general framework.

Throughout this section, $M_1(\Sigma)$ is equipped with the τ-topology and the cylinder σ-field $\mathcal{B}^{cy}$. (For the definitions see Section 6.2.) For any $\mu \in M_1(\Sigma)$, let $\mu^n \in M_1(\Sigma^n)$ denote the induced product measure on Σ^n and let Q_n be the measure induced by μ^n in $(M_1(\Sigma), \mathcal{B}^{cy})$ through $L_n^{\mathbf{Y}}$. Let $A_\delta \in \mathcal{B}^{cy}$, $\delta > 0$ be nested measurable sets, i.e., $A_\delta \subseteq A_{\delta'}$ if $\delta < \delta'$. Let F_δ be nested *closed* sets such that $A_\delta \subseteq F_\delta$. Define $F_0 = \cap_{\delta>0} F_\delta$ and $A_0 = \cap_{\delta>0} A_\delta$ (so that $A_0 \subseteq F_0$). The following assumption prevails in this section.

Assumption (A-1)   *There exists a $\nu_* \in A_0$ (not necessarily unique) satisfying*

$$H(\nu_*|\mu) = \inf_{\nu \in F_0} H(\nu|\mu) \triangleq I_F < \infty \, ,$$

and for all $\delta > 0$,

$$\lim_{n \to \infty} \nu_*^n(\{L_n^{\mathbf{Y}} \in A_\delta\}) = 1 \, . \tag{7.3.2}$$

Think of the following situation as representative: $A_\delta = \{\nu : |\Phi(\nu)| \leq \delta\}$, where $\Phi : M_1(\Sigma) \to [-\infty, \infty]$ is only lower semicontinuous, and thus A_δ is neither open nor closed. (For example, the energy functional $\Phi(\nu) = \int_\Sigma (\| x \|^2 - 1)\nu(dx)$ when Σ is a separable Banach space.) The nested, closed sets F_δ are then chosen as $F_\delta = \{\nu : \Phi(\nu) \leq \delta\}$ with $F_0 = \{\nu : \Phi(\nu) \leq 0\}$, while $A_0 = \{\nu : \Phi(\nu) = 0\}$. We are then interested in the conditional distribution of Y_1 under a constraint of the form $\Phi(L_n^{\mathbf{Y}}) = 0$ (for example, a specified average energy).

Theorem 7.3.3 *Assume (A-1). Then $\mathcal{M} \triangleq \{\nu \in F_0 : H(\nu|\mu) = I_F\}$ is a non-empty, compact set. Further, for any $\Gamma \in \mathcal{B}^{cy}$ with $\mathcal{M} \subset \Gamma^o$,*

$$\limsup_{\delta \to 0} \limsup_{n \to \infty} \frac{1}{n} \log \mu^n(L_n^{\mathbf{Y}} \notin \Gamma \,|\, L_n^{\mathbf{Y}} \in A_\delta) < 0 \, .$$

Proof: Note that $A_0 \subseteq F_0$, so $\nu_* \in \mathcal{M}$ by Assumption (A-1). Moreover, $I_F < \infty$ implies that $\mathcal{M}$ being the intersection of the closed set F_0 and the compact set $\{\nu : H(\nu|\mu) \leq I_F\}$ (see Lemma 6.2.12), is a compact set. Clearly,

$$\limsup_{\delta \to 0} \limsup_{n \to \infty} \frac{1}{n} \log \mu^n(L_n^{\mathbf{Y}} \notin \Gamma | L_n^{\mathbf{Y}} \in A_\delta)$$

$$\leq \lim_{\delta \to 0} \limsup_{n \to \infty} \frac{1}{n} \log Q_n(\Gamma^c \cap A_\delta) - \lim_{\delta \to 0} \liminf_{n \to \infty} \frac{1}{n} \log Q_n(A_\delta).$$

Let $G \triangleq \Gamma^o$. Then, since $\Gamma^c \cap A_\delta \subset G^c \cap F_\delta$, with $G^c \cap F_\delta$ being a closed set, the upper bound of Sanov's theorem (Theorem 6.2.10) yields

$$\lim_{\delta \to 0} \limsup_{n \to \infty} \frac{1}{n} \log Q_n(\Gamma^c \cap A_\delta)$$

$$\leq \lim_{\delta \to 0} \left\{ -\inf_{\nu \in G^c \cap F_\delta} H(\nu|\mu) \right\} = -\inf_{\nu \in G^c \cap F_0} H(\nu|\mu) < -I_F \,,$$

where the preceding equality follows by applying Lemma 4.1.6 to the nested, closed sets $G^c \cap F_\delta$, and the strict inequality follows from the closedness of $G^c \cap F_0$ and the definition of $\mathcal{M}$. The proof is now completed by the following lemma.

Lemma 7.3.4 *Assume (A-1). Then, for all $\delta > 0$,*

$$\liminf_{n \to \infty} \frac{1}{n} \log Q_n(A_\delta) \geq -I_F \,.$$

Proof of Lemma 7.3.4: Since A_δ in general may contain no neighborhood of points from $\mathcal{M}$, the lower bound of Sanov's theorem cannot be used directly. Instead, a direct computation of the lower bound via the change of measure argument will be used in conjunction with (7.3.2).

Let ν_* be as in Assumption (A-1). Since $H(\nu_*|\mu) < \infty$, the Radon–Nikodym derivative $f = d\nu_*/d\mu$ exists. Fix $\delta > 0$ and define the sets

$$\Gamma_n \triangleq \left\{ \mathbf{y} \in \Sigma^n : f_n(\mathbf{y}) \triangleq \prod_{i=1}^n f(y_i) > 0, L_n^{\mathbf{y}} \in A_\delta \right\} \,.$$

It follows by (7.3.2) that $\nu_*^n(\Gamma_n) \xrightarrow[n \to \infty]{} 1$. Hence,

$$\liminf_{n \to \infty} \frac{1}{n} \log Q_n(A_\delta) \geq \liminf_{n \to \infty} \frac{1}{n} \log \int_{\Gamma_n} \frac{1}{f_n(\mathbf{y})} \nu_*^n(dy)$$

$$= \liminf_{n \to \infty} \frac{1}{n} \log \left(\frac{1}{\nu_*^n(\Gamma_n)} \int_{\Gamma_n} \frac{1}{f_n(\mathbf{y})} \nu_*^n(dy) \right) \,.$$

Therefore, by Jensen's inequality,

$$\liminf_{n \to \infty} \frac{1}{n} \log Q_n(A_\delta) \geq -\limsup_{n \to \infty} \frac{1}{n \nu_*^n(\Gamma_n)} \int_{\Gamma_n} \log(f_n(\mathbf{y})) \nu_*^n(dy)$$

$$= -H(\nu_*|\mu) + \liminf_{n \to \infty} \frac{1}{n} \int_{(\Gamma_n)^c} \log(f_n(\mathbf{y})) \nu_*^n(dy) \,.$$

Note that

$$\int_{(\Gamma_n)^c} \log(f_n(\mathbf{y})) \nu_*^n(dy) = \int_{(\Gamma_n)^c} f_n(\mathbf{y}) \log(f_n(\mathbf{y})) \mu^n(dy) \geq C,$$

where $C = \inf_{x \geq 0}\{x \log x\} > -\infty$. Since $H(\nu_*|\mu) = I_F$, the proof is complete. □

The following corollary shows that if ν_* of (A-1) is unique, then $\mu^n_{\mathbf{Y}^k|A_\delta}$, the law of $\mathbf{Y}^k = (Y_1, \ldots, Y_k)$ conditional upon the event $\{L^{\mathbf{Y}}_n \in A_\delta\}$, is approximately a product measure.

Corollary 7.3.5 If $\mathcal{M} = \{\nu_*\}$ then $\mu^n_{\mathbf{Y}^k|A_\delta} \to (\nu_*)^k$ weakly in $M_1(\Sigma^k)$ for $n \to \infty$ followed by $\delta \to 0$.

Proof: Assume $\mathcal{M} = \{\nu_*\}$ and fix $\phi_j \in C_b(\Sigma)$, $j = 1, \ldots, k$. By the invariance of $\mu^n_{\mathbf{Y}^n|A_\delta}$ with respect to permutations of $\{Y_1, \ldots, Y_n\}$,

$$\langle \prod_{j=1}^k \phi_j, \mu^n_{\mathbf{Y}^k|A_\delta} \rangle = \frac{(n-k)!}{n!} \sum_{i_1 \neq \cdots \neq i_k} \int_{\Sigma^n} \prod_{j=1}^k \phi_j(y_{i_j}) \mu^n_{\mathbf{Y}^n|A_\delta}(d\mathbf{y}) .$$

Since,

$$E(\prod_{j=1}^k \langle \phi_j, L^{\mathbf{Y}}_n \rangle \,|\, L^{\mathbf{Y}}_n \in A_\delta) = \frac{1}{n^k} \sum_{i_1, \ldots, i_k} \int_{\Sigma^n} \prod_{j=1}^k \phi_j(y_{i_j}) \mu^n_{\mathbf{Y}^n|A_\delta}(d\mathbf{y}) ,$$

and ϕ_j are bounded functions, it follows that

$$|\langle \prod_{j=1}^k \phi_j, \mu^n_{\mathbf{Y}^k|A_\delta} \rangle - E(\prod_{j=1}^k \langle \phi_j, L^{\mathbf{Y}}_n \rangle \,|\, L^{\mathbf{Y}}_n \in A_\delta)| \leq C(1 - \frac{n!}{n^k(n-k)!}) \xrightarrow[n \to \infty]{} 0 .$$

For $\mathcal{M} = \{\nu_*\}$, Theorem 7.3.3 implies that for any $\eta > 0$,

$$\mu^n(|\langle \phi_j, L^{\mathbf{Y}}_n \rangle - \langle \phi_j, \nu_* \rangle| > \eta \,|\, L^{\mathbf{Y}}_n \in A_\delta) \to 0$$

as $n \to \infty$ followed by $\delta \to 0$. Since $\langle \phi_j, L^{\mathbf{Y}}_n \rangle$ are bounded,

$$E(\prod_{j=1}^k \langle \phi_j, L^{\mathbf{Y}}_n \rangle \,|\, L^{\mathbf{Y}}_n \in A_\delta) \to \langle \prod_{j=1}^k \phi_j, (\nu_*)^k \rangle ,$$

so that

$$\limsup_{\delta \to 0} \limsup_{n \to \infty} \langle \prod_{j=1}^k \phi_j, \mu^n_{\mathbf{Y}^k|A_\delta} - (\nu_*)^k \rangle = 0 .$$

Recall that $C_b(\Sigma)^k$ is convergence determining for $M_1(\Sigma^k)$, hence it follows that $\mu^n_{\mathbf{Y}^k|A_\delta} \to (\nu_*)^k$ weakly in $M_1(\Sigma^k)$. □

Having stated a general conditioning result, it is worthwhile checking Assumption (A-1) and the resulting set of measures $\mathcal{M}$ for some particular choices of the functional Φ. Two options are considered in detail in the following sections; non-interacting particles, in which case $n^{-1} \sum_{i=1}^{n} U(Y_i)$ is specified, and interacting particles, in which case $n^{-2} \sum_{i,j=1}^{n} U(Y_i, Y_j)$ is specified. We return to refinements of Corollary 7.3.5 in Section 7.3.3.

7.3.1   The Non-Interacting Case

Let $U : \Sigma \to [0, \infty)$ be a Borel measurable function. Define the functional $\Phi : M_1(\Sigma) \to [-1, \infty]$ by

$$\Phi(\nu) = \langle U, \nu \rangle - 1 \ ,$$

and consider the constraint

$$\{L_n^{\mathbf{Y}} \in A_\delta\} \triangleq \{|\Phi(L_n^{\mathbf{Y}})| \leq \delta\} = \{|\frac{1}{n} \sum_{i=1}^{n} U(Y_i) - 1| \leq \delta\}.$$

By formally solving the optimization problem

$$\inf_{\{\nu:\ \langle U, \nu \rangle = 1\}} H(\nu|\mu),$$

one is led to conjecture that ν_* of Assumption (A-1) should be a *Gibbs measure*, namely, one of the measures γ_β, where

$$\frac{d\gamma_\beta}{d\mu} = \frac{e^{-\beta U(x)}}{Z_\beta},$$

and Z_β, the *partition function*, is the normalizing constant

$$Z_\beta = \int_\Sigma e^{-\beta U(x)} \mu(dx).$$

Throughout this section, $\beta \in (\beta_\infty, \infty)$, where $\beta_\infty \triangleq \inf\{\beta :\ Z_\beta < \infty\}$.

The following lemma (whose proof is deferred to the end of this section) is key to the verification of this conjecture.

Lemma 7.3.6 *Assume that* $\mu(\{x : U(x) > 1\}) > 0$, $\mu(\{x : U(x) < 1\}) > 0$, *and either* $\beta_\infty = -\infty$ *or*

$$\lim_{\beta \searrow \beta_\infty} \langle U, \gamma_\beta \rangle > 1. \tag{7.3.7}$$

Then there exists a unique $\beta^* \in (\beta_\infty, \infty)$ *such that* $\langle U, \gamma_{\beta^*} \rangle = 1$.

The main result of this section is the following.

Theorem 7.3.8 *Let U, μ and β^* be as in the preceding lemma. If either U is bounded or $\beta^* \geq 0$, then Theorem 7.3.3 applies, with $\mathcal{M}$ consisting of a unique Gibbs measure γ_{β^*}.*

In particular, Theorem 7.3.8 states that the conditional law of Y_1 converges, as $n \to \infty$, to the Gibbs measure γ_{β^*}.

Proof: Note that by the monotone convergence theorem, $\langle U, \cdot \rangle = \sup_n \langle U \wedge, n, \cdot \rangle$. Since $U \wedge n \in B(\Sigma)$, it follows that $\Phi(\cdot) = \langle U, \cdot \rangle - 1$ is a τ-lower semicontinuous functional. Hence, $F_\delta \triangleq \{\nu : \langle U, \nu \rangle \leq 1 + \delta\}$, $\delta > 0$, are nested closed sets, whereas $F_0 = \{\nu : \langle U, \nu \rangle \leq 1\}$ is a convex, closed set. By the preceding lemma, $\gamma_{\beta^*} \in F_0$, and by a direct computation,

$$H(\gamma_{\beta^*} | \mu) = -\beta^* \langle U, \gamma_{\beta^*} \rangle - \log Z_{\beta^*} < \infty \,,$$

implying that $I_F < \infty$. Since $H(\cdot | \mu)$ is strictly convex within its I_F level set, it follows that $\mathcal{M}$ contains precisely one probability measure, denoted ν_0. A direct computation, using the equivalence of μ and γ_{β^*}, yields that

$$
\begin{aligned}
-H(\nu_0 | \gamma_{\beta^*}) &\geq -H(\nu_0 | \gamma_{\beta^*}) + [H(\nu_0 | \mu) - H(\gamma_{\beta^*} | \mu)] \\
&= \beta^*(\langle U, \gamma_{\beta^*} \rangle - \langle U, \nu_0 \rangle) = \beta^*(1 - \langle U, \nu_0 \rangle),
\end{aligned}
$$

where the preceding inequality is implied by $\nu_0 \in \mathcal{M}$ and $\gamma_{\beta^*} \in F_0$. For $\beta^* \geq 0$, it follows that $H(\nu_0 | \gamma_{\beta^*}) \leq 0$, since $\langle U, \nu_0 \rangle \leq 1$. Hence, $\nu_0 = \gamma_{\beta^*}$, and consequently, $\mathcal{M} = \{\gamma_{\beta^*}\}$. Now, Assumption (A-1) holds for $\nu_* = \gamma_{\beta^*} \in A_0$ as the limit (7.3.2) follows by the weak law of large numbers. Consequently, Theorem 7.3.3 holds. When U is bounded, then A_δ are closed sets. Therefore, in this case, $F_\delta = A_\delta$ can be chosen to start with, yielding $\langle U, \nu_0 \rangle = 1$. Consequently, when U is bounded, $\nu_0 = \gamma_{\beta^*} = \nu_*$ even for $\beta^* < 0$. $\qquad \square$

Proof of Lemma 7.3.6: Recall that by Exercise 2.2.24, $\log Z_\beta$ is a C^∞ function in (β_∞, ∞). By dominated convergence,

$$\langle U, \gamma_\beta \rangle = -\frac{d}{d\beta} \log Z_\beta$$

and is finite for all $\beta > \beta_\infty$, as follows from the proof of (2.2.9). By a similar argument,

$$\frac{d}{d\beta} \langle U, \gamma_\beta \rangle = -\int_\Sigma (U - \langle U, \gamma_\beta \rangle)^2 d\gamma_\beta < 0 \,,$$

where the strict inequality follows, since by our assumptions, U cannot be constant μ a.e. Hence, $\langle U, \gamma_\beta \rangle$ is strictly decreasing and continuous as a function of $\beta \in (\beta_\infty, \infty)$. Thus, it suffices to show that

$$\lim_{\beta \to \infty} \langle U, \gamma_\beta \rangle < 1, \qquad (7.3.9)$$

and that when $\beta_\infty = -\infty$,

$$\lim_{\beta \to -\infty} \langle U, \gamma_\beta \rangle > 1. \tag{7.3.10}$$

To see (7.3.9), note that by assumption, there exists a $0 < u_0 < 1$ such that $\mu(\{x : U(x) < u_0\}) > 0$. Now, for $\beta > 0$,

$$\int_\Sigma e^{-\beta U(x)} \mu(dx) \geq e^{-\beta u_0} \mu(\{x : U(x) \in [0, u_0)\})$$

and

$$\int_\Sigma (U(x) - u_0) e^{-\beta U(x)} \mu(dx)$$

$$\leq e^{-\beta u_0} \int_\Sigma (U(x) - u_0) 1_{\{U(x) > u_0\}} e^{-\beta(U(x) - u_0)} \mu(dx)$$

$$\leq \frac{e^{-\beta u_0}}{\beta} \sup_{y \geq 0} \{y e^{-y}\}.$$

Hence, for some $C < \infty$,

$$\langle U, \gamma_\beta \rangle = u_0 + \frac{\int_\Sigma (U(x) - u_0) e^{-\beta U(x)} \mu(dx)}{\int_\Sigma e^{-\beta U(x)} \mu(dx)} \leq u_0 + \frac{C}{\beta},$$

and (7.3.9) follows by considering the limit $\beta \to \infty$.

To see (7.3.10) when $\beta_\infty = -\infty$, choose $u_2 > u_1 > 1$ such that $\mu(\{x : U(x) \in [u_2, \infty)\}) > 0$. Note that for all $\beta \leq 0$,

$$\int_\Sigma 1_{\{x : U(x) \in [0, u_1)\}} e^{-\beta U(x)} \mu(dx) \leq e^{-\beta u_1}$$

and

$$\int_\Sigma 1_{\{x : U(x) \in [u_1, \infty)\}} e^{-\beta U(x)} \mu(dx) \geq e^{-\beta u_2} \mu(\{x : U(x) \in [u_2, \infty)\}).$$

Hence, for all $\beta \leq 0$,

$$\frac{1}{\gamma_\beta(\{x : U(x) \in [u_1, \infty)\})} = 1 + \frac{\int_\Sigma 1_{\{x : U(x) \in [0, u_1)\}} e^{-\beta U(x)} \mu(dx)}{\int_\Sigma 1_{\{x : U(x) \in [u_1, \infty)\}} e^{-\beta U(x)} \mu(dx)}$$

$$\leq 1 + \frac{e^{\beta(u_2 - u_1)}}{\mu(\{x : U(x) \in [u_2, \infty)\})},$$

implying that

$$\liminf_{\beta \to -\infty} \langle U, \gamma_\beta \rangle \geq u_1 \liminf_{\beta \to -\infty} \gamma_\beta(\{x : U(x) \in [u_1, \infty)\}) \geq u_1,$$

and consequently (7.3.10) holds. $\qquad \square$

Exercise 7.3.11 Let $\Sigma = [0, \infty)$ and $\mu(dx) = C \frac{e^{-x}}{x^3 + 1} dx$, where $C < \infty$ is a normalization constant. Let $U(x) = \epsilon x$ with $\epsilon > 0$.
(a) Check that for all $\epsilon > 0$, both $\mu(\{x : U(x) > 1\}) > 0$ and $\mu(\{x : U(x) < 1\}) > 0$, and that $\beta_\infty = -1/\epsilon$.
(b) Show that for $\epsilon > 0$ small enough, (7.3.7) fails to hold.
(c) Verify that for $\epsilon > 0$ small enough, there is no Gibbs measure in A_0.

7.3.2 The Interacting Case

The previous section deals with the case where there is no interaction present, i.e., when the "particles" $Y_1, \ldots, Y_n$ do not affect each other. The case where interaction is present is interesting from a physical point of view. To build a model of such a situation, let $M > 1$ be given, let $U : \Sigma^2 \to [0, M]$ be a continuous, symmetric, bounded function, and define $\Phi(\nu) = \langle U\nu, \nu \rangle - 1$ and $A_\delta = \{\nu : |\Phi(\nu)| \le \delta\}$ for $\delta \ge 0$. Throughout, $U\nu$ denotes the bounded, continuous function

$$U\nu(x) = \int_\Sigma U(x, y)\nu(dy) .$$

The restriction that U be bounded is made here for the sake of simplicity, as it leads to $\Phi(\cdot)$ being a continuous functional, implying that the sets A_δ are closed.

Lemma 7.3.12 *The functional $\nu \mapsto \langle U\nu, \nu \rangle$ is continuous with respect to the τ-topology on $M_1(\Sigma)$.*

Proof: Clearly, it suffices to prove the continuity of the functional $\nu \mapsto \langle U\nu, \nu \rangle$ with respect to the weak topology on $M_1(\Sigma)$. For $U(x, y) = f(x)g(y)$ with $f, g \in C_b(\Sigma)$, the continuity of $\nu \mapsto \langle U\nu, \nu \rangle$ is trivial. With Σ Polish, the collection $\{f(x)g(y)\}_{f, g \in C_b(\Sigma)}$ is convergence determining for the weak topology on $M_1(\Sigma^2)$ and hence the continuity of $\nu \mapsto \langle U\nu, \nu \rangle$ holds for all $U \in C_b(\Sigma^2)$. $\square$

Remark: For measurable $U(x, y)$, the map $\nu \mapsto \langle U\nu, \nu \rangle$ is sequentially continuous with respect to the τ-topology. However, the continuity of $U(\cdot, \cdot)$ is essential for the above lemma to hold true (see Exercise 7.3.18 for an example of $A \in \mathcal{B}_{[0,1]^2}$ for which the map $\nu \mapsto \nu \times \nu(A)$ is discontinuous with respect to the τ-topology on $M_1([0, 1])$).

As in the non-interacting case, a formal computation reveals that Theorem 7.3.3, if applicable, would lead to $\mathcal{M}$ consisting of Gibbs measures γ_β such that

$$\frac{d\gamma_\beta}{d\mu} = \frac{e^{-\beta U \gamma_\beta(x)}}{Z_\beta} , \tag{7.3.13}$$

where Z_β, the *partition function*, would be the normalizing constant

$$Z_\beta = \int_\Sigma e^{-\beta U \gamma_\beta(x)} \mu(dx).$$

It is the goal of this section to check that, under the following three assumptions, this is indeed the case.

Assumption (A-2)　*For any ν_i such that $H(\nu_i|\mu) < \infty$, $i = 1, 2$,*

$$\langle U\nu_1, \nu_2 \rangle \le \frac{1}{2} \left(\langle U\nu_1, \nu_1 \rangle + \langle U\nu_2, \nu_2 \rangle \right).$$

Assumption (A-3)　$\int_{\Sigma^2} U(x, y) \mu(dx) \mu(dy) \ge 1$.

Assumption (A-4)　*There exists a probability measure ν with $H(\nu|\mu) < \infty$ and $\langle U\nu, \nu \rangle < 1$.*

Note that, unlike the non-interacting case, here even the existence of Gibbs measures needs to be proved. For that purpose, it is natural to define the following *Hamiltonian*:

$$H_\beta(\nu) = H(\nu|\mu) + \frac{\beta}{2} \langle U\nu, \nu \rangle,$$

where $\beta \in [0, \infty)$. The following lemma (whose proof is deferred to the end of this section) summarizes the key properties of the Gibbs measures that are related to $H_\beta(\cdot)$.

Lemma 7.3.14 *Assume (A-2). Then:*
(a) For each $\beta \ge 0$, there exists a unique minimizer of $H_\beta(\cdot)$, denoted γ_β, such that (7.3.13) holds.
(b) The function $g(\beta) \triangleq \langle U\gamma_\beta, \gamma_\beta \rangle$ is continuous on $[0, \infty)$.
(c) Assume (A-2), (A-3), and (A-4), and define

$$\beta^* \triangleq \inf\{\beta \ge 0 : g(\beta) \le 1\}. \tag{7.3.15}$$

Then $\beta^* < \infty$ and $g(\beta^*) = 1$.

Equipped with Lemma 7.3.14, the characterization of $\mathcal{M}$ and the proof of Theorem 7.3.3 is an easy matter.

Theorem 7.3.16 *Assume (A-2)–(A-4). Then Theorem 7.3.3 applies, with $\mathcal{M}$ consisting of a unique Gibbs measure γ_{β^*}, where β^* is as defined in (7.3.15).*

Proof: Since $\Phi(\cdot)$ is a continuous functional (see Lemma 7.3.12), $F_\delta = A_\delta$ may be taken here, yielding $F_0 = A_0 = \{\nu : \langle U\nu, \nu \rangle = 1\}$. Recall that μ

and γ_{β^*} are equivalent with a globally bounded Radon–Nikodym derivative. (See (7.3.13).) Since $\gamma_{\beta^*} \in F_0$, $I_F < \infty$, and $\mathcal{M}$ is non-empty. Moreover, as in the proof of Theorem 7.3.8, observe that by a direct computation, whenever $\nu \in F_0$ and $H(\nu|\mu) < \infty$,

$$H(\nu|\mu) - H(\nu|\gamma_{\beta^*}) - H(\gamma_{\beta^*}|\mu) = \beta^* \left(\langle U\gamma_{\beta^*}, \gamma_{\beta^*} \rangle - \langle U\gamma_{\beta^*}, \nu \rangle \right)$$
$$\geq \frac{\beta^*}{2} \left(\langle U\gamma_{\beta^*}, \gamma_{\beta^*} \rangle - \langle U\nu, \nu \rangle \right) = 0,$$

where the inequality follows by Assumption (A-2). Thus, for all $\nu \in \mathcal{M}$,

$$-H(\nu|\gamma_{\beta^*}) \geq -H(\nu|\gamma_{\beta^*}) + H(\nu|\mu) - H(\gamma_{\beta^*}|\mu) \geq 0,$$

which is clearly possible only if $\nu = \gamma_{\beta^*}$. Therefore, $\mathcal{M} = \{\gamma_{\beta^*}\}$. Finally, note that Assumption (A-1) holds for $\nu_* = \gamma_{\beta^*}$, and consequently, Theorem 7.3.3 holds as well. Indeed, here,

$$\{L_n^{\mathbf{Y}} \in A_\delta\} \equiv \left\{ \left| \frac{1}{n^2} \sum_{i,j=1}^{n} (U(Y_i, Y_j) - E_{\nu_*^2}[U(X, Y)]) \right| \leq \delta \right\}.$$

Therefore, by Chebycheff's inequality,

$$\nu_*^n(\{L_n^{\mathbf{Y}} \notin A_\delta\})$$
$$\leq \frac{1}{n^4 \delta^2} \sum_{i,j,k,\ell=1}^{n} E_{\nu_*^n}(U(Y_i, Y_j) - E_{\nu_*^2}[U(X, Y)])(U(Y_k, Y_\ell) - E_{\nu_*^2}[U(X, Y)])$$
$$\leq \frac{6M^2}{n\delta^2},$$

and (7.3.2) follows by considering $n \to \infty$. $\qquad\qquad\square$

Proof of Lemma 7.3.14: (a) Fix $\beta \geq 0$. Note that $H(\nu|\mu) \leq H_\beta(\nu)$ for all $\nu \in M_1(\Sigma)$. Further, $H(\cdot|\mu)$ is a good rate function, and hence by Lemma 7.3.12, so is $H_\beta(\cdot)$. Since U is bounded, $H_\beta(\mu) < \infty$, and therefore, there exists a $\tilde{\nu}$ such that

$$H_\beta(\tilde{\nu}) = \inf_{\nu \in M_1(\Sigma)} H_\beta(\nu) < \infty.$$

Assumption (A-2) and the convexity of $H(\cdot|\mu)$ imply that if $H_\beta(\nu_i) < \infty$ for $i = 1, 2$, then

$$H_\beta \left(\frac{\nu_1 + \nu_2}{2} \right) \leq \frac{1}{2} H_\beta(\nu_1) + \frac{1}{2} H_\beta(\nu_2).$$

The preceding inequality extends by iterations to cover $\frac{k}{2^n}\nu_1 + (1 - \frac{k}{2^n})\nu_2$ for all integers k, n with $1 \leq k \leq 2^n$, and the convexity of $H_\beta(\cdot)$ within its level sets follows as a consequence of its lower semicontinuity. Further, $H(\cdot|\mu)$ is strictly convex within its level sets (see Lemma 6.2.12), and hence so is $H_\beta(\cdot)$. Therefore, let γ_β denote the unique minimizer of $H_\beta(\cdot)$. Let $f = d\gamma_\beta/d\mu$. First check that $\mu(\{x : f(x) = 0\}) = 0$. Assume otherwise, let $z = \mu(\{x : f(x) = 0\}) > 0$, and define the probability measure

$$\nu(dx) = \frac{1_{\{x:f(x)=0\}}}{z}\mu(dx) \, .$$

Note that $\nu_t = t\nu + (1-t)\gamma_\beta$ is a probability measure for all $t \in [0,1]$. Since the supports of ν and γ_β are disjoint, by a direct computation,

$$0 \leq \frac{1}{t}[H_\beta(\nu_t) - H_\beta(\gamma_\beta)]$$

$$= H_\beta(\nu) - H_\beta(\gamma_\beta) + \log t + \frac{1-t}{t}\log(1-t) - \frac{\beta}{2}(1-t)\langle U(\gamma_\beta - \nu), \gamma_\beta - \nu\rangle.$$

Since $H(\nu|\mu) = -\log z < \infty$, $H_\beta(\nu) < \infty$, and the preceding inequality results with a contradiction when considering the limit $t \searrow 0$. It remains, therefore, to check that (7.3.13) holds. To this end, fix $\phi \in B(\Sigma)$, $\phi \neq 0$ and $\delta = 2/||\phi|| > 0$. For all $t \in (-\delta, \delta)$, define $\nu_t \in M_1(\Sigma)$ via

$$\frac{d\nu_t}{d\gamma_\beta} = 1 + t(\phi - \langle\phi, \gamma_\beta\rangle).$$

Since $H_\beta(\nu_t)$ is differentiable in t and possesses a minimum at $t = 0$, it follows that $dH_\beta(\nu_t)/dt = 0$ at $t = 0$. Hence,

$$0 = \int_\Sigma (\phi - \langle\phi, \gamma_\beta\rangle)(\log f + \beta U\gamma_\beta)d\gamma_\beta \qquad (7.3.17)$$

$$= \int_\Sigma \phi f\left(\log f + \beta U\gamma_\beta - H(\gamma_\beta|\mu) - \beta\langle U\gamma_\beta, \gamma_\beta\rangle\right) d\mu.$$

Since ϕ is arbitrary, $f > 0$ μ-a.e., and $H(\gamma_\beta|\mu)$ and $\beta\langle U\gamma_\beta, \gamma_\beta\rangle$ are finite constants, (7.3.13) follows.

(b) Suppose that $\beta_n \to \beta \in [0, \infty)$. Then β_n is a bounded sequence, and hence, $\{\gamma_{\beta_n}\}_{n=1}^\infty$ are contained in some compact level set of $H(\cdot|\mu)$. Consequently, this sequence of measures has at least one limit point, denoted ν. Passing to a convergent subsequence, it follows by Lemma 7.3.12 and the characterization of γ_{β_n} that

$$H_\beta(\nu) \leq \liminf_{n\to\infty} H_{\beta_n}(\gamma_{\beta_n}) \leq \liminf_{n\to\infty} H_{\beta_n}(\gamma_\beta) = H_\beta(\gamma_\beta).$$

Hence, by part (a) of the lemma, the sequence $\{\gamma_{\beta_n}\}$ converges to γ_β. The continuity of $g(\beta)$ now follows by Lemma 7.3.12.

(c) Let ν be as in Assumption (A-4), i.e., $\langle U\nu, \nu \rangle < 1$ and $H(\nu|\mu) < \infty$. Observe that by part (a) of the lemma and the nonnegativity of $H(\cdot|\mu)$,

$$g(\beta) \leq \frac{2}{\beta} H_\beta(\gamma_\beta) \leq \frac{2}{\beta} H(\nu|\mu) + \langle U\nu, \nu \rangle, \quad \forall \beta > 0.$$

Thus,

$$\limsup_{\beta \to \infty} g(\beta) \leq \langle U\nu, \nu \rangle < 1.$$

Clearly, $\gamma_0 = \mu$ and $g(0) \geq 1$ by Assumption (A-3). Hence, by part (b) of the lemma, $\beta^* < \infty$ and $g(\beta^*) = 1$. $\square$

Remark: Note that (7.3.17) implies in particular that

$$\log Z_\beta = -H(\gamma_\beta|\mu) - \beta \langle U\gamma_\beta, \gamma_\beta \rangle.$$

Exercise 7.3.18 [Suggested by Y. Peres] In this exercise, you show that Lemma 7.3.12 cannot be extended to general bounded measurable functions $U(\cdot, \cdot)$.
(a) Let m be Lesbegue measure on $[0, 1]$. Check that for any $B_i, i = 1, \ldots, N$, disjoint subsets of $[0, 1]$ of positive Lesbegue measure, there exist $y_i \in B_i$ such that $y_i - y_j$ is rational for all $i, j \in \{1, \ldots, N\}$.
Hint: Let $f(x_1, \ldots, x_{N-1}) = \int_0^1 1_{B_N}(z) \prod_{i=1}^{N-1} 1_{B_i}(z - x_i)dz$. Check that $f(\cdot)$ is continuous on $[-1, 1]^{N-1}$, and that

$$f(x_1, \ldots, x_{N-1}) = m\left(\bigcap_{i=1}^{N-1} (x_i + B_i) \cap B_N \right),$$

while

$$\int_{-1}^1 \cdots \int_{-1}^1 f(x_1, \ldots, x_{N-1})dx_1 \cdots dx_{N-1} = \prod_{i=1}^N m(B_i) > 0.$$

(b) Consider the measurable set $A = \{(x, y) : x, y \in [0, 1], x - y \text{ is rational}\}$. Show that for every finite measurable partition $B_1, \ldots, B_N$ of $[0, 1]$, there is a measure ν with $\nu(B_i) = m(B_i)$ but $\nu \times \nu(A) = 1$.
Hint: Let $\nu = \sum_{i=1}^N m(B_i)\delta_{y_i}$, with $\{y_i\}$ as in part (a).
(c) Conclude that at $\nu = m$, the map $\nu \mapsto \langle 1_A \nu, \nu \rangle$ is discontinuous with respect to the τ-topology of $M_1([0, 1])$.

Exercise 7.3.19 Assume (A-2). Check that when U is unbounded, but $\langle U\mu, \mu \rangle < \infty$, the existence of a unique minimizer of $H_\beta(\cdot)$ asserted in part (a) of Lemma 7.3.14 still holds true, that Lemma 7.3.12 (and hence part (b) of Lemma 7.3.14) is replaced by a lower semicontinuity statement, and that the minimizer of $H_\beta(\cdot)$ satisfies

$$\frac{d\gamma_\beta}{d\mu} = 1_{\{x: U\gamma_\beta(x) < \infty\}} \exp(H(\gamma_\beta|\mu) + \beta \langle U\gamma_\beta, \gamma_\beta \rangle - \beta U\gamma_\beta(x)).$$

Exercise 7.3.20 Prove that if U is unbounded, Theorem 7.3.16 still holds true, provided that β^* defined in (7.3.15) is nonzero, that $\langle U\gamma_\beta, \gamma_\beta \rangle$ is continuous at $\beta = \beta^*$, and that $\int_\Sigma U(x,x)\gamma_{\beta^*}(dx) < \infty$.

7.3.3 Refinements of the Gibbs Conditioning Principle

We return in this section to the general setup discussed in Theorem 7.3.3. Our goal is to explore the structure of the conditional law $\mu^n_{\mathbf{Y}^k|A_\delta}$ when $k = k(n) \to_{n\to\infty} \infty$. The motivation is clear: we wish to consider the effect of Gibbs conditioning on subsets of the system whose size increases with the size of the system.

The following simplifying assumption prevails in this section.
Assumption (A-5) $F_\delta = A_\delta \equiv A \in \mathcal{B}^{cy}$ is a closed, convex set of probability measures on a compact metric space (Σ, d) such that

$$I_F \stackrel{\triangle}{=} \inf_{\nu \in A} H(\nu|\mu) = \inf_{\nu \in A^\circ} H(\nu|\mu) < \infty.$$

With $H(\cdot|\mu)$ strictly convex on the compact convex sets $\{\nu : H(\nu|\mu) \leq \alpha\}$ (see Lemma 6.2.12), there exists a unique $\nu_* \in A$ such that $H(\nu_*|\mu) = I_F$. The main result of this section is the following refinement of Corollary 7.3.5.

Theorem 7.3.21 *Assume (A-5), and further that*

$$\mu^n(L^{\mathbf{Y}}_n \in A)e^{nI_F} \geq g_n > 0. \tag{7.3.22}$$

Then, for any $k = k(n)$,

$$H\left(\mu^n_{\mathbf{Y}^k|A}\Big|(\nu_*)^k\right) \leq \frac{1}{\lfloor \frac{n}{k(n)} \rfloor} \log(1/g_n). \tag{7.3.23}$$

Remarks:
(a) By Exercise 6.2.17 and (7.3.23), if $n^{-1}k(n)\log(1/g_n) \to 0$ then

$$\left\|\mu^n_{\mathbf{Y}^{k(n)}|A} - (\nu_*)^{k(n)}\right\|_{var} \xrightarrow[n\to\infty]{} 0. \tag{7.3.24}$$

(b) By Sanov's theorem (Theorem 6.2.10) and Assumption (A-5),

$$\lim_{n\to\infty} \frac{1}{n} \log \mu^n(L^{\mathbf{Y}}_n \in A) = -I_F,$$

implying that (7.3.22) always holds for some g_n such that $n^{-1}\log g_n \to 0$. Hence, (7.3.24) holds for any fixed $k \in \mathbb{Z}_+$, already an improvement over Corollary 7.3.5. More can be said about g_n for certain choices of the conditioning set A. Corollary 7.3.34 provides one such example.

Key to the proof of Theorem 7.3.21 are the properties of the relative entropy $H(\cdot|\cdot)$ shown in the following two lemmas.

Lemma 7.3.25 *For $\mathcal{X}$ a Polish space, any $P \in M_1(\mathcal{X}^m)$, $m \in \mathbb{Z}_+$ and any $Q \in M_1(\mathcal{X})$,*

$$H(P|Q^m) = H(P|P_1 \times \cdots \times P_m) + \sum_{i=1}^{m} H(P_i|Q) , \qquad (7.3.26)$$

where $P_i \in M_1(\mathcal{X})$ denotes the i-th marginal of P.

Proof: Suppose $P_i(B) > Q(B) = 0$ for some $B \subset \mathcal{X}$ and $i = 1, \ldots, m$. Then, $P(\tilde{B}) > Q^m(\tilde{B}) = 0$ for $\tilde{B} = \mathcal{X}^{i-1} \times B \times \mathcal{X}^{m-i}$ in which case both sides of (7.3.26) are infinite. Thus, we may and shall assume that $f_i = dP_i/dQ$ exist for every $i = 1, \ldots, m$. Since $P(\{\mathbf{y} : \prod_{i=1}^{m} f_i(y_i) = 0\}) \leq \sum_{i=1}^{m} P_i(\{y_i : f_i(y_i) = 0\}) = 0$, if $P(\tilde{B}) > P_1 \times \cdots \times P_m(\tilde{B}) = 0$ for some $\tilde{B} \subset \mathcal{X}^m$ then also $Q^m(\tilde{B}) = 0$ and again both sides of (7.3.26) are infinite. Thus, we may and shall assume also that $g = dP/d(P_1 \times \cdots \times P_m)$ exists, in which case $dP/dQ^m(\mathbf{y}) = g(\mathbf{y}) \prod_{i=1}^{m} f_i(y_i)$, implying that

$$H(P|Q^m) = \int_{\mathcal{X}^m} \log g \, dP + \sum_{i=1}^{m} \int_{\mathcal{X}} \log f_i \, dP_i ,$$

and (7.3.26) follows. $\qquad\square$

Lemma 7.3.27 (Csiszàr) *Suppose that $H(\nu_0|\mu) = \inf_{\nu \in A} H(\nu|\mu)$ for a convex set $A \subset M_1(\Sigma)$ and some $\nu_0 \in A$. Then, for any $\nu \in A$,*

$$H(\nu|\mu) \geq H(\nu|\nu_0) + H(\nu_0|\mu) . \qquad (7.3.28)$$

Proof: Fix $\nu \triangleq \nu_1 \in A$ and let $\nu_\alpha = \alpha\nu + (1 - \alpha)\nu_0 \in A$ for $\alpha \in [0, 1)$. There is nothing to prove unless $H(\nu_0|\mu) \leq H(\nu|\mu) < \infty$, in which case $f_\alpha = d\nu_\alpha/d\mu$ exists for every $\alpha \in [0, 1]$ and $f_\alpha = \alpha f_1 + (1 - \alpha)f_0$. Since $\phi(\alpha) \triangleq f_\alpha \log f_\alpha$ is convex, it follows that $h(\alpha) \triangleq \phi(1) - \phi(0) - \alpha^{-1}(\phi(\alpha) - \phi(0))$ is non-negative and monotone non-decreasing on $(0, 1]$ with

$$\lim_{\alpha \searrow 0} h(\alpha) = \phi(1) - \phi(0) - \phi'(0^+) = f_1 \log(f_1/f_0) + f_0 - f_1 . \qquad (7.3.29)$$

Since $H(\nu_\alpha|\mu) \geq H(\nu_0|\mu)$, it follows that

$$H(\nu|\mu) - H(\nu_0|\mu) \geq H(\nu|\mu) - H(\nu_0|\mu) - \frac{1}{\alpha}(H(\nu_\alpha|\mu) - H(\nu_0|\mu)) = \int_{\Sigma} h(\alpha) d\mu .$$

Considering $\alpha \to 0$, by the monotone convergence theorem (Theorem C.11) and (7.3.29), it follows that

$$H(\nu|\mu) - H(\nu_0|\mu) \geq \int_\Sigma (\lim_{\alpha \searrow 0} h(\alpha))d\mu = \int_\Sigma f_1 \log(f_1/f_0)d\mu. \qquad (7.3.30)$$

In particular, $\mu(\{y : f_1(y) > 0, f_0(y) = 0\}) = 0$, implying that $d\nu/d\nu_0 = f_1/f_0$ exists ν_0 almost surely. Hence, $H(\nu|\nu_0) = \int_\Sigma f_1 \log(f_1/f_0)d\mu$, with (7.3.30) implying (7.3.28). □

Proof of Theorem 7.3.21: First note that,

$$H\left(\mu_{\mathbf{Y}^n|A}^n \middle| \mu^n\right) = -\log \mu^n(L_n^{\mathbf{Y}} \in A) < \infty.$$

Since all marginals of $P = \mu_{\mathbf{Y}^n|A}^n \in M_1(\mathcal{X}^n)$ on $\mathcal{X} = \Sigma$ are identical, applying Lemma 7.3.25 for P, once with $Q = \mu \in M_1(\Sigma)$ and once with $Q = \nu_* \in M_1(\Sigma)$, it follows that

$$-\log \mu^n(L_n^{\mathbf{Y}} \in A) = H\left(\mu_{\mathbf{Y}^n|A}^n \middle| \mu^n\right) \qquad (7.3.31)$$

$$= H\left(\mu_{\mathbf{Y}^n|A}^n \middle| (\nu_*)^n\right) + n\left(H(\mu_{\mathbf{Y}^1|A}^n|\mu) - H(\mu_{\mathbf{Y}^1|A}^n|\nu_*)\right).$$

Recall that $M(\Sigma)$, equipped with the $B(\Sigma)$-topology, is a locally convex, Hausdorff topological vector space, whose topological dual is $B(\Sigma)$ (see Theorem B.8). Therefore, with A a closed, convex subset of this space, if $\mu_{\mathbf{Y}^1|A}^n \notin A$, by the Hahn-Banach theorem (Theorem B.6), there exist $f \in B(\Sigma)$ and $\gamma \in \mathbb{R}$, such that

$$E(f(Y_1)|L_n^{\mathbf{Y}} \in A) = \langle f, \mu_{\mathbf{Y}^1|A}^n \rangle < \gamma \leq \inf_{\nu \in A} \langle f, \nu \rangle \leq E(\langle f, L_n^{\mathbf{Y}} \rangle | L_n^{\mathbf{Y}} \in A),$$

in contradiction with (7.3.1). Hence, $\mu_{\mathbf{Y}^1|A}^n \in A$, and by Lemma 7.3.27,

$$H\left(\mu_{\mathbf{Y}^1|A}^n \middle| \mu\right) - H\left(\mu_{\mathbf{Y}^1|A}^n \middle| \nu_*\right) \geq H(\nu_*|\mu). \qquad (7.3.32)$$

Combining (7.3.31) and (7.3.32) leads to the bound

$$-\log\left(\mu^n(L_n^{\mathbf{Y}} \in A)e^{nI_F}\right) \geq H\left(\mu_{\mathbf{Y}^n|A}^n \middle| (\nu_*)^n\right). \qquad (7.3.33)$$

Apply Lemma 7.3.25 for $\mathcal{X} = \Sigma^k$, $Q = (\nu_*)^k$ and $P = \mu_{\mathbf{Y}^n|A}^n \in M_1(\mathcal{X}^m)$, where $n = km$ for some $m \in \mathbb{Z}_+$. Since $P_i = \mu_{\mathbf{Y}^k|A}^n$ for $i = 1, \ldots, m$, it follows that

$$H\left(\mu_{\mathbf{Y}^n|A}^n \middle| (\nu_*)^n\right) = H\left(\mu_{\mathbf{Y}^n|A}^n \middle| (\mu_{\mathbf{Y}^k|A}^n)^m\right) + mH\left(\mu_{\mathbf{Y}^k|A}^n \middle| (\nu_*)^k\right).$$

Consequently, for any $1 \leq k(n) \leq n$

$$H\left(\mu_{\mathbf{Y}^n|A}^n \middle| (\nu_*)^n\right) \geq \left\lfloor \frac{n}{k(n)} \right\rfloor H\left(\mu_{\mathbf{Y}^k|A}^n \middle| (\nu_*)^k\right),$$

which by (7.3.22) and (7.3.33) completes the proof of Theorem 7.3.21. $\square$

The following is a concrete application of Theorem 7.3.21. See the historical notes for other applications and extensions.

Corollary 7.3.34 Let $A = \{\nu \in M_1[0,1] : \langle U, \nu \rangle \leq 1\}$ for a bounded non-negative Borel function $U(\cdot)$, such that $\mu \circ U^{-1}$ is a non-lattice law, $\int_0^1 U(x) d\mu(x) > 1$ and $\mu(\{x : U(x) < 1\}) > 0$. Then (A-5) holds with $\nu_* = \gamma_{\beta^*}$ of Theorem 7.3.8 and for $n^{-1}k(n) \log n \to_{n \to \infty} 0$,

$$H\left(\mu_{\mathbf{Y}^{k(n)}|A}^n \middle| (\nu_*)^{k(n)}\right) \xrightarrow[n \to \infty]{} 0. \tag{7.3.35}$$

Proof: It is shown in the course of proving Theorem 7.3.8 that $\nu_* = \gamma_{\beta^*}$ is such that $\langle U, \nu_* \rangle = 1$ and

$$H(\nu_*|\mu) = \inf_{\nu \in A} H(\nu|\mu) < \infty.$$

Note that $\langle U, \gamma_{\beta^*} \rangle = \Lambda'(-\beta^*) = 1$ for $\Lambda(\lambda) = \log \int_0^1 e^{\lambda U(x)} \mu(dx)$. Moreover, $\Lambda(\cdot)$ is differentiable on $\mathbb{R}$, and with β^* finite, it follows that $\Lambda'(-\beta^*) = 1$ is in the interior of $\{\Lambda'(\lambda) : \lambda \in \mathbb{R}\}$, so that $\Lambda^*(\cdot)$, the Fenchel–Legendre transform of $\Lambda(\cdot)$, is continuous at 1 (see Exercise 2.2.24). Therefore,

$$\inf_{x \in [0,1]} \Lambda^*(x) = \inf_{x \in [0,1)} \Lambda^*(x),$$

which by comparing Cramér's theorem (Theorem 2.2.3) for $\hat{S}_n = \langle U, L_n^{\mathbf{Y}} \rangle$, with Sanov's theorem (Theorem 6.2.10), and using the the contraction principle (Theorem 4.2.1) for $\nu \mapsto \langle U, \nu \rangle$ implies that (A-5) holds. Moreover, by Theorem 3.7.4, for some constant $C > 0$,

$$\lim_{n \to \infty} \mu^n(L_n^{\mathbf{Y}} \in A)\sqrt{n}\, e^{nI_F} = C.$$

Hence, Theorem 7.3.21 applies with $g_n = \frac{C}{\sqrt{n}}$, implying (7.3.35). $\square$

7.4 Historical Notes and References

Applications of the LDP to statistics and statistical mechanics abound, and it is impossible to provide an extensive bibliography of such applications here. We have chosen to provide in this chapter some applications that seem both representative and interesting. We limit ourselves to bringing some of the related references, running the risk of being inaccurate and unjust towards authors whose work we do not cite.

The results on universal hypothesis testing are based on the results in [ZG91]. For references to the finite dimensional case, see the historical notes of Chapter 3. Some recent extensions and an application to Gaussian processes, based on the results of [DV85] and [DeuSZ91], may be found in [SZ92].

Section 7.2 follows [DeZ96a], which also contains the corresponding moderate deviations results. The approach taken here leads to stronger results than those in the first edition of this book, where projective limits were used. For Theorem 7.2.3 and some of its applications in statistics and in statistical physics, see also [GG97] and [ElGP93], respectively.

The Gibbs conditioning question is one of the motivations of Ruelle's and Lanford's studies, which as seen earlier were influential in the development of the large deviations "theory." Most of the analysis here is taken from [StZ91], where more general functions U are considered. Note however that in [StZ91], Lemma 7.3.12 is incorrectly used for measurable functions, and that this difficulty may be circumvented by considering the LDP for the product empirical measure $L_n^Y \times L_n^Y$, as done for example in [EicS97]. Corollary 7.3.5 is adapted from Proposition 2.2 of [Szn91]. The analysis of Section 7.3.3 is a simplified version of [DeZ96b], which in turn is based on [Cs84], with Lemma 7.3.27 taken from [Cs75]. For related work, see [Bol90]. See also [Scr93] for the discrete parameter Markov chain case. For a discussion of the multidimensional (field) situation, see [DeuSZ91], and for similar results in the context of mean field models, see [BeZ98]. It should be mentioned that one may treat the same question from a CLT and not LDP point of view, as is done in [DT77].

Large deviations techniques have been used extensively in recent years in connection with statistical mechanics and interacting particles systems. For an introduction to the LDP for classical statistical mechanics and spin models, see [Ell85], whereas for a sample of more recent publications, see [KuT84, Dur85, CS87, DV87, Leo87, BCG88, DuS88, LS88, Ore88, Deu89, DV89, BeB90, CD90, KO90, Pap90, StZg92, BeG95, SchS95] and references therein. A particularly interesting application of large deviations techniques is in the construction of refined large deviations at the surface level; see, for a sample of results, [Sch87, Pfi91, DKS92, Io95, Pis96].

Appendix

Good references for the material in the appendices are [Roc70] for Appendix A, [DunS58] for Appendices B and C, [Par67] for Appendix D, and [KS88] for Appendix E.

A Convex Analysis Considerations in $\mathbb{R}^d$

This appendix completes the convexity analysis preliminaries needed in the proof of the Gärtner–Ellis theorem in Section 2.3, culminating with the proof of the two lemmas on which the proof of Lemma 2.3.12 is based.

First, we recall properties of the relative interior of a set and some continuity results concerning convex functions. Let $C \subseteq \mathbb{R}^d$ be a non-empty, convex set. Then $\operatorname{ri} C$ is non-empty, and

$$x \in C, y \in \operatorname{ri} C \Rightarrow (1 - \alpha)x + \alpha y \in \operatorname{ri} C, \ \forall \alpha \in (0,1]. \qquad \text{(A.1)}$$

Let $f : \mathbb{R}^d \to (-\infty, \infty]$ be a convex function and denote its domain by $\mathcal{D}_f$. Then f is continuous in $\operatorname{ri} \mathcal{D}_f$, i.e., for every sequence $x_n \to x$ with $x_n, x \in \operatorname{ri} \mathcal{D}_f$, $f(x_n) \to f(x)$. Moreover, f is Lipschitz continuous on compact subsets of $\operatorname{ri} \mathcal{D}_f$. Finally, let $x, y \in \mathcal{D}_f$; then

$$\lim_{\alpha \searrow 0} f((1 - \alpha)x + \alpha y) \le f(x).$$

Lemma A.2 If $f : \mathbb{R}^d \to [0, \infty]$ is a convex, lower semicontinuous function, with $\inf_{\lambda \in \mathbb{R}^d} f(\lambda) = 0$ and $0 \in \operatorname{ri} \mathcal{D}_{f^*}$, then $f(\eta) = 0$ for some $\eta \in \mathbb{R}^d$.

Proof: Note that $f^*(0) = -\inf_{\lambda \in \mathbb{R}^d} f(\lambda) = 0$. Define the function

$$g(y) \stackrel{\triangle}{=} \inf_{\delta > 0} \frac{f^*(\delta y)}{\delta} = \lim_{\delta \searrow 0} \frac{f^*(\delta y)}{\delta}, \qquad \text{(A.3)}$$

where the convexity of f^* results with a monotonicity in δ, which in turn implies the preceding equality (and that the preceding limit exists). The

A. Dembo, O. Zeitouni, *Large Deviations Techniques and Applications*, 341
Stochastic Modelling and Applied Probability 38,
DOI 10.1007/978-3-642-03311-7,
© Springer-Verlag Berlin Heidelberg 1998, corrected printing 2010

function $g(\cdot)$ is the pointwise limit of convex functions and hence is convex. Further, $g(\alpha y) = \alpha g(y)$ for all $\alpha \geq 0$ and, in particular, $g(0) = 0$. As $0 \in \mathrm{ri}\, \mathcal{D}_{f^*}$, either $f^*(\delta y) = \infty$ for all $\delta > 0$, in which case $g(y) = \infty$, or $f^*(-\epsilon y) < \infty$ for some $\epsilon > 0$. In the latter case, by the convexity of f^*,

$$\frac{f^*(\delta y)}{\delta} + \frac{f^*(-\epsilon y)}{\epsilon} \geq 0, \qquad \forall \delta > 0.$$

Thus, considering the limit $\delta \searrow 0$, it follows that $g(y) > -\infty$ for every $y \in \mathbb{R}^d$. Moreover, since $0 \in \mathrm{ri}\, \mathcal{D}_{f^*}$, by (A.3), also $0 \in \mathrm{ri}\, \mathcal{D}_g$. Hence, since $g(y) = |y|g(y/|y|)$, it follows from (A.1) that $\mathrm{ri}\, \mathcal{D}_g = \mathcal{D}_g$. Consequently, by the continuity of g on $\mathrm{ri}\, \mathcal{D}_g$, $\liminf_{y \to 0} g(y) \geq 0$. In particular, the convex set $\mathcal{E} \triangleq \{(y, \xi) : \xi \geq g(y)\} \subseteq \mathbb{R}^d \times \mathbb{R}$ is non-empty (for example, $(0,0) \in \mathcal{E}$), and $(0, -1) \notin \overline{\mathcal{E}}$. Therefore, there exists a hyperplane in $\mathbb{R}^d \times \mathbb{R}$ that strictly separates the point $(0, -1)$ and the set $\mathcal{E}$. (This is a particular instance of the Hahn–Banach theorem (Theorem B.6).) Specifically, there exist a $\lambda \in \mathbb{R}^d$ and a $\rho \in \mathbb{R}$ such that, for all $(y, \xi) \in \mathcal{E}$,

$$\langle \lambda, 0 \rangle + \rho = \rho > \langle \lambda, y \rangle - \xi \rho. \qquad (A.4)$$

Considering $y = 0$, it is clear that $\rho > 0$. With $\eta = \lambda/\rho$ and choosing $\xi = g(y)$, the inequality (A.4) implies that $1 > [\langle \eta, y \rangle - g(y)]$ for all $y \in \mathbb{R}^d$. Since $g(\alpha y) = \alpha g(y)$ for all $\alpha > 0$, $y \in \mathbb{R}^d$, it follows by considering $\alpha \to \infty$, while y is fixed, that $g(y) \geq \langle \eta, y \rangle$ for all $y \in \mathbb{R}^d$. Consequently, by (A.3), also $f^*(y) \geq \langle \eta, y \rangle$ for all $y \in \mathbb{R}^d$. Since f is a convex, lower semicontinuous function such that $f(\cdot) > -\infty$ everywhere,

$$f(\eta) = \sup_{y \in \mathbb{R}^d} \{\langle \eta, y \rangle - f^*(y)\}.$$

(See the duality lemma (Lemma 4.5.8) or [Roc70], Theorem 12.2, for a proof.) Thus, necessarily $f(\eta) = 0$. $\qquad \square$

Lemma A.5 [Suggested by A. Ioffe] *Let f be an essentially smooth, convex function. If $f(0) = 0$ and $f^*(x) = 0$ for some $x \in \mathbb{R}^d$, then $0 \in \mathcal{D}_f^o$.*

Proof: Since $f(0) = 0$, it follows by convexity of f that

$$f(t\lambda) \leq t f(\lambda), \qquad \forall t \in [0, 1], \quad \forall \lambda \in \mathbb{R}^d.$$

Moreover, since $f^*(x) = 0$,

$$f(t\lambda) \geq \langle t\lambda, x \rangle \geq -t|\lambda||x|.$$

Because f is essentially smooth, $\mathcal{D}_f^o$ contains a closed ball, e.g., $\overline{B}_{z,r}$, $r > 0$, in which f is differentiable. Hence,

$$M = \sup_{\lambda \in \overline{B}_{z,r}} \{f(\lambda) \vee |\lambda||x|\} < \infty.$$

Note that $t\lambda \in \overline{B}_{tz,tr}$ iff $\lambda \in \overline{B}_{z,r}$, and so by the preceding inequalities, for any $t \in (0,1]$,

$$\sup_{\theta \in \overline{B}_{tz,tr}} |f(\theta)| \leq tM .$$

For any $\theta \in \overline{B}_{tz,tr}$, $\theta \neq tz$, by the convexity of f,

$$f(\theta) - f(tz) \leq \frac{|\theta - tz|}{tr}(f(y) - f(tz)) \leq \frac{2tM}{tr}|\theta - tz| ,$$

where $y = tz + \frac{tr}{|\theta - tz|}(\theta - tz) \in \overline{B}_{tz,tr}$.

By a similar convexity argument, also $f(tz) - f(\theta) \leq \frac{2tM}{tr}|\theta - tz|$. Thus,

$$|f(\theta) - f(tz)| \leq \frac{2M}{r}|\theta - tz| \quad \forall \theta \in \overline{B}_{tz,tr} .$$

Observe that $tz \in \mathcal{D}_f^o$ because of the convexity of $\mathcal{D}_f$. Hence, by assumption, $\nabla f(tz)$ exists, and by the preceding inequality, $|\nabla f(tz)| \leq 2M/r$. Since f is steep, it follows by considering $t \to 0$, in which case $tz \to 0$, that $0 \in \mathcal{D}_f^o$. $\qquad\qquad\square$

B   Topological Preliminaries

B.1   Generalities

A family τ of subsets of a set $\mathcal{X}$ is a *topology* if $\emptyset \in \tau$, if $\mathcal{X} \in \tau$, if any union of sets of τ belongs to τ, and if any finite intersection of elements of τ belongs to τ. A topological space is denoted $(\mathcal{X}, \tau)$, and this notation is abbreviated to $\mathcal{X}$ if the topology is obvious from the context. Sets that belong to τ are called *open sets*. Complements of open sets are *closed sets*. An open set containing a point $x \in \mathcal{X}$ is a *neighborhood* of x. Likewise, an open set containing a subset $A \subset \mathcal{X}$ is a neighborhood of A. The *interior* of a subset $A \subset \mathcal{X}$, denoted A^o, is the union of the open subsets of A. The *closure* of A, denoted $\bar{A}$, is the intersection of all closed sets containing A. A point p is called an *accumulation point* of a set $A \subset \mathcal{X}$ if every neighborhood of p contains at least one point in A. The closure of A is the union of its accumulation points.

A *base* for the topology τ is a collection of sets $\mathcal{A} \subset \tau$ such that any set from τ is the union of sets in $\mathcal{A}$. If τ_1 and τ_2 are two topologies on $\mathcal{X}$, τ_1 is called stronger (or finer) than τ_2, and τ_2 is called weaker (or coarser) than τ_1 if $\tau_2 \subset \tau_1$.

A topological space is *Hausdorff* if single points are closed and every two distinct points $x, y \in \mathcal{X}$ have disjoint neighborhoods. It is *regular* if, in addition, any closed set $F \subset \mathcal{X}$ and any point $x \notin F$ possess disjoint neighborhoods. It is *normal* if, in addition, any two disjoint closed sets F_1, F_2 possess disjoint neighborhoods.

If $(\mathcal{X}, \tau_1)$ and $(\mathcal{Y}, \tau_2)$ are topological spaces, a function $f : \mathcal{X} \to \mathcal{Y}$ is a *bijection* if it is one-to-one and onto. It is *continuous* if $f^{-1}(A) \in \tau_1$ for any $A \in \tau_2$. This implies also that the inverse image of a closed set is closed. Continuity is preserved under compositions, i.e., if $f : \mathcal{X} \to \mathcal{Y}$ and $g : \mathcal{Y} \to \mathcal{Z}$ are continuous, then $g \circ f : \mathcal{X} \to \mathcal{Z}$ is continuous. If both f and f^{-1} are continuous, then f is a *homeomorphism*, and spaces $\mathcal{X}, \mathcal{Y}$ are called homeomorphic if there exists a homeomorphism $f : \mathcal{X} \to \mathcal{Y}$.

A function $f : \mathcal{X} \to \mathbb{R}$ is *lower semicontinuous* (*upper semicontinuous*) if its level sets $\{x \in \mathcal{X} : f(x) \leq \alpha\}$ (respectively, $\{x \in \mathcal{X} : f(x) \geq \alpha\}$) are closed sets. Clearly, every continuous function is lower (upper) semicontinuous and the pointwise supremum of a family of lower semicontinuous functions is lower semicontinuous.

Theorem B.1 *A lower (upper) semicontinuous function f achieves its minimum (respectively, maximum) over any compact set K.*

A Hausdorff topological space is *completely regular* if for any closed set $F \subset \mathcal{X}$ and any point $x \notin F$, there exists a continuous function $f : \mathcal{X} \to [0, 1]$ such that $f(x) = 1$ and $f(y) = 0$ for all $y \in F$.

A *cover* of a set $A \subset \mathcal{X}$ is a collection of open sets whose union contains A. A set is *compact* if every cover of it has a finite subset that is also a cover. A continuous image of a compact set is compact. A continuous bijection between compact spaces is a homeomorphism. Every compact subset of a Hausdorff topological space is closed. A set is *pre-compact* if its closure is compact.

Let $(\mathcal{X}, \tau)$ be a topological space, and let $A \subset \mathcal{X}$. The *relative* (or induced) topology on A is the collection of sets $A \bigcap \tau$. The Hausdorff, normality, and regularity properties are preserved under the relative topology. Furthermore, the compactness is preserved, i.e., $B \subset A$ is compact in the relative topology iff it is compact in the original topology τ. Note, however, that the "closedness" property is *not* preserved. ·

A nonnegative real function $d : \mathcal{X} \times \mathcal{X} \to \mathbb{R}$ is called a *metric* if $d(x, y) = 0 \Leftrightarrow x = y$, $d(x, y) = d(y, x)$, and $d(x, y) \leq d(x, z) + d(z, y)$. The last property is referred to as the *triangle* inequality. The set $B_{x, \delta} = \{y : d(x, y) < \delta\}$ is called the *ball* of center x and radius δ. The metric topology of $\mathcal{X}$ is the weakest topology which contains all balls. The set $\mathcal{X}$ equipped with the metric topology is a *metric* space $(\mathcal{X}, d)$. A topological space whose topology is the same as some metric topology is called *metrizable*. Every metrizable space is normal. Every regular space that possesses a countable base is metrizable.

A sequence $x_n \in \mathcal{X}$ *converges* to $x \in \mathcal{X}$ (denoted $x_n \to x$) if every neighborhood of x contains all but a finite number of elements of the sequence $\{x_n\}$. If $\mathcal{X}, \mathcal{Y}$ are metric spaces, then $f : \mathcal{X} \to \mathcal{Y}$ is continuous iff $f(x_n) \to f(x)$ for any convergent sequence $x_n \to x$. A subset $A \subset \mathcal{X}$ of a topological space is *sequentially compact* if every sequence of points in A has a subsequence converging to a point in $\mathcal{X}$.

Theorem B.2 *A subset of a metric space is compact iff it is closed and sequentially compact.*

A set $A \subset \mathcal{X}$ is *dense* if its closure is $\mathcal{X}$. A topological space is *separable* if it contains a countable dense set. Any topological space that possesses a countable base is separable, whereas any separable metric space possesses a countable base.

Even if a space is not metric, the notion of convergence on a sequence may be extended to convergence on *filters* such that compactness, "closedness," etc. may be checked by convergence. Filters are not used in this book. The interested reader is referred to [DunS58] or [Bou87] for details.

Let J be an arbitrary set. Let $\mathcal{X}$ be the Cartesian product of topological spaces $\mathcal{X}_j$, i.e., $\mathcal{X} = \prod_j \mathcal{X}_j$. The *product topology* on $\mathcal{X}$ is the topology generated by the base $\prod_j U_j$, where U_j are open and equal to $\mathcal{X}_j$ except for a finite number of values of j. This topology is the weakest one which makes all projections $p_j : \mathcal{X} \to \mathcal{X}_j$ continuous. The Hausdorff property is preserved under products, and any countable product of metric spaces (with metric $d_n(\cdot, \cdot)$) is metrizable, with the metric on $\mathcal{X}$ given by

$$d(x,y) = \sum_{n=1}^{\infty} \frac{1}{2^n} \frac{d_n(p_n x, p_n y)}{1 + d_n(p_n x, p_n y)}.$$

Theorem B.3 (Tychonoff) *A product of compact spaces is compact.*

Let $(J, \leq)$ be a partially ordered right-filtering set, i.e., for every $i, j \in J$, there exists a $k \in J$ with $i \leq k$ and $j \leq k$. The *projective system* of Hausdorff topological spaces $\mathcal{Y}_j$ consists of these spaces and for each $i \leq j$, a continuous map $p_{ij} : \mathcal{Y}_j \to \mathcal{Y}_i$, satisfying the consistency conditions $p_{ik} = p_{ij} \circ p_{jk}$ if $i \leq j \leq k$, where p_{ii} is the identity map on $\mathcal{Y}_i$. The *projective limit* $\mathcal{X}$ of the system $(\mathcal{Y}_j, p_{ij})$, denoted $\mathcal{X} = \varprojlim \mathcal{Y}_j$, is the subset of the topological product space $\mathcal{Y} = \Pi_{j \in J} \mathcal{Y}_j$, consisting of the elements $\mathbf{x} = (y_j)_{j \in J}$, which, for all $i \leq j$, satisfy the consistency conditions $y_i = p_{ij}(y_j)$. The topology on $\mathcal{X}$ is the topology induced by the product topology on $\mathcal{Y}$. The canonical projections $p_j : \mathcal{X} \to \mathcal{Y}_j$ are the restrictions of the coordinate projections from $\mathcal{Y}$ to $\mathcal{Y}_j$, and are continuous. $\mathcal{X}$ is a closed subset of $\mathcal{Y}$ and is Hausdorff. The collection $\{p_j^{-1}(U_j) : U_j \subset \mathcal{Y}_j \text{ is open}\}$ is a base for the topology on $\mathcal{X}$.

The notion of projective limits is inherited by closed sets. Thus, a closed set $F \subseteq \mathcal{X}$ is the *projective limit* of $F_j \subseteq \mathcal{Y}_j$ (denoted: $F = \varprojlim F_j$), if $p_{ij}(F_j) \subseteq F_i$ for all $i \leq j$ and $F = \bigcap_{j \in J} p_j^{-1}(F_j)$.

Theorem B.4 *A projective limit of non-empty compact sets is non-empty.*

B.2   Topological Vector Spaces and Weak Topologies

A *vector space* over the reals is a set $\mathcal{X}$ that is closed under the operations of addition and multiplication by scalars, i.e., if $x, y \in \mathcal{X}$, then $x + y \in \mathcal{X}$ and $\alpha x \in \mathcal{X}$ for all $\alpha \in \mathbb{R}$. All vector spaces in this book are over the reals. A *topological vector space* is a vector space equipped with a Hausdorff topology that makes the vector space operations continuous. The *convex hull* of a set A, denoted $\mathrm{co}(A)$, is the intersection of all convex sets containing A. The closure of $\mathrm{co}(A)$ is denoted $\overline{\mathrm{co}}(A)$. $\mathrm{co}(\{x_1, \ldots, x_N\})$ is compact, and, if K_i are compact, convex sets, then the set $\mathrm{co}(\cup_{i=1}^N K_i)$ is closed. A *locally convex* topological vector space is a vector space that possesses a convex base for its topology.

Theorem B.5 *Every (Hausdorff) topological vector space is regular.*

A *linear functional* on the vector space $\mathcal{X}$ is a function $f : \mathcal{X} \to \mathbb{R}$ that satisfies $f(\alpha x + \beta y) = \alpha f(x) + \beta f(y)$ for any 'scalars $\alpha, \beta \in \mathbb{R}$ and any $x, y \in \mathcal{X}$. The *algebraic dual* of $\mathcal{X}$, denoted $\mathcal{X}'$, is the collection of all linear functionals on $\mathcal{X}$. The *topological dual* of $\mathcal{X}$, denoted $\mathcal{X}^*$, is the collection of all continuous linear functionals on the *topological* vector space $\mathcal{X}$. Both the algebraic dual and the topological dual are vector spaces. Note that whereas the algebraic dual may be defined for any vector space, the topological dual may be defined only for a topological vector space. The product of two topological vector spaces is a topological vector space, and is locally convex if each of the coordinate spaces is locally convex. The topological dual of the product space is the product of the topological duals of the coordinate spaces. A set $\mathcal{H} \subset \mathcal{X}'$ is called *separating* if for any point $x \in \mathcal{X}$, $x \neq 0$, one may find an $h \in \mathcal{H}$ such that $h(x) \neq 0$. It follows from its definition that $\mathcal{X}'$ is separating.

Theorem B.6 (Hahn–Banach) *Suppose A and B are two disjoint, non-empty, closed, convex sets in the locally convex topological vector space $\mathcal{X}$.*

If A is compact, then there exists an $f \in \mathcal{X}^*$ and scalars $\alpha, \beta \in \mathbb{R}$ such that, for all $x \in A$, $y \in B$,

$$f(x) < \alpha < \beta < f(y). \qquad (B.7)$$

It follows in particular that if $\mathcal{X}$ is locally convex, then $\mathcal{X}^*$ is separating. Now let $\mathcal{H}$ be a separating family of linear functionals on $\mathcal{X}$. The $\mathcal{H}$-*topology* of $\mathcal{X}$ is the weakest (coarsest) one that makes all elements of $\mathcal{H}$ continuous. Two particular cases are of interest:

(a) If $\mathcal{H} = \mathcal{X}^*$, then the $\mathcal{X}^*$-topology on $\mathcal{X}$ obtained in this way is called the *weak topology* of $\mathcal{X}$. It is always weaker (coarser) than the original topology on $\mathcal{X}$.

(b) Let $\mathcal{X}$ be a topological vector space (not necessarily locally convex). Every $x \in \mathcal{X}$ defines a linear functionals f_x on $\mathcal{X}^*$ by the formula $f_x(x^*) = x^*(x)$. The set of all such functionals is separating in $\mathcal{X}^*$. The $\mathcal{X}$-topology of $\mathcal{X}^*$ obtained in this way is referred to as the *weak* topology* of $\mathcal{X}^*$.

Theorem B.8 *Suppose $\mathcal{X}$ is a vector space and $\mathcal{Y} \subset \mathcal{X}'$ is a separating vector space. Then the $\mathcal{Y}$-topology makes $\mathcal{X}$ into a locally convex topological vector space with $\mathcal{X}^* = \mathcal{Y}$.*

It follows in particular that there may be different topological vector spaces with the same topological dual. Such examples arise when the original topology on $\mathcal{X}$ is strictly finer than the weak topology.

Theorem B.9 *Let $\mathcal{X}$ be a locally convex topological vector space. A convex subset of $\mathcal{X}$ is weakly closed iff it is originally closed.*

Theorem B.10 (Banach–Alaoglu) *Let V be a neighborhood of 0 in the topological vector space $\mathcal{X}$. Let $K = \{x^* \in \mathcal{X}^* : |x^*(x)| \leq 1 , \forall x \in V\}$. Then K is weak* compact.*

B.3   Banach and Polish Spaces

A *norm* $||\cdot||$ on a vector space $\mathcal{X}$ is a metric $d(x,y) = ||x - y||$ that satisfies the scaling property $||\alpha(x - y)|| = \alpha||x - y||$ for all $\alpha > 0$. The metric topology then yields a topological vector space structure on $\mathcal{X}$, which is referred to as a *normed* space. The standard norm on the topological dual of a normed space $\mathcal{X}$ is $||x^*||_{\mathcal{X}^*} = \sup_{||x|| \leq 1} |x^*(x)|$, and then $||x|| = \sup_{||x^*||_{\mathcal{X}^*} \leq 1} x^*(x)$, for all $x \in \mathcal{X}$.

A *Cauchy sequence* in a metric space $\mathcal{X}$ is a sequence $x_n \in \mathcal{X}$ such that for every $\epsilon > 0$, there exists an $N(\epsilon)$ such that $d(x_n, x_m) < \epsilon$ for any

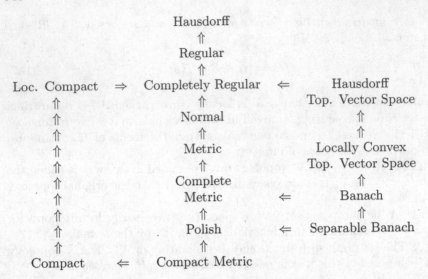

Figure B.1: Dependencies between topological spaces.

$n > N(\epsilon)$ and $m > N(\epsilon)$. If every Cauchy sequence in $\mathcal{X}$ converges to a point in $\mathcal{X}$, the metric in $\mathcal{X}$ is called *complete*. Note that completeness is not preserved under homeomorphism. A complete separable metric space is called a *Polish* space. In particular, a compact metric space is Polish, and an open subset of a Polish space (equipped with the induced topology) is homeomorphic to a Polish space.

A complete normed space is called a *Banach* space. The natural topology on a Banach space is the topology defined by its norm.

A set B in a topological vector space $\mathcal{X}$ is *bounded* if, given any neighborhood V of the origin in $\mathcal{X}$, there exists an $\epsilon > 0$ such that $\{\alpha x : x \in B, |\alpha| \leq \epsilon\} \subset V$. In particular, a set B in a normed space is bounded iff $\sup_{x \in B} \|x\| < \infty$. A set B in a metric space $\mathcal{X}$ is *totally bounded* if, for every $\delta > 0$, it is possible to cover B by a finite number of balls of radius δ centered in B. A totally bounded subset of a complete metric space is pre-compact.

Unlike in the Euclidean setup, balls need not be convex in a metric space. However, in normed spaces, all balls are convex. Actually, the following partial converse holds.

Theorem B.11 *A topological vector space is normable, i.e., a norm may be defined on it that is compatible with its topology, iff its origin has a convex bounded neighborhood.*

Weak topologies may be defined on Banach spaces and their topological duals. A striking property of the weak topology of Banach spaces is the fact that compactness, apart from closure, may be checked using sequences.

Theorem B.12 (Eberlein–Šmulian) *Let $\mathcal{X}$ be a Banach space. In the weak topology of $\mathcal{X}$, a set is sequentially compact iff it is pre-compact.*

B.4 Mazur's Theorem

In this section, the statement and proof of Mazur's theorem as used in Section 6.1 are provided. Since this particular variant of Mazur's theorem is not readily available in standard textbooks, a complete proof is presented.

Theorem B.13 (Mazur) *Let $\mathcal{X}$ be a Hausdorff topological vector space, and let $\mathcal{E}$, be a closed, convex subset of $\mathcal{X}$ such that $(\mathcal{E}, d)$ is a complete metric space, whose metric topology is compatible with the topology induced by $\mathcal{X}$. Further assume that for all $\alpha \in [0,1]$, $x_1, x_2, y_1, y_2 \in \mathcal{E}$,*

$$d(\alpha x_1 + (1-\alpha)x_2, \alpha y_1 + (1-\alpha)y_2) \leq \max\{d(x_1,y_1), d(x_2,y_2)\}. \quad (\text{B.14})$$

Then the closed convex hull of every compact subset of $\mathcal{E}$ is compact.

Remark: The proof is adapted from Theorem V.2.6 in [DunS58], where $\mathcal{X} = \mathcal{E}$ is a Banach space. A related proof may be found in [DeuS89b], Lemma 3.1.1.

Proof: Fix a compact $K \subset \mathcal{E}$ and $\delta > 0$. By compactness, $K \subset \cup_{i=1}^{N} B_{x_i,\delta}$ for some $x_i \in K$ and finite N. By (B.14), balls in $\mathcal{E}$ are convex. Hence, if $y \in \text{co}(\cup_{i=1}^{N} B_{x_i,\delta})$, then $y = \sum_{i=1}^{N} a_i y_i$, for some $a_i \geq 0$ with $\sum_{i=1}^{N} a_i = 1$ and $y_i \in B_{x_i,\delta}$. Thus, again by (B.14),

$$d(\sum_{i=1}^{N} a_i y_i, \sum_{i=1}^{N} a_i x_i) \leq \max_{i=1}^{N} d(y_i, x_i) < \delta.$$

Consequently,

$$\text{co}(K) \subset \text{co}(\bigcup_{i=1}^{N} B_{x_i,\delta}) \subset (\text{co}(\{x_1, \ldots, x_N\}))^{\delta}.$$

By the compactness of $\text{co}(\{x_1, \ldots, x_N\})$ in $\mathcal{E}$, the set $(\text{co}(\{x_1, \ldots, x_N\}))^{\delta}$ can be covered by a finite number of balls of radius 2δ. Thus, by the arbitrariness of $\delta > 0$, it follows that $\text{co}(K)$ is a totally bounded subset of $\mathcal{E}$. By the completeness of $\mathcal{E}$, $\text{co}(K)$ is pre-compact, and $\overline{\text{co}}(K) = \overline{\text{co}(K)}$ is compact, since $\mathcal{E}$ is a closed subset of $\mathcal{X}$. □

C   Integration and Function Spaces

C.1   Additive Set Functions

Let $\mathcal{B}$ be a family of sets. A *set function* μ on $\mathcal{B}$ is a function that assigns an extended real value (possibly ∞ or $-\infty$, but not both) to each set $b \in \mathcal{B}$. A *positive* set function is a set function that assigns only nonnegative values to sets $b \in \mathcal{B}$. Similarly, a *finite* set function assigns only finite values to sets, and a *bounded* set function has a bounded range. Hereafter, unless specifically mentioned otherwise, all set functions are either positive or bounded. A set function is *additive* if $\mathcal{B}$ contains $\emptyset$, if $\mu(\emptyset) = 0$, and if for any finite family of disjoint sets $b_1, \ldots, b_n \in \mathcal{B}$, $\mu(\bigcup_{i=1}^n b_i) = \sum_{i=1}^n \mu(b_i)$ whenever $\bigcup_{i=1}^n b_i \in \mathcal{B}$. It is *countably additive* if the union in the preceding definition may be taken over a countable number of disjoint sets b_i.

Associated with the notion of an additive set function is the notion of field; namely, let $\mathcal{X}$ be some space, then a family of subsets $\mathcal{B}$ of $\mathcal{X}$ is a *field* if $\emptyset \in \mathcal{B}$, if the complement of any element of $\mathcal{B}$ (with respect to $\mathcal{X}$) belongs to $\mathcal{B}$ and if the union of two elements of $\mathcal{B}$ belongs to $\mathcal{B}$. The *total variation* of a set function μ with respect to the field $\mathcal{B}$ is defined as $v(\mu) = \sup \sum_{i=1}^n |\mu(b_i)|$, where the supremum is taken over all finite collections of disjoint elements of $\mathcal{B}$. For any *simple* function of the form $f(x) = \sum_{i=1}^n a_i 1_{b_i}(x)$, where b_i are disjoint elements of $\mathcal{B}$ and a_i are arbitrary constants, one may define the *integral* of f with respect to the set function μ as $\int f d\mu = \sum_{i=1}^n a_i \mu(b_i)$. Every additive set function defines an integral over the class of real valued functions $\mathcal{H}$ that are obtained by monotone limits of simple functions. This integral is a linear functional on $\mathcal{H}$.

A *σ-field* is a field that is closed under countable unions. A *σ-additive* set function μ is a countably additive set function on a σ-field $\mathcal{B}$. σ-additive set functions are also called *measures*, and the triplet $(\mathcal{X}, \mathcal{B}, \mu)$ is called a *measure space*, with the sets in $\mathcal{B}$ called *measurable sets*. As before, integrals with respect to μ are defined for monotone limits of simple functions, i.e., for *measurable functions*. A measure is bounded iff it is finite. For any measure space $(\mathcal{X}, \mathcal{B}, \mu)$ there exists $B \in \mathcal{B}$ such that both $\mu(\cdot \cap B)$ and $-\mu(\cdot \cap B^c)$ are positive measures. The measure $\mu_+(\cdot) = \mu(\cdot \cap B)$ is then called the positive part of $\mu(\cdot)$. A measure ν is *absolutely continuous* with respect to a positive measure μ if both are defined on the same σ-field $\mathcal{B}$ and $\nu(A) = 0$ for every $A \in \mathcal{B}$ such that $\mu(A) = 0$. Two measures that are mutually absolutely continuous are called *equivalent*.

For every given space $\mathcal{X}$ and family of subsets $\mathcal{B} \subset \mathcal{X}$, there exist a smallest field (and a smallest σ-field) that contain $\mathcal{B}$, also called the field (respectively, σ-field) generated by $\mathcal{B}$. If $\mathcal{B}$ is taken as the set of all closed subsets of the topological Hausdorff space $\mathcal{X}$, the resulting σ-field is the *Borel* σ-field of $\mathcal{X}$, denoted $\mathcal{B}_\mathcal{X}$. Functions that are measurable with respect to the Borel σ-field are called *Borel functions*, and form a convenient family of functions suitable for integration. An alternative characterization of Borel functions is as functions f such that $f^{-1}(A)$ belongs to the Borel σ-field for any Borel measurable subset A of $\mathbb{R}$. Similarly, a map $f : \mathcal{X} \to \mathcal{Y}$ between the measure spaces $(\mathcal{X}, \mathcal{B}, \mu)$ and $(\mathcal{Y}, \mathcal{B}', \nu)$ is a *measurable map* if $f^{-1}(A) \in \mathcal{B}$ for every $A \in \mathcal{B}'$. All lower (upper) semicontinuous functions are Borel measurable. The space of all bounded, real valued, Borel functions on a topological space $\mathcal{X}$ is denoted $B(\mathcal{X})$. This space, when equipped with the supremum norm $\|f\| = \sup_{x \in \mathcal{X}} |f(x)|$, is a Banach space.

Theorem C.1 (Hahn) *Every countably additive, positive (bounded) set function on a field $\mathcal{B}$ possesses a unique extension to a σ-additive, positive (respectively, bounded), set function on the smallest σ-field containing $\mathcal{B}$.*

Another useful characterization of countably additive set functions is the following.

Theorem C.2 *Let μ be an additive set function that is defined on the Borel σ-field of the Hausdorff topological space $\mathcal{X}$. If, for any sequence of sets E_n that decrease to the empty set, $\mu(E_n) \to 0$, then μ is countably additive.*

Let μ be an additive set function defined on some field $\mathcal{B}$ of subsets of a Hausdorff topological space $\mathcal{X}$. μ is *regular* if for each $b \in \mathcal{B}$ and each $\epsilon > 0$, there is a set $A \in \mathcal{B}$ with $\bar{A} \subset b$ and a set $C \in \mathcal{B}$ with $b \subset C^o$ such that $|\mu(D)| < \epsilon$ for each $D \subset C \backslash A$, $D \in \mathcal{B}$.

Theorem C.3 *Let $\mathcal{X}$ be a Hausdorff topological space. Then $B(\mathcal{X})^*$ may be represented by the space of all bounded additive set functions on the Borel σ-field of $\mathcal{X}$.*

The space of bounded continuous functions on a Hausdorff topological space is denoted $C_b(\mathcal{X})$.

Theorem C.4 *Let $\mathcal{X}$ be a normal topological space. Then $C_b(\mathcal{X})^*$ may be represented by the space of all regular, bounded additive set functions on the field generated by the closed subsets of $\mathcal{X}$.*

Theorem C.5 *Every bounded measure on the Borel σ-field of a metric space is regular.*

A bounded regular measure μ on the Borel σ-field of $\mathcal{X}$ is uniquely determined by the values it assigns to the closed subsets of $\mathcal{X}$, and when $\mathcal{X}$ is metrizable, it is also uniquely determined by the integrals $\{\int f d\mu\}_{f \in C_b(\mathcal{X})}$.

C.2   Integration and Spaces of Functions

Let μ be a positive measure on the Borel σ-field of $\mathcal{X}$. A Borel function is μ *integrable* if $\int_{\mathcal{X}} |f| d\mu < \infty$. The space of functions whose pth power is integrable is denoted $L_p(\mathcal{X}, \mu)$, or simply $L_p(\mathcal{X})$, $L_p(\mu)$, or even L_p if no confusion as to the space $\mathcal{X}$ or the measure μ arises. To be precise, L_p is the space of *equivalence classes* of functions whose pth power is μ integrable, where the equivalence is with respect to the relation $\int_{\mathcal{X}} |f - g|^p d\mu = 0$. It is a fact that two functions f, g are in the same equivalence class, regardless of p, if they differ on a set of zero μ measure. We denote this fact by $f = g$ (a.e.) (almost everywhere) or, when μ is a probability measure, by $f = g$ (a.s.) (almost surely).

Theorem C.6 L_p, $1 \le p < \infty$ *are Banach spaces when equipped with the norm* $||f||_p = (\int_{\mathcal{X}} |f^p| d\mu)^{1/p}$. *Moreover*, $(L_p)^* = L_q$, *where* $1 < p < \infty$ *and q is the conjugate of p, i.e.*, $1/p + 1/q = 1$.

If $\mathcal{X}$ is metrizable, then $C_b(\mathcal{X})$ is a dense subset of $L_p(\mathcal{X}, \mu)$ for every finite measure μ, and for all $1 \le p < \infty$. Similarly, if $\mathcal{X}$ is Polish and μ is finite, then $C_u(\mathcal{X})$, the space of uniformly continuous bounded functions on $\mathcal{X}$, is dense in $L_p(\mathcal{X}, \mu)$ for $1 \le p < \infty$.

The space $L_\infty(\mathcal{X}, \mu)$ denotes the space of equivalence classes in $B(\mathcal{X})$, where two functions are equivalent if they are equal a.e (μ). L_∞, when equipped with the essential supremum norm, is a Banach space.

A set $K \subset L_1(\mu)$ is *uniformly integrable* if, for any $\epsilon > 0$, one may find a constant c such that $\int_{\mathcal{X}} 1_{\{|f| > c\}} |f| d\mu < \epsilon$ for all $f \in K$.

Theorem C.7 *Let μ be a finite measure. Then* $L_1^*(\mu) = L_\infty(\mu)$. *Moreover, $L_1(\mu)$ is weakly complete, and a subset K of $L_1(\mu)$ is weakly sequentially compact if it is bounded and uniformly integrable.*

A compactness criterion in $C_b(\mathcal{X})$ can be established if $\mathcal{X}$ is compact. A set $K \subset C_b(\mathcal{X})$ is *equicontinuous* if, for all $\epsilon > 0$ and all $x \in \mathcal{X}$, there exists a neighborhood $N_\epsilon(x)$ of x such that

$$\sup_{f \in K} \sup_{y \in N_\epsilon(x)} |f(x) - f(y)| < \epsilon.$$

Theorem C.8 (Arzelà–Ascoli) *Assume $\mathcal{X}$ is compact. Then $K \subset C_b(\mathcal{X})$ is pre-compact iff it is bounded and equicontinuous.*

Finally, the following theorems are elementary.

Theorem C.9 (Radon–Nikodym) *Let μ be a finite measure. The following statements are equivalent:*
(1) ν is absolutely continuous with respect to μ.
(2) There is a unique $h \in L_1(\mu)$ such that $\nu(E) = \int_E h \, d\mu$ for every measurable set E.
(3) To every $\epsilon > 0$, there corresponds a $\delta > 0$ such that $|\nu(E)| < \epsilon$ as soon as $\mu(E) < \delta$.
The function h that occurs in (2) is called the Radon–Nikodym derivative of ν with respect to μ, and is denoted $d\nu/d\mu$.

Theorem C.10 (Lebesgue's dominated convergence) *Let $\{f_n\}$ be a sequence in $L_p(\mu)$, $1 \le p < \infty$. Assume that f_n converges a.e. (μ) to a function f. Suppose that for some $g \in L_p(\mu)$, $|f_n| \le |g|$ a.e.. Then $f \in L_p(\mu)$ and $\{f_n\}$ converges to f in L_p.*

Theorem C.11 (Monotone convergence theorem) *Let $\{f_n\}$ be a sequence of nonnegative, monotonically increasing (in n) Borel functions, which converges a.e. to some function f. Then, regardless of the finiteness of both sides of the equality,*

$$\lim_{n \to \infty} \int_{\mathcal{X}} f_n \, d\mu = \int_{\mathcal{X}} f \, d\mu.$$

Theorem C.12 (Fatou) *Let $\{f_n\}$ be a sequence of nonnegative Borel functions. Then*

$$\int_{\mathcal{X}} (\liminf_{n \to \infty} f_n) \, d\mu \le \liminf_{n \to \infty} \int_{\mathcal{X}} f_n \, d\mu.$$

A sequence of sets $E_i \subset \mathbb{R}^d$ shrinks nicely to x if there exists $\alpha > 0$ such that each E_i lies in a ball B_{x,r_i}, $r_i \to 0$ and $m(E_i) \ge \alpha m(B_{x,r_i})$, where m denotes Lebesgue's measure on $\mathbb{R}^d$, i.e., the countably additive extension of volume to the Borel σ-field.

Theorem C.13 (Lebesgue) *Let $f \in L_1(\mathbb{R}^d, m)$. Then for almost all x,*

$$\lim_{i \to \infty} \frac{1}{m(E_i)} \int_{E_i} |f(y) - f(x)| m(dy) = 0$$

for every sequence $\{E_i\}$ that shrinks nicely to x.

D   Probability Measures on Polish Spaces

D.1   Generalities

The following indicates why Polish spaces are convenient when handling measurability issues. Throughout, unless explicitly stated otherwise, Polish spaces are equipped with their Borel σ-fields.

Theorem D.1 (Kuratowski) *Let Σ_1, Σ_2 be Polish spaces, and let $f : \Sigma_1 \to \Sigma_2$ be a measurable, one-to-one map. Let $E_1 \subset \Sigma_1$ be a Borel set. Then $f(E_1)$ is a Borel set in Σ_2.*

A *probability measure* on the Borel σ-field $\mathcal{B}_\Sigma$ of a Hausdorff topological space Σ is a countably additive, positive set function μ with $\mu(\Sigma) = 1$. The space of (Borel) probability measures on Σ is denoted $M_1(\Sigma)$. A probability measure is regular if it is regular as an additive set function. When Σ is separable, the structure of $M_1(\Sigma)$ becomes simpler, and conditioning becomes easier to handle; namely, let Σ, Σ_1 be two separable Hausdorff spaces, and let μ be a probability measure on $(\Sigma, \mathcal{B}_\Sigma)$. Let $\pi : \Sigma \to \Sigma_1$ be measurable, and let $\nu = \mu \circ \pi^{-1}$ be the measure on $\mathcal{B}_{\Sigma_1}$ defined by $\nu(E_1) = \mu(\pi^{-1}(E_1))$.

Definition D.2 *A regular conditional probability distribution given π (referred to as r.c.p.d.) is a mapping $\sigma_1 \in \Sigma_1 \mapsto \mu^{\sigma_1} \in M_1(\Sigma)$ such that:*
(1) There exists a set $N \in \mathcal{B}_{\Sigma_1}$ with $\nu(N) = 0$, and for each $\sigma_1 \in \Sigma_1 \backslash N$,

$$\mu^{\sigma_1}(\{\sigma : \pi(\sigma) \neq \sigma_1\}) = 0.$$

(2) For any set $E \in \mathcal{B}_\Sigma$, the map $\sigma_1 \mapsto \mu^{\sigma_1}(E)$ is $\mathcal{B}_{\Sigma_1}$ measurable and

$$\mu(E) = \int_{\Sigma_1} \mu^{\sigma_1}(E) \nu(d\sigma_1).$$

It is property (2) that allows for the decomposition of measures. In Polish spaces, the existence of an r.c.p.d. follows from:

Theorem D.3 *Let Σ, Σ_1 be Polish spaces, $\mu \in M_1(\Sigma)$, and $\pi : \Sigma \to \Sigma_1$ a measurable map. Then there exists an r.c.p.d. μ^{σ_1}. Moreover, it is unique in the sense that any other r.c.p.d. $\overline{\mu}^{\sigma_1}$ satisfies*

$$\nu(\{\sigma_1 : \overline{\mu}^{\sigma_1} \neq \mu^{\sigma_1}\}) = 0.$$

Another useful property of separable spaces is their behavior under products.

Theorem D.4 *Let N be either finite or $N = \infty$.*
(a) $\prod_{i=1}^{N} \mathcal{B}_{\Sigma_i} \subset \mathcal{B}_{\prod_{i=1}^{N} \Sigma_i}$.
(b) If Σ_i are separable, then $\prod_{i=1}^{N} \mathcal{B}_{\Sigma_i} = \mathcal{B}_{\prod_{i=1}^{N} \Sigma_i}$.

Additional properties of r.c.p.d. when products of Polish spaces are involved are collected in Appendix D.3 below.

We now turn our attention to the particular case where Σ is metric (and, whenever needed, Polish).

Theorem D.5 *Let Σ be a metric space. Then any $\mu \in M_1(\Sigma)$ is regular.*

Theorem D.6 *Let Σ be Polish, and let $\mu \in M_1(\Sigma)$. Then there exists a unique closed set C_μ such that $\mu(C_\mu) = 1$ and, if D is any other closed set with $\mu(D) = 1$, then $C_\mu \subseteq D$. Finally,*

$$ C_\mu = \{ \sigma \in \Sigma : \sigma \in U^o \ \Rightarrow \ \mu(U^o) > 0 \} . $$

The set C_μ of Theorem D.6 is called the *support* of μ.

A probability measure μ on the metric space Σ is *tight* if for each $\eta > 0$, there exists a compact set $K_\eta \subset \Sigma$ such that $\mu(K_\eta^c) < \eta$. A family of probability measures $\{\mu_\alpha\}$ on the metric space Σ is called a *tight family* if the set K_η may be chosen independently of α.

Theorem D.7 *Each probability measure on a Polish space Σ is tight.*

D.2 Weak Topology

Whenever Σ is Polish, a topology may be defined on $M_1(\Sigma)$ that possesses nice properties; namely, define the *weak topology* on $M_1(\Sigma)$ as the topology generated by the sets

$$ U_{\phi,x,\delta} = \{ \nu \in M_1(\Sigma) : | \int_\Sigma \phi d\nu - x | < \delta \}, $$

where $\phi \in C_b(\Sigma)$, $\delta > 0$ and $x \in \mathbb{R}$.

Hereafter, $M_1(\Sigma)$ always denotes $M_1(\Sigma)$ equipped with the weak topology. The following are some basic properties of this topological space.

Theorem D.8 *Let Σ be Polish.*
(1) $M_1(\Sigma)$ is Polish.
(2) A metric compatible with the weak topology is the Lévy metric:

$$ d(\mu, \nu) = \inf \{ \delta : \ \mu(F) \leq \nu(F^\delta) + \delta \ \ \forall F \subset \Sigma \ closed \} . $$

(3) $M_1(\Sigma)$ is compact iff Σ is compact.

(4) Let $E \subset \Sigma$ be a dense countable subset of Σ. The set of all probability measures whose supports are finite subsets of E is dense in $M_1(\Sigma)$.

(5) Another metric compatible with the weak topology is the Lipschitz bounded metric:

$$d_{LU}(\mu, \nu) = \sup_{f \in \mathcal{F}_{LU}} \left| \int_\Sigma f d\nu - \int_\Sigma f d\mu \right|,$$

where $\mathcal{F}_{LU}$ is the class of Lipschitz continuous functions $f : \Sigma \to \mathbb{R}$, with Lipschitz constant at most 1 and uniform bound 1.

$M_1(\Sigma)$ possesses a useful criterion for compactness.

Theorem D.9 (Prohorov) Let Σ be Polish, and let $\Gamma \subset M_1(\Sigma)$. Then $\overline{\Gamma}$ is compact iff Γ is tight.

Since $M_1(\Sigma)$ is Polish, convergence may be decided by sequences. The following lists some useful properties of converging sequences in $M_1(\Sigma)$.

Theorem D.10 (Portmanteau theorem) Let Σ be Polish. The following statements are equivalent:

(1) $\mu_n \to \mu$ as $n \to \infty$.

(2) $\forall g$ bounded and uniformly continuous, $\lim_{n\to\infty} \int_\Sigma g \, d\mu_n = \int_\Sigma g \, d\mu$.

(3) $\forall F \subset \Sigma$ closed, $\limsup_{n\to\infty} \mu_n(F) \le \mu(F)$.

(4) $\forall G \subset \Sigma$ open, $\liminf_{n\to\infty} \mu_n(G) \ge \mu(G)$.

(5) $\forall A \in \mathcal{B}_\Sigma$, which is a continuity set, i.e., such that $\mu(\overline{A} \backslash A^\circ) = 0$, $\lim_{n\to\infty} \mu_n(A) = \mu(A)$.

A collection of functions $\mathcal{G} \subset B(\Sigma)$ is called *convergence determining* for $M_1(\Sigma)$ if

$$\lim_{n\to\infty} \int_\Sigma g d\mu_n = \int_\Sigma g d\mu, \quad \forall g \in \mathcal{G} \Rightarrow \mu_n \xrightarrow{n\to\infty} \mu.$$

For Σ Polish, there exists a countable convergence determining collection of functions for $M_1(\Sigma)$ and the collection $\{f(x)g(y)\}_{f,g \in C_b(\Sigma)}$ is convergence determining for $M_1(\Sigma^2)$.

Theorem D.11 Let Σ be Polish. If K is a set of continuous, uniformly bounded functions on Σ that are equicontinuous on compact subsets of Σ, then $\mu_n \to \mu$ implies that

$$\limsup_{n\to\infty} \sup_{\phi \in K} \left\{ \left| \int_\Sigma \phi d\mu_n - \int_\Sigma \phi d\mu \right| \right\} = 0.$$

The following theorem is the analog of Fatou's lemma for measures. It is proved from Fatou's lemma either directly or by using the Skorohod representation theorem.

Theorem D.12 *Let Σ be Polish. Let $f : \Sigma \to [0, \infty]$ be a lower semicontinuous function, and assume $\mu_n \to \mu$. Then*

$$\liminf_{n \to \infty} \int_\Sigma f d\mu_n \geq \int_\Sigma f d\mu.$$

D.3 Product Space and Relative Entropy Decompositions

A particularly important situation requiring measure decompositions is as follows. Let $\Sigma = \Sigma_1 \times \Sigma_2$ where Σ_i are Polish spaces equipped with their Borel σ-field. Throughout, we let $\sigma = (\sigma_1, \sigma_2)$ denote a generic point in Σ, and use $\pi : \Sigma \mapsto \Sigma_1$ with $\pi(\sigma) = \sigma_1$. Then, π is measurable by Theorem D.4. With $\mu \in M_1(\Sigma)$, by Theorem D.3, the r.c.p.d. of μ given π denoted $\mu^{\sigma_1}(\cdot)$ exists, with $\mu_1 = \mu \circ \pi^{-1} \in M_1(\Sigma_1)$. Because $\mu^{\sigma_1}(\{x \in \Sigma : \pi(x) \neq \sigma_1\}) = 0$, one has for any $A \in \mathcal{B}_\Sigma$,

$$\mu^{\sigma_1}(A) = \mu^{\sigma_1}(\{\sigma_2 : (\sigma_1, \sigma_2) \in A\}),$$

and hence $\mu^{\sigma_1}(\cdot)$ can be considered also as a measure on Σ_2. Further, for any $g : \Sigma \to \mathbb{R}$,

$$\int_\Sigma g(\sigma)\mu(d\sigma) = \int_{\Sigma_1} \int_\Sigma g(\sigma)\mu^{\sigma_1}(d\sigma)\mu_1(d\sigma_1).$$

The *relative entropy* of $\nu \in M_1(\Sigma)$ with respect to $\mu \in M_1(\Sigma)$ is

$$H(\nu|\mu) \triangleq \begin{cases} \int_\Sigma f \log f d\mu & \text{if } f \triangleq \frac{d\nu}{d\mu} \text{ exists} \\ \infty & \text{otherwise}. \end{cases}$$

The following decomposition of the relative entropy functional in the particular case $\Sigma = \Sigma_1 \times \Sigma_2$ discussed above is used extensively in Sections 2.4.2 and 6.6. Its proof is reproduced here for completeness.

Theorem D.13 *With $\Sigma = \Sigma_1 \times \Sigma_2$ and Σ_i Polish, let μ_1, ν_1 denote the restrictions of $\mu, \nu \in M_1(\Sigma)$ to Σ_1, with $\mu^{\sigma_1}(\cdot)$ and $\nu^{\sigma_1}(\cdot)$ the r.c.p.d. corresponding to the projection map $\pi : \Sigma \to \Sigma_1$. Then, the map*

$$\sigma_1 \mapsto H\left(\nu^{\sigma_1}(\cdot)|\mu^{\sigma_1}(\cdot)\right) : \Sigma_1 \to [0, \infty] \tag{D.14}$$

is measurable, and

$$H(\nu|\mu) = H(\nu_1|\mu_1) + \int_\Sigma H\Big(\nu^{\sigma_1}(\cdot)|\mu^{\sigma_1}(\cdot)\Big)\nu_1(d\sigma_1). \tag{D.15}$$

Proof: The function $H(\cdot|\cdot) : M_1(\Sigma) \times M_1(\Sigma) \to [0,\infty]$ is lower semicontinuous since

$$H(\nu|\mu) = \sup_{\phi \in C_b(\Sigma)} \{\int_\Sigma \phi d\nu - \log \int_\Sigma e^\phi d\mu\}$$

(see for example, Lemma 6.2.13). With $M_1(\Sigma)$ Polish, Definition D.2 implies that $\sigma_1 \mapsto \nu^{\sigma_1}(\cdot)$ and $\sigma_1 \mapsto \mu^{\sigma_1}(\cdot)$ are both measurable maps from Σ_1 to $M_1(\Sigma)$. The measurability of the map in (D.14) follows.

Suppose that the right side of (D.15) is finite. Then $g \triangleq d\nu_1/d\mu_1$ exists and $f(\sigma) \triangleq d\nu^{\sigma_1}/d\mu^{\sigma_1}(\sigma)$ exists for any $\sigma_1 \in \tilde{\Sigma}_1$ and some Borel set $\tilde{\Sigma}_1$ such that $\nu_1(\tilde{\Sigma}_1) = 1$. If $\Gamma \subset \Sigma$ is such that $\mu(\Gamma) = 0$, then by Definition D.2, $\mu^{\sigma_1}(\Gamma) = 0$ for all $\sigma_1 \notin B$, some $B \subset \Sigma_1$ with $\mu_1(B) = 0$. Since $d\nu_1/d\mu_1$ exists, also $\nu_1(B) = 0$, and for any $\sigma_1 \in \tilde{\Sigma}_1 \cap B^c$, $\nu^{\sigma_1}(\Gamma) = 0$, leading to

$$\nu(\Gamma) = \int_\Sigma \nu^{\sigma_1}(\Gamma)\nu_1(d\sigma_1) = \int_{\tilde{\Sigma}_1 \cap B^c} \nu^{\sigma_1}(\Gamma)\nu_1(d\sigma_1) = 0,$$

that is, $\rho \triangleq d\nu/d\mu$ exists. In particular, if $d\nu/d\mu$ does not exist, then both sides of (D.15) are infinite. Hence, we may and will assume that $\rho \triangleq d\nu/d\mu$ exists, implying that $g \triangleq d\nu_1/d\mu_1$ exists too.

Fix $A_i \in \mathcal{B}_{\Sigma_i}$, $i = 1,2$ and $A \triangleq A_1 \times A_2$. Then,

$$\nu(A) = \nu(A_1 \times A_2) = \int_{\Sigma_1} \nu^{\sigma_1}(A_1 \times A_2)\nu_1(d\sigma_1)$$

$$= \int_{\Sigma_1} 1_{\{\sigma_1 \in A_1\}}\nu^{\sigma_1}(\Sigma_1 \times A_2)\nu_1(d\sigma_1) = \int_{A_1} \nu^{\sigma_1}(\Sigma_1 \times A_2)g(\sigma_1)\mu_1(d\sigma_1).$$

On the other hand,

$$\nu(A) = \int_A \rho(\sigma)\mu(d\sigma) = \int_{A_1} \left(\int_{\Sigma_1 \times A_2} \rho(\sigma)\mu^{\sigma_1}(d\sigma)\right)\mu_1(d\sigma_1).$$

Thus, for each such A_2, there exists a μ_1-null set Γ_{A_2} such that, for all $\sigma_1 \notin \Gamma_{A_2}$,

$$\nu^{\sigma_1}(\Sigma_1 \times A_2)g(\sigma_1) = \int_{\Sigma_1 \times A_2} \rho(\sigma)\mu^{\sigma_1}(d\sigma). \tag{D.16}$$

Let $\Gamma \subset \Sigma_1$ be such that $\mu_1(\Gamma) = 0$ and, for any A_2 in the countable base of Σ_2, and all $\sigma_1 \notin \Gamma$, (D.16) holds. Since (D.16) is preserved under both countable unions and monotone limits of A_2 it thus holds for any Borel set A_2 and all $\sigma_1 \notin \Gamma$. Enlarging Γ if necessary, still with $\mu_1(\Gamma) = 0$, one concludes that, for all $\sigma_1 \notin \Gamma$, for any $A_1 \in \mathcal{B}_{\Sigma_1}$, and all $A_2 \in \mathcal{B}_{\Sigma_2}$,

$$\nu^{\sigma_1}(A_1 \times A_2)g(\sigma_1) = 1_{\sigma_1 \in A_1} \nu^{\sigma_1}(\Sigma_1 \times A_2)g(\sigma_1)$$

$$= 1_{\sigma_1 \in A_1} \int_{\Sigma_1 \times A_2} \rho(\sigma)\mu^{\sigma_1}(d\sigma) = \int_{A_1 \times A_2} \rho(\sigma)\mu^{\sigma_1}(d\sigma).$$

Letting $\tilde{\Gamma} = \Gamma \cup \{g = 0\}$, by Theorem D.4 suffices to consider such product sets in order to conclude that for $\sigma_1 \notin \tilde{\Gamma}$, $f(\sigma) \triangleq d\nu^{\sigma_1}/d\mu^{\sigma_1}(\sigma)$ exists and satisfies $f(\sigma) = \rho(\sigma)/g(\sigma_1)$. Hence, using the fact that $\nu_1(\tilde{\Gamma}) = 0$,

$$
\begin{aligned}
H(\nu|\mu) &= \int_{\Sigma} \log\big(\rho(\sigma)\big)\nu(d\sigma) = \int_{\Sigma \cap \{\sigma:\sigma_1 \notin \tilde{\Gamma}\}} \log \rho(\sigma)\nu(d\sigma) \\
&= \int_{\Sigma \cap \{\sigma:\sigma_1 \notin \tilde{\Gamma}\}} \big(\log g(\sigma_1) + \log f(\sigma)\big)\nu(d\sigma) \\
&= \int_{\Sigma \cap \{\sigma:\sigma_1 \notin \tilde{\Gamma}\}} \log g(\sigma_1)\nu(d\sigma) + \int_{\Sigma} \log f(\sigma)\nu(d\sigma) \\
&= \int_{\Sigma_1} \log g(\sigma_1)\nu_1(d\sigma_1) + \int_{\Sigma_1} \nu_1(d\sigma_1) \int_{\Sigma} \log f(\sigma)\nu^{\sigma_1}(d\sigma) \\
&= H(\nu_1|\mu_1) + \int_{\Sigma_1} H\big(\nu^{\sigma_1}(\cdot)|\mu^{\sigma_1}(\cdot)\big)\nu_1(d\sigma_1).
\end{aligned}
$$

$\square$

E   Stochastic Analysis

It is assumed that the reader has a working knowledge of Itô's theory of stochastic integration, and therefore an account of it is not presented here. In particular, we will be using the facts that a stochastic integral of the form $M_t = \int_0^t \sigma_s dw_s$, where $\{w_t, t \geq 0\}$ is a Brownian motion and σ_s is square integrable, is a continuous martingale with increasing process $\langle M \rangle_t = \int_0^t \sigma_s^2 ds$. An excellent modern account of the theory and the main reference for the results quoted here is [KS88].

Let $\{\mathcal{F}_t, 0 \leq t \leq \infty\}$ be an increasing family of σ-fields.

Theorem E.1 (Doob's optional sampling) *Let M_t be an $\mathcal{F}_t$ continuous martingale, and let τ be a bounded stopping time. Then $E(M_\tau) = E(M_0)$.*

Theorem E.2 (Time change for martingales) *Let M_t be an $\mathcal{F}_t$ continuous martingale. Let $\langle M \rangle_t$ denote the increasing process associated with M_t. For each $0 \leq s < \langle M \rangle_\infty$, define*

$$\tau(s) = \inf\{t \geq 0 : \langle M \rangle_t > s\},$$

and let $\tau(s) = \infty$ for $s \geq \langle M \rangle_\infty$. Define $w_s \triangleq M_{\tau(s)}, \mathcal{G}_s = \mathcal{F}_{\tau(s)}$. Then $\{w_s\}$ is a $\mathcal{G}$ adapted Brownian motion (up to $\langle M \rangle_\infty$), and a.s.

$$M_t = w_{\langle M \rangle_t}; \ 0 \leq t < \infty.$$

Theorem E.3 (Burkholder–Davis–Gundy maximal inequality) *Let M_t be an $\mathcal{F}_t$ continuous martingale, with increasing process $\langle M \rangle_t$. Then, for every $m > 0$, there exist universal constants $k_m, K_m > 0$ such that, for every stopping time τ,*

$$k_m \, E\left(\langle M \rangle_\tau^m\right) \leq E\left[\left(\sup_{0 \leq t \leq \tau} |M_t|\right)^{2m}\right] \leq K_m E\left(\langle M \rangle_\tau^m\right).$$

Let w_t be a one-dimensional Brownian motion.

Theorem E.4 (Désiré André) *For any $b \geq 0$,*

$$P(\sup_{0 \leq t \leq 1} w_t \geq b) = 2P(w_1 \geq b).$$

Theorem E.5 (Itô) *Let $x.$ be a semi-martingale admitting the decomposition $x_t = \int_0^t b_s ds + \int_0^t \sigma_s dw_s$, with $b., \sigma.$ adapted square integrable processes. Let $f(\cdot)$ be a C^2 function (i.e., possesses continuous derivatives up to second order). Then*

$$f(x_t) = f(x_0) + \int_0^t f'(x_s) b_s ds + \int_0^t f'(x_s) \sigma_s dw_s + \frac{1}{2} \int_0^t f''(x_s) \sigma_s^2 ds.$$

A fundamental tool in the analysis of deterministic and stochastic differential equations is the following.

Lemma E.6 (Gronwall) *Let $g(t) \geq 0$ be a nonnegative, continuous function satisfying*

$$g(t) \leq \alpha(t) + \beta \int_0^t g(s) \, ds,$$

where $0 \leq t \leq T$, $\alpha \in L_1([0,T])$ and $\beta \geq 0$. Then

$$g(t) \leq \alpha(t) + \beta \int_0^t \alpha(s) e^{\beta(t-s)} \, ds, \qquad 0 \leq t \leq T.$$

The following existence result is the basis for our treatment of diffusion processes in Section 5.6. Let w_t be a d-dimensional Brownian motion of law P, adapted to the filtration $\mathcal{F}_t$, which is augmented with its P null sets as usual and thus satisfies the "standard conditions." Let $f(t, x) : [0, \infty) \times \mathbb{R}^d \to \mathbb{R}^d$ and $\sigma(t, x) : [0, \infty) \times \mathbb{R}^d \to \mathbb{R}^{d \times d}$ be jointly measurable, and satisfy the uniform Lipschitz type growth condition

$$|f(t, x) - f(t, y)| + |\sigma(t, x) - \sigma(t, y)| \leq K|x - y|$$

and

$$|f(t, x)|^2 + |\sigma(t, x)|^2 \leq K(1 + |x|^2)$$

for some $K > 0$. where $|\cdot|$ denotes both the Euclidean norm in $\mathbb{R}^d$ and the matrix norm.

Theorem E.7 (Strong existence) *The stochastic differential equation*

$$dX_t = f(t, X_t)\, dt + \sigma(t, X_t)\, dw_t, \; X_0 = x_0 \qquad (E.8)$$

possesses a unique strong solution, i.e., there exists a unique $\mathcal{F}_t$ adapted stochastic process X_t such that

$$P\left(\sup_{0 \leq t < \infty} \left| X_t - x_0 - \int_0^t f(s, X_s)\, ds - \int_0^t \sigma(s, X_s)\, dw_s \right| > 0 \right) = 0.$$

This solution is a strong Markov process.

Remark: This solution is referred to as the *strong solution* of (E.8). There is another concept of solutions to (E.8), called weak solutions, which involve only the construction of the probability law associated with (E.8) and avoid the path picture of the strong solutions. Weak solutions are not used in this book.

Bibliography

[AD95]   D. Aldous and P. Diaconis. Hammersley's interacting particle process
           and longest increasing subsequences. *Prob. Th. Rel. Fields*, 103:199–
           213, 1995.

[AH98]   M. Alanyali and B. Hajek. On large deviations of Markov processes
           with discontinuous statistics. To appear, *Ann. Appl. Probab.*, 1998.

[AM92]   P. H. Algoet and B. M. Marcus. Large deviation theorems for empirical
           types of Markov chains constrained to thin sets. *IEEE Trans. Inf.
           Theory*, IT-38:1276–1291, 1992.

[Ana88]   V. Anantharam. How large delays build up in a $GI/G/1$ queue. *Queue-
           ing Systems*, 5:345–368, 1988.

[ArGW90]   R. Arratia, L. Gordon and M. S. Waterman. The Erdös-Rényi law
           in distribution for coin tossing and sequence matching. *Ann. Stat.*,
           18:539–570, 1990.

[ArMW88]   R. Arratia, P. Morris and M. S. Waterman. Stochastic scrabble: large
           deviations for sequences with scores. *J. Appl. Prob.*, 25:106–119, 1988.

[Arr89]   S. Arrhenius. Über die reaktionsgeschwindigkeit bei der inversion von
           rohrzucker durch säuren. *Z. Phys. Chem.*, 4:226–242, 1889.

[As79]   Grande deviations et applications statistiques. *Astérisque*, 68, 1979.

[As81]   Geodésique et diffusions en temps petit. *Astérisque*, 84–85, 1981.

[AS91]   N. Alon and J. H. Spencer. *The probabilistic method*. Wiley, New
           York, 1991.

[Asm82]   S. Asmussen. Conditional limit theorems relating the random walk
           to its associate, with applications to risk processes and the $GI/G/1$
           queue. *Advances in Applied Probability*, 14:143–170, 1982.

[Aze80]   R. Azencott. Grandes déviations et applications. In P. L. Hennequin,
           editor, *Ecole d'Été de Probabilités de Saint-Flour VIII–1978*, Lecture
           Notes in Math. 774, pages 1–176. Springer-Verlag, Berlin, 1980.

[AzR77]   R. Azencott and G. Ruget. Mélanges d'equations differentielles et
           grands écarts a la loi des grands nombres. *Z. Wahrsch verw. Geb.*,
           38:1–54, 1977.

A. Dembo, O. Zeitouni, *Large Deviations Techniques and Applications*,   363
Stochastic Modelling and Applied Probability 38,
DOI 10.1007/978-3-642-03311-7,
© Springer-Verlag Berlin Heidelberg 1998, corrected printing 2010

[Azu67]   K. Azuma. Weighted sums of certain dependent random variables. *Tohoku Math. J.*, 3:357–367, 1967.

[Bah67]   R. R. Bahadur. Rates of convergence of estimates and test statistics. *Ann. Math. Statist.*, 38:303–324, 1967.

[Bah71]   R. R. Bahadur. *Some limit theorems in statistics*, volume 4 of *CBMS–NSF regional conference series in applied mathematics*. SIAM, Philadelphia, 1971.

[Bah83]   R. R. Bahadur. Large deviations of the maximum likelihood estimate in the Markov chain case. In J. S. Rostag, M. H. Rizvi and D. Siegmund, editors, *Recent advances in statistics*, pages 273–283. Academic Press, Boston, 1983.

[BaR60]   R. R. Bahadur and R. Ranga Rao. On deviations of the sample mean. *Ann. Math. Statis.*, 31:1015–1027, 1960.

[BaZ79]   R. R. Bahadur and S. L. Zabell. Large deviations of the sample mean in general vector spaces. *Ann. Probab.*, 7:587–621, 1979.

[BaZG80]   R. R. Bahadur, S. L. Zabell and J. C. Gupta. Large deviations, tests and estimates. In I. M. Chakravarti, editor, *Asymptotic theory of statistical tests and estimation*, pages 33–64. Academic Press, Boston, 1980.

[Bal86]   P. Baldi. Large deviations and functional iterated logarithm law for diffusion processes. *Prob. Th. Rel. Fields*, 71:435–453, 1986.

[Bal88]   P. Baldi. Large deviations and stochastic homogenization. *Ann. Mat. Pura Appl.*, 151:161–177, 1988.

[Bal91]   P. Baldi. Large deviations for diffusion processes with homogenization and applications. *Ann. Probab.*, 19:509–524, 1991.

[Barn78]   O. Barndorff-Nielsen. *Information and Exponential Families in Statistical Theory*. Wiley, Chichester, 1978.

[Barr85]   A. R. Barron. The strong ergodic theorem for densities: generalized Shannon–McMillan–Breiman theorem. *Ann. Probab.*, 13:1292–1303, 1985.

[BB92]   B. Bollobàs and G. Brightwell. The height of a random partial order: concentration of measure. *Ann. Appl. Probab.*, 2:1009–1018, 1992.

[BCG88]   M. Bramson, J. T. Cox and D. Griffeath. Occupation time large deviations of the voter model. *Prob. Th. Rel. Fields*, 77:401–413, 1988.

[BD90]   R. M. Burton and H. Dehling. Large deviations for some weakly dependent random processes. *Stat. and Prob. Lett.*, 9:397–401, 1990.

[BeB90]   G. Ben Arous and M. Brunaud. Methode de Laplace: etude variationnelle des fluctuations de diffusions de type "champs moyens". *Stochastics*, 31:79–144, 1990.

[BeC96]   G. Ben Arous and F. Castelle. Flow decomposition and large deviations. *J. Funct. Anal.*, 140:23–67, 1996.

[BeG95]   G. Ben Arous and A. Guionnet. Large deviations for Langevin spin glass dynamics. *Prob. Th. Rel. Fields*, 102:455–509, 1995.

[BeLa91]   G. Ben Arous and R. Léandre. Décroissance exponentielle du noyau de la chaleur sur la diagonale. *Prob. Th. Rel. Fields*, 90:175–202, 1991.

[BeLd93]   G. Ben Arous and M. Ledoux. Schilder's large deviations principle without topology. In *Asymptotic problems in probability theory: Wiener functionals and asymptotics (Sanda/Kyoto, 1990)*, pages 107–121. Pitman Res. Notes Math. Ser., 284, Longman Sci. Tech., Harlowi, 1993.

[BeZ98]   G. Ben Arous and O. Zeitouni. Increasing propagation of chaos for mean field models. To appear, *Ann. Inst. H. Poincaré Probab. Stat.*, 1998.

[Ben88]   G. Ben Arous. Methodes de Laplace et de la phase stationnaire sur l'espace de Wiener. *Stochastics*, 25:125–153, 1988.

[Benn62]   G. Bennett. Probability inequalities for the sum of independent random variables. *J. Am. Stat. Assoc.*, 57:33–45, 1962.

[Ber71]   T. Berger. *Rate Distortion Theory*. Prentice Hall, Englewood Cliffs, NJ, 1971.

[Bez87a]   C. Bezuidenhout. A large deviations principle for small perturbations of random evolution equations. *Ann. Probab.*, 15:646–658, 1987.

[Bez87b]   C. Bezuidenhout. Singular perturbations of degenerate diffusions. *Ann. Probab.*, 15:1014–1043, 1987.

[BGR97]   B. Bercu, F. Gamboa and A. Rouault. Large deviations for quadratic forms of stationary Gaussian processes. *Stoch. Proc. and Appl.*, 71:75–90, 1997.

[BH59]   D. Blackwell and J. L. Hodges. The probability in the extreme tail of a convolution. *Ann. Math. Statist.*, 30:1113–1120, 1959.

[BhR76]   R. N. Bhattacharya and R. Ranga Rao. *Normal Approximations and Asymptotic Expansions*. Wiley, New York, 1976.

[Big77]   J. D. Biggins. Chernoff's theorem in the branching random walk. *J. Appl. Prob.*, 14:630–636, 1977.

[Big79]   J. D. Biggins. Growth rates in the branching random walk. *Z. Wahrsch verw. Geb.*, 48:17–34, 1979.

[Bis84]   J. M. Bismut. *Large deviations and Malliavin calculus*. Birkhäuser, Basel, Switzerland, 1984.

[BlD94]   V. M. Blinovskii and R. L. Dobrushin. Process level large deviations for a class of piecewise homogeneous random walks. In M. I. Freidlin, editor, *The Dynkin Festschrift, Progress in Probability*, volume 34, pages 1–59, Birkhäuser, Basel, Switzerland, 1994.

[BoL97]   S. Bobkov and M. Ledoux. Poincaré's inequalities and Talagrand's concentration phenomenon for the exponential distribution. *Prob. Th. Rel. Fields*, 107:383–400, 1997.

[Bol84]   E. Bolthausen. On the probability of large deviations in Banach spaces. *Ann. Probab.*, 12:427–435, 1984.

[Bol86]   E. Bolthausen. Laplace approximations for sums of independent random vectors. *Prob. Th. Rel. Fields*, 72:305–318, 1986.

[Bol87a]   E. Bolthausen. Laplace approximations for sums of independent random vectors, II. *Prob. Th. Rel. Fields*, 76:167–206, 1987.

[Bol87b]   E. Bolthausen. Markov process large deviations in the τ-topology. *Stoch. Proc. and Appl.*, 25:95–108, 1987.

[Bol90]   E. Bolthausen. Maximum entropy principles for Markov processes. *Stochastic processes and applications in mathematical physics. Proceeding of the Bielefeld Conference, 1985.* Mathematics and its applications, Vol. 61. Kluwer, 1990.

[BolD93]   E. Bolthausen and J. D Deuschel. Critical large deviations for Gaussian fields in the phase transition regime, I. *Ann. Probab.*, 21:1876–1920, 1993.

[BolDT95]   E. Bolthausen, J. D. Deuschel and Y. Tamura. Laplace approximations for large deviations of nonreversible Markov processes. Part I: The nondegenerate case. *Ann. Probab.*, 23:236–267, 1995.

[BolDT96]   E. Bolthausen, J. D. Deuschel and Y. Tamura. Laplace approximations for large deviations of nonreversible Markov processes. Part II: The degenerate case. Preprint, 1996.

[BolS89]   E. Bolthausen and U. Schmock. On the maximum entropy principle for uniformly ergodic Markov chains. *Stoch. Proc. and Appl.*, 33:1–27, 1989.

[Boo75]   S. A. Book. Convergence rates for a class of large deviation probabilities. *Ann. Probab.*, 3:516–525, 1975.

[BoM78]   A. A. Borovkov and A. A. Mogulskii. Probabilities of large deviations in topological spaces, I. *Siberian Math. J.*, 19:697–709, 1978.

[BoM80]   A. A. Borovkov and A. A. Mogulskii. Probabilities of large deviations in topological spaces, II. *Siberian Math. J.*, 21:653–664, 1980.

[Bor65]   A. A. Borovkov. New limit theorems in boundary problems for sums of independent terms. *Sel. Transl. Math. Stat.*, 5:315–372, 1965.

[Bor67]   A. A. Borovkov. Boundary–value problems for random walks and large deviations in function spaces. *Th. Prob. Appl.*, 12:575–595, 1967.

[BoR65]   A. A. Borovkov and B. A. Rogozin. On the multidimensional central limit theorem. *Th. Prob. Appl.*, 10:55–62, 1965.

[Bore75]   C. Borell. The Brunn–Minkowski inequality in Gauss space. *Invent. Math.*, 30:205–216, 1975.

[Bou87]   N. Bourbaki. *Elements of Mathematics – general topology.* Springer, Berlin, 1987.

[Bra86]   R. C. Bradley. Basic properties of strong mixing conditions. In E. Eber-
           lein and M. Taqqu, editors, *Dependence in Probability and Statistics*,
           pages 165–192. Birkhäuser, Basel, Switzerland, 1986.

[BrM91]   D. C. Brydges and I. M. Maya. An application of Berezin integration
           to large deviations. *J. Theoretical Prob.*, 4:371–389, 1991.

[BrR65]   A. Brønsted and R. T. Rockafellar. On the subdifferentiability of con-
           vex functions. *Proc. Amer. Math. Soc.*, 16:605–611, 1965.

[Bry90]   W. Bryc. Large deviations by the asymptotic value method. In M.
           Pinsky, editor, *Diffusion Processes and Related Problems in Analysis*,
           pages 447–472. Birkhäuser, Basel, Switzerland, 1990.

[Bry92]   W. Bryc. On the large deviation principle for stationary weakly de-
           pendent random fields. *Ann. Probab.*, 20:1004–1030, 1992.

[Bry93]   W. Bryc. A remark on the connection between the large deviation
           principle and the central limit theorem. *Stat. and Prob. Lett.*, 18:253–
           256, 1993.

[BryD95]   W. Bryc and A. Dembo. On large deviations of empirical measures
           for stationary Gaussian processes. *Stoch. Proc. and Appl.*, 58:23–34,
           1995.

[BryD96]   W. Bryc and A. Dembo. Large deviations and strong mixing. *Ann.
           Inst. H. Poincaré Probab. Stat.*, 32:549–569, 1996.

[BryD97]   W. Bryc and A. Dembo. Large deviations for quadratic functionals of
           Gaussian processes. *J. Theoretical Prob.*, 10:307–332, 1997.

[BryS93]   W. Bryc and W. Smolenski. On the convergence of averages of mixing
           sequences. *J. Theoretical Prob.*, 6:473–483, 1993.

[BS82]   B. Z. Bobrovsky and Z. Schuss. A singular perturbation method for
           the computation of the mean first passage time in a nonlinear filter.
           SIAM J. Appl. Math., 42:174–187, 1982.

[BS88]   B. Z. Bobrovsky and Z. Schuss. Jamming and maneuvering induced
           loss of lock in a range tracking loop. Unpublished, 1988.

[Buc90]   J. A. Bucklew. *Large Deviations Techniques in Decision, Simulation,
           and Estimation*. Wiley, New York, 1990.

[BxJ88]   J. R. Baxter and N. C. Jain. A comparison principle for large devia-
           tions. *Proc. Amer. Math. Soc.*, 103:925–960, 1988.

[BxJ96]   J. R. Baxter and N. C. Jain. An approximation condition for large
           deviations and some applications. In *Convergence in ergodic theory
           and probability (Columbus, OH 1993)*, pages 63–90. Ohio State Univ.
           Math. Res. Inst. Publ. 5, de Gruyter, Berlin, 1996.

[BxJV91]   J. R. Baxter, N. C. Jain and S. R. S. Varadhan. Some familiar examples
           for which the large deviation principle does not hold. *Comm. Pure
           Appl. Math.*, 34:911–923, 1991.

[CaC81]   J. M. Van Campenhout and T. M. Cover. Maximum entropy and con-
           ditional probability. *IEEE Trans. Inf. Theory*, IT-27:483–489, 1981.

[Car98]   L. Caramellino. Strassen law of the iterated logarithm for diffusion processes for small time. To appear, *Stoch. Proc. and Appl.*, 1998.

[Cas93]   F. Castelle. Asymptotic expansion of stochastic flows. *Prob. Th. Rel. Fields*, 96:225–239, 1993.

[CatC97]   O. Catoni. and R. Cerf. The exit path of a Markov chain with rare transitions. *ESAIM: Probability and Statistics*, 1:95–144, 1997.

[Cat92]   O. Catoni. Rough large deviation estimates for simulated annealing: application to exponential schedules. *Ann. Probab.*, 20:1109–1146, 1992.

[CD90]   J. T. Cox and R. Durrett. Large deviations for independent random walks. *Prob. Th. Rel. Fields*, 84:67–82, 1990.

[Che52]   H. Chernoff. A measure of asymptotic efficiency for tests of a hypothesis based on the sum of observations. *Ann. Math. Statist.*, 23:493–507, 1952.

[Che56]   H. Chernoff. Large sample theory: parametric case. *Ann. Math. Statist.*, 27:1–22, 1956.

[CK88]   T. Chiyonobu and S. Kusuoka. The large deviation principle for hypermixing processes. *Prob. Th. and Rel. Fields*, 78:627–649, 1988.

[CM90]   P. L. Chow and J. L. Menaldi. Exponential estimates in exit probability for some diffusion processes in Hilbert spaces. *Stochastics and Stoch. Rep.*, 29:377–393, 1990.

[Com89]   F. Comets. Large deviations estimates for a conditional probability distribution. Applications to random interacting Gibbs measures. *Prob. Th. Rel. Fields*, 80:407–432, 1989.

[Com94]   F. Comets. Erdös-Rényi laws for Gibbs measures. *Comm. Math. Physics*, 162:353–369, 1994.

[Cra38]   H. Cramér. Sur un nouveau théorème–limite de la théorie des probabilités. In *Actualités Scientifiques et Industrielles*, number 736 in Colloque consacré à la théorie des probabilités, pages 5–23. Hermann, Paris, 1938.

[CS87]   N. R. Chaganty and J. Sethuraman. Limit theorems in the area of large deviations for some dependent random variables. *Ann. Probab.*, 15:628–645, 1987.

[Cs75]   I. Csiszár. *I*–divergence geometry of probability distributions and minimization problems. *Ann. Probab.*, 3:146–158, 1975.

[Cs84]   I. Csiszár. Sanov property, generalized *I*–projection and a conditional limit theorem. *Ann. Probab.*, 12:768–793, 1984.

[CsCC87]   I. Csiszár, T. M. Cover and B. S. Choi. Conditional limit theorems under Markov conditioning. *IEEE Trans. Inf. Theory*, IT-33:788–801, 1987.

[CsK81]   I. Csiszár and J. Körner. *Information Theory: Coding Theorems for Discrete Memoryless Systems*. Academic Press, New York, 1981.

[CT91]   T. M. Cover and J. A. Thomas. *Elements of Information Theory*. Wiley, New York, 1991.

[DaG87]   D. A. Dawson and J. Gärtner. Large deviations from the McKean-Vlasov limit for weakly interacting diffusions. *Stochastics*, 20:247–308, 1987.

[Dan54]   H. E. Daniels. Saddlepoint approximations in statistics. *Ann. Math. Statist.*, 25:631–650, 1954.

[DaPZ92]   G. Da Prato and J. Zabczyk. *Stochastic equations in infinite dimensions*. Cambridge University Press, Cambridge, 1992.

[Day87]   M. V. Day. Recent progress on the small parameter exit problem. *Stochastics*, 20:121–150, 1987.

[Day89]   M. V. Day. Boundary local time and small parameter exit problems with characteristic boundaries. *SIAM J. Math. Anal.*, 20:222–248, 1989.

[Day90a]   M. V. Day. Large deviations results for the exit problem with characteristic boundary. *J. Math Anal. and Appl.*, 147:134–153, 1990.

[Day90b]   M. V. Day. Some phenomena of the characteristic boundary exit problem. In M. Pinsky, editor, *Diffusion processes and related problems in analysis*, pages 55–72. Birkhäuser, Basel, Switzerland, 1990.

[Day92]   M. V. Day. Conditional exits for small noise diffusions with characteristic boundary. *Ann. Probab.*, 20:1385–1419, 1992.

[deA85a]   A. de Acosta. Upper bounds for large deviations of dependent random vectors. *Z. Wahrsch. verw. Geb.*, 69:551–565, 1985.

[deA85b]   A. de Acosta. On large deviations of sums of independent random variables. In *Lecture Notes in Math. 1153*, pages 1–14. Springer-Verlag, New York, 1985.

[deA88]   A. de Acosta. Large deviations for vector-valued functionals of a Markov chain: lower bounds. *Ann. Probab.*, 16:925–960, 1988.

[deA90]   A. de Acosta. Large deviations for empirical measures of Markov chains. *J. Theoretical Prob.*, 3:395–431, 1990.

[deA92]   A. de Acosta. Moderate deviations and associated Laplace approximations for sums of independent random variables. *Trans. Amer. Math. Soc.*, 329:357–375, 1992.

[deA94a]   A. de Acosta. Large deviations for vector-valued Lévy processes. *Stoch. Proc. and Appl.*, 51:75–115, 1994.

[deA94b]   A. de Acosta. On large deviations of empirical measures in the τ-topology. Studies in applied probability. *J. Appl. Prob.*, 31A:41–47, 1994

[deA94c]   A. de Acosta. Projective systems in large deviation theory, II: Some applications. In J. Hoffmann-Jørgensen, J. Kuelbs and M. B. Marcus, editors, *Probability in Banach Spaces, 9*, Progress in Probability, volume 35, pages 241–250. Birkhäuser, Boston, MA, 1994.

[deA97a]   A. de Acosta. Exponential tightness and projective systems in large deviation theory. In D. Pollard and E. Torgersen, editors, *Festschrift for Lucien Le Cam*, pages 143–156. Springer-Verlag, New York, 1997.

[deA97b]   A. de Acosta. Moderate deviations for empirical measures of Markov chains: lower bounds. *Ann. Probab.*, 25:259–284, 1997.

[DeK91a]   A. Dembo and S. Karlin. Strong limit theorems of empirical distributions for large segmental exceedances of partial sums of Markov variables. *Ann. Probab.*, 19:1755–1767, 1991.

[DcK91b]   A. Dembo and S. Karlin. Strong limit theorems of empirical functionals for large exceedances of partial sums of i.i.d. variables. *Ann. Probab.*, 19:1737–1755, 1991.

[Dem96]   A. Dembo. Moderate deviations for martingales with bounded jumps. *Elect. Commun. in Probab.*, 1:11–17, 1996.

[Dem97]   A. Dembo. Information inequalities and concentration of measure. *Ann. Probab.*, 25:927–939, 1997.

[DeS98]   A. Dembo and Q. Shao. Self-normalized large deviations in vector spaces. In M. G. Hahn, M. Talagrand and E. Eberlein, editors, *High dimensional probability*, Birkhäuser, Basel, Switzerland, to appear, 1998.

[DeZa95]   A. Dembo and T. Zajic. Large deviations: from empirical mean and measure to partial sums process. *Stoc. Proc. and Appl.*, 57:191–224, 1995.

[DeZa97]   A. Dembo and T. Zajic. Uniform large and moderate deviations for functional empirical processes. *Stoch. Proc. and Appl.*, 67:195–211, 1997.

[DeZ94]   A. Dembo and O. Zeitouni. A large deviations analysis of range tracking loops. *IEEE Trans. Aut. Control*, AC-39:360–364, 1994.

[DeZ95]   A. Dembo and O. Zeitouni. Large deviations via parameter dependent change of measure and an application to the lower tail of Gaussian processes. In E. Bolthausen, M. Dozzi and F. Russo, editors, *Progress in Probability, volume 36*, pages 111–121. Birkhäuser, Basel, Switzerland, 1995.

[DeZ96a]   A. Dembo and O. Zeitouni. Large deviations of sub-sampling from individual sequences. *Stat. and Prob. Lett.*, 27:201–205, 1996.

[DeZ96b]   A. Dembo and O. Zeitouni. Refinements of the Gibbs conditioning principle. *Prob. Th. Rel. Fields*, 104:1–14, 1996.

[DeZ96c]   A. Dembo and O. Zeitouni. Transportation approach to some concentration inequalities in product spaces. *Elect. Commun. in Probab.*, 1:83–90, 1996.

[DGZ95]   A. Dembo, V. Galperin and O. Zeitouni. Exponential rates for error probabilities in DMPSK systems. *IEEE Trans. Commun.*, 43:915–921, 1995.

[DKZ94a] A. Dembo, S. Karlin and O. Zeitouni. Critical phenomena for sequence matching with scoring. *Ann. Probab.*, 22:1993–2021, 1994.

[DKZ94b] A. Dembo, S. Karlin and O. Zeitouni. Limit distribution of maximal non-aligned two-sequence segmental scoring. *Ann. Probab.*, 22:2022–2039, 1994.

[DKZ94c] A. Dembo, S. Karlin and O. Zeitouni. Large exceedances for multidimensional Lévy processes. *Ann. Appl. Probab.*, 4:432–447, 1994.

[Deu89] J. D. Deuschel. Invariance principle and empirical mean large deviations of the critical Orenstein-Uhlenbeck process. *Ann. Probab.*, 17:74–90, 1989.

[DeuS89a] J. D. Deuschel and D. W. Stroock. A function space large deviation principle for certain stochastic integrals. *Prob. Th. Rel. Fields*, 83:279–307, 1989.

[DeuS89b] J. D. Deuschel and D. W. Stroock. *Large Deviations*. Academic Press, Boston, 1989.

[DeuSZ91] J. D. Deuschel, D. W. Stroock and H. Zessin. Microcanonical distributions for lattice gases. *Comm. Math. Physics*, 139:83–101, 1991.

[DeuZ98] J. D. Deuschel and O. Zeitouni. On increasing subsequences of i.i.d. samples. To appear, *Combinatorics, Probability and Computing*, 1998.

[DiaS91] P. Diaconis and D. W. Stroock. Geometric bounds for eigenvalues of Markov chains. *Ann. Appl. Probab.*, 1:36–61, 1991.

[DiaSa93] P. Diaconis and L. Saloff-Coste. Comparison theorems for reversible Markov chains. *Ann. Appl. Probab.*, 3:696–730, 1993.

[Din91] I. H. Dinwoodie. A note on the upper bound for i.i.d. large deviations. *Ann. Probab.*, 19:1732–1736, 1991.

[Din92] I. H. Dinwoodie. Measures dominantes et théorème de Sanov. *Ann. Inst. H. Poincaré Probab. Stat.*, 28:365–373, 1992.

[Din93] I. H. Dinwoodie. Identifying a large deviation rate function. *Ann. Probab.*, 21:216–231, 1993.

[DiN95] I. H. Dinwoodie and P. Ney. Occupation measures for Markov chains. *J. Theoretical Prob.*, 8:679–691, 1995.

[DiZ92] I. H. Dinwoodie and S. L. Zabell. Large deviations for exchangeable random vectors. *Ann. Probab.*, 20:1147–1166, 1992.

[DoD86] H. Doss and M. Dozzi. Estimations de grande deviations pour les processus a parametre multidimensionnel. In *Lecture Notes in Math. 1204*, pages 68–80. Springer-Verlag, New York, 1986.

[DoS91] H. Doss and D. W. Stroock. Nouveaux résultats concernant les petites perturbations de systèmes dynamiques. *J. Funct. Anal.*, 101:370–391, 1991.

[DKS92] R. L. Dobrushin, R. Kotecky and S. Shlosman. *Wulff construction: a global shape from local interactions.* AMS Translation Series, Providence, 1992.

[DT77]   R. L. Dobrushin and B. Tirozzi. The central limit theorem and the
           problem of equivalence of ensembles. *Comm. Math. Physics*, 54:173–
           192, 1977.

[DuE92]   P. Dupuis and R. S. Ellis. Large deviations for Markov processes
           with discontinuous statistics, II. Random walks. *Prob. Th. Rel. Fields*,
           91:153–194, 1992.

[DuE95]   P. Dupuis and R. S. Ellis. The large deviation principle for a general
           class of queueing systems, I. *Trans. Amer. Math. Soc.*, 347:2689–2751,
           1995.

[DuE97]   P. Dupuis and R. S. Ellis. *A Weak Convergence Approach to the The-
           ory of Large Deviations*. Wiley, New York, 1997.

[DuK86]   P. Dupuis and H. J. Kushner. Large deviations for systems with small
           noise effects, and applications to stochastic systems theory. *SIAM J.
           Cont. Opt.*, 24:979–1008, 1986.

[DuK87]   P. Dupuis and H. J. Kushner. Stochastic systems with small noise,
           analysis and simulation; a phase locked loop example. *SIAM J. Appl.
           Math.*, 47:643–661, 1987.

[DuZ96]   P. Dupuis and O. Zeitouni. A nonstandard form of the rate function
           for the occupation measure of a Markov chain. *Stoch. Proc. and Appl.*,
           61:249–261, 1996.

[DunS58]   N. Dunford and J. T. Schwartz. *Linear Operators, Part I*. Interscience
           Publishers Inc., New York, 1958.

[Dur85]   R. Durrett. Particle Systems, Random Media and Large Deviations.
           Contemp. Math. 41. Amer. Math. Soc., Providence, RI, 1985.

[DuS88]   R. Durrett and R. H. Schonmann. Large deviations for the contact pro-
           cess and two dimensional percolation. *Prob. Th. Rel. Fields*, 77:583–
           603, 1988.

[DV75a]   M. D. Donsker and S. R. S. Varadhan. Asymptotic evaluation of cer-
           tain Markov process expectations for large time, I. *Comm. Pure Appl.
           Math.*, 28:1–47, 1975.

[DV75b]   M. D. Donsker and S. R. S. Varadhan. Asymptotic evaluation of cer-
           tain Markov process expectations for large time, II. *Comm. Pure Appl.
           Math.*, 28:279–301, 1975.

[DV76]   M. D. Donsker and S. R. S. Varadhan. Asymptotic evaluation of cer-
           tain Markov process expectations for large time, III. *Comm. Pure
           Appl. Math.*, 29:389–461, 1976.

[DV83]   M. D. Donsker and S. R. S. Varadhan. Asymptotic evaluation of cer-
           tain Markov process expectations for large time, IV. *Comm. Pure
           Appl. Math.*, 36:183–212, 1983.

[DV85]   M. D. Donsker and S. R. S. Varadhan. Large deviations for stationary
           Gaussian processes. *Comm. Math. Physics*, 97:187–210, 1985.

[DV87]   M. D. Donsker and S. R. S. Varadhan. Large deviations for noninter-
           acting particle systems. *J. Stat. Physics*, 46:1195–1232, 1987.

[DV89]   M. D. Donsker and S. R. S. Varadhan. Large deviations from a hydro-
           dynamic scaling limit. *Comm. Pure Appl. Math.*, 42:243–270, 1989.

[EicG98]   P. Eichelsbacher and M. Grunwald. Exponential tightness can fail in
           the strong topology. Preprint, 1998.

[EicS96]   P. Eichelsbacher and U. Schmock. Exponential approximations in com-
           pletely regular topological spaces and extensions of Sanov's theorem.
           Preprint 96-109, Univesität Bielefeld, SFB 343 Diskrete Strukturen in
           der Mathematik, 1996.

[EicS97]   P. Eichelsbacher and U. Schmock. Large and moderate deviations of
           products of empirical measures and U-empirical measures in strong
           topologies. Preprint 97-087, Univesität Bielefeld, SFB 343 Diskrete
           Strukturen in der Mathematik, 1997.

[EinK96]   U. Einmahl and J. Kuelbs. Dominating points and large deviations for
           random vectors. *Prob. Th. Rel. Fields*, 105:529–543, 1996.

[Eiz84]   A. Eizenberg. The exit distribution for small random perturbations of
           dynamical systems with a repulsive type stationary point. *Stochastics*,
           12:251–275, 1984.

[EK87]   A. Eizenberg and Y. Kifer. The asymptotic behavior of the principal
           eigenvalue in a singular perturbation problem with invariant bound-
           aries. *Prob. Th. Rel. Fields*, 76:439–476, 1987.

[EKW94]   A. Eizenberg, Y. Kifer and B. Weiss. Large deviations for $\mathbf{Z}^d$-actions.
           Comm. Math. Physics, 164:433–454, 1994.

[Ell84]   R. S. Ellis. Large deviations for a general class of random vectors.
           Ann. Probab., 12:1–12, 1984.

[Ell85]   R. S. Ellis. *Entropy, Large Deviations and Statistical Mechanics.*
           Springer-Verlag, New York, 1985.

[Ell88]   R. S. Ellis. Large deviation for the empirical measure of a Markov
           chain with an application to the multivariate empirical measure. *Ann.
           Probab.*, 16:1496–1508, 1988.

[ElW89]   R. S. Ellis and A. D. Wyner. Uniform large deviation property of
           the empirical process of a Markov chain. *Ann. Probab.*, 17:1147–1151,
           1989.

[ElGP93]   R. S. Ellis, J. Gough and J. V. Puli. The large deviation principle for
           measures with random weights. *Rev. Math. Phys.*, 5:659–692, 1993.

[ET76]   I. Ekeland and R. Temam. *Convex Analysis and Variational Problems.*
           North-Holland, Amsterdam, 1976.

[Fel71]   W. Feller. *An Introduction to Probability Theory and Its Applications*,
           volume II. Wiley, New York, second edition, 1971.

[Fle78]   W. H. Fleming. Exit probabilities and optimal stochastic control.
           Appl. Math. Opt., 4:329–346, 1978.

[FO88]   H. Föllmer and S. Orey. Large deviations for the empirical field of a
           Gibbs measure. *Ann. Probab.*, 16:961–977, 1988.

[Föl73]　　H. Föllmer. On entropy and information gain in random fields. *Z. Wahrsch. verw. Geb.*, 26:207–217, 1973.

[Frd75]　　D. Freedman. On tail probabilities for martingales. *Ann. Probab.*, 3:100–118, 1975.

[Fre85a]　　M. I. Freidlin. *Functional Integration and Partial Differential Equations*. Princeton University Press, Princeton, 1985.

[Fre85b]　　M. I. Freidlin. Limit theorems for large deviations and reaction-diffusion equations. *Ann. Probab.*, 13:639–675, 1985.

[FW84]　　M. I. Freidlin and A. D. Wentzell. *Random Perturbations of Dynamical Systems*. Springer-Verlag, New York, 1984.

[Ga68]　　R. G. Gallager. *Information Theory and Reliable Communication*. Wiley, New York, 1968.

[Gan93]　　N. Gantert. An inversion of Strassen's law of the iterated logarithm for small time. *Ann. Probab.*, 21:1045–1049, 1993.

[Gär77]　　J. Gärtner. On large deviations from the invariant measure. *Th. Prob. Appl.*, 22:24–39, 1977.

[GG97]　　F. Gamboa and E. Gassiat. Bayesian methods and maximum entropy for ill-posed inverse problems. *Ann. Stat.*, 25:328–350, 1997.

[GOR79]　　P. Groeneboom, J. Oosterhoff and F. H. Ruymgaart. Large deviation theorems for empirical probability measures. *Ann. Probab.*, 7:553–586, 1979.

[Gro80]　　P. Groeneboom. Large deviations and asymptotic efficiencies. In *Mathematical Centre Tracts 118*. Mathematisch Centrum, Amsterdam, 1980.

[GrS81]　　P. Groeneboom and G. R. Shorack. Large deviations of goodness of fit statistics and linear combinations of order statistics. *Ann. Probab.*, 9:971–987, 1981.

[GS58]　　U. Grenander and G. Szegö. *Toeplitz forms and their applications*. University of California Press, Berkeley, 1958.

[Haj88]　　B. Hajek. Cooling schedules for optimal annealing. *Math. Oper. Res.*, 13:311–329, 1988.

[Ham62]　　J. M. Hammersley. Generalization of the fundamental theorem on subadditive functions. *Math. Proc. Camb. Philos. Soc.*, 58:235–238, 1962.

[Hea67]　　C. R. Heathcote. Complete exponential convergence and some related topics. *J. Appl. Prob.*, 4:218–256, 1967.

[Hoa67]　　A. B. Hoadley. On the probability of large deviations of functions of several empirical cdf's. *Ann. Math. Statist.*, 38:360–382, 1967.

[Hoe63]　　W. Hoeffding. Probability inequalities for sums of bounded random variables. *J. Am. Stat. Assoc.*, 58:13–30, 1963.

[Hoe65] W. Hoeffding. On probabilities of large deviations. In *Proceedings of the Fifth Berkeley Symposium on Mathematical Statistics and Probability*, pages 203–219, Berkeley, 1965. University of California Press.

[HoKS89] R. A. Holley, S. Kusuoka and D. W. Stroock. Asymptotics of the spectral gap with applications to the theory of simulated annealing. *J. Funct. Anal.*, 83:333–347, 1989.

[HoS88] R. A. Holley and D. W. Stroock. Simulated annealing via Sobolev inequalities. *Comm. Math. Physics*, 115:553–569, 1988.

[HTB90] P. Hänggi, P. Talkner and M. Borkovec. Reaction rate theory: fifty years after Kramers. *Reviews of Modern Physics*, 62:251–341, 1990.

[Igl72] D. L. Iglehart. Extreme values in the $GI/G/1$ queue. *Ann. Math. Statist.*, 43:627–635, 1972.

[Ilt95] M. Iltis. Sharp asymptotics of large deviations in $\mathbb{R}^d$. *J. Theoretical Prob.*, 8:501–524, 1995.

[IMS94] I. A. Ignatyuk, V. Malyshev and V. V. Scherbakov. Boundary effects in large deviation problems. *Russian Math. Surveys*, 49:41–99, 1994.

[INN85] I. Iscoe, P. Ney and E. Nummelin. Large deviations of uniformly recurrent Markov additive processes. *Adv. in Appl. Math.*, 6:373–412, 1985.

[Io91a] D. Ioffe. On some applicable versions of abstract large deviations theorems. *Ann. Probab.*, 19:1629–1639, 1991.

[Io91b] D. Ioffe. *Probabilistic methods in PDE*. PhD thesis, Dept. of Math., Technion, Haifa, Israel, 1991.

[Io95] D. Ioffe. Exact large deviation bounds up to T_c for the Ising model in two dimensions. *Prob. Th. Rel. Fields*, 102:313–330, 1995.

[Jai90] N. C. Jain. Large deviation lower bounds for additive functionals of Markov processes. *Ann. Probab.*, 18:1071–1098, 1990.

[Jen91] J. L. Jensen. Saddlepoint expansions for sums of Markov dependent variables on a continuous state space. *Prob. Th. Rel. Fields*, 89:181–199, 1991.

[Jia94] T. Jiang. Large deviations for renewal processes. *Stoch. Proc. and Appl.*, 50:57–71, 1994.

[Jia95] T. Jiang. The metric of large deviation convergence. Preprint, Dept. of Statistics, Stanford University, Stanford, 1995.

[JiRW95] T. Jiang, M. B. Rao and X. Wang. Large deviations for moving average processes. *Stoch. Proc. and Appl.*, 59:309–320, 1995.

[JiWR92] T. Jiang, X. Wang and M. B. Rao. Moderate deviations for some weakly dependent random processes. *Stat. and Prob. Lett.*, 15:71–76, 1992.

[Joh70] J. A. Johnson. Banach spaces of Lipschitz functions and vector valued Lipschitz functional. *Trans. Amer. Math. Soc.*, 148:147–169, 1970.

[JS89]   M. Jerrum and A. Sinclair. Approximating the permanent. *SIAM J. Comput.*, 18:1149–1178, 1989.

[JS93]   M. Jerrum and A. Sinclair. Polynomial time approximation algorithms for the Ising model. *SIAM J. Comput.*, 22:1087–1116, 1993.

[KaDK90] S. Karlin, A. Dembo, and T. Kawabata. Statistical composition of high scoring segments from molecular sequences. *Ann. Stat.*, 18:571–581, 1990.

[KaS66]   S. Karlin and W. J. Studden. *Tchebycheff systems: with applications in analysis and statistics.* Interscience Publishers Inc., New York, 1966.

[Kam78] S. Kamin. Elliptic perturbations of a first order operator with a singular point of attracting type. *Indiana Univ. Math. J.*, 27:935–952, 1978.

[Kem73] J. H. B. Kemperman. Moment problems for sampling without replacement. *Indag. Math.*, 35:149–188, 1973.

[Khi29]   A. I. Khinchin. Über einen neuen grenzwertsatz der wahrscheinlichkeitsrechnung. *Math. Annalen*, 101:745–752, 1929.

[Kif81]   Y. Kifer. The exit problem for small random perturbations of dynamical systems with a hyperbolic fixed point. *Israel J. Math.*, 40:74–96, 1981.

[Kif90a]   Y. Kifer. Large deviations in dynamical systems and stochastic processes. *Trans. Amer. Math. Soc.*, 321:505–524, 1990.

[Kif90b]   Y. Kifer. A discrete-time version of the Wentzell-Freidlin theory. *Ann. Probab.*, 18:1676–1692, 1990.

[Kif92]   Y. Kifer. Averaging in dynamical systems and large deviations. *Invent. Math.*, 110:337–370, 1992.

[Kif95]   Y. Kifer. Multidimensional random subshifts of finite type and their large deviations. *Prob. Th. Rel. Fields*, 103:223–248, 1995.

[Kif96]   Y. Kifer. Perron-Frobenius theorem, large deviations, and random perturbations in random environments. *Math. Z.*, 222:677–698, 1996.

[KK86]   A. D. M. Kester and W. C. M. Kallenberg. Large deviations of estimators. *Ann. Stat.*, 14:648–664, 1986.

[KL51]   S. Kullback and R. A. Leibler. On information and sufficiency. *Ann. Math. Statist.*, 22:79–86, 1951.

[Ko92]   Y. Kofman. *Combined coding and modulation.* PhD thesis, Dept. of Electrical Engineering, Technion, Haifa, Israel, 1992.

[KO90]   C. Kipnis and S. Olla. Large deviations from the hydrodynamical limit for a system of independent Brownian particles. *Stochastics and Stoch. Rep.*, 33:17–25, 1990.

[Kra40]   H. A. Kramers. Brownian motion in a field of force and the diffusion model of chemical reaction. *Physica*, 7:284–304, 1940.

[KS88]   I. Karatzas and S. E. Shreve. *Brownian Motion and Stochastic Calculus*. Springer-Verlag, New York, 1988.

[KucC91] T. Kuczek and K. N. Crank. A large deviation result for regenerative processes. *J. Theoretical Prob.*, 4:551–561, 1991.

[KuS91]   S. Kusuoka and D. W. Stroock. Precise asymptotics of certain Wiener functionals. *J. Funct. Anal.*, 1:1–74, 1991.

[KuS94]   S. Kusuoka and D. W. Stroock. Asymptotics of certain Wiener functionals with degenerate extrema. *Comm. Pure Appl. Math.*, 47:477–501, 1994.

[KuT84]   S. Kusuoka and Y. Tamura. The convergence of Gibbs measures associated with mean fields potentials. *J. Fac. Sci. Univ. Tokyo*, 31:223–245, 1984.

[KZ94]   F. Klebaner and O. Zeitouni. The exit problem for a class of period doubling systems. *Ann. Appl. Probab.*, 4:1188–1205, 1994.

[Lan73]   O. E. Lanford. Entropy and equilibrium states in classical statistical mechanics. In A. Lenard, editor, *Statistical Mechanics and Mathematical Problems*, Lecture Notes in Physics 20, pages 1–113. Springer-Verlag, New York, 1973.

[Lanc69] P. Lancaster. *Theory of Matrices*. Academic Press, New York, 1969.

[Land89] R. Landauer. Noise activated escape from metastable states: an historical view. In F. Moss and P. V. E. McClintock, editors, *Noise in nonlinear dynamical systems*, volume I, pages 1–16. Cambridge University Press, 1989.

[LaS88]   G. F. Lawler and A. D. Sokal Bounds on the L^2 spectrum for Markov chains and Markov processes: a generalization of Cheeger's inequality. *Trans. Amer. Math. Soc.*, 309:557–580, 1988.

[Led92]   M. Ledoux. On moderate deviations of sums of i.i.d. vector random variables. *Ann. Inst. H. Poincaré Probab. Stat.*, 28:267–280, 1992.

[Led96]   M. Ledoux. On Talagrand's deviation inequalities for product measures. *ESAIM: Probability and Statistics*, 1:63–87, 1996.

[Leh59]   E. L. Lehmann. *Testing Statistical Hypotheses*. Wiley, New York, 1959.

[Leo87]   C. Leonard. Large deviations and law of large numbers for a mean field type interacting particle system. *Stoch. Proc. and Appl.*, 25:215–235, 1987.

[Lin61]   Yu. V. Linnik. On the probability of large deviations for the sums of independent variables. In J. Neyman, editor, *Proceedings of the Fourth Berkeley Symposium on Mathematical Statistics and Probability*, pages 289–306, Berkeley, 1961. University of California Press.

[LoS77]   B. F. Logan and L. A. Shepp. A variational problem for random Young tableaux. *Adv. in Math.*, 26:206–222, 1977.

[LP92]   R. S. Liptser and A. A. Pukhalskii. Limit theorems on large deviations
            for semimartingales. *Stochastics and Stoch. Rep.*, 38:201–249, 1992.

[LS88]   J. L. Lebowitz and R. H. Schonmann. Pseudo free energies and large
            deviations for non Gibssian FKG measures. *Prob. Th. Rel. Fields*,
            77:49–64, 1988.

[LyS87]   J. Lynch and J. Sethuraman. Large deviations for processes with in-
            dependent increments. *Ann. Probab.*, 15:610–627, 1987.

[Mal82]   A. Martin-Löf. A Laplace approximation for sums of independent ran-
            dom variables. *Z. Wahrsch. verw. Geb.*, 59:101–116, 1982.

[Mar96a]   K. Marton. Bounding $\bar{d}$-distance by information divergence: a method
            to prove measure concentration. *Ann. Probab.*, 24:857–866, 1996.

[Mar96b]   K. Marton. A measure concentration inequality for contracting
            Markov chains. *Geom. Func. Anal.*, 6:556–571, 1996. Erratum, 7:609–
            613, 1997.

[Mart98]   F. Martinelli. Glauber Dynamics for Discrete Spin Models. In P. L.
            Hennequin, editor, *Ecole d'Été de Probabilités de Saint-Flour XXV–
            1997*, Lecture Notes in Math., to appear, 1998.

[McD89]   C. McDiarmid. On the method of bounded differences. *Surveys in
            Combinatorics*, 141:148–188, Cambridge University Press, Cambridge,
            1989.

[Mic92]   L. Miclo. Recuit simulé sans potentiel sur une variété riemannienne
            compacte. *Stochastics and Stoch. Rep.*, 41:23–56, 1992.

[Mic95]   L. Miclo. Comportement de spectres d'opérateurs de Schrödinger a
            basse temperature. *Bull. Sci. Math.*, 119:529–533, 1995.

[Mik88]   T. Mikami. Some generalizations of Wentzell's lower estimates on large
            deviations. *Stochastics*, 24:269–284, 1988.

[Mil61]   H. Miller. A convexity property in the theory of random variables
            defined on a finite Markov chain. *Ann. Math. Statist.*, 32:1260–1270,
            1961.

[MiS86]   V. D. Milman and G. Schechtman. Asymptotic theory of finite dimen-
            sional normed spaces. In *Lecture Notes in Math. 1200*. Springer-Verlag,
            New York, 1986.

[MNS92]   A. Millet, D. Nualart and M. Sanz. Large deviations for a class of
            anticipating stochastic differential equations. *Ann. Probab.*, 20:1902–
            1931, 1992.

[Mog74]   A. A. Mogulskii. Large deviations in the space $C(0,1)$ for sums given
            on a finite Markov chain. *Siberian Math. J.*, 15:43–53, 1974.

[Mog76]   A. A. Mogulskii. Large deviations for trajectories of multi dimensional
            random walks. *Th. Prob. Appl.*, 21:300–315, 1976.

[Mog93]   A. A. Mogulskii. Large deviations for processes with independent in-
            crements. *Ann. Probab.*, 21:202–215, 1993.

[MoS89] G. J. Morrow and S. Sawyer. Large deviation results for a class of Markov chains arising from population genetics. *Ann. Probab.*, 17:1124–1146, 1989.

[Moy61] S. C. Moy. Generalizations of Shannon–McMillan theorem. *Pac. J. Math.*, 11:705–714, 1961.

[MS77] B. J. Matkowsky and Z. Schuss. The exit problem for randomly perturbed dynamical systems. *SIAM J. Appl. Math.*, 33:365–382, 1977.

[MS82] B. J. Matkowsky and Z. Schuss. Diffusion across characteristic boundaries. *SIAM J. Appl. Math.*, 42:822–834, 1982.

[MST83] B. J. Matkowsky, Z. Schuss and C. Tier. Diffusion across characteristic boundaries with critical points. *SIAM J. Appl. Math.*, 43:673–695, 1983.

[MWZ93] E. Mayer-Wolf and O. Zeitouni. On the probability of small Gaussian ellipsoids and associated conditional moments. *Ann. Probab.*, 21:14–24, 1993.

[Na95] I. Nagot. Grandes deviations pour les processus d'apprentissage lent a statistique discontinues sur une surface. Ph.D. thesis, University Paris XI, Orsay, 1995.

[Nag79] S. V. Nagaev. Large deviations of sums of independent random variables. *Ann. Probab.*, 7:745–789, 1979.

[Ney83] P. Ney. Dominating points and the asymptotics of large deviations for random walk on $\mathbb{R}^d$. *Ann. Probab.*, 11:158–167, 1983.

[NN87a] P. Ney and E. Nummelin. Markov additive processes, I. Eigenvalues properties and limit theorems. *Ann. Probab.*, 15:561–592, 1987.

[NN87b] P. Ney and E. Nummelin. Markov additive processes, II. Large deviations. *Ann. Probab.*, 15:593–609, 1987.

[Num90] E. Nummelin. Large deviations for functionals of stationary processes. *Prob. Th. Rel. Fields*, 86:387–401, 1990.

[OBr96] G. L. O'Brien. Sequences of capacities, with connections to large-deviation theory. *J. Theoretical Prob.*, 9:19–35, 1996.

[OBS96] G. L. O'Brien and J. Sun. Large deviations on linear spaces. *Probab. Math. Statist.*, 16:261–273, 1996.

[OBV91] G. L. O'Brien and W. Vervaat. Capacities, large deviations and loglog laws. In S. Cambanis, G. Samorodnitsky and M. Taqqu, editors, *Stable Processes and Related Topics*, pages 43–84. Birkhäuser, Basel, Switzerland, 1991.

[OBV95] G. L. O'Brien and W. Vervaat. Compactness in the theory of large deviations. *Stoch. Proc. and Appl.*, 57:1–10, 1995.

[Oll87] S. Olla. Large deviations for almost Markovian processes. *Prob. Th. Rel. Fields*, 76:395–409, 1987.

[Oll88] S. Olla. Large deviations for Gibbs random fields. *Prob. Th. Rel. Fields*, 77:343–357, 1988.

[OP88]   S. Orey and S. Pelikan. Large deviation principles for stationary pro-
           cesses. *Ann. Probab.*, 16:1481–1495, 1988.

[Ore85]   S. Orey. Large deviations in ergodic theory. In K. L. Chung, E. Çinlar
           and R. K. Getoor, editors, *Seminar on Stochastic Processes 12*, pages
           195–249. Birkhäuser, Basel, Switzerland, 1985.

[Ore88]   S. Orey. Large deviations for the empirical field of Curie–Weiss models.
           Stochastics, 25:3–14, 1988.

[Pap90]   F. Papangelou. Large deviations and the internal fluctuations of crit-
           ical mean field systems. *Stoch. Proc. and Appl.*, 36:1–14, 1990.

[Par67]   K. R. Parthasarathy. *Probability Measures on Metric Spaces*. Aca-
           demic Press, New York, 1967.

[PAV89]   L. Pontryagin, A. Andronov and A. Vitt. On the statistical treatment
           of dynamical systems. In F. Moss and P. V. E. McClintock, editors,
           Noise in Nonlinear Dynamical Systems, volume I, pages 329–348. Cam-
           bridge University Press, Cambridge, 1989. Translated from *Zh. Eksp.
           Teor. Fiz. (1933)*.

[Pet75]   V. V. Petrov. *Sums of Independent Random Variables*. Springer-
           Verlag, Berlin, 1975.

[Pfi91]   C. Pfister. Large deviations and phase separation in the two dimen-
           sional Ising model. *Elv. Phys. Acta*, 64:953–1054, 1991.

[Pi64]   M. S. Pinsker. *Information and Information Stability of Random Vari-
           ables*. Holden-Day, San Francisco, 1964.

[Pie81]   I. F. Pinelis. A problem on large deviations in the space of trajectories.
           Th. Prob. Appl., 26:69–84, 1981.

[Pin85a]   R. Pinsky. The I-function for diffusion processes with boundaries. *Ann.
           Probab.*, 13:676–692, 1985.

[Pin85b]   R. Pinsky. On evaluating the Donsker–Varadhan I-function. *Ann.
           Probab.*, 13:342–362, 1985.

[Pis96]   A. Pisztora. Surface order large deviations for Ising, Potts and perco-
           lation models. *Prob. Th. Rel. Fields*, 104:427–466, 1996.

[Pro83]   J. G. Proakis. *Digital Communications*. McGraw-Hill, New York, first
           edition, 1983.

[PS75]   D. Plachky and J. Steinebach. A theorem about probabilities of large
           deviations with an application to queuing theory. *Periodica Mathe-
           matica Hungarica*, 6:343–345, 1975.

[Puk91]   A. A. Pukhalskii. On functional principle of large deviations. In V.
           Sazonov and T. Shervashidze, editors, *New Trends in Probability and
           Statistics*, pages 198–218. VSP Moks'las, Moskva, 1991.

[Puk94a]   A. A. Pukhalskii. The method of stochastic exponentials for large de-
           viations. *Stoch. Proc. and Appl.*, 54:45–70, 1994.

[Puk94b] A. A. Pukhalskii. Large deviations of semimartingales via convergence of the predictable characteristics. *Stochastics and Stoch. Rep.*, 49:27–85, 1994.

[PuW97]   A. A. Pukhalskii and W. Whitt. Functional large deviation principles for first-passage-time processes. *Ann. Appl. Probab.*, 7:362–381, 1997.

[Rad83]   M. E. Radavichyus. On the probability of large deviations of maximum likelihood estimators. *Sov. Math. Dokl.*, 27:127–131, 1983.

[Rei95]   R. Reidi. An improved multifractal formalism and self-similar measures. *J. Math. Anal. Applic.*, 189:462–490, 1995.

[Roc70]   R. T. Rockafellar. *Convex Analysis*. Princeton University Press, Princeton, 1970.

[Roz86]   L. V. Rozovskii. Asymptotic expansions for probabilities of large deviations. *Prob. Th. Rel. Fields*, 73:299–318, 1986.

[Rue65]   D. Ruelle. Correlation functionals. *J. Math. Physics*, 6:201–220, 1965.

[Rue67]   D. Ruelle. A variational formulation of equilibrium statistical mechanics and the Gibbs phase rule. *Comm. Math. Physics*, 5:324–329, 1967.

[Sal97]   L. Saloff-Coste. Markov Chains. In P. L. Hennequin, editor, *Ecole d'Été de Probabilités de Saint-Flour XXIV–1996*, Lecture Notes in Math. 1665, pages 301–413. Springer-Verlag, New York, 1997.

[San57]   I. N. Sanov. On the probability of large deviations of random variables. In Russian, 1957. (English translation from *Mat. Sb. (42)*) in Selected Translations in Mathematical Statistics and Probability I (1961), pp. 213–244).

[Sc66]   M. Schilder. Some asymptotic formulae for Wiener integrals. *Trans. Amer. Math. Soc.*, 125:63–85, 1966.

[Scd98]   A. Schied. Cramér's condition and Sanov's theorem. To appear, *Stat. and Prob. Lett.*, 1998.

[Sch87]   R. Schonmann. Second order large deviation estimates for ferromagnetic systems in the phase coexistence region. *Comm. Math. Physics*, 112:409–422, 1987.

[Sch89]   R. H. Schonmann. Exponential convergence under mixing. *Prob. Th. Rel. Fields*, 81:235–238, 1989.

[SchS95] R. H. Schonmann and S. B. Shlosman. Complete analyticity for 2D Ising model completed. *Comm. Math. Physics*, 170:453–482, 1995.

[Scr93]   C. Schroeder. *I*-projection and conditional limit theorems for discrete parameter Markov processes. *Ann. Probab.*, 21:721–758, 1993.

[Sen81]   E. Seneta. *Non Negative Matrices and Markov Chains*. Springer-Verlag, New York, second edition, 1981.

[Sep93]   T. Seppäläinen. Large deviations for lattice systems, I, II. *Prob. Th. Rel. Fields*, 96:241–260, 97:103–112, 1993.

[Sep97]   T. Seppäläinen. Large deviations for increasing sequences on the plane.
            Preprint, 1997.

[Set64]   J. Sethuraman. On the probability of large deviations of families of
            sample means. *Ann. Math. Statist.*, 35:1304–1316, 1964. Corrections:
            Ann. Math. Statist., 41:1376–1380, 1970.

[SeW97]   R. Serfling and W. Wang. Large deviation results for U- and V- statis-
            tics, -empiricals, and -processes. Preprint, 1995.

[Sha48]   C. Shannon. A mathematical theory of communication. *Bell Sys. Tech.
            Journal*, 27:379–423, 623–656, 1948.

[Sha59]   C. Shannon. Coding theorems for a discrete source with a fidelity
            criterion. *IRE National Convention Record, Part 4*, pages 142–163,
            1959.

[Sie75]   D. Siegmund. The time until ruin in collective risk theory. *Schweiz-
            erischer Ver. der Versich. Mitteilungen*, 2, 1975.

[Sie85]   D. Siegmund. *Sequential Analysis: Tests and Confidence Intervals.*
            Springer-Verlag, New York, 1985.

[Sio58]   M. Sion. On general minimax theorems. *Pac. J. Math.*, 8:171–176,
            1958.

[Sla88]   M. Slaby. On the upper bound for large deviations for sums of i.i.d.
            random vectors. *Ann. Probab.*, 16:978–990, 1988.

[Smi33]   N. Smirnoff. Über wahrscheinlichkeiten grosser abweichungen. *Rec.
            Soc. Math. Moscou*, 40:441–455, 1933.

[SOSL85]   M. K. Simon, J. K. Omura, R. A. Scholtz, and B. K. Levitt.
            Spread Spectrum Communications, Vol. 1. Computer Science Press,
            Rockville, MD, 1985.

[Str64]   V. Strassen. An invariance principle for the law of the iterated loga-
            rithm. *Z. Wahrsch. verw. Geb.*, 3:211–226, 1964.

[Str65]   V. Strassen. The existence of probability measures with given
            marginals. *Ann. Math. Statist.*, 36:423–439, 1965.

[St84]   D. W. Stroock. *An Introduction to the Theory of Large Deviations.*
            Springer-Verlag, Berlin, 1984.

[StZ91]   D. W. Stroock and O. Zeitouni. Microcanonical distributions, Gibbs
            states, and the equivalence of ensembles. In R. Durrett and H. Kesten,
            editors, *Festschrift in honour of F. Spitzer*, pages 399–424. Birkhäuser,
            Basel, Switzerland, 1991.

[StZg92]   D. W. Stroock and B. Zegarlinski. The logarithmic Sobolev inequality
            for discrete spin systems on a lattice. *Comm. Math. Physics*, 149:175–
            193, 1992.

[SW95]   A. Shwartz and A. Weiss. *Large deviations for performance analysis.*
            Chapman and Hall, London, 1995.

[SZ92]   Y. Steinberg and O. Zeitouni. On tests for normality. *IEEE Trans. Inf.
            Theory*, IT-38:1779–1787, 1992.

[Szn91]   A. Sznitman. *Topics in propagation of chaos.* In *Lecture Notes in Math. 1464.* Springer-Verlag, New York, 1991.

[Tak82]   Y. Takahashi. Entropy functional (free energy) for dynamical systems and their random perturbations. In K. Itô, editor, *Proc. Taniguchi Symp. on Stoch. Anal.*, pages 437–467, Tokyo, 1982. Kinokuniya North-Holland.

[Tal95]   M. Talagrand. Concentration of measure and isoperimetric inequalities in product spaces. *Publ. Mathématiques de l'I.H.E.S.*, 81:73–205, 1995.

[Tal96a]   M. Talagrand. New concentration inequalities in product spaces. *Invent. Math.*, 126:505–563, 1996.

[Tal96b]   M. Talagrand. Transportation cost for Gaussian and other product measures. *Geom. Func. Anal.*, 6:587–600, 1996.

[Var66]   S. R. S. Varadhan. Asymptotic probabilities and differential equations. *Comm. Pure Appl. Math.*, 19:261–286, 1966.

[Var84]   S. R. S. Varadhan. *Large Deviations and Applications.* SIAM, Philadelphia, 1984.

[Vas80]   O. A. Vasicek. A conditional law of large numbers. *Ann. Probab.*, 8:142–147, 1980.

[VF70]   A. D. Ventcel and M. I. Freidlin. On small random perturbations of dynamical systems. *Russian Math. Surveys*, 25:1–55, 1970.

[VF72]   A. D. Ventcel and M. I. Freidlin. Some problems concerning stability under small random perturbations. *Th. Prob. Appl.*, 17:269–283, 1972.

[Vit66]   A. J. Viterbi. *Principles of Coherent Communication.* McGraw-Hill, New York, 1966.

[VK77]   A. M. Vershik and S. V. Kerov. Asymptotics of the Plancherel measure of the symmetric group and the limiting form of Young tables. *Dokl. Acad. Nauk.*, 233:1024–1028, 1977. See also *Funct. Anal. Appl.*, 19:21–31, 1985.

[Wei76]   H. V. Weizsäcker. A note on finite dimensional convex sets. *Math. Scand.*, 38:321–324, 1976.

[Wen90]   A. D. Wentzell. *Limit Theorems on Large Deviations for Markov Stochastic Processes.* Kluwer, Dordrecht, Holland, 1990.

[Wu94]   L. Wu. Large deviations, moderate deviations and LIL for empirical processes. *Ann. Probab.*, 22:17–27, 1994.

[Wu95]   L. Wu. Moderate deviations of dependent random variables related to CLT. *Ann. Probab.*, 23:420–445, 1995.

[Zab92]   S. L. Zabell. Mosco convergence and large deviations. In M. G. Hahn, R. M. Dudley and J. Kuelbs, editors, *Probability in Banach Spaces, 8*, pages 245–252. Birkhäuser, Basel, Switzerland, 1992.

[Zaj95]   T. Zajic. Large exceedances for uniformly recurrent Markov additive processes and strong mixing stationary processes. *J. Appl. Prob.*, 32:679–691, 1995.

[Zei88] O. Zeitouni. Approximate and limit results for nonlinear filters with small observation noise: the linear sensor and constant diffusion coefficient case. *IEEE Trans. Aut. Cont.*, AC-33:595–599, 1988.

[ZG91] O. Zeitouni and M. Gutman. On universal hypotheses testing via large deviations. *IEEE Trans. Inf. Theory*, IT-37:285–290, 1991.

[ZK95] O. Zeitouni and S. Kulkarni. A general classification rule for probability measures. *Ann. Stat.*, 23:1393–1407, 1995.

[ZZ92] O. Zeitouni and M. Zakai. On the optimal tracking problem. *SIAM J. Cont. Opt.*, 30:426–439, 1992. Erratum, 32:1194, 1994.

[Zoh96] G. Zohar. Large deviations formalism for multifractals. Preprint, Dept. of Electrical Engineering, Technion, Haifa, Israel, 1996.

General Conventions

a.s., a.e.	almost surely, almost everywhere
μ^n	product measure
$\Sigma^n(\mathcal{X}^n), \mathcal{B}^n$	product topological spaces and product σ-fields
$\mu \circ f^{-1}$	composition of measure and a measurable map
ν, μ, ν'	probability measures
$\emptyset$	the empty set
$\wedge, \vee$	(pointwise) minimum, maximum
$\overset{\text{Prob}}{\to}$	convergence in probability
$1_A(\cdot), 1_a(\cdot)$	indicator on A and on $\{a\}$
$\overline{A}, A^o, A^c$	closure, interior and complement of A
$A \backslash B$	set difference
$\subset$	contained in (not necessarily properly)
$d(\cdot, \cdot), d(x, A)$	metric and distance from point x to a set A
$(\mathcal{Y}, d)$	metric space
F, G, K	closed, open, compact sets
$f', f'', \nabla f$	first and second derivatives, gradient of f
$f(A)$	image of A under f
f^{-1}	inverse image of f
$f \circ g$	composition of functions
$I(\cdot)$	generic rate function
$\log(\cdot)$	logarithm, natural base
N(0,I)	zero mean, identity covariance standard multivariate normal
O	order of
$P(\cdot), E(\cdot)$	probability and expectation, respectively
$\mathbb{R}, \mathbb{R}^d$	real line, d-dimensional Euclidean space, (d positive integer)
$[t]$	integer part of t
Trace	trace of a matrix
v'	transpose of the vector (matrix) v
$\{x\}$	set consisting of the point x
$\mathbb{Z}_+$	positive integers
$\langle \cdot, \cdot \rangle$	scalar product in $\mathbb{R}^d$; duality between $\mathcal{X}$ and $\mathcal{X}'$
$\langle f, \mu \rangle$	integral of f with respect to μ
C^k	functions with continuous derivatives up to order k
C^∞	infinitely differentiable functions

A. Dembo, O. Zeitouni, *Large Deviations Techniques and Applications*,   385
Stochastic Modelling and Applied Probability 38,
DOI 10.1007/978-3-642-03311-7,
© Springer-Verlag Berlin Heidelberg 1998, corrected printing 2010

Glossary

α_n, 91, 312

β^*, 327, 331

β_∞, 327

β_n, 91, 312

Γ_δ, 130

$\Gamma_{\epsilon,\delta}$, 133

γ, γ_n, 92

γ_β, 327, 330

γ_k, 194

$\Delta^\epsilon, \Delta_j^\epsilon$, 189

δ_{X_i}, 3, 260

Θ, 274, 315

Θ', 274

$\theta_{cr}, \theta_{cr}^x, \theta_{cr}^y$, 239

θ_m, 228

θ_t, 238

$\Lambda(\cdot)$, 19, 26, 43, 149, 252, 261, 290

$\overline{\Lambda}(\cdot)$, 149

$\Lambda^*(\cdot)$, 26, 149, 261

$\Lambda_\lambda, \Lambda_\lambda^*$, 154, 155

Λ_ϵ^*, 151

Λ_{μ_ϵ}, 148

$\Lambda_0(\cdot), \Lambda_0^*(\cdot)$, 92

Λ_f, 142

$\Lambda_n(\lambda)$, 43

μ, 12, 26, 252

$\mu\pi$, 290

$\mu \otimes_k \pi$, 296

$\mu^{\mathbb{Z}_+}$, 14, 253

$\{\mu_\epsilon\}$, 5

$\tilde{\mu}_\epsilon$, 131

$\mu_{\epsilon,\sigma}$, 121

$\mu_{\epsilon,m}$, 131

$\mu_{\theta,n}$, 315

μ_ϕ, 190

$\mu_{0,n}, \mu_{1,n}$, 312

μ_0^*, μ_1^*, 301

$\{\mu_n\}$, 6, 176, 252

$\tilde{\mu}_n$, 49, 177, 190, 273

μ_n^*, 87

$\mu_{n,\sigma}$, 273

$\{\nu_\epsilon\}$, 128, 185

ν_{k-1}, 296

ν_*, 324

$\nu_0^* \otimes \pi$, 301

ω_o, 194

$\mathbf{\Pi} = \{\pi(i,j)\}$, 72

$\mathbf{\Pi}_\lambda$, 74

$\pi(\sigma, \cdot)$, 272

$\pi^{(2)} = \pi^{(2)}(k \times \ell, i \times j)$, 78

$\pi^m(\tau, \cdot)$, 275

π_f, 290

$\pi_k(\cdot, \cdot)$, 295

ρ, 101

$\rho^{(J)}$, 105

$\rho(\mathbf{\Pi}_\lambda)$, 74

ρ_{C_n}, 101

ρ_{max}, 101

$\overline{\rho}_n$, 103

ρ_Q, 102

$\rho_Q^{(J)}$, 106

Σ, 11, 260

Σ_μ, 12

$\sigma(\cdot), \sigma(\cdot, \cdot)$, 213, 361

A. Dembo, O. Zeitouni, *Large Deviations Techniques and Applications*,   387
Stochastic Modelling and Applied Probability 38,
DOI 10.1007/978-3-642-03311-7,
© Springer-Verlag Berlin Heidelberg 1998, corrected printing 2010

σ_t, 217

σ_ρ, 226

$\tau, \hat{\tau}$, 240

τ^ϵ, 221

τ_ϵ, 201

τ_1, 217

τ_k, 242

τ_m, 228

$\hat{\tau}_k$, 244

$\Phi(\cdot)$, 323

$\Phi(x), \phi(x)$, 111

$\Psi_I(\alpha)$, 4

ϕ_t^*, 202

Ω, 101, 253

$\mathcal{A}$, 120, 343

$\mathcal{AC}$, 176

$\mathcal{AC}_0$, 190

$\mathcal{AC}^T$, 184

$|A|$, 11

$(A+B)/2$, 123

A^δ, 119

$A^{\delta,o}$, 168

A_δ, 274, 324

a_n, 7

$\mathcal{B}$, 4, 116, 350

$\mathcal{B}^\omega$, 261

$\mathcal{B}_\epsilon$, 130

$\mathcal{B}^{cy}$, 263

$\mathcal{B}_\mathcal{X}$, 4, 351

$\mathbf{B} = \{b(i,j)\}$, 72

$B(\Sigma)$, 165, 263

$B(\mathcal{X})$, 351

B_ρ, 223

$B_{x,\delta}$, 9, 38, 344

$\tilde{B}_{z,2\delta}$, 313

$b(\cdot)$, 213

b_t, 217

$\mathcal{C}^o$, 257

$\mathcal{C}_n$, 103

$\mathbf{C}$, 109

$co(A)$, 87, 346

C_η, 208

$C_b(\mathcal{X})$, 141, 351

C_n, 101

$C([0,1])$, 176

$C_0([0,1])$, 177

$C_0([0,1]^d)$, 190

$C_u(\mathcal{X})$, 352

$\mathcal{D}_\Lambda$, 26, 43

$\mathcal{D}_{\Lambda^*}$, 26

$\mathcal{D}_f$, 341

$\mathcal{D}_I$, 4

D_1, D_2, 239

$d_\alpha(\cdot,\cdot)$, 61

$d_k(\cdot,\cdot), d_\infty(\cdot,\cdot)$, 299

$d_{LU}(\cdot,\cdot)$, 320 ,356

$d_V(\cdot,\cdot), \|\cdot\|_{var}$, 12, 266

$\frac{d\nu}{d\mu}$, 353

$\mathcal{E}$, 155, 252

$\mathcal{E}^*$, 155

E_σ^π, 72

E_σ, 277

E_P, 285

E_x, 221

$\mathcal{F}$, 44

$\mathcal{F}_n$, 273

F_δ, F_0, 119, 324

$\|f\|$, 176

$\|f\|_\alpha$, 188

f^*, 149

f_ϵ, 133

f_m, 133

$\mathcal{G}$, 144

$\mathcal{G}^+$, 147

$G, \partial G$, 220, 221

$g(\cdot)$, 243

$H(\nu)$, 13

$H(\nu|\mu)$, 13, 262

$H_\beta(\cdot)$, 331

H_0, H_1, 90, 91, 311

$H_1([0,T]), H_1$, 185, 187

H_1^x, 214

$\| \cdot \|_{H_1}$, 185

$I(\cdot)$, 4

$I_\theta(\cdot)$, 315

$I'(\cdot)$, 126

$I(\nu|\beta, \mu)$, 21, 318

$I^\delta(x)$, 6

$I^U(\cdot)$, 292

I_Γ, 5

I_A, 83

I_F, 324

$I_\infty(\nu)$, 300

I_0, 196

$I_k(\cdot), I_k^U(\cdot)$, 296

I_k, 82, 194

$I_m(\cdot)$ 131, 134

I_t, 202

$I_w(\cdot)$, 196

$I_x(\cdot)$, 214

$I_\mathcal{X}(\cdot)$, 180, 191

$I_{y,t}(\cdot), I_y(\cdot), I_t(\cdot)$, 223

$(J, \leq)$, 162, 345

$J_\delta(\cdot)$, 313, 315

J_B, J_Π, 77

J, 76

J_n, 110

K_α, 8, 193

K_L, 262

$\mathcal{L}_A$, 120, 121

$\mathcal{L}_n$, 12, 82

Lip_α, 188

L^ϵ, 222

$L_\infty([0,1]), L_\infty([0,1]^d)$, 176, 190

$L_1(\cdot)$, 190

L_p, L_∞, 352

$L_{0\|1}, L_{1\|0}$, 91

$L_\epsilon^\mathbf{Y}, L_{\epsilon,m}^\mathbf{Y}$, 135

$L_m^\mathbf{Y}$, 21, 318

$L_n^\mathbf{Y}$, 12

$L_n^{\mathbf{Y},(2)}$, 79

$L_n^\mathbf{Y}$, 260

$L_{n,k}^\mathbf{Y}$, 295

$L_n^{\mathbf{Y},m}$, 273

$L_{n,\infty}^\mathbf{Y}$, 298

$\varprojlim$, 162, 345

$\mathcal{M}$, 87, 324

$M(\lambda)$, 26

$M(\Sigma)$, 260

$M_1(\Sigma)$, 12, 260, 354

$m(\cdot)$, 190, 353

$m(A, \delta)$, 151

$N(t), \hat{N}(t)$, 188

N_0, 243

$\mathcal{P}$, 101

$\mathcal{P}_J$, 106

P, P_n, 285

P_ϵ, 130

$P_{\epsilon, m}$, 131

P_μ, 14

P_σ^π, 72

P_σ, 273

$P_{n,\sigma}$, 272

P_x, 221

$p_{\lambda_1, \lambda_2, \ldots, \lambda_d}, p_V$, 164

p_j, 162, 345

$p_k, p_{m,k}$, 299

$\mathcal{Q}$, 189

$\mathcal{Q}(\mathcal{X})$, 169

Q_X, Q_Y, 102

Q_k, 194

$q_1(\cdot), q_2(\cdot)$, 79

$q_f(\cdot|\cdot)$, 79

$q_b(\cdot|\cdot)$, 80

$R_1(D)$, 102

$R_J(D), R(D)$, 106

R_{C_n}, 102

R_m, 83

r.c.p.d., 354

ri C, 47, 341

$\mathcal{S}$, 91, 311

$\mathcal{S}^*$, 98

$\mathcal{S}^n$, 91, 312

$\mathcal{S}^{n,\delta}, \mathcal{S}_0^{n,\delta}, \mathcal{S}_1^{n,\delta}, \mathcal{S}_\delta^*$, 313

$\mathcal{S}_1^n, \mathcal{S}_0^n$, 97, 312

S_ρ, 223

$\hat{S}_n$, 2, 26, 252

$\hat{S}_n^m$, 252

$s(t)$, 243

T_ϵ, 201

$T_n(\nu)$, 13

T_r, 83

$\bar{l}$, 203

U, 327

$U\nu$, 330

$U_{\phi,x,\delta}$, 261, 355

$U_{f,x,\delta}, \hat{U}_{\hat{f},x,\delta,k}$, 299, 300

u_t, 239

$\mathbf{u} \gg 0$, 76

$\overline{V}$, 223

$V(x,t)$, 202

$V(y,z,t), V(y,z)$, 223

V_A, 202

$V(t)$, 194

$V_{t,x,\delta}$, 178

v_k, 242

v_t, 239

$\mathcal{W}$, 164

$\mathcal{W}'$, 164

$W_{\phi,x,\delta}$, 263

weak*, 153, 160, 347

$w_\epsilon(t)$, 185

$\hat{w}_\epsilon(t)$, 185

w_t, 42

w_k, 244

$\mathcal{X}$, 4, 115, 116, 148, 252, 343

$\tilde{\mathcal{X}}$, 180

$\mathcal{X}'$, 346

$X_t^{\epsilon,y}$, 216

X_i, 2, 18, 252

$|x|$, 26

$\overline{x}$, 26

$\overline{x}_0, \overline{x}_1$, 91

$x^*(x)$, 347

$\mathcal{X}^*$, 148, 346

$\| \cdot \|, \| \cdot \|_{\mathcal{X}^*}$, 347

$(\mathcal{X}, \tau_1), (\mathcal{X}, \tau_2)$, 129

$x_t^{\epsilon,m}$, 214

x_t^ϵ, 213

$\mathcal{Y}$, 130

$(\mathcal{Y}_j, p_{ij})$, 162, 345

Y_t^ϵ, 201, 208

$Y_\epsilon(t)$, 183, 245

Y_i, 12, 18, 260

$\hat{Y}_s^\epsilon$, 203

y_t, 239

$\mathbb{Z}_0, \mathbb{Z}_1$, 300

$Z_\epsilon, \tilde{Z}_\epsilon$, 130

$Z_\epsilon(\cdot)$, 245

$\hat{Z}_\epsilon(\cdot)$, 245

$Z_{\epsilon,m}$, 131

$Z_n(t)$, 176, 189

$\tilde{Z}_n(t)$, 190

Z_β, 327, 331

Z_n, 43, 312

Z_k, 244

z_m, 228

z_t, 240

$\hat{z}_t$, 241

Index

Absolutely continuous functions, **176**, 184, 190, 202

Absolutely continuous measures, **350**

Additive functional of Markov chain, *see Markov, additive functional*

Additive set function, 189, **350**

Affine regularization, **153**

Algebraic dual, **164–168**, 263, **346**

Alphabet, **11**, 72, 87, 99

Approximate derivative, **189**

Arzelà–Ascoli, theorem, 182, **352**

Asymptotic expansion, 114, 307, 309

Azencott, 173, 249

$B(\Sigma)$-topology, 263

Bahadur, **110**, 174, 306

Baldi's theorem, **157**, 267

Ball, **344**

Banach space, 69, 148, 160, 248, 253, 268, 324, **348**

Bandwidth, 242

Base of topology, 120–126, 274–276, 280, **343**

Bayes, 92, 93

Berry–Esséen expansion, 111, 112

Bessel process, 241

Bijection, 127, **344**

Bin packing, 59, 67

Birkhoff's ergodic theorem, 103

Bootstrap, 323

Borel–Cantelli lemma, 71, 84

Borel probability measures, 10, 137, 162, 168, 175, 252, 272, 311

Borel σ-field, 5, 116, 190, 261, 299, **350– 351**, 354

Bounded variation, **189**

Branching process, 310

Brønsted–Rockafellar, theorem, 161

Brownian motion, 43, 175, **185–188**, 194, 208, 211, 212–242, 249, 360

Brownian sheet, **188–193**, **192**

Bryc's theorem, **141–148**, 278

Cauchy sequence, 268, **348**

Central Limit Theorem (CLT), 34, 111

Change of measure, 33, 53, 248, 260, 308, 325

Change point, 175

Characteristic boundary, 224–225, 250

Chebycheff's bound
  see Inequalities, Chebycheff

Chernoff's bound, **93**

Closed set, **343**

Code, **101**–108

Coding gain, **102**

Compact set, 7, 8, 37, 88, 144, 150, **344**, 345

Complete space, **348**

Concave function, 42, 143, 151

Concentration inequalities, **55**, 60, 69, 307

Contraction principle, 20, **126**, 163, 173, 179, 212
  approximate, 133–134, 173
  inverse, **127**, 174

Controllability, 224

Controlled process, 302

A. Dembo, O. Zeitouni, *Large Deviations Techniques and Applications*, 391
Stochastic Modelling and Applied Probability 38,
DOI 10.1007/978-3-642-03311-7,
© Springer-Verlag Berlin Heidelberg 1998, corrected printing 2010

Convergence, **345**

Convergence determining, 330, **356**

Convex analysis, 45, 157, 174, **341**–343

Convex hull, 87, 252, **346**

Countably additive, 190, **350**

Coupling, **61**, 64

Covariance, 109, 192

Cover, **344**

Cramér's theorem, 2, 18–20, 26–42, 44, 68, 71, 83, 108, 248, **251**–260, 306

Cumulant generating function, **26**

CUSUM, 250

Dawson–Gärtner, theorem, **162**, 180, 300

Decision test
  see Test

δ-rate function, **6**

δ-smoothed rate function, **313**

Désiré André
  see Reflection principle

Diffusion process, 212–213

Digital communication, 45, 53, 193, 250

Distortion, **101**, 102, 106

Divergence, 262

DMPSK, **193**–200, 250

DNA sequences, 83, 175

Doëblin recurrence, 308

Domain, **4**

Dominated convergence, **353**

Donsker, 9, 173

Doob, 218, **359**

Duality lemma, **152**–157, 264, 342

Dupuis–Ellis, 302

Dynamical systems, 175, 225, 310

Eberlein–Šmulian, theorem, 266, **349**

Effective domain, **4**

Ellis, 9, 339
  see Dupuis–Ellis, Gärtner–Ellis

Elliptic, 238

Empirical,
  measure, 3, 12, 21, 79, 96, **260**–339
  mean, 26, 83, 253

Energy, 89, **323**

Ensemble, 105, 323

Entropy, **13**, 107

Epigraph, 157

Equilibrium point, **221**

Equicontinuous, 193, **352**, 356

Equivalent measures, 91, **350**

Ergodic, 76, 102, 103, 106, 285

Error probability, 53, 71, **91**, 97, **195**, **311**–313

Essentially smooth, **44**, 47, 52, 166, 342

Euler–Lagrange, 200

Exact asymptotics, 108, **110**
  of Bahadur and Rao, **110**, 114

Exchangeable, 89, 100, 323

Exit from a domain, **220**–238, 249–250

Exponentially equivalent, **130**, 175, 177, 190

Exponentially good approximations, **131**, 133, 135–136, 200, 214

Exponentially tight, **8**, 38, 49, 120, 128, 140, 144, 154, 157, 167, 173, 253, 260–261, 269, 277, 316

Exposed point, **44**, 50, **157**

Exposing hyperplane, **44**, 50, **157**

Fatou's lemma, 35, 51, 180, 191, 254, **353**

Feller continuous, 293–294, 309

Fenchel–Legendre transform, **26**, 41, 51, 115, **152**–157, 176, 179, 189, 237, 252, 268, 279, 289–290

Field, 190, 310, **350**

Freidlin, 3, 9, **212**–220, 247

Gateaux differentiable, **160**, 167, 265

Gaussian process, 53, 188, 192, 310, 339

Gärtner–Ellis, theorem, 43, **44**–55, 71, 109, 141, 148, **157**–161, 179, 251, 341

Generator, 135

Gibbs, 3, 71, **87**–90, 310, **323**–338
  field, 114, 310
  measure, **327**, 330, 331
  parameter, **89**
Girsanov, 248
Green's function, 222, 238
Gronwall's lemma, 213, 216, 218, 233, 235, 246, **360**

Hahn's theorem, **351**
Hahn–Banach theorem, 156, 167, 342, **347**
Hamilton–Jacobi, 237
Hamiltonian, 331
Hammersley, 282
Hoeffding, **56**, 69, 96, **98**
Hölder continuous, 188
Homeomorphism, **344**
Hypercontractive, 288, 309
Hypermixing
  see Mixing
Hypothesis testing, 71, **90**–100, **311**–317, 339

I continuity set, **5**
Inequalities,
  Azuma, 69
  Bennett, **56**, 59, 69
  Borell, **188**, 193, 259
  Burkholder–Davis–Gundy, 235, **360**
  Cauchy–Schwartz, 197, 215
  Chebycheff, 30, 33, 37, 42, 44, 48, 150, 151, 159, 182, 187, 218, 230, 262, 268, 292, 332
  Csiszár, **336**
  Fernique, 248
  Hoeffding, **56**, 69
  Hölder, 28, 37, 288
  Isoperimetric, 248
  Jensen, 13, 19, 28, 40, 93, 108, 202, 264, 267, 287, 294, 325
  Logarithmic Sobolev, 309
  Talagrand, 60, **62**, 69
Initial

conditions, 216, 275, 277
measure, 292
Integral, **350**
Interacting particles, 327, 330, 339
Interior, **343**
Invariant measure, **290**, 295
Inverse contraction principle,
  see Contraction principle, inverse
Ioffe, 342
Irreducible, **72**, 74, 77, 79, 86, 275
Itô, 218, **360**

Kofman, 53
Kolmogorov's extension, 299, 301
Kullback–Leibler distance, 262, 308

Lanford, 306, 339
Langevin's equation, 238
Laplace's method, 137, 174, 307
Large Deviation Principle (LDP), **5**–9
  behavior under transformations, 118, **126**–137
  existence, 120
  for Banach space, **268**, 306
  for continuous time Markov processes, 135, 272
  for diffusion processes, 212–220
  for empirical mean, 2, **44**, 174
  for empirical measure, 3, **16**, 76, 81, **261**–263, 308
  for empirical process, 298–302, **300**
  for i.i.d. empirical sum, 2, **18**, **27**, **36**, **252**
  for Markov chains, 2, **72**-82, **74**, **272**–278
  for Markov occupation time, **289**–295
  for multivariate random walk, **188**, 189–193
  for projective limits, **161**–168
  for random walk, **176**–184
  for sampling without replacement, **23**, **318**
  for topological vector spaces, 148–161
  uniqueness, **117**

Large exceedances, **200**–212, 250

Lattice law, **110**

Law of large numbers, 20, 39, 45, 49, 94, 251

Lebesgue measure, 190, 192, 277

Lebesgue's theorem, 180, **353**

Level sets, 37, 139 , 163

Lévy metric, 261, 267, 272, **355**

Lévy process, 221, 225

Likelihood ratio, **91**, 96, 200, 311–312

Lipschitz bounded metric, 320, **356**

Lipschitz continuous function, 184, 212–214, 220, 235, 240, 242, 281, 341

Logarithmic moment generating function, **26**, 40, **43**, 83, 103, 110, **149**, 252, 254, **290**, 306

Log-likelihood ratio, 91

Longest increasing subsequence, 64, 69

Lower bound, **6**, 7, 39, 45, 99, 142

Lower semicontinuous, **4**, 42, 117, 138, **344**

Markov,
  additive functional, **73**, 110, 308
  chain, 71, 74, 86, 173, 277, 294, 308
  kernel, 294, 295
  process, continuous time, 135, 175
  semigroup, 9, 287
  see LDP for Markov chain

Martingale, 218, 235, 359
  difference, **57**, 60, 69, 306

Maximal length segments, **83**

Mazur's theorem, 253, **349**

Mean, of a Banach space valued variable, **268**

Measurable function, **350**

Measure space, **350**

Metric entropy, **151**, 267

Metric space, 4, 9, 116

Micro-canonical, 323

Min–Max theorem, 42, **151**, 206, 211

Mixing, 259, 278–288, 309, 310
  hypermixing, **286**, 288, 289, 292, 309

ψ-*mixing*, 279, 287

Moderate deviations, 108, **109**, 110

Mogulskii, 3, **176**, 190, 204

Monotone convergence, **353**

Monotone class theorem, 293

Moving average, 55, 69

Multi-index, 189

Mutual information, **102**

Neighborhood, **343**

Neyman–Pearson, **91**–96, 113, 311–312

Noise, 194, 225

Non-coherent detection, 45, 53

Norm, **347**

Occupation time, **289**–295, 302

Open set, **343**

Partition function, **327**, 331

Partially ordered set, 162, **345**

Peres, 285, 334

Perron–Frobenius, **72**, 77

Phase Lock Loop, 237, 250

Pinsker, 301

Pointwise convergence, 162, 177, 180, 191, 293, 342

Poisson process, 51, 187, 221

Polygonal approximation and interpolation, **177**, 190, 209, 245

Polish space
  see Topological, Polish space

Portmanteau theorem, 262, 299, **356**

Pre-compact set, 120, 151, 188, 315, **344**, **352**

Pre-exponent, 71

Probability of error,
  see Error probability,

Process level LDP, 298–302

Product measure, 71, 90, 106, 108, 129, 252, 324

Product topology, 253, 256, **345**

Prohorov's criterion, 124, 262, **356**

Projective limit, **161**–168, 174, 180, 191, 260, 265, 299, 309, **345**

Projective topology, **162**–168, 180, 299, 302, 339, **345–346**

Pulse, 242–243

Quadrature noise, 53

Quadrature signal, 53, 194

Quasi-potential, **202, 223**, 238

Queueing, 225, 250

Radar, 238, 243

Radon–Nikodym, 62, 107, 187, 190, 262, 325, 332, **353**

Random variables,
Bernoulli, 35, 88
Exponential, 35, 45, 113, 184
Geometric, 113
Normal, 2, 35, 40, 45, 52, 108, 185, 208, 211, 242, 244, 288
Poisson, 35, 51, 188

Random walk, 83, **176**–184
multivariate, **188**-193

Range gate, 243–244

Rao, **110**

Rate distortion, 101–108, **102, 106**

Rate function, **4**, 7
good, **4**, 34, 37, 118, 133, 160, 162, 253, 261–266, 279, 292, 296
convex, 27, 37, 123, 149 , 167, 253, 261–266, 279

Refinements of LDP, 250
see Exact asymptotics

Regular conditional probability distribution (r.c.p.d.), 61, 297, **354**

Regular measure, **351**, 355

D. André's reflection principle, 187, **360**

Relative entropy, **13**, 61, 76, **79**, 95, 260, **262**, 308, **357**

Relative interior, **47**, 341

Relative topology, **344**

Right filtering set, 162, **345**

Rockafellar's lemma, **47**, 52

Ruelle, 306, 339

Sample path, 173

Sampling without replacement, **20**–25, 89, 100, **318**–323

Sanov's theorem, 12, **16**, 18, 36, 41, 76–78, 88, 98, 251, **260**–272, 306, 307, 317, 325

k-scan process, **81**

Schilder's theorem, **185**, 187, 196, 213, 215, 248, 259

Separating
ℓ-separated, **286**
space, set, **346**
well-separating, **143**

Sequentially compact, 266, **345**

Set function, 169, **350**
Additive, Countably additive, **350**

Shannon's theorem, 71, **103**

Shift invariant, 79, 296, 300

Shrinks nicely, 191, **353**

σ-field, 5, 190, 252, **350**
Borel, see Borel σ-field
cylinder, **263**, 324

Signal, 53, 194

Signal-to-Noise Ratio, 53

Source coding, 71, **106**

Spectral radius, 71, 78, 81

Stationary, 76, 278, 285, 290

Statistical mechanics
see Gibbs

Steep, **44**, 51, 253, 343

Stein's lemma, **94**

Stirling's approximation, 14, 22, 68

Stochastic matrix, **72**

Stochastic differential equation, 212–213

Strassen, 249

Strictly convex, 35

Strong solutions, **361**

Strong topology, **343**
see τ-topology

Sub-additivity, 36, 108, 174, 251, **255**–256, 259, 272–278, 282, 306

Subdifferential, **47**
Sufficient statistics, **96**, 317
Super multiplicative, 273
Sup-measure, **169**
Support, **12**, **355**
Supremum norm, **176**, 185, 191, 261, 285, **351**

Test, **91**–96, **96**–100, 113, 311–317
Tight, 124, 170, 261, 269, **355**
  see Exponentially tight
Topology, **343**
Topological,
  space, **343**
  completely regular space, **141**, 147
  dual, 143, **148**, 164, 167, **346**
  Hausdorff space, **116**, 141, **343**
  locally compact space, 117
  locally convex space, 120, 143, 152, 167, 252, 261, 264, **346**
  metric space, 116, 117, 124, 312, **344**
  normal space, **344**, 351
  Polish space, 106, 119, 137, 165, 251–252, 258, 261, 267, 272, 285, 299, 307, 318, 323, **348**, 354–357
  regular space, **116**, 117, **343**
  separable space, 129, 157, **345**
  vector space, 116, **148**, 164, **346**
Total variation, **350**
Totally bounded, **348**
Tracking loop, 221, **238**–248, 250
Transition probability, **272**, 275, 289, 294–296
Tychonoff's theorem, 163, **345**
Type, **12**, 13, 26, 72, 82, 97, 99
τ-topology, 260, 263, 269–271, 286, 289, 307, 324

Ulam's problem, 69
Uniform Markov (U), **272**–273, **275**–278, 289, 290, 292, 308
Uniformly integrable, 266, **352**
Uniformly continuous, 300, 356

Upper bound, **6**, 7, 37, 44, 120, 124, 142, **149**–151, 174
Upper semicontinuous, 42, 138, **344**

Varadhan, 9, **137**–141, 142, 147, 153, 173, 174
Variational distance (norm), **12**, **266**–**267**
Viterbi, 237

Weak convergence, **168**–173, 174, **302**–306, 309
Weak LDP, **7**, 120–126, 154, 168, 252–254, 258, 261
Weak topology, 174, 261, 263, 267, 292, 299, 300, 307, 318, 330, **343**, **347**, **355**
Weak* topology, 153, 160, **347**
W-topology, **164**, 347
Well-separating
  see Separating
Wentzell, 3, 9, **212**–220, 247
Wiener measure, 259

Zabell, 174, 306

Stochastic Modelling and Applied Probability
formerly: Applications of Mathematics

1 Fleming/Rishel, **Deterministic and Stochastic Optimal Control** (1975)
2 Marchuk, **Methods of Numerical Mathematics** (1975, 2nd ed. 1982)
3 Balakrishnan, **Applied Functional Analysis** (1976, 2nd ed. 1981)
4 Borovkov, **Stochastic Processes in Queueing Theory** (1976)
5 Liptser/Shiryaev, **Statistics of Random Processes I: General Theory** (1977, 2nd ed. 2001)
6 Liptser/Shiryaev, **Statistics of Random Processes II: Applications** (1978, 2nd ed. 2001)
7 Vorob'ev, **Game Theory: Lectures for Economists and Systems Scientists** (1977)
8 Shiryaev, **Optimal Stopping Rules** (1978, 2008)
9 Ibragimov/Rozanov, **Gaussian Random Processes** (1978)
10 Wonham, **Linear Multivariable Control: A Geometric Approach** (1979, 2nd ed. 1985)
11 Hida, **Brownian Motion** (1980)
12 Hestenes, **Conjugate Direction Methods in Optimization** (1980)
13 Kallianpur, **Stochastic Filtering Theory** (1980)
14 Krylov, **Controlled Diffusion Processes** (1980, 2009)
15 Prabhu, **Stochastic Storage Processes: Queues, Insurance Risk, and Dams** (1980)
16 Ibragimov/Has'minskii, **Statistical Estimation: Asymptotic Theory** (1981)
17 Cesari, **Optimization: Theory and Applications** (1982)
18 Elliott, **Stochastic Calculus and Applications** (1982)
19 Marchuk/Shaidourov, **Difference Methods and Their Extrapolations** (1983)
20 Hijab, **Stabilization of Control Systems** (1986)
21 Protter, **Stochastic Integration and Differential Equations** (1990)
22 Benveniste/Métivier/Priouret, **Adaptive Algorithms and Stochastic Approximations** (1990)
23 Kloeden/Platen, **Numerical Solution of Stochastic Differential Equations** (1992, corr. 2nd printing 1999)
24 Kushner/Dupuis, **Numerical Methods for Stochastic Control Problems in Continuous Time** (1992)
25 Fleming/Soner, **Controlled Markov Processes and Viscosity Solutions** (1993)
26 Baccelli/Brémaud, **Elements of Queueing Theory** (1994, 2nd ed. 2003)
27 Winkler, **Image Analysis, Random Fields and Dynamic Monte Carlo Methods** (1995, 2nd ed. 2003)
28 Kalpazidou, **Cycle Representations of Markov Processes** (1995)
29 Elliott/Aggoun/Moore, **Hidden Markov Models: Estimation and Control** (1995, corr. 3rd printing 2008)
30 Hernández-Lerma/Lasserre, **Discrete-Time Markov Control Processes** (1995)
31 Devroye/Györfi/Lugosi, **A Probabilistic Theory of Pattern Recognition** (1996)
32 Maitra/Sudderth, **Discrete Gambling and Stochastic Games** (1996)
33 Embrechts/Klüppelberg/Mikosch, **Modelling Extremal Events for Insurance and Finance** (1997, corr. 4th printing 2008)
34 Duflo, **Random Iterative Models** (1997)
35 Kushner/Yin, **Stochastic Approximation Algorithms and Applications** (1997)
36 Musiela/Rutkowski, **Martingale Methods in Financial Modelling** (1997, 2nd ed. 2005, corr. 3rd printing 2009)
37 Yin, **Continuous-Time Markov Chains and Applications** (1998)
38 Dembo/Zeitouni, **Large Deviations Techniques and Applications** (1998, corr. 2nd printing 2010)
39 Karatzas, **Methods of Mathematical Finance** (1998)
40 Fayolle/Iasnogorodski/Malyshev, **Random Walks in the Quarter-Plane** (1999)
41 Aven/Jensen, **Stochastic Models in Reliability** (1999)
42 Hernandez-Lerma/Lasserre, **Further Topics on Discrete-Time Markov Control Processes** (1999)
43 Yong/Zhou, **Stochastic Controls. Hamiltonian Systems and HJB Equations** (1999)
44 Serfozo, **Introduction to Stochastic Networks** (1999)

Stochastic Modelling and Applied Probability
formerly: Applications of Mathematics

45 Steele, **Stochastic Calculus and Financial Applications** (2001)
46 Chen/Yao, **Fundamentals of Queuing Networks: Performance, Asymptotics, and Optimization** (2001)
47 Kushner, **Heavy Traffic Analysis of Controlled Queueing and Communications Networks** (2001)
48 Fernholz, **Stochastic Portfolio Theory** (2002)
49 Kabanov/Pergamenshchikov, **Two-Scale Stochastic Systems** (2003)
50 Han, **Information-Spectrum Methods in Information Theory** (2003)
51 Asmussen, **Applied Probability and Queues** (2nd ed. 2003)
52 Robert, **Stochastic Networks and Queues** (2003)
53 Glasserman, **Monte Carlo Methods in Financial Engineering** (2004)
54 Sethi/Zhang/Zhang, **Average-Cost Control of Stochastic Manufacturing Systems** (2005)
55 Yin/Zhang, **Discrete-Time Markov Chains** (2005)
56 Fouque/Garnier/Papanicolaou/Sølna, **Wave Propagation and Time Reversal in Randomly Layered Media** (2007)
57 Asmussen/Glynn, **Stochastic Simulation: Algorithms and Analysis** (2007)
58 Kotelenez, **Stochastic Ordinary and Stochastic Partial Differential Equations: Transition from Microscopic to Macroscopic Equations** (2008)
59 Chang, **Stochastic Control of Hereditary Systems and Applications** (2008)
60 Bain/Crisan, **Fundamentals of Stochastic Filtering** (2009)
61 Pham, **Continuous-time Stochastic Control and Optimization with Financial Applications** (2009)
62 Guo/Hernández-Lerma, **Continuous-Time Markov Decision Processes** (2009)
63 Yin/Zhu, **Hybrid Switching Diffusions** (2010)